Birds of Europe, North Africa and the Middle East

An Annotated Checklist

Dominic Mitchell

Birds of Europe, North Africa and the Middle East

An Annotated Checklist

Dominic Mitchell

Lynx

Cover (left to right): Monteiro's Storm-petrel (*Oceanodroma monteiroi*), Scottish Crossbill (*Loxia scotica*), Moussier's Redstart (*Phoenicurus moussieri*), Rock Partridge (*Alectoris graeca*), European Herring Gull (*Larus argentatus*), Caucasian Grouse (*Lyrurus mlokosiewiczi*), Arabian Grosbeak (*Rhynchostruthus percivali*) and Pleske's Ground Jay (*Podoces pleskei*).

First Edition: February 2017

Printed by: Ingoprint, S.A.
Legal Deposit: B-1192-2017
ISBN: 978-84-941892-9-6

CONTENTS

Preface ... 9

Acknowledgements ... 11

About this book ... 13
Scientific and English names ... 13
Taxonomy ... 13
Distribution ... 13
General definitions, abbreviations and terms ... 15
Geographical definitions ... 15

Introduction ... 17
Regional boundaries ... 17
Avifauna ... 20

SYSTEMATIC LIST
Family Struthionidae – Ostriches ... 25
Family Anatidae – Ducks, geese and swans ... 25
Family Numididae – Guineafowl ... 41
Family Odontophoridae – New World quail ... 41
Family Phasianidae – Grouse, partridges, francolins and pheasants ... 42
Family Gaviidae – Divers ... 46
Family Oceanitidae – Austral storm-petrels ... 47
Family Diomedeidae – Albatrosses ... 48
Family Hydrobatidae – Northern storm-petrels ... 49
Family Procellariidae – Petrels and shearwaters ... 50
Family Podicipedidae – Grebes ... 56
Family Phoenicopteridae – Flamingos ... 57
Family Phaethontidae – Tropicbirds ... 58
Family Ciconiidae – Storks ... 58
Family Threskiornithidae – Ibises and spoonbills ... 60
Family Ardeidae – Bitterns, herons and egrets ... 61
Family Scopidae – Hamerkop ... 66
Family Pelecanidae – Pelicans ... 66
Family Fregatidae – Frigatebirds ... 67
Family Sulidae – Gannets and boobies ... 68
Family Phalacrocoracidae – Cormorants and shags ... 69
Family Anhingidae – Anhingas and darters ... 70
Family Pandionidae – Ospreys ... 70
Family Accipitridae – Kites, hawks and eagles ... 71
Family Otididae – Bustards ... 81
Family Rallidae – Rails, crakes and coots ... 83
Family Gruidae – Cranes ... 86
Family Turnicidae – Buttonquail ... 87
Family Burhinidae – Stone-curlews and thick-knees ... 87
Family Haematopodidae – Oystercatchers ... 88
Family Dromadidae – Crab-plover ... 88
Family Recurvirostridae – Stilts and avocets ... 89
Family Charadriidae – Lapwings and plovers ... 89
Family Pluvianidae – Egyptian Plover ... 93

Family Rostratulidae – Painted-snipes 94
Family Jacanidae – Jacanas 94
Family Scolopacidae – Sandpipers and snipes 94
Family Glareolidae – Coursers and pratincoles 106
Family Laridae – Gulls, terns and skimmers 107
Family Stercorariidae – Skuas 119
Family Alcidae – Auks 120
Family Pteroclidae – Sandgrouse 122
Family Columbidae – Pigeons and doves 123
Family Cuculidae – Cuckoos 127
Family Tytonidae – Barn owls 129
Family Strigidae – Typical owls 129
Family Caprimulgidae – Nightjars 133
Family Apodidae – Swifts 135
Family Coliidae – Mousebirds 137
Family Coraciidae – Rollers 137
Family Alcedinidae – Kingfishers 138
Family Meropidae – Bee-eaters 139
Family Upupidae – Hoopoes 140
Family Bucerotidae – Hornbills 140
Family Picidae – Woodpeckers 140
Family Falconidae – Falcons 143
Family Psittacidae – African and New World parrots 146
Family Psittaculidae – Old World parrots 146
Family Pittidae – Pittas 147
Family Tyrannidae – Tyrant flycatchers and allies 147
Family Malaconotidae – Bush-shrikes 148
Family Laniidae – Shrikes 148
Family Vireonidae – Vireos 151
Family Oriolidae – Old World orioles 152
Family Dicruridae – Drongos 152
Family Monarchidae – Monarch flycatchers and allies 153
Family Corvidae – Crows, jays and magpies 153
Family Bombycillidae – Waxwings 157
Family Hypocoliidae – Grey Hypocolius 157
Family Paridae – Tits and chickadees 157
Family Remizidae – Penduline tits 160
Family Panuridae – Bearded Reedling 161
Family Alaudidae – Larks 161
Family Pycnonotidae – Bulbuls 166
Family Hirundinidae – Swallows and martins 167
Family Cettiidae – Bush warblers 170
Family Scotocercidae – Streaked Scrub Warbler 170
Family Aegithalidae – Bushtits 171
Family Phylloscopidae – Leaf warblers 171
Family Acrocephalidae – Reed warblers and allies 175
Family Locustellidae – Grass and bush warblers 180
Family Cisticolidae – Cisticolas and allies 181
Family Leiothrichidae – Laughingthrushes and allies 181
Family Sylviidae – Sylviid babblers 182
Family Zosteropidae – White-eyes 187
Family Regulidae – Kinglets 188
Family Troglodytidae – Wrens 188

Family Sittidae – Nuthatches 189
Family Tichodromadidae – Wallcreeper 190
Family Certhiidae – Treecreepers 190
Family Mimidae – Mockingbirds and thrashers 191
Family Sturnidae – Starlings and allies 191
Family Buphagidae – Oxpeckers 193
Family Turdidae – Thrushes 193
Family Muscicapidae – Chats and Old World flycatchers 197
Family Cinclidae – Dippers 209
Family Nectariniidae – Sunbirds 209
Family Passeridae – Old World sparrows and snowfinches 210
Family Ploceidae – Weavers and allies 213
Family Estrildidae – Waxbills, munias and allies 214
Family Prunellidae – Accentors 215
Family Motacillidae – Wagtails and pipits 216
Family Fringillidae – Finches 221
Family Parulidae – New World warblers 229
Family Icteridae – New World orioles, blackbirds and allies 232
Family Emberizidae – Buntings, New World sparrows and allies 233
Family Calcariidae – Longspurs and snow buntings 239
Family Cardinalidae – Cardinals, grosbeaks and allies 240

Annex A: Indeterminate species recorded in the region 241

Appendices
Appendix 1: Endemic species 243
Appendix 2: Extinct species 247
Appendix 3: Omitted species 249
Appendix 4: National lists 253

Checklist 255

References 289

Index of species 323

PREFACE

To birders of a certain generation, the acronym *BWP* will immediately strike a chord. I am one of them. In the 1970s, after cutting my birding teeth on Heinzel, Fitter and Parslow's *The Birds of Britain and Europe, with North Africa and the Middle East*, I moved on to the first volume of Oxford University Press's *Handbook of the Birds of Europe, the Middle East and North Africa: the Birds of the Western Palearctic*. I was instantly hooked: it was not so much a progression as an eye-opening revelation. Doing for the region what Witherby's *Handbook* had done for Britain 40 years before, *BWP* presented its birds in succinct but scholarly detail. A standard had been set, a region defined and an avifauna described.

How time flies. It is already 22 years since the final volume of *BWP* was published, and 18 since that 11-volume series was condensed into *Concise BWP*. This two-part distillation appeared in 1998, and by that time it was already apparent that the species list for a work launched two decades earlier was in need of update. That same year, as a side project at *Birdwatch* magazine, we produced a simple updated list of Western Palearctic birds and distributed it to the readership; its popularity was such that it resulted in a new edition being published just two years later.

My involvement with those checklists reflected not only a strong personal interest in the region's birds, but also the realisation that there remained a need for accurate, up-to-date information which was not always accessible to the wider birding community. With this in mind I founded the WestPal-Birds email group in 2003, while also contemplating a new checklist – not least because of debate over the region's boundaries, and the question of whether or not the Arabian Peninsula and Iran, both excluded from *BWP*, should really be included because of their avifaunal affinities.

Over time, my idea of producing a more detailed modern checklist for an enlarged region developed into a format which briefly summarised key information for every species. It soon became clear that a couple of lines of annotated text would be insufficient to say anything very meaningful about taxonomy, status and distribution, especially for a new suite of Middle Eastern species which might be unfamiliar to the intended audience. Consequently, what started out as a relatively straightforward plan evolved into a significantly more complex and long-term project.

The end result is the book you are now reading. There is so much more that could have been said for many species, especially in terms of taxonomy and conservation status, but space constraints and the need to keep the final product to a manageable and practical size have necessarily limited the length of the species accounts.

By its very nature, any book such as this is a snapshot in time. Across Europe, North Africa and the Middle East, discoveries in the field – especially around the fringes of the region – and in the laboratory, as well as decisions in committee meetings, mean that status, distribution and even the species themselves are being constantly redefined. That will always be the case, of course, but I hope that this annotated checklist provides a fair and accurate account of all the birds which have occurred in this fascinating region up to the present time.

ACKNOWLEDGEMENTS

I am very grateful to Josep del Hoyo and his team at Lynx Edicions, particularly Susanna Silva, for their interest in and support for this book, and for patiently tolerating its long gestation.

I have benefited from many productive exchanges with Marcel Haas about Western Palearctic birds, and from some of his research into the occurrence of extremely rare species in the region which he generously shared. Marcel also made useful comments on a draft of the text, as did David Christie, Marc Duquet, Ray Galea, Yann Kolbeinsson and Jan Södersved. From the early stages, Richard Porter was another source of valuable feedback and information, especially on the birds of the Middle East. Gianluigi Castelli similarly responded positively to my wide-ranging queries, as did Joost Brouwer on the birds of the southern Sahara and Ljubomir Profirov on those of the Balkans. David Callahan helped to source numerous obscure taxonomic papers and references, and also provided useful feedback on the introductory text.

Countless questions arose during the compilation of the distribution sections of the species accounts. On more than one occasion I had reason to be grateful to the following for their help with records, queries or other information relating to particular recording areas, countries or regions: Mike Jennings (Arabia), Leander Khil (Austria), Michael Heiss (Azerbaijan), Peter Alfrey (Azores), Howard King (Bahrain), Dr Alexandre Vintchevski (Belarus), Peter Adriaens (Belgium), Nigel Hudson (Britain), Bojidar Ivanov (Bulgaria), Eduardo García-del-Rey (Canary Islands), Jelena Kralj (Croatia), Colin Richardson (Cyprus), Martin Vavřík (Czech Republic), Morten Bentzon Hansen (Denmark), Frédéric Jiguet (Egypt), Silas Olofson (Faroe Islands), Jan Södersved (Finland), Marc Duquet and Pierre Le Marechal (France), Stefan Pfützke, Jochen Dierschke and Christopher König (Germany), Ernest Garcia (Gibraltar), Nikos Probonas (Greece), Rolland Kern (Hungary), Yann Kolbeinsson (Iceland), Derek Scott (Iran), Osama Mohammed Kadhim (Iraq), Kieran Fahy and Mark Carmody (Ireland), Yoav Perlman (Israel), Dr Fares Khoury and Tareq Qaneer (Jordan), Maris Jaunzemis (Latvia), Liutauras Raudonikis (Lithuania), Catarina Fagundes and Hugo Romano (Madeira), Ray Galea (Malta), Paul Isenmann (Mauritania and Tunisia), Darko Saveljić (Montenegro), Patrick Bergier (Morocco), Arnoud van den Berg and Pim Wolf (Netherlands), Oddvar Heddøy, Georg Bangjord and Frode Falkenberg (Norway, Svalbard and Jan Mayen), Jens Eriksen (Oman), Marcin Faber and Antoni Marczewski (Poland), Neil Morris (Qatar), Vaughan Ashby (Russia), Jeremy Babbington (Saudi Arabia), Marko Šćiban and Dragan Simić (Serbia), Ricard Gutiérrez and Dani López Velasco (Spain), Erik Hirschfeld (Sweden), Michael Schaad (Switzerland), David Murdoch (Syria), Kerem Ali Boyla (Turkey), Oscar Campbell, Tommy Pedersen and Mark Smiles (UAE), and Dr Vitaly Grishchenko (Ukraine). While their assistance with data and fact-checking was essential, responsibility for any errors is of course mine.

I am also grateful to the following for assisting in various ways (with apologies for any omissions): Ian Andrews, Raffael Ayé, Peter Barthel, Dr Leo Batten, Mike Blair, Joël Bried, Graham Bundy, Dr Stuart Butchart, Dr J Martin Collinson, José Luis Copete, Pierre-André Crochet, Hugues Dufourny, Nils van Duivendijk, Laith El-Moghrabi, Richard Ek, Dave Farrow, Martin Fowlie, Boyko Georgiev, Gerard Gorman, Roland Graf, Hein van Grouw, Marcello Grussu, Trinus Haitjema, Tom van der Have, Tim Hesselballe Hansen, Chris Harbard, Morten Heegaard, Magnus Hellström, Joe Hobbs, Jan Hullen, Tim Inskipp, Måns Karlsson, Peter Kennerley, Guy Kirwan, Steve Madge, Dr Clive Mann, Veronica Neves, Staffan Rodebrand, Steve Rooke, Edward Rickson, Hadoram Shirihai, Roy Slaterus, Laurens Steijn, Dr Voislav Vasić, Brecht Verhelst, Ahmed Waheed, Ingo Weiss, Tom Whiley, Georg Willi, Edwin Winkel, Martin Woodcock, Mark Zekhuis, and the many members of the WestPalBirds Yahoo and Facebook groups.

Finally, it would not have been possible to write this book without such an understanding and patient family. I am especially grateful to Hazel, Ed and Ava for their constant support, encouragement and understanding during the long periods when I was immersed in researching and writing the text.

ABOUT THIS BOOK

This checklist details all 1,148 extant species recorded in Europe, North Africa and the Middle East by 31 December 2015. Species are listed in systematic order, and each species entry details scientific and English names, taxonomy, distribution and range/occurrences within the region. The following paragraphs explain more about how this information is presented in the main systematic list.

Scientific and English names

The scientific name for each species is listed first, followed by the principal English name. These closely follow those published by the IOC (v4.4, 2014), but in a few instances a preferred English name is given where this is already in widespread use in the region, or where for good reason the IOC name is considered less appropriate. For ease of use and for cross-referencing to other works, alternative common names are also given. These are generally listed in alphabetical order, first by species and then, if applicable, by subspecies known to occur in the region. Some species, especially those with multiple distinctive forms (such as Canada and Cackling Geese), have a considerable list of alternative names, but for others (for example Lesser Spotted Eagle and Garden Warbler) there are few or no options in common usage. Minor differences in names, for example American or Old English spellings (*eg* Gray-cheeked Thrush and Ring Ousel respectively) or those with different hyphenation, are not listed, and neither are colloquial or folk names generally included.

Taxonomy

The briefest summary of taxonomic information includes the species' author and year, together with the number of subspecies for polytypic species and, if relevant, a very brief indication of key taxonomic issues. These are mainly confined to those affecting taxa in a regional context, particularly where changes to taxonomic treatment have been made recently or are expected; other sources should be consulted for more background information (see References).

Distribution

Primary world range is very briefly summarised in square brackets, along with BirdLife International's current population-trend assessment and the species' IUCN Red List category. There are nine such categories (see www.iucnredlist.org/static/categories_criteria_3_1 for a detailed explanation): Extinct (EX), Extinct in the Wild (EW), Critically Endangered (CE), Endangered (EN), Vulnerable (VU), Near Threatened (NT), Least Concern (LC), Data Deficient (DD) and Not Evaluated (NE).

In the present work, this last category also includes taxa which have been elevated to species level since the last version of the IUCN Red List was published, or which are separate-

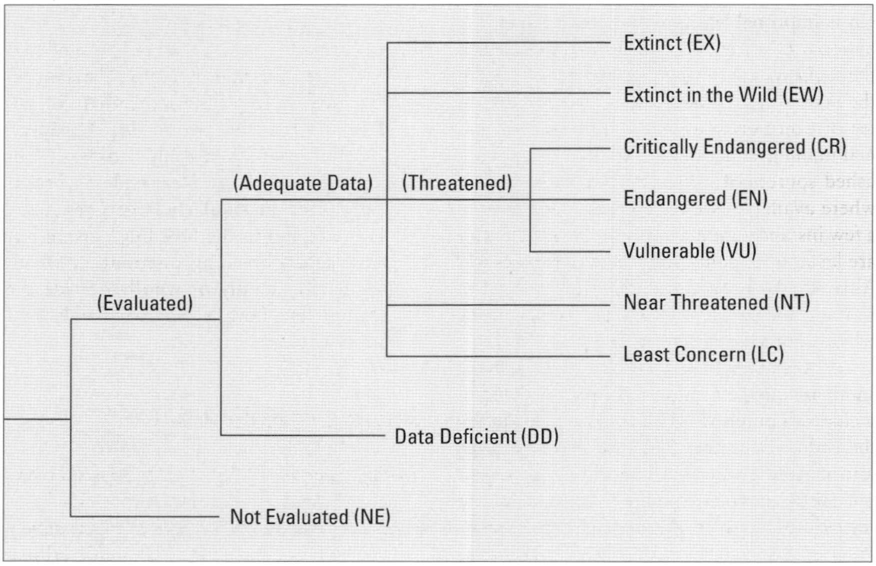

ly treated as species here, and which therefore lack assessment.

This is followed by a regional summary of range and status, by necessity broad, including distribution by subspecies, and an overview of occurrences in the case of vagrants – the 306 species that are genuinely rare and unpredictable in a regional context. They are defined as such by the total number of accepted/documented records up to 31 December 2013, or 31 December 2015 in the case of three species (Eastern Wood Pewee, Large-billed Leaf Warbler and Dark-sided Flycatcher) which have only occurred or been identified retrospectively since then. The following definitions are adapted from those used in Mitchell and Young (1999):

- Extreme vagrant: 1–10 regional records, all occurrences listed in full.
- Very rare vagrant: 11–50 regional records, occurrences summarised.
- Rare vagrant: 51–100 regional records, recorded annually (or almost so) in the region, occurrences/distribution summarised.
- Vagrant: more than 100 regional records, occurrences unpredictable but recorded annually in the region, distribution summarised.

The treatment of the number of records listed for rarer species is unavoidably inconsistent, because some national rarities committees and comparable authorities differ in the way in which such occurrences are recorded at a national level. For example, a flock of 10 birds of one rare species will usually be treated as a single record, but sometimes – as in the Azores and Iceland, for example – the data usually relate to the number of individual birds, not records. In the case of the Azores I have also frequently based occurrence data on records from the database of the islands' birds, as the processing of these thousands of records by the Portuguese Rarities Committee is currently far from complete. All such rarity records from the Azores and elsewhere in the region remain subject to official acceptance by the appropriate committee or authority, and where it is clear this is yet to happen these records are marked in the text with an asterisk.

In the main, however, most records published herein have been accepted by the appropriate authorities and placed in Categories A, B and C of their respective national lists, usually in line with practice set out by the Association of European Records and Rarities Committees (AERC). In a few instances, national committees have amended this categorisation model to suit local preferences; for example, in Hungary and Romania Category D has been divided into D_A (species of apparently wild origin for which a non-wild origin cannot be fully excluded) and D_E (species for which a wild origin is improbable), and here, as in Hungary's national checklist, D_A species are included in the main systematic list. For those countries in the region that do not have a rarities committee or other organisation vetting records and maintaining a national checklist, other published sources of information have been used where available (see References). However, in a few instances authoritative or accessible data are lacking for some areas, and in this respect Albania, Moldova, the Palestinian territories and the Salvage Islands are among those where data are incomplete.

In compiling the records of vagrants or summarising them into totals, I have aimed to include all published data up to and including the end of December 2013. In some cases the information given here may differ from that published elsewhere, partly because of recent reviews of some national lists (*eg* Austria, Poland and Romania) or re-examination of records documentation or specimen material for certain species. Despite the care taken in compiling this information, in the course of researching and listing tens of thousands of records it is inevitable that errors and omissions may have made it into print – please notify any corrections to the author.

General definitions, abbreviations and terms

AOU: American Ornithologists' Union.

BOURC: British Ornithologists' Union's Records Committee.

Breeding endemic: Species whose breeding distribution falls entirely within the region (*eg* Scopoli's Shearwater, Audouin's Gull and Masked Shrike), but whose non-breeding distribution is partly or wholly outside the region.

EBA: Endemic Bird Area.

Endemic: Species whose year-round distribution falls entirely within the region (*eg* Rock Partridge, Marmora's Warbler and Corsican Nuthatch). Rarely, some endemics may occur as vagrants outside the region (*eg* Fea's Petrel).

IOC: International Ornithological Congress.

IBA: Important Bird Area.

IUCN: International Union for the Conservation of Nature.

mtDNA: Mitochondrial DNA.

Near-endemic: Species whose combined breeding and non-breeding distribution falls largely (typically more than 90 per cent) within the region (*eg* European Green Woodpecker, Ring Ouzel and European Crested Tit). Near-endemic breeding species are similarly largely restricted to the region in the breeding season, but have a non-breeding distribution which lies partly, mainly or wholly outside the region (*eg* Levant Sparrowhawk and Sooty Falcon).

Geographical definitions

N Africa: Specifically Egypt and Libya, but variably extending west into NW Africa.

NE Africa: The Horn of Africa (alternatively the Somali Peninsula) – Djibouti, Eritrea, Ethiopia and Somalia.

NW Africa: Morocco, Algeria and Tunisia.

Arabia: The Arabian Peninsula, as defined by Jennings (2010) – Bahrain, Kuwait, Oman, Qatar, Saudi Arabia, UAE, Yemen and their associated islands (but excluding the Socotra archipelago, which biogeographically is part of Africa). Here, Arabia equates to the southern (S) Middle East.

S Asia: The Indian Subcontinent, as defined by Rasmussen and Anderton (2012) – Afghanistan, Bhutan, India, Nepal, Pakistan and Sri Lanka.

Asia Minor: The Asian part of Turkey.

Atlantic islands: The Azores, Canary Islands, Cape Verde Islands, Madeira (including the Desertas and Porto Santo), Salvage Islands and other associated islands and islets. Also known as Macaronesia.

Balkans: The Balkan peninsula, including Greece and European Turkey.

Baltic States: Estonia, Latvia and Lithuania.

Fennoscandia: Norway, Sweden, Finland and the Kola Peninsula of Russia.

NW Kazakhstan: The area of the province of West Kazakhstan lying west of the Ural River (so inside the regional boundary), thus the most north-westerly part of Kazakhstan.

Levant: Collective name for Lebanon, Syria, Israel and Jordan (but here excluding Cyprus), equating to what used to be known as the Near East.

Middle East: Definition follows Porter and Aspinall (2010) – Turkey, Iran, Iraq, Cyprus, Lebanon, Syria, Jordan, Palestine and Israel in the north, and Arabia (*qv*) in the south.

Morocco: Includes the disputed Maghreb territory of Western Sahara, which has been largely under Moroccan control since 1991.

Palestine: The Palestinian territories of Gaza and West Bank.

Sahel: Transitional zone in Africa between the Sahara to the north and the savanna to the south, extending from southern Mauritania in the west to Sudan in the east.

Scandinavia: Denmark, Norway and Sweden.

INTRODUCTION

The concept of biogeographical regions was first developed by Philip Sclater, H G Adolf Engler and other zoogeographers during the mid-19th century. Sclater crystallised his thinking in an 1858 review of the distribution of passerine birds and, among other things, proposed six regions in two series and first coined the name Palearctic. Encompassing Europe, North Africa and south-west, central and northern Asia, the Palearctic region meets the Afrotropical (formerly Ethiopian) region at its southern boundary in Africa and the Oriental (formerly Indian or Indo-Malay) region in Asia.

The evolutionary pioneer Alfred Russel Wallace (1876) went further than Sclater by analysing other groups of terrestrial animals and defining four sub-regions for each of the six regions. Wallace's sub-regions for the Palearctic were European, Mediterranean, Siberian and Manchurian, and the first two represent much of the landmass of what is today known as the Western Palearctic.

Regional boundaries

Expanded to include the Sahara Desert and the Middle East, the Western Palearctic is now a widely recognised and broadly understood zoogeographical province, even if precise delineation of some of its boundaries has been problematic (see, for example, Vaurie 1959, Udvardy 1975, Cramp 1977, Voous 1977 and Beaman 1994 for differing treatments). The most widely adopted modern interpretation is that outlined by Cramp (1977) who, when discussing the Palearctic as a whole, observed that "like all faunal regions, [it] is not susceptible to precise definition". The relatively arbitrary limits he adopted have since been contested, the main issue being the southern and south-eastern boundaries, especially south of the Sahara and in the Middle East.

In Africa, Cramp (1977) followed Vaurie (1959) by including the Cape Verde Islands, and added those of the Banc d'Arguin in Mauritania, though not the adjacent mainland. Here, the regional border began slightly farther north at 21°N and continued east in the southernmost Sahara as far as northern Chad, where it ran south of the principal areas of the Tibesti massif. Beaman (1994) used an identical boundary west of north-western Chad, but extended his area of coverage southwards to include all of the Tibesti, Libya and Chad north of 19°N, and even northernmost Sudan.

Others have taken a different approach. A more restrained limit adopted by the Dutch Birding Association (2008) constitutes the smallest variant of the Western Palearctic, and has been referred to as 'greater Europe'; it is defined simply as Europe with Macaronesia and all countries bordering the Black, Dead or Mediterranean Seas. In this respect it perhaps better reflects Wallace's original view, and also that of botanists who generally favour a boundary along the southern fringe of the Mediterranean sub-region (*eg* Takhtadzhyan 1978). While geographically convenient it would, however, mean excluding a significant part of the Sahara, which has an essentially Palearctic avifauna as far south as the northern limits of the Sahel, and most of the Middle East except its north-west fringe.

For these key reasons, and taking into account current knowledge of bird distributions, it can be argued that Cramp (1977) was too conservative in positioning the core boundary in Africa at 21°N. Indeed, before and since *BWP*, others have determined a more southerly limit. Hall and Moreau (1970), Snow (1978) and Sinclair and Ryan (2003) all set the northern limit of the Afrotropical region, and thus the southern limit of the Western Palearctic, at 20°N, while Eck (1996) positioned his boundary farther south still, at 19°N.

To the east in the Arabian Peninsula, Cramp (1977) located the boundary much farther north than in Africa, at 28°N, excluding the Arabian Peninsula south of the northern fringe of Saudi Arabia – a more conservative view even than Wallace, who over a century earlier had included the upper half of the peninsula. Notably, Harrison (1982) later set his baseline boundary through both Arabia and Africa at 20°N. Cramp did not give reasons for such a northerly division in the Middle East, stating only that "If the boundary of the Palearctic region eludes any logically unquestionable demarcation, the determination of the eastern limits of the western part is clearly largely arbitrary". The problem he perceived would have included the Afrotropical influences in the mountains of south-west Arabia and Oriental elements which penetrate the south-east, despite the otherwise largely Palearctic character of Arabia's avifauna.

Noting that knowledge of the distribution and status of Arabian birds had substantially advanced since *BWP* more than 20 years earlier, Martins and Hirschfeld (1998) argued for the inclusion of the great majority of the Arabian Peninsula within the Western Palearctic. They omitted only mountainous south-west Saudi Arabia and west and south-west Yemen, and the easternmost Yemeni province of Mahrah and the adjacent Dhofar region of south-west Oman, because of the Afrotropical affinities of those areas. Collectively, these omitted areas were estimated to amount to only 5 per cent or so of the total landmass of Arabia. Martins and Hirschfeld also included much of the western half of Iran within their regional definition, though their boundary meandered widely between different biomes and deliberately petered out north of the Elburz Mountains, rendering any full practical application effectively impossible.

In response to Martins and Hirschfeld and the Arabian question, Ullmann (2000) went further still and demonstrated convincingly that the affinities of the species breeding in south-west Arabia lay more with the avifauna of the rest of the peninsula, which is Palearctic in character, rather than with tropical Africa. He reasoned that while a relatively large percentage of the birds of south-west Arabia were "of Afrotropical origin, the large numbers of such species whose distributions do not straddle the Red Sea belies any strong affinity between the two regions".

A more precise southern border of the Palearctic was proposed by Roselaar (2006), based on the mapped distribution of Palearctic, Afrotropical and Oriental passerine species and the areas in which different faunal elements were dominant. With the Sahara included in the Palearctic, Roselaar's line took a more southerly route than did Cramp's in Africa, more or less at 19°N, but moved north to Gebel Elba at about 22°N in the extreme east. It also included much of the Arabian Peninsula except the mountains of the south and west (considered Afrotropical) and the coastal plain of northern Oman (Oriental), and continued on to include all of Iran except for the country's southern coastal plain (Oriental).

Following a theoretical and somewhat erratic mapped mid-point division between avifaunas, however, would render any such boundary impractical to determine with ease in the field. The avian distributions it reflects – this particular study was restricted to passerines – are themselves fluid, especially at the edges of breeding ranges. In harsh environments such as the southern Sahara, the variable northward shift and southward return of the Intertropical Convergence Zone and the presence or absence of drought or rainfall from one year to the next have a significant effect on the extent of vegetation and availability of food and water, and thus avian distributions. These may differ significantly from year to year as a result of such factors, especially in the case of more nomadic species.

Taking all these factors and Cramp's original caveats about the difficulty of determining regional limits into account, it is clear that an effective boundary must be both justifiable and workable. In the words of Ullmann (2000), "Boundaries between faunal regions should be kept, where possible (in many cases it is not), relatively simple". The region covered by this checklist therefore utilises a core southern boundary in Africa at 20°N, and includes the entire Arabian Peninsula, and also the whole of Iran. The term 'greater Western Palearctic' seems appropriate for this enlarged interpretation of the biogeographical province which encompasses Europe, North Africa and the Middle East. More precisely, the limits of this region can be defined as follows:

> The northern boundary extends into the high Arctic to 82°N, to include Novaya Zemlya, all of Franz Josef Land and Svalbard, then west to 10°W and south to the Arctic Circle, so excluding Greenland but including Jan Mayen. It follows the Arctic Circle west across northern Iceland, extending north slightly to include Icelandic territorial waters (12 nautical miles from shore), then at 30°W turns south to the Azores, with a minor westward extension to include Corvo, Flores and surrounding territorial waters in that archipelago, before continuing south along 30°W. At 14°N, the boundary runs south of and thus includes the Cape Verde Islands, before heading north at 20°W and then east again at 19°N to the Mauritanian coast. To the north, the islands of the Banc d'Arguin fall within the region but the adjacent mainland coast does not, the boundary lying on the low tide midpoint between the two. From the Mauritanian coast, the boundary runs inland east at 20°N through the southern Sahara to northern Chad, thereby including the

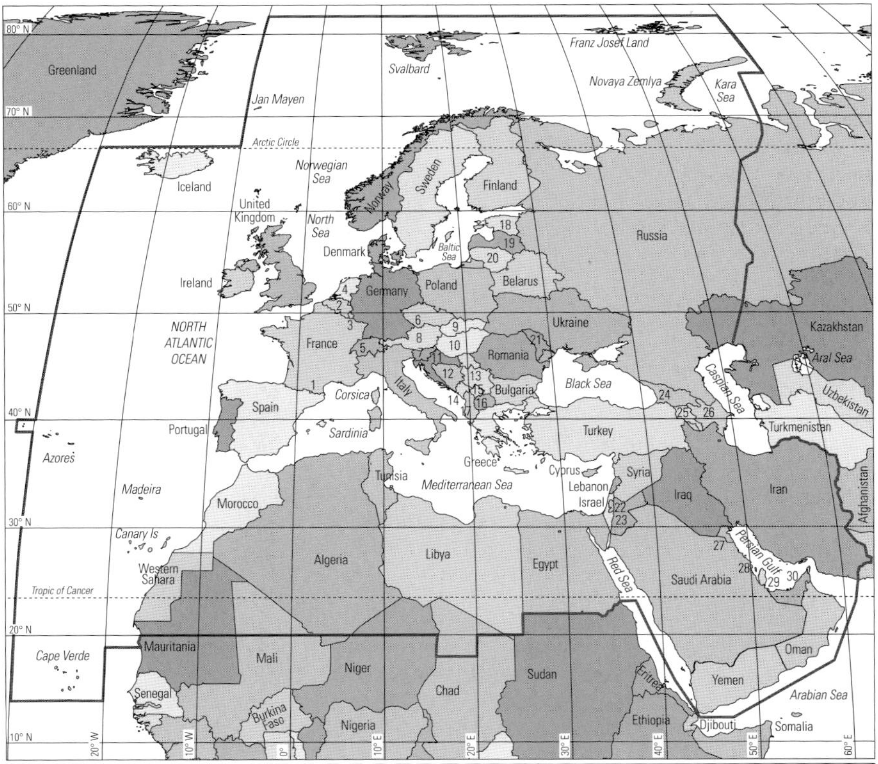

1. Andorra	6. Czech Republic	11. Croatia	16. Macedonia	21. Moldova	26. Azerbaijan
2. Belgium	7. Slovakia	12. Bosnia-Herzegovina	17. Albania	22. Palestine	27. Kuwait
3. Luxembourg	8. Austria	13. Serbia	18. Estonia	23. Jordan	28. Bahrain
4. Netherlands	9. Slovenia	14. Montenegro	19. Latvia	24. Georgia	29. Qatar
5. Switzerland	10. Hungary	15. Kosovo	20. Lithuania	25. Armenia	30. United Arab Emirates

Hoggar but not the extreme south of Algeria nor the Aïr Massif in north-west Niger. Between 16° and 20°E in northern Chad the boundary runs farther south at 18°N to include the Tibesti but not the Ennedi, then east at 20°N again to the Libya–Sudan border and north to the Egyptian border. The regional boundary follows this border east along 22°N as far as Gebel Elba, where at about 34°E it moves north-east, thus excluding this Sudan-administered Egyptian territory, and reaches the Red Sea at about 22°N. All of the Arabian Peninsula is included within the region, as is all of Iran. The eastern border of the region extends in the north from the Kara Sea east of Novaya Zemlya, the north-eastern tip of which represents the most easterly point of Europe, south along the Ural ridge (following administrative boundaries) to 58°30'N, thence by an arbitrary straight line to a point 50 km east of Yekaterinburg, and by another arbitrary straight line to the head waters of the Ural River south of Zlatoust, and then south along the Ural River to the Caspian Sea, thus including the European section of West Kazakhstan to the west. In the Caspian Sea, the boundary continues south along a theoretical meandering line midway between the west and east shores until it meets the Iranian border with Turkmenistan to the south-east.

Avifauna

Precise definition and practical application of regional boundaries are essential starting points for any work aiming to document all species which have occurred in a given area. Using the above geographical limits, a total of 1,148 extant bird species of 104 families has so far been reliably recorded in Europe, North Africa and the Middle East. The net effect of the enlarged regional boundaries used here is to add a further 129 species – an increase of some 12 per cent – to the total recorded in the region as defined by Cramp (1977). The additions include 14 endemic species (13 in Arabia and one in Iran), as well as a diverse range of regular breeding species and migrants whose ranges are otherwise largely Afrotropical or Oriental; naturalised introductions and vagrants make up the rest.

Though it hosts several of the world's most range-restricted species, the Palearctic faunal region is the least rich in terms of endemic birds (Stattersfield *et al* 1998). An endemic can be defined as "A taxon naturally found in any specific area and nowhere else; this is a relative term in that a taxon can be endemic to a small island, to a country, or to a continent" (IUCN 2012). In the context of this checklist, the word 'endemic' is applied to a species whose regular year-round distribution is restricted to Europe, North Africa and the Middle East, as opposed to breeding endemics which nest solely in the region but regularly occur elsewhere at other times of the year.

The greater Western Palearctic as a whole is home to 87 'true' endemic bird species (Appendix 1). Among them, gamebirds (seven species), tubenoses (six species), woodpeckers (five species), nuthatches (four species) and finches (14 species) are particularly well represented. Aside from seven seabirds and two gulls, no other endemic species are associated with marine or freshwater habitats. The region has just six Endemic Bird Areas (essentially defined as locations which encompass the complete ranges of two or more range-restricted breeding landbird species): Madeira and the Canary Islands (treated together), the Cape Verde Islands, south-west Arabian mountains, Mesopotamian Marshes, Cyprus and the Caucasus. With taxonomic changes resulting in more taxa being confirmed as endemic species, the number of EBAs may need revising upwards. Important Bird Areas – sites identified as critical for conservation of the world's birds – are much more strongly represented, Europe alone having 3,619 IBAs, equivalent to 7 per cent of its land area (Heath and Evans 2004).

Perhaps as many as 37 species feature in this checklist only by virtue of their status as naturalised introductions. Purists may not regard such birds as part of the regional avifauna, and periodically the population levels of some seemingly well-established introduced species fall steadily and become no longer viable, proving the point. This may be in the process of happening to Golden and Lady Amherst's Pheasants in Britain, and has already done so in the case of Fischer's Lovebird in France. Nonetheless, other 'exotics' appear to have found a permanent niche, established themselves in numbers – sometimes to the detriment of native species which they outnumber and out-compete – and are now distributed in multiple countries with populations numbering tens or even hundreds of thousands of individuals. Canada Goose and Ring-necked Parakeet are two obvious examples of introduced species that are here to stay, and, as the AERC's widely adopted Category C model for national lists was specifically designed to classify such established introductions, it is appropriate to include them here.

Another key factor which will continue to impact the number of species in the region is ongoing taxonomic reassessment. *BWP* began life in an age pre-dating the genetic revolution, its last volumes being released as the landmark work of Sibley and Monroe (1990) introduced a new era of DNA-based research. Avian taxonomy continues to be radically overhauled and, whether through laboratory work or field studies, it is certain that more species, perhaps including cryptic taxa, will be described in the region in due course. A separate and long-awaited work on Western Palearctic birds (Shirihai and Svensson in prep) will, when completed, likely recognise new taxa at species level, necessitating further updates to this checklist.

Such rapid developments in taxonomy mean that, almost inevitably, there are differences in checklists at both national and regional levels as committees and authors adopt alternative treatments of certain taxa or react to change at different paces. In Europe the AERC has attempted to introduce taxonomic stability but, while its recommendations have gained significant traction in some countries, there is no region-wide consensus. Rather than exacerbate this problem and publish another checklist with its own perspective on splits, lumps and species boundaries, I have chosen instead to adhere closely to the authoritative IOC World Bird List, which provides a stable and ongoing platform based on an international consensus view. The IOC's list has already been widely adopted and is regularly revised to reflect changes in taxonomy and nomenclature. Only in a few instances does it differ from views prevailing in this region and, where this checklist has had to make a small number of compromises in taxonomic treatment, these differences from the IOC – some of which that body is already considering and may adopt in future – are duly noted in the text.

Finally, the IOC list is also the primary basis for the English names used here. Although a very small minority have been modified, again principally to reflect prevailing regional usage, I hope that in time this initiative to standardise the English names of birds in use around the world gains even greater support. It certainly deserves it, as the long list of alternative English names for some species featured here illustrates.

SYSTEMATIC LIST

ORDER STRUTHIONIFORMES

FAMILY STRUTHIONIDAE

Ostriches · one species

Struthio camelus
Common Ostrich

OTHER NAMES Ostrich; North African (Red-necked) Ostrich (*S c camelus*).
TAXONOMY Linnaeus, 1758. Four subspecies (two in region, one of which extinct).
DISTRIBUTION [Sub-Saharan Africa. Decreasing. LC] Formerly widespread N Africa (*camelus*) and Middle East (*syriacus*†), but disappeared from most of regional range by 1900 and latter population hunted to extinction in 1930s. Last confirmed 1988 in S Morocco (where bred 1963); may have occurred more recently NW Niger (known from Ténéré desert, which extends N just into region from Aïr Massif, but population severely impacted by civil conflict in area). In late 19th century Middle East range included disjunct areas in N and S Arabia, latter population possibly involving extralimital Somali Ostrich *S molybdophanes* (NE Africa); lack of specimen evidence prevents firm identification. Reintroduction attempted (mainly unsuccessfully) in Arabia since 1997.

ORDER ANSERIFORMES

FAMILY ANATIDAE

Ducks, geese and swans · 75 species

Dendrocygna bicolor
Fulvous Whistling Duck (V)

OTHER NAMES Fulvous (Tree-)Duck, Large Whistling-teal.
TAXONOMY (Vieillot, 1816). Monotypic.
DISTRIBUTION [S USA to Argentina, sub-Saharan Africa, S Asia. Decreasing. LC] Extreme vagrant (most records Apr-Oct): Azores (29 Aug 2000); Morocco (3 sets of records: 7-26 Apr 1977, 11-12 birds, then one at same site 22 Jul and 3 Sep, two there 5 Oct; 6 May 1977, two at different site; 5-6 Sep 1980, two at original site); Oman (2: 1 Sep-18 Oct 1999; 26 Sep-2 Nov 2000); Yemen (19-21 Feb 2000). Also considered 'semi-introduced' Canary Is (Tenerife), and birds of unknown origin or escapes recorded elsewhere (*eg* Britain, France, Hungary and Spain).

Dendrocygna javanica
Lesser Whistling Duck (V)

OTHER NAMES Indian (or Lesser) Tree-duck, Indian (or Javan) Whistling-duck, (Lesser) Whistling-teal.
TAXONOMY (Horsfield, 1821). Monotypic.
DISTRIBUTION [S, SE and E Asia. Decreasing. LC] Extreme vagrant (five records, all from Middle East and most in winter/spring): Israel (15 Nov 1966-mid Mar 1967); Oman (4: 24 Dec 1999-19 Mar 2000, eight, four of which remaining to 14 Apr; 20 Nov 2008-5 Mar 2009, four; 4 Jan-29 Apr 2013, three; 11-18 Aug 2013).

Anser fabalis
Taiga Bean Goose

OTHER NAMES Bean Goose; Middendorff's (Bean) Goose (*A f middendorffii*), Sushkin's Goose (*A f 'neglectus'*).
TAXONOMY (Latham, 1787). Three subspecies (all in region). Formerly included Tundra Bean Goose and Pink-footed Goose (*qv*); *middendorffii* recently also supported at species level in one genetic analysis, but status of W Siberian *johanseni* may be questionable.
DISTRIBUTION [N Eurasia > NW Europe and E Asia. Population trend not assessed. NE] Breeds N Fennoscandia and NW Russia (*fabalis*). Migrates S and SW to winter S Sweden, Denmark, Germany and Netherlands; scarcer C Scotland and E England, rare Ireland. Regular on passage Hungary. Regional population in winter includes some birds from E of Ural Mtns (subspecies uncertain). Status requires confirmation Iran, where rare but possibly reg-

ular (*fabalis* and sole regional *middendorffii* specimens, *johanseni* also postulated). *Fabalis* vagrant *eg* Azerbaijan, Azores, Balearic Is, Belgium, Channel Is, Faroe Is, France, Madeira, Portugal, Spain and Svalbard, but many extralimital (and often pre-split) bean geese records not specifically assigned, including those from Algeria, Corsica, Egypt, Iceland, Israel, Jan Mayen, Jordan, Lebanon, Liechtenstein, Malta, Moldova, Morocco, Tunisia and Turkey; *johanseni* vagrant Czech Republic.

Anser serrirostris
Tundra Bean Goose

OTHER NAME Bean Goose.
TAXONOMY Gould, 1852. Two subspecies (one in region). Formerly considered conspecific with Taiga Bean Goose, but separated on basis of morphology, behaviour and lack of hybridisation.
DISTRIBUTION [N and NE Siberia > Europe, C and E Asia. Population trend not assessed. NE] Breeds N Norway; moult gatherings of birds breeding outside region in NW Siberia also occur Novaya Zemlya (*rossicus*). Migrates SW via Finland/S Baltic, occurring passage/winter from Poland and Germany W to S Sweden, S Norway and Denmark, S to NE France and N Italy, and through parts of C and E Europe to Black Sea; scarce Britain. Formerly also numerous Iberia, where now very rare, and – presumed this bean goose form – NW Africa in 19th century. Vagrant *rossicus* have occurred *eg* Balearic Is, Faroe Is, Channel Is, Greece, Ireland, Portugal, Romania, Spain and Svalbard, but many pre-split records of extralimital bean geese do not discriminate between taxa – see summary under Taiga Bean Goose.

Anser brachyrhynchus
Pink-footed Goose

TAXONOMY Baillon, 1834. Monotypic. At one time treated with both bean geese as a single species, but morphological and genetic differences justify separate treatment.
DISTRIBUTION [E Greenland and NW Europe > NW Europe. Increasing. LC] Breeds Iceland and Svalbard. Icelandic population, together with extralimital Greenland breeders, migrates SE to winter Britain; scarce Ireland, rare Channel Is. Svalbard population migrates S via W Scandinavia (some also to E via NW Russia, Finland and Baltic) to winter Denmark, Germany, Netherlands and Belgium. Irregular Poland. Vagrant Albania, Austria, Azores, Bosnia-Herzegovina, Bulgaria, Canary Is*, Croatia, Czech Republic, France, Italy, Hungary, Lithuania, Luxembourg, Madeira, Portugal, Romania, Sardinia, Serbia, Slovakia, Slovenia, Spain, Switzerland and Ukraine.

Anser anser
Greylag Goose

OTHER NAMES Greylag; Eastern (or Siberian) Greylag Goose (*A a rubrirostris*), Western (or European) Greylag Goose (*A a anser*).
TAXONOMY (Linnaeus, 1758). Two subspecies (both in region).
DISTRIBUTION [Europe, W, C and E Asia > NW Africa, W and S Europe, SW to E Asia. Increasing. LC] Breeds from Iceland discontinuously E across parts of NW, N, W and C Europe (*anser*), and from E and SE Europe to Turkey, NW Iran and SE Iraq (*rubrirostris*); breeding range has increased significantly, partly through widespread introductions in W Europe, and feral birds in *eg* Britain now greatly outnumber wild population. Mainly resident W Europe but otherwise largely migratory, wintering parts of W and S Europe, N Middle East S to E Arabia, and locally coastal NW Africa. Vagrant Azores, Canary Is, Cyprus, Egypt, Gibraltar, Lebanon, Libya, Liechtenstein, Madeira and Malta; *rubrirostris* vagrant Iceland.

Anser albifrons
Greater White-fronted Goose

OTHER NAMES (Eurasian) White-fronted Goose, White-front; European (or Russian/Siberian) White-fronted Goose (*A a albifrons*), Greenland White-fronted Goose (*A a flavirostris*).
TAXONOMY (Scopoli, 1769). Five subspecies (two in region).
DISTRIBUTION [N North America and N Eurasia > W and S North America, Europe, SW and E Asia. Population trend unknown.

LC] Breeds NW Russia and Novaya Zemlya (*albifrons*). Migrates SW and S to winter Britain, NW Europe S to France, Spain (uncommon) and Italy, C and SE Europe, Egypt, N Middle East and Oman. Separate population breeding W Greenland (*flavirostris*) migrates via Iceland to winter Ireland, W Scotland and W Wales; rare England and Norway. Vagrant Algeria, Azores, Bahrain, Balearic Is, Canary Is, Iceland (*albifrons*), Jan Mayen, Jordan, Kuwait, Libya, Liechtenstein, Madeira, Malta, N Mauritania, Morocco, Portugal, Saudi Arabia, Svalbard, Tunisia and Yemen; *flavirostris* vagrant *eg* Belgium, Channel Is, Faroe Is, Estonia, Finland, France, Netherlands, Norway, Poland (records under review) and Spain.

Anser erythropus
Lesser White-fronted Goose

OTHER NAME Lesser White-front.
TAXONOMY (Linnaeus, 1758). Monotypic.
DISTRIBUTION [N Eurasia > SE Europe, SW and E Asia. Decreasing. VU] Breeding range extends from Norway, Sweden and Finland to NW Russia; dramatic population decline since mid 20th century, resulting in absence from large areas of former range. On migration some stage *eg* Estonia and Poland, with wintering grounds mainly around Evros Delta (Greece/Turkey), Black and Caspian Seas, Azerbaijan, Iraq, and possibly Iran and Syria. Small numbers on passage or in winter Denmark, Germany, Hungary, Slovakia, SE Europe and Armenia. Reintroduced Sweden, this population wintering Netherlands. Vagrant Albania, Austria, Belarus, Belgium, Bosnia-Herzegovina, Britain, Croatia, Cyprus, Czech Republic, France, Egypt, Ireland, Israel, Italy, Jordan, Latvia, Libya, Lithuania, Macedonia, Moldova, Montenegro, Oman, Romania, Serbia, Slovenia, Spain, Switzerland, Turkey and UAE.

Anser indicus
Bar-headed Goose (I)

TAXONOMY (Latham, 1790). Monotypic.
DISTRIBUTION [C and CE Asia > S and SE Asia. Decreasing. LC] Introduced Netherlands, where small naturalised population now self-sustaining, and also considered Cat C vagrant to Spain. Not yet treated as fully established elsewhere but birds of feral origin/escapes also recorded (and sometimes breed) *eg* Belgium, Britain, Croatia, Finland, France, Germany, Hungary, Iceland, Italy, Malta, Svalbard, Switzerland and Ukraine.

Anser caerulescens
Snow Goose (V, I)

OTHER NAMES Greater Snow Goose (*A c atlantica*), Lesser Snow Goose (*A c caerulescens*); Blue Goose (blue morphs of both subspecies).
TAXONOMY (Linnaeus, 1758). Two subspecies (both in region).
DISTRIBUTION [N North America and NE Asia > S North America. Increasing. LC] Rare winter visitor Britain (mainly *caerulescens*, two specimens *atlantica*) and Netherlands; records no longer assessed nationally in either country, but involve wild individuals (*eg* among Pink-footed Geese from Greenland/Iceland, and also Dutch flock of 18 including Canadian-ringed bird) as well as some of captive/feral origin. Tiny naturalised feral population Britain (Cat C) is declining and mainly sedentary; feral breeding also reported Finland, Iceland, Norway and Sweden. Feral/escaped birds may wander widely (*eg* ringed bird from British collection shot Morocco), sometimes with wild geese, thus difficult to determine scale of genuine vagrancy. Classified as vagrant Azerbaijan, Azores*, Belgium, Bulgaria, Czech Republic, Finland, France, Greece, Iceland, Ireland (*caerulescens* and *atlantica*), Italy, NW Kazakhstan, Morocco, Romania and Ukraine; records from Estonia, Germany, Hungary, Latvia, Norway, Svalbard, Switzerland and Tunisia all considered to involve birds of unknown, dubious or captive origin.

Anser rossii
Ross's Goose (V)

OTHER NAME Ross's Snow Goose.
TAXONOMY (Cassin, 1861). Monotypic.
DISTRIBUTION [N Canada > S and SW USA, N Mexico. Increasing. LC] Extreme vagrant (eight records of 14 individuals): Iceland (20 Oct 2013*); Netherlands (six records of 12

individuals); Russia (9 Jun-12 Aug 2006, male paired with Barnacle Goose and three hybrid young fledged; same pairing also recorded 29 Jun 2007, with four eggs; bird then shot 2008). Other records elsewhere (*eg* Britain, Denmark, Faroe Is, Finland, Germany, Norway, Poland and Sweden) generally considered to involve birds of unknown origin or escapes, but some may relate to wild individuals.

Branta canadensis
Canada Goose (V, I)

OTHER NAMES Greater Canada Goose; Atlantic Canada Goose (*B c canadensis*), Dusky Canada Goose (*B c occidentalis*), Giant Canada Goose (*B c maxima*), Intermediate (or Lesser) Canada Goose (*B c parvipes*), Moffitt's (or Great Basin) Canada Goose (*B c moffitti*), Todd's (or Interior/Hudson Bay) Canada Goose (*B c interior*), Vancouver Canada Goose (*B c fulva*).
TAXONOMY (Linnaeus, 1758). Seven subspecies (at least two in region). Formerly also included Cackling Goose (*qv*).
DISTRIBUTION [N North America > S North America. Increasing. LC] Widely introduced and naturalised resident NW Europe, especially Britain and Sweden, also Belgium, Finland, Netherlands, Norway and Germany (*canadensis*), range recently expanding S and E to include *eg* France, Austria and Poland; feral birds have occurred as vagrants (and occasionally bred) in neighbouring countries/territories, including Channel Is, Czech Republic, Hungary, Italy, Slovakia, Slovenia and Spain. Geunuine vagrancy of wild individuals among other geese also established, *eg* Azores (5*), Britain (5, *interior/parvipes*, including neck-collared probable *interior* individual from Maryland, USA, shot Scotland 26 Jan 1993), Iceland (neck-collared individual recorded Oct 2010 marked Maryland, USA, 1 Mar 1994), Ireland (33, *interior*) and Norway (11-12 Oct 2001, *interior*); exact racial identity of vagrants often difficult to confirm and subspecific assessment work ongoing.

Branta hutchinsii
Cackling Goose (V)

OTHER NAMES Lesser Canada Goose; Aleutian (Canada) Goose (*B h leucopareia*), Cackling Canada (or Ridgway's Canada/Dark Cackling) Goose (*B h minima*), Richardson's (or Hutchins's) (Cackling) Goose (*B h hutchinsii*), Taverner's (or Alaska) (Cackling) Goose (*B h taverneri*).
TAXONOMY (Richardson, 1832). Four extant subspecies (at least one in region); fifth taxon, extralimital Bering Canada Goose *B h asiatica*† (NE Russia), now extinct. Formerly considered conspecific with Canada Goose, but separated on basis of morphology, genetics, ecology and behaviour. Limited zone of hybridisation recently reported in Hudson Bay, Canada.
DISTRIBUTION [N North America > W and S North America. Population trend not assessed. NE] Rare vagrant (mainly autumn and winter): Britain (official assessment of records ongoing, but *hutchinsii*, *minima* and *taverneri* all reported among dozens of occurrences); Iceland (16, *hutchinsii*); Ireland (25, *hutchinsii*); Netherlands (9, *hutchinsii*); Svalbard. Other records elsewhere (*eg* Belgium, Finland, France, Germany, Latvia and Lithuania) generally considered to involve birds of unknown origin or escapes, but some occurrences may relate to wild individuals.

Branta bernicla
Brent Goose

OTHER NAMES Brant Goose (IOC), Brant; Black Brant (*B b nigricans*), Dark-bellied (or Russian) Brent Goose (*B b bernicla*), Grey-bellied Brant (form breeding Parry Is, arctic Canada), Pale-bellied (or Light-bellied/Atlantic) Brent Goose (*B b hrota*).
TAXONOMY (Linnaeus, 1758). Three subspecies (all in region). Fourth form from Canadian Arctic, known as Grey-bellied Brant, is distinctive and may be valid subspecies, but lacks formal description. All three recognised subspecies separated at species level by some authorities; research into species limits ongoing.
DISTRIBUTION [Holarctic > NW, W and E North America, NW Europe and E Asia. Population trend unknown. LC] Breeds Svalbard and Franz Josef Land (*hrota*) and very locally inside regional boundary in arctic Russia (*bernicla*). *Hrota* migrates S via

W Norway to Denmark, wintering there and NE England, also Netherlands, with arctic Canada and Greenland population migrating via Iceland to winter Ireland, some also NW France and Channel Is. *Bernicla* migrates SW via White Sea, Gulf of Finland and Baltic to Denmark, wintering S along North Sea coasts to S England and W France, irregularly reaching Spain and Italy. Siberian-breeding *nigricans* scarce but regular among *bernicla* in Britain (annual mean 19, 2000-09), France and Netherlands, and also recorded *eg* Denmark, Finland, Germany, Iceland, Ireland, Svalbard and Sweden. Vagrant Albania, Algeria, Austria, Azores, Belarus, Bulgaria, Canary Is, Croatia, Czech Republic, Egypt, Greece, Hungary, Libya, Luxembourg, Madeira, N Mauritania, Morocco, Portugal, Romania, Salvage Is, Sardinia, Serbia, Slovakia, Slovenia, Switzerland, Tunisia, Turkey and Ukraine; Grey-bellied Brant provisionally documented Britain, Iceland and Ireland.

Branta leucopsis
Barnacle Goose

TAXONOMY (Bechstein, 1803). Monotypic.
DISTRIBUTION [Greenland, N Europe and NW Siberia > NW Europe. Increasing. LC] Breeds Svalbard, Jan Mayen, NW Russia and Novaya Zemlya, and since 1970s locally also SE Iceland, Faroe Is, Fennoscandia, Baltic, Germany, Netherlands and Belgium; latter population partly, perhaps mostly, of captive origin. Populations have largely discrete wintering areas: Russian breeders migrate SW via White Sea, Gulf of Finland and Baltic to *eg* Netherlands, those from Svalbard via W Norway to SW Scotland and NW England, and extralimital Greenland population via Iceland to W Scotland and Ireland. Non-Arctic breeders winter chiefly Netherlands, with these or Arctic birds occasionally S to *eg* France, C Europe and Italy. Difficult to determine whether individual records elsewhere involve genuine vagrants, birds from established feral populations or escapes, but considered as vagrant Algeria, Azores, Belarus, Bulgaria, Canary Is, Egypt, Hungary, Iran, Luxembourg, Morocco, Portugal, Romania, Serbia, Slovakia, Switzerland, Turkey and Ukraine.

Branta ruficollis
Red-breasted Goose

TAXONOMY (Pallas, 1769). Monotypic.
DISTRIBUTION [NW Siberia > SE Europe and SW Asia. Decreasing. VU] May nest arctic Russia just inside regional boundary. Winter visitor Bulgaria, Romania and Ukraine, with smaller numbers regularly *eg* Greece, SW Russia and Turkey (rare); scarce Finland, Sweden, Netherlands, Poland and C Europe. Prior to 1950s wintered farther E, where now rare, notably Azerbaijan, also Iran and Iraq. On migration stages Hungary, also Manych depression, SW Russia, and NW Kazakhstan. Stragglers reach N and W Europe typically in flocks of Brent and White-fronted Geese migrating from Siberian breeding grounds. Vagrant Albania, Algeria, Armenia, Austria, Belarus, Belgium, Bosnia-Herzegovina, Britain, Channel Is, Croatia, Cyprus, Czech Republic, Denmark, Egypt, Estonia, France, Germany, Georgia, Iceland, Israel, Italy, Latvia, Lithuania, Macedonia, Malta, Montenegro, Oman, Norway, Sardinia, Serbia, Slovakia, Slovenia, Spain, Svalbard, Switzerland and Syria. Feral/escaped birds also occur and have bred *eg* Britain and Netherlands; Cat D in Ireland.

Cygnus atratus
Black Swan (I)

TAXONOMY (Latham, 1790). Monotypic.
DISTRIBUTION [Australia and New Zealand. Stable. LC] Introduced Netherlands, where small population now self-sustaining. Not yet fully established elsewhere but has also bred Belgium, Britain, France, Germany, Italy, Switzerland, Ukraine and formerly Austria. Occasional feral/escaped individuals also recorded *eg* Faroe Is, Iceland, Croatia, Czech Republic, Slovakia and UAE.

Cygnus olor
Mute Swan

TAXONOMY (Gmelin, J F, 1789). Monotypic.
DISTRIBUTION [Europe to C Asia. Increasing. LC] Widespread W and C Europe, breeding N to C Sweden and Finland, S to Spain

(scarce) and E through C, E and SE Europe to NW Kazakhstan and Turkey. Wild population augmented by introductions in many areas. Largely migratory in NE and E of range, wintering W and C Europe, Balkans, Turkey, Cyprus, Egypt (rare), Israel (rare) and NW Iran. Vagrant Algeria, Armenia, Azores, Bahrain, Balearic Is, Corsica, Faroe Is, Iceland, Jordan, Kuwait, Lebanon, Malta, Morocco, Oman, Saudi Arabia, Svalbard, Tunisia and UAE; Cat D Portugal.

Cygnus columbianus
Tundra Swan

OTHER NAMES Bewick's Swan (*C c bewickii*), Whistling Swan (*C c columbianus*).
TAXONOMY (Ord, 1815). Two subspecies (both in region).
DISTRIBUTION [N North America and N Eurasia > S North America, Europe and E Asia. Population trend unknown. LC] Breeds NW Russia (*bewickii*). Migrates SW through White Sea and Baltic to winter chiefly Britain, Netherlands and Germany, also Ireland, Denmark, Belgium and N France; some migrate S across continental Europe to winter *eg* S France, W and N Black Sea coasts, Turkey, SE Iraq and W Iran. Irregular Italy. Vagrant Albania, Algeria, Azerbaijan, Balearic Is, Belarus, Channel Is, Corsica, Croatia, Cyprus, Czech Republic, Hungary, Iceland, Israel, Jordan, Libya, Oman, Portugal, Romania, Sardinia, Saudi Arabia, Serbia, Slovakia, Slovenia, Spain, Svalbard and UAE; *columbianus* (North America) vagrant Britain (2), Estonia (4), Ireland (5), Netherlands (5) and Sweden (2).

Cygnus cygnus
Whooper Swan

TAXONOMY (Linnaeus, 1758). Monotypic.
DISTRIBUTION [N and C Eurasia > Europe and E Asia. Population trend unknown. LC] Breeds Iceland and from Scandinavia and N Germany E through Baltic region to N and E Russia. Majority of Icelandic population winters Britain and Ireland; continental European breeders winter mainly from Norway S to C Europe, and from NE Italy and parts of Balkans E to W Turkey and Black and Caspian Seas, notably NW Iran. Vagrant Algeria, Channel Is, Cyprus, Egypt, Iraq, Israel, Jordan, Liechtenstein, Morocco, Oman, Portugal, Saudi Arabia, Slovenia, Spain, Tunisia and UAE.

Plectropterus gambensis
Spur-winged Goose (V)

OTHER NAMES Black (or Gambian) Spur-winged (or Spurwing) Goose, Spur-winged Duck.
TAXONOMY (Linnaeus, 1766). Two subspecies (one in region).
DISTRIBUTION [Sub-Saharan Africa. Increasing. LC] Extreme vagrant (two records, *gambensis*): N Mauritania (11-14 Dec 2004, 14 birds); Morocco (12 and 23 Mar, 23 Apr-12 Dec 1984).

Sarkidiornis melanotos
Knob-billed Duck (V)

OTHER NAMES African Comb Duck, Knob-billed (or Black-backed) Goose.
TAXONOMY (Pennant, 1769). Monotypic. Sometimes considered to include extralimital Comb Duck *S sylvicola* (South America).
DISTRIBUTION [Sub-Saharan Africa, S and SE Asia. Decreasing. LC] Extreme vagrant (two records): N Mauritania (7 Dec 1984); Oman (21 Aug 1990-8 Nov 1998, male, paired with feral female Mallard in 1992; produced single hybrid young).

Alopochen aegyptiaca
Egyptian Goose (I)

OTHER NAMES African Sheldgoose, Nile Goose.
TAXONOMY (Linnaeus, 1766). Monotypic.
DISTRIBUTION [Sub-Saharan Africa. Decreasing. LC] Naturalised self-sustaining populations established Belgium, Britain, Germany, Netherlands, UAE and possibly France, Israel and Italy. Former native breeding range included S Egypt; rare winter visitor Cyprus 1906-28 (three later occurrences possibly involving feral birds), and few records Alge-

ria mainly 19th century and Tunisia early 20th century. Single records from Malta in 1914 and N Mauritania in 2000, and classified as vagrant Hungary, Israel, Jordan, Saudi Arabia, Turkey and Yemen, but in latter countries and elsewhere recent records more likely to refer to wandering or colonising feral individuals and escapes.

Tadorna tadorna
Common Shelduck

OTHER NAMES (Eurasian or Northern) Shelduck.
TAXONOMY (Linnaeus, 1758). Monotypic.
DISTRIBUTION [Europe and N Africa to E Asia > S and SE Asia. Increasing. LC] Breeds NW Europe (mainly coastally) from Iceland and N Scandinavia S to Biscay, SW Iberia, Tunisia, parts of Mediterranean Europe E to Black Sea, SE European Russia, NW Kazakhstan and Caspian Sea S to scattered areas of N Middle East; scarce but increasing C and E Europe. In winter more widespread S of main breeding areas, including N Africa and throughout Middle East, but rarer Arabia. Vagrant Azores, Canary Is, Faroe Is, Liechtenstein, Madeira, N Mauritania, Moldova and Svalbard.

Tadorna ferruginea
Ruddy Shelduck

OTHER NAME Brahminy Duck.
TAXONOMY (Pallas, 1764). Monotypic.
DISTRIBUTION [NW and E Africa, SE Europe to E Asia > S Asia. Population trend unknown. LC] Breeds Canary Is, NW Africa and from S Balkans E to NW Kazakhstan, W Caspian region, Asia Minor, Iraq and Iran. Winter range includes Cyprus and scattered areas of N Middle East; rare or irregular S Arabia and Italy, and formerly also wintered S Iberia. Wild population augmented by reintroduced/feral stock *eg* Bulgaria, Russia and Ukraine, and escapes from captivity also frequent in region (and occasionally breeding). True pattern of movements and vagrancy status thus obscured, but significant periodic influxes linked to drought in E of range (*eg* in 1892 and 1994) known to have triggered widespread records across Europe (including recovery in Poland of bird ringed 4,000 km E in Kyrgyzstan). Extralimital records from more than 25 countries in region of birds variously classified as wild, feral, captive or unknown origin; see Vinicombe and Harrop 1999 for historical summary. Away from Europe, vagrant Libya and N Mauritania.

Cairina moschata
Muscovy Duck (I)

OTHER NAMES Muscovy, Musk Duck.
TAXONOMY (Linnaeus, 1758). Monotypic.
DISTRIBUTION [S North America to N South America. Decreasing. LC] Naturalised population established Italy. Other small/localised populations originating from domestic stock also established elsewhere in region (*eg* in Britain known from 53 sites in 2000), but generally considered semi-feral rather than self-sustaining.

Aix sponsa
Wood Duck (V)

OTHER NAME Carolina Duck.
TAXONOMY (Linnaeus, 1758). Monotypic.
DISTRIBUTION [North America > S North America. Increasing. LC] Very rare vagrant (mainly autumn-winter): Azores (12*, including one ringed in North Carolina, USA); Iceland (7*). Other records attributed to unknown origin, escapes from captivity or failed introduction attempts, with occasional breeding reported, especially Britain but also *eg* Austria, Belgium, Germany and Ireland.

Aix galericulata
Mandarin Duck (I)

OTHER NAME Mandarin.
TAXONOMY (Linnaeus, 1758). Monotypic.
DISTRIBUTION [E Asia; introduced elsewhere. Decreasing. LC] Main naturalised population Britain, but has also bred Austria, Belgium, France, Germany, Italy, Netherlands, Norway, Poland, Slovakia, Spain and Switzerland. Essentially non-migratory; some dispersal recorded, including notable recoveries of British-ringed birds in Hungary and W Russia,

and records documented in at least 20 further countries.

Nettapus coromandelianus
Cotton Pygmy Goose

OTHER NAMES Cotton Teal, Goose Teal, Indian Pygmy-goose, Rice Teal, White Pygmy-goose.
TAXONOMY (Gmelin, J F, 1789). Two subspecies (one in region).
DISTRIBUTION [S, SE and E Asia to NE Australia. Stable. LC] Winters Oman (*coromandelianus*). Vagrant Bahrain, Jordan, Iran, Turkey*, UAE and Yemen.

Anas capensis
Cape Teal (V)

OTHER NAME Cape Wigeon.
TAXONOMY Gmelin, J F, 1789. Monotypic.
DISTRIBUTION [Sub-Saharan Africa. Increasing. LC] Extreme vagrant: Israel (3 records of 4 birds: 26 Jun-16 Oct 1978; 15-19 May 1982; 28 July 1984, two); Libya (3 records of 6 birds: Apr 1961, remains found; 3 Apr 1961; 31 Mar-5 Apr 1968, two pairs).

Anas strepera
Gadwall

OTHER NAME Grey Duck.
TAXONOMY Linnaeus, 1758. Two subspecies recognised; Pacific island *couesii* now extinct.
DISTRIBUTION [N North America, Europe, W and SW to E Asia > W and S North America, N and NE Africa, W and S Europe, SW to E Asia. Population trend unknown. LC] Breeds Iceland, Ireland, Britain, patchily in SW, W, C and SE Europe N to S Fennoscandia, and from E Europe E to Ural Mts, NW Kazakhstan and Caucasus, locally also Turkey and NW Iran. Mainly resident W and S Europe and Asia Minor, but otherwise largely migratory, occurring across much of Europe and Middle East on passage; fragmented winter range includes discontinuous areas of W, S and SE Europe, Middle East, Nile Valley and locally coastal N Africa (irregular W to Canary Is). Rare migrant Faroe Is. Vagrant Azores, Gibraltar, Madeira, Malta and Svalbard.

Anas falcata
Falcated Duck (V)

OTHER NAMES Falcated (or Bronze-capped) Teal.
TAXONOMY Georgi, 1775. Monotypic.
DISTRIBUTION [E Asia > S, SE and E Asia. Decreasing. NT] Very rare vagrant (autumn-spring, often among flocks of wintering Eurasian Wigeon and Gadwall): Austria (1839, shot); Bulgaria (23 Feb 1978); Iraq (2: 19 Mar 1916; winter 1968-1969); Jordan (10 Jan 1969, shot); Malta (10 Oct 1979, shot); Netherlands (12); Spain (3). Possibility of occurrence of escaped birds from wildfowl collections means that many other regional records are treated as of unknown origin (*eg* Cat D in Finland, France, Ireland, Portugal and Sweden), involving escapes (*eg* Cat E in Germany), or placed in both categories (Belgium and Britain); planned stable isotope analysis of drake found shot in Britain (24 Nov 2000) may help determine vagrancy likelihood.

Anas penelope
Eurasian Wigeon

OTHER NAMES (European) Wigeon.
TAXONOMY Linnaeus, 1758. Monotypic.
DISTRIBUTION [N Eurasia > Sahel, W, N and NE Africa, S Eurasia. Decreasing. LC] Breeds Iceland, Faroe Is, N Britain (formerly also Ireland) and from Fennoscandia and Baltic States E through NW Russia, locally S also to *eg* N Germany, Poland and N Ukraine. Largely migratory, occurring across much of Europe and parts of Middle East on migration, but winters principally in North Sea countries of W Europe, also S Scandinavia, France and Iberia; Siberian breeding populations winter from NW Africa and Mediterranean basin to Black and Caspian Seas and Middle East. Scarce but regular Azores, irregular Canary Is. Vagrant Cape Verde Is, Gibraltar, Jan Mayen and Madeira.

Anas americana
American Wigeon

OTHER NAME Baldpate.
TAXONOMY Gmelin, J F, 1789. Monotypic.

DISTRIBUTION [N North America > W and S North America to N South America. Stable. LC] Scarce autumn and winter visitor, mainly Oct-May, Azores (annual mean 15*, 2005-2012) and Britain (annual mean 17, 2000-09; total 433, 1958-2007). Significant majority of records relate to more readily identifiable males, thus clearly under-recorded. Vagrant Austria, Belgium, Canary Is, Cape Verde Is, Channel Is, Czech Republic, Denmark, Estonia, Faroe Is, Finland, France, Germany, Hungary, Iceland, Ireland, Italy, Lithuania, Madeira, Morocco, Netherlands, Norway, Poland, Portugal, Spain and Sweden.

Anas rubripes
American Black Duck (V)

OTHER NAMES (North American) Black Duck.
TAXONOMY Brewster, 1902. Monotypic.
DISTRIBUTION [NC and NE North America > SE North America. Decreasing. LC] Vagrant, but many records involve long-staying or returning individuals, some also breeding: Azores (33 records of 58 birds*; resident in very small numbers Flores/Corvo since 1999, with breeding of pure pair and subsequent hybridisation with Mallards establishing mixed population of hybrids and pure individuals); Britain (40; hybridised with Mallard 1978-1983, producing 22 young); Canary Is (5 Jan-3 Apr 1994); Faroe Is (series of records 2003-2010 may have involved only 2 individuals; hybrids with Mallard also recorded, suggesting local breeding); Finland (26 May-30 Jun 2004); France (5, including Canadian-ringed individual); Iceland (39); Ireland (19); Norway (4); Spain (12); Sweden (2).

Anas platyrhynchos
Mallard

TAXONOMY Linnaeus, 1758. Two subspecies (one in region).
DISTRIBUTION [North America, NW Africa and Eurasia > North America, West Indies, Europe, NW to NE Africa and SW to NE Asia. Decreasing. LC] Breeds across Europe from Mediterranean N to Iceland and N Fennoscandia, W to Britain and Ireland and E to Ural Mtns and NW Kazakhstan, also Caucasus, Turkey, NW Iran, isolated areas of Levant and Arabia, and locally NW Africa, Canary Is and Azores (*platyrhynchos*); populations in some areas introduced (*eg* Arabia, Canary Is) or with domestic influence. Largely migratory in N and E of range, wintering S to N Africa, W to Azores (rare on passage Madeira) and through Middle East. Vagrant Gibraltar, Jan Mayen and N Mauritania.

Anas discors
Blue-winged Teal

TAXONOMY Linnaeus, 1766. Monotypic.
DISTRIBUTION [N North America > S North America to N South America. Decreasing. LC] Very scarce autumn (mostly) and winter visitor Azores, where occurs annually in variable but sometimes high numbers (*eg* 66* in 2009 and 56* in 2010), and occasionally in small flocks (max 12, 18 Jan 2010). Vagrant Algeria, Austria, Belgium, Britain, Canary Is, Cape Verde Is, Channel Is, Corsica, Czech Republic, Denmark, Egypt, Estonia, Faroe Is, Finland, France, Germany, Greece, Hungary, Iceland, Ireland, Italy, Libya, Malta, Morocco, Netherlands, Norway, Poland, Portugal, Spain, Svalbard, Sweden, Switzerland and Tunisia.

Anas clypeata
Northern Shoveler

OTHER NAMES (Common or European) Shoveler.
TAXONOMY Linnaeus, 1758. Monotypic.
DISTRIBUTION [N North America and N Eurasia > S North America to N South America, Sahel, N and NE Africa, S Eurasia. Decreasing. LC] Fragmented breeding range in W of region extends from S Iberia N to Britain, Ireland, Iceland and S Scandinavia, and E to parts of C and SE Europe; much more widespread from E Europe N to Finland and NW Russia and E to N Caspian and Caucasus, with isolated small populations also Turkey, NW Iran and Syria; introduced Saudi Arabia. Migratory in N and E of range, occurring across Europe and Middle East on passage, and wintering S to coastal N Africa (irregular W to Canary Is) and Nile Valley. Vagrant Azores, Cape Verde Is, Madeira and Svalbard.

Anas erythrorhyncha
Red-billed Teal (V)

OTHER NAMES African Red-billed Teal, Red-billed Pintail.
TAXONOMY Gmelin, J F, 1789. Monotypic.
DISTRIBUTION [S, SW, C, E and NE Africa. Decreasing. LC] Extreme vagrant: Israel (20 Jun-12 July 1958).

Anas acuta
Northern Pintail

OTHER NAMES (Common) Pintail.
TAXONOMY Linnaeus, 1758. Monotypic.
DISTRIBUTION [N North America and N Eurasia > W and S North America to N South America, S Eurasia, Sahel, N and NE Africa. Decreasing. LC] Breeds Iceland, Ireland, Britain and across much of N Europe E to Ural Mtns, with scattered breeding areas to S in C and E Europe and Turkey. Largely migratory, occurring widely on passage W to Azores (scarce but regular), with fragmented winter range extending S to N Africa (irregular Canary Is) and Middle East. Vagrant Cape Verde Is, Gibraltar, Jan Mayen, Madeira and Moldova.

Anas querquedula
Garganey

TAXONOMY Linnaeus, 1758. Monotypic.
DISTRIBUTION [W Europe to E Asia > Sahel, W, C, E and NE Africa, S and SE Asia. Decreasing. LC] Summer visitor from Ireland (rare) and Britain (scarce) E across much of Europe to Ural Mtns, NW Kazakhstan and Caucasus, N to C Fennoscandia and patchily S to Iberia and Mediterranean, Turkey and NW Iran, occasionally also Israel and Lebanon; may breed Iraq and Syria. Occurs widely on passage S Europe, N Africa (W to Canary Is) and Middle East; some winter Arabia, N Africa and occasionally also S Europe. Vagrant Azores, Cape Verde Is, Faroe Is, Gibraltar, Iceland, Madeira and Svalbard.

Anas formosa
Baikal Teal (V)

OTHER NAME Formosa Teal.
TAXONOMY Georgi, 1775. Monotypic.
DISTRIBUTION [NC and NE Asia > E Asia. Increasing. LC] Very rare vagrant (chiefly in winter): Belgium (4); Britain (6); Denmark (24 Nov 2005, shot); Finland (2); France (4 pre-1950 records, others in Cat D); Ireland (19-23 Feb 2010); Italy (11); Malta (16 Apr 1912, three); Netherlands (12); Norway; Spain (2); Svalbard; Ukraine (29 Aug 1954). Further records considered to be of uncertain origin (Cat D) or involving escapes from captivity (Cat E); stable isotope analysis of British and Danish specimens has confirmed capability of species to reach region from NE Asia.

Anas crecca
Eurasian Teal

OTHER NAMES (Common) Teal; Aleutian Green-winged Teal (*A c nimia*).
TAXONOMY Linnaeus, 1758. Two subspecies (one in region). Formerly included Green-winged Teal (*qv*).
DISTRIBUTION [W Europe to E and NE Asia, NW North America > N and NE Africa, Sahel, W and S Europe, SW to E Asia. Population trend unknown. LC] Breeds across much of N Europe from Britain, Ireland and W France E to Caucasus, SW Russia and NW Kazakhstan, and N to Iceland, Fennoscandia and NW Russia, with isolated populations Italy and Turkey (*crecca*); has bred Svalbard. Migratory across much of N and E of range, occurring on passage across Europe and wintering S to N Africa and through Middle East; regular W to Azores, Canary Is and perhaps N Mauritania (rare). Vagrant Cape Verde Is, Gibraltar, Jan Mayen and Madeira.

Anas carolinensis
Green-winged Teal

OTHER NAMES (North) American Green-winged Teal.
TAXONOMY Gmelin, J F, 1789. Monotypic. Previously considered conspecific with Eurasian Teal, but morphological and behavioural differences and genetic evidence justify separate treatment at species level.
DISTRIBUTION [N North America > S North America and West Indies. Population trend not assessed. NE] Scarce autumn and winter

visitor Britain, mainly Nov-May (annual mean 46, 2000-09), but significantly under-recorded in region as only adult male plumage is readily identifiable. Vagrant Austria, Azores, Belgium, Bulgaria, Canary Is, Cape Verde Is, Channel Is, Denmark, Estonia, Faroe Is, Finland, France, Germany, Hungary, Iceland, Ireland, Italy, Jan Mayen, Madeira, Morocco, Netherlands, Norway, Poland, Portugal, Slovenia, Spain, Svalbard, Sweden and Switzerland.

Marmaronetta angustirostris
Marbled Duck

OTHER NAME Marbled Teal.
TAXONOMY (Ménétriés, 1832). Monotypic.
DISTRIBUTION [NW Africa, SW Europe, SW to C Asia > N Africa, Sahel, SW Europe and SW Asia. Decreasing. VU] Breeds locally NW Africa (recently also Libya), S and E Spain and occasionally Canary Is and Mallorca, but main populations scattered through N Middle East and S Caucasus; also formerly or irregularly bred *eg* Balkans, Crete, Cyprus, Egypt, France and Italy. Populations in Turkey and S Caucasus principally migratory; occurs more widely through N Middle East on passage and in winter, especially SE Iraq. Spanish population also prone to dispersal, especially in droughts; some winter Morocco. Vagrant Albania, Bahrain, Bulgaria, Bosnia-Herzegovina, Cape Verde Is, Cyprus, Czech Republic, France, Georgia, Germany, Greece, Hungary, Kuwait, Macedonia, Madeira, Malta, Netherlands, Oman, Portugal, Romania, Qatar, Sardinia, Saudi Arabia, Serbia and UAE. Further records from *eg* Belgium, Britain, Denmark, Poland and Switzerland treated as uncertain origin (Cat D) or escapes (Cat E).

Netta rufina
Red-crested Pochard

OTHER NAME Red-crested Duck.
TAXONOMY (Pallas, 1773). Monotypic. European and Asian populations genetically distinct.
DISTRIBUTION [SW Europe to C Asia > N Africa, S Europe, SW and S Asia. Population trend unknown. LC] Breeds SW, S and C Europe N to Netherlands, E Europe and S Baltic, and from Danube Delta discontinuously E to N Caucasus and NW Kazakhstan, also N Morocco, Turkey, Syria and Iraq; has bred Cyprus and Iran, and feral population established Britain. Migratory in N and E of range, occurring on passage and in winter more widely C and S Europe, also Black and Caspian Seas, N Middle East, Oman (rare) and locally N Africa. Vagrant Algeria (formerly bred), Bahrain, Belarus, Corsica, Estonia, Finland, Ireland, Kuwait, Latvia, Lebanon, Liechtenstein, Lithuania, Malta, Moldova, Montenegro, Norway, Qatar, Saudi Arabia and UAE.

Netta erythrophthalma
Southern Pochard (V)

OTHER NAMES Red-eyed Pochard; (South) African Pochard (*N e brunnea*), South American Pochard (*N e erythrophthalma*).
TAXONOMY (zu Wied-Neuwied, 1833). Two subspecies (one in region).
DISTRIBUTION [NW, N and E South America, NE, E, C and S Africa. Decreasing. LC] Extreme vagrant: Israel (22 Apr-8 May 1998, *brunnea*).

Aythya valisineria
Canvasback (V)

TAXONOMY (Wilson, A, 1814). Monotypic.
DISTRIBUTION [N North America > S North America. Decreasing. LC] Extreme vagrant (eight records, two involving returning wintering individuals): Britain (7: 7 Dec 1996; 18 Jan-10 Mar 1997, returning 23-24, 29 and 30 Nov, 3 Dec 1997-9 Mar 1998, 6-16 Apr 1999, 8 Nov 1999-15 Feb 2000 and 12 Nov 2000-13 Feb 2001; 28 May 1999; 29 Jan-14 Mar 2000, returning 18 Nov-8 Mar 2001; 19 Apr 2000; 21-23 Jun 2000; 11-30 July 2002); Iceland (11 Apr 1977, collected). Previously accepted record from Netherlands (initially 9 Jan 2003 then recorded most winters until Nov 2009) rejected after review when bird discovered to have wing-tag.

Aythya americana
Redhead (V)

OTHER NAMES American (or Red-headed) Pochard.

TAXONOMY (Eyton, 1838). Monotypic.
DISTRIBUTION [N North America > S North America and West Indies. Increasing. LC] Extreme vagrant (four records, of which three in midsummer contrast with typical vagrancy pattern for Nearctic wildfowl): Britain (8-27 Mar 1996 and again 4-24 Feb 1997); Iceland (2: 15 Jun-10 Jul 1998; 11-12 Jul 1998); Ireland (12-15 Jul 2003).

Aythya ferina
Common Pochard

OTHER NAMES (Eurasian or European) Pochard, Northern Pochard.
TAXONOMY (Linnaeus, 1758). Monotypic.
DISTRIBUTION [W Europe to CE Asia > W and S Europe, SW to E Asia. Decreasing. VU] Breeds from Britain and Ireland E across much of Europe to Ural Mtns, N to S Fennoscandia and more patchily S to Iberia and W Mediterranean, with small outlying populations *eg* S Caucasus, NW Iran, Turkey and Tunisia. Largely migratory in E of range, occurring widely to W, S to Mediterranean and across Middle East on passage and in winter, also locally coastal N Africa (irregular Canary Is). Vagrant Azores, Cape Verde Is, Faroe Is, Iceland (has bred), Madeira, N Mauritania and Svalbard.

Aythya nyroca
Ferruginous Duck

OTHER NAMES Ferruginous (or White-eyed) Pochard, Ferruginous (or Common) White-eye.
TAXONOMY (Güldenstädt, 1770). Monotypic.
DISTRIBUTION [NW Africa, SW Europe to C Asia > N Africa, Sahel, SW Europe to E Asia. Decreasing. NT] Breeds mainly from EC Europe and N Balkans through E Europe to NW Kazakhstan and Caucasus, also NW Africa and scattered localities in Spain (very rare), Italy and occasionally France, and N Middle East S to C Arabia; has bred Belgium, Corsica, Cyprus, Czech Republic, Egypt, Germany, Netherlands and Switzerland. Mainly migratory in E of European range, occurring through C and SE Europe and Middle East on passage; winters more widely N Mediterranean and Middle East, also N Africa. Scarce visitor NW to *eg* Britain, Benelux and Germany. Vagrant Azores, Bahrain, Canary Is, Cape Verde Is, Channel Is, Denmark, Estonia, Finland, Ireland, Kuwait (has bred), Latvia, Norway, Portugal (has bred) and Sweden.

Aythya collaris
Ring-necked Duck

OTHER NAME Ring-billed Duck.
TAXONOMY (Donovan, 1809). Monotypic.
DISTRIBUTION [N North America > W and S North America, Central America and West Indies. Increasing. LC] Scarce autumn and winter visitor, mainly Oct-Mar, to Azores (average winter population 21, 2004-2009*), Britain (annual mean 32, 2000-09) and France (annual mean 12, 2003-2008). Vagrant Algeria, Austria, Belgium, Canary Is, Cape Verde Is, Channel Is, Denmark, Estonia, Faroe Is, Finland, Germany, Hungary, Iceland, Ireland, Italy, Madeira, Morocco, Netherlands, Norway, Poland, Portugal, Slovakia, Spain, Svalbard, Sweden and Switzerland.

Aythya fuligula
Tufted Duck

OTHER NAME Tufted Pochard.
TAXONOMY (Linnaeus, 1758). Monotypic.
DISTRIBUTION [Europe and N Asia > W and S Europe, N, W and E Africa, SW to E Asia. Stable. LC] Breeds mainly W, C, N and E Europe, with scattered populations S to *eg* N Spain, S France, N Italy and W Balkans. Mainly resident in W of range but migratory in N and E, occurring widely on passage and in winter S to Mediterranean, N Africa (W to Canary Is, rare N Mauritania) and across Middle East. Vagrant Azores, Cape Verde Is, Gibraltar, Madeira and Malta.

Aythya marila
Greater Scaup

OTHER NAME Scaup.
TAXONOMY (Linnaeus, 1761). Two subspecies (one in region).

DISTRIBUTION [N North America and N Eurasia > NW, W and E North America, Europe, SW and E Asia. Decreasing. LC] Breeds Iceland, Fennoscandia, Estonia and far NW Russia (*marila*); has also bred *eg* Britain, Denmark, Germany and Svalbard. Migrates S to winter chiefly Baltic, North Sea and Atlantic coasts S to Biscay, also C, E and SE Europe, Mediterranean, Adriatic, Black and Caspian Seas, Turkey and N Iran. Vagrant Azores, Algeria, Balearic Is, Bear Is, Canary Is, Channel Is, Corsica, Cyprus, Gibraltar, Iraq, Israel, Jan Mayen, Liechtenstein, Madeira, Malta, Montenegro, Morocco, Portugal, Svalbard, Tunisia and UAE.

Aythya affinis
Lesser Scaup (V)

TAXONOMY (Eyton, 1838). Monotypic.
DISTRIBUTION [N North America > S North America to extreme N South America. Decreasing. LC] Vagrant (almost all records from countries on W seaboard of region): Azores (55 to 2011*); Belgium (3); Britain (181); Canary Is (17); Cape Verde Is (2); Channel Is (2); Denmark (8); Faroe Is (2); Finland (21 Mar-10 Apr 2004); France (31); Germany (18); Hungary (19-20 Apr 2003); Iceland (14*); Ireland (27); Morocco (3); Netherlands (9); Norway (6); Poland (9 Dec 2001); Portugal (4); Spain (19); Sweden (6).

Polysticta stelleri
Steller's Eider

TAXONOMY (Pallas, 1769). Monotypic.
DISTRIBUTION [Alaska and arctic Siberia > Alaska, N Europe, NE Asia and Bering Sea. Decreasing. VU] Winters extreme NW Russia, N Norway and E Baltic, chiefly Estonia and Lithuania, also Sweden (now scarce); regular visitor Svalbard and rare but regular Finland, mostly on migration early May. Some non-breeders oversummer in N of winter range, and has bred NW Russia. Vagrant Belarus, Britain, Czech Republic, Denmark, France, Germany, Hungary, Iceland, Latvia (perhaps more regular than records suggest), Netherlands, Norway and Poland.

Somateria fischeri
Spectacled Eider (V)

OTHER NAME Fischer's Eider.
TAXONOMY (von Brandt, J F, 1847). Monotypic.
DISTRIBUTION [Alaska and NE Siberia > Bering Sea. Decreasing. LC] Extreme vagrant: Norway (3: 12 Dec 1933, collected; 23-24 Feb 1988, three; 15 Jun 1997); Svalbard (2: 30 Apr 2002, at sea, three; 20 Jun 2012, pair).

Somateria spectabilis
King Eider

TAXONOMY (Linnaeus, 1758). Monotypic.
DISTRIBUTION [Circumpolar > NW and NE North America, N Europe. Decreasing. LC] Breeds Svalbard, NW Russia and Novaya Zemlya. Migrates W to winter primarily from White Sea to arctic Norway, small numbers regularly W to Iceland and Jan Mayen, S to Sweden, Denmark and Finland (rare, mostly Apr-May); regular on passage Estonia, and some non-breeders oversummer in N of winter range. Vagrant Azores, Belgium, Britain, Czech Republic, France, Germany, Hungary, Ireland, Italy, Latvia, Lithuania, Netherlands, Poland and Spain.

Somateria mollissima
Common Eider

OTHER NAMES Eider (Duck); Dresser's (or American) Eider (*S m dresseri*), European Eider (*S m mollissima*), Northern Eider (*S m borealis*), Pacific Eider (*S m v-nigra*).
TAXONOMY (Linnaeus, 1758). Six subspecies (four in region). Relationships determined by genetic studies do not correspond with currently accepted subspecies limits, but are weakly supported; more research needed.
DISTRIBUTION [N North America, NW and NE Eurasia > NW and NE North America, NW and W Europe, NE Asia. Population trend unknown. NT] Breeds N Ireland, N Britain except Northern Isles, and from Netherlands N along North Sea, Baltic and Scandinavian coasts to extreme NW Russia and Novaya Zemlya, also Ukraine and N Italy (*mollissima*), Shetland, Ork-

ney and Faroe Is (*faeroeensis*) and Iceland, Jan Mayen, Svalbard and Franz Josef Land (*borealis*); has bred or does so irregularly *eg* Austria, Bulgaria, France, Poland and Switzerland. Mainly resident in W, but Baltic and High Arctic breeders move S and W in winter, when species' range extends S to Biscay coasts, small numbers also C Europe, NW Mediterranean, Adriatic and parts of Black Sea. Rare Montenegro. Vagrant Azores, Balearic Is, Belarus, Bosnia-Herzegovina, Corsica, Croatia, Czech Republic, Greece, Hungary, Israel, Liechtenstein, Luxembourg, Portugal, Romania, Serbia, Slovakia, Tunisia and Turkey; *borealis* vagrant *eg* Denmark*, Ireland and Scotland; *dresseri* vagrant Ireland.

Histrionicus histrionicus
Harlequin Duck

OTHER NAME Harlequin.
TAXONOMY (Linnaeus, 1758). Monotypic.
DISTRIBUTION [NW and NE North America, Iceland and NE Asia. Increasing. LC] Resident Iceland, breeding on inland rivers and wintering coastally. Sedentary nature of this small population suggests that some extralimital records (*eg* continental Europe) more likely to involve more migratory NE Asian population, and conceivably some British vagrants originate from North America. Vagrant Austria, Britain, Faroe Is, Finland, Germany, Italy, Jan Mayen, Lithuania, Netherlands, Norway, Poland, Russia, Slovakia, Svalbard, Sweden and Ukraine.

Melanitta perspicillata
Surf Scoter

TAXONOMY (Linnaeus, 1758). Monotypic.
DISTRIBUTION [N North America > W and E North America. Decreasing. LC] Scarce autumn and winter visitor Britain (mainly Oct-May but noted all months), chiefly in N, W and SW (annual mean 18, 2000-09, including many returning individuals). Vagrant Azores, Belgium, Channel Is, Denmark, Estonia, Faroe Is, Finland, France, Germany, Iceland, Ireland, Madeira, Netherlands, Norway, Portugal, Spain and Sweden.

Melanitta fusca
Velvet Scoter

TAXONOMY (Linnaeus, 1758). Monotypic. Formerly included White-winged Scoter (*qv*).
DISTRIBUTION [N Europe, W Siberia and SW Asia > N and W Europe, SW Asia. Decreasing. VU] Breeds from Norway E through Fennoscandia, N Baltic and Estonia to far NW Russia and Ural Mtns, with small disjunct population in Caucasus and E Turkey. Migrates S and SW to winter principally S Baltic, W Scandinavia and from North Sea and Atlantic coasts of Britain, Ireland and continental Europe S to Biscay coasts, small numbers also reaching Mediterranean and (especially in cold winters) C Europe; in E, winters locally *eg* Balkans, Turkey, Black and Caspian Seas. Vagrant Albania, Algeria, Balearic Is, Belarus, Bosnia-Herzegovina, Corsica, Egypt, Faroe Is, Greece, Iceland, Iran, Israel, Jan Mayen, Lebanon, Luxembourg, Macedonia, Morocco, Portugal, Slovenia and Svalbard.

Melanitta deglandi
White-winged Scoter (V)

OTHER NAMES American (or Degland's) White-winged Scoter (*M d deglandi*), Asian White-winged (or Asiatic/Stejneger's) Scoter (*M d stejnegeri*).
TAXONOMY (Bonaparte, 1850). Two subspecies (both in region). Previously considered conspecific with Velvet Scoter, but separated on basis of diagnostic differences in bill structure and pattern, male plumage, tracheal structure and vocalisations; some of these distinctions also apply between the two White-winged subspecies, indicating potential further split, but more study required.
DISTRIBUTION [N North America, C and E Siberia > W and E North America, E and NE Asia. Decreasing. LC] American *deglandi* and Asian *stejnegeri* both very rare vagrants to region, former outnumbering latter *c* 2:1; occurrences listed here separately by subspecies. *Deglandi* (30 records of *c* 19 individuals, all NW Europe): Britain (11-23 Jun 2011); Denmark (23 Jan-3 Mar 2013); Faroe Is (1 Jul 2011); Finland (11-13 Jun 2012); Iceland (26 records thought to involve 15 individuals, including 2 returning males 1998-2007). *Stejnegeri* (11

records, scattered across E, N, W and SW Europe; hypothesised occurrence in Caspian Sea as yet unsubstantiated): Britain (26 Dec 2013*); Denmark (12-18 Oct 2009, returning 19 Mar 2010 & 18 Feb-26 Mar 2011); Finland (2: 27 May-8 Jun 1996, trapped in fishing net; 27 May-12 Jun 2012); France (4 Dec 1886, collected); Iceland (6 Apr-2 May 2003); Ireland (1 Feb-11 Apr 2011); Norway (21-22 Jun 2011); Poland (10 Mar 2007); Spain (13-23 and 27-30 Dec 2011); Sweden (3 May 2012).

Melanitta nigra
Common Scoter

OTHER NAMES (Black) Scoter.
TAXONOMY (Linnaeus, 1758). Monotypic. Formerly included Black Scoter (*qv*).
DISTRIBUTION [N Europe to N Asia > W Europe and NW Africa. Population trend unknown. LC] Breeds Iceland, Ireland, N Britain, N Fennoscandia and far NW Russia, occasionally also Svalbard; has bred Faroe Is and Latvia. Migrates S and SW to winter principally S Baltic, W Scandinavia and from North Sea and Atlantic coasts of Britain, Ireland and continental Europe S to N Morocco, small numbers reaching N Mauritania and W Mediterranean. A few migrants/winterers regularly occur also C Europe and Black and Caspian Seas. Vagrant Albania, Algeria, Azerbaijan, Azores, Balearic Is, Belarus, Bosnia-Herzegovina, Canary Is, Corsica, Croatia, Cyprus, Greece, Hungary, Iran, Jan Mayen, Libya, Luxembourg, Madeira, Malta, Romania, Serbia, Slovakia, Slovenia, Tunisia and Ukraine.

Melanitta americana
Black Scoter (V)

OTHER NAMES American (Black) Scoter.
TAXONOMY (Swainson, 1832). Monotypic. Formerly treated as conspecific with Common Scoter, but now separated at species level on basis of differences in bill pattern and shape and in vocalisations, which indicate different evolutionary lineages and long-term reproductive isolation.
DISTRIBUTION [NE Asia, NW and NE North America > N Pacific and NW Atlantic coasts. Decreasing. NT] Rare vagrant (all records of identifiable adult males, so very likely significantly more numerous than totals imply): Britain (11); Denmark (9); Estonia (7); Finland (3); France (12); Germany (13-16 Feb 2008); Netherlands (3); Norway (15 May 2013); Poland (4); Spain (3); Sweden (17).

Clangula hyemalis
Long-tailed Duck

OTHER NAME Oldsquaw.
TAXONOMY (Linnaeus, 1758). Monotypic.
DISTRIBUTION [Holarctic > NW, NC and NE North America, NW Europe, NC and NE Asia. Decreasing. VU] Breeds Iceland, N Fennoscandia, Jan Mayen, Svalbard and extreme NW Russia. Migrates SW to winter mainly in Baltic and off Scandinavian coasts, also North Sea W to Faroe Is, Iceland and Ireland, very small numbers regularly S to N Spain, more rarely WC Europe and Italy; also migrant and winter visitor Hungary, Ukraine and formerly Azerbaijan. Vagrant Albania, Armenia, Azores, Balearic Is, Bosnia-Herzegovina, Bulgaria, Sardinia, Corsica, Croatia, Greece, Iran, Israel, Jordan, Kuwait, Luxembourg, Macedonia, Madeira, Montenegro, Portugal, Romania, Serbia, Slovakia, Slovenia and Turkey.

Bucephala albeola
Bufflehead (V)

TAXONOMY (Linnaeus, 1758). Monotypic.
DISTRIBUTION [N North America > W and S North America. Increasing. LC] Very rare vagrant: Azores (4); Britain (16); Channel Is (8-11 Nov 1967, shot); France (records under review); Iceland (4); Ireland (2); Netherlands (3); Portugal (24 Jan-7 Feb 1993); Romania (16 Jan 2006); Spain (3). Further records treated as unknown origin and placed in Cat D (*eg* Czech Republic, Denmark, Finland, Lithuania, Norway, Poland and Slovakia), or as escapes in Cat E (*eg* France and Germany), or in both categories (Belgium).

Bucephala clangula
Common Goldeneye

OTHER NAME Goldeneye.

TAXONOMY (Linnaeus, 1758). Two subspecies (one in region).
DISTRIBUTION [N North America and N Eurasia > W and S North America, W and S Europe, SW to NE Asia. Stable. LC] Breeds Scotland, Fennoscandia S to Netherlands, NE Germany, Poland and Belarus, and E through Baltic States and NW Russia (*clangula*). Migrates W and SW to winter chiefly W Baltic, North Sea and Atlantic coasts of NW Europe, also C, S and SE Europe, Black and Caspian Seas and parts of N Middle East; irregular Spain. Vagrant Algeria, Azores, Balearic Is, Corsica, Cyprus, Israel, Jan Mayen, Lebanon, Libya, Madeira, Malta, Morocco, Portugal, Svalbard, Syria and Tunisia.

Bucephala islandica
Barrow's Goldeneye

TAXONOMY (Gmelin, J F, 1789). Monotypic.
DISTRIBUTION [W and E North America, Iceland. Increasing. LC] Resident Iceland. Vagrant Britain, Czech Republic, Faroe Is, France, Germany, Ireland, Norway, Spain and Svalbard.

Mergellus albellus
Smew

TAXONOMY (Linnaeus, 1758). Monotypic.
DISTRIBUTION [N Eurasia > Europe, SW to E Asia. Decreasing. LC] Breeds N Fennoscandia and NW Russia, with isolated population to S in Belarus; formerly more widespread, breeding S to Romania, Ukraine and SW Russia, and has also bred Czech Republic and Netherlands. Migrates SW and S to winter from S Baltic W to Britain, Ireland (rare) and N France, S to C and SE Europe, and in Black Sea, Caspian Sea and N Middle East; winter population significantly augmented by Asian breeders. Vagrant Algeria, Balearic Is, Channel Is, Corsica, Cyprus, Egypt, Faroe Is, Iceland, Israel, Libya, Liechtenstein, Malta, Portugal, Saudi Arabia, Spain and Tunisia.

Lophodytes cucullatus
Hooded Merganser (V)

TAXONOMY (Linnaeus, 1758). Monotypic.
DISTRIBUTION [SW and SE Canada, NW and E USA > W Canada and USA. Increasing. LC] Very rare vagrant (mostly late autumn and winter, all in Atlantic and North Sea sectors): Azores (5*); Britain (7); Canary Is (11 Dec 2001-16 Feb 2002); Iceland (9); Ireland (3 records of 4 individuals, last in 1957); Netherlands (2); Spain (8-24 Dec 2009). True distribution of records clouded by occasional escapes from captivity and varying treatment of records within and between countries; occurrences in some placed in Cat D (*eg* in Estonia, Finland, Norway, Poland and in France, where 25 records of 31 individuals), Cat E (*eg* Denmark, Germany and Slovakia) or both D and E (*eg* Belgium).

Mergus merganser
Common Merganser

OTHER NAME Goosander.
TAXONOMY Linnaeus, 1758. Three subspecies (one in region).
DISTRIBUTION [N North America and N Eurasia > S North America, W and S Eurasia. Increasing. LC] Breeds Iceland, Britain, Fennoscandia and Baltic region E through NW Russia, also WC Europe and very locally in S Balkans (*merganser*). N and E populations mainly migratory, wintering from W and S Fennoscandia through much of NW Europe, also parts of C and E Europe S to Balkans, Black and Caspian Seas, Turkey and N Iran. Vagrant Algeria, Armenia, Azores, Balearic Is, Corsica, Cyprus, Faroe Is, Iraq, Israel, Madeira, Malta, Morocco, Portugal, Spain, Svalbard and Tunisia.

Mergus serrator
Red-breasted Merganser

TAXONOMY Linnaeus, 1758. Monotypic.
DISTRIBUTION [N North America and N Eurasia > North America, W and S Eurasia. Stable. LC] Breeds Iceland, Faroe Is, Ireland, Britain, Fennoscandia and Baltic E through N Ukraine and NW Russia. N and E populations mainly migratory, wintering NW Europe on W Baltic, Atlantic and North Sea coasts S to W France, small numbers also reaching Iberia and N Africa, C Europe, N and C Mediter-

ranean, Aegean, Black Sea, S Caspian and SE Iran. Vagrant Algeria, Armenia, Azores, Bahrain, Canary Is, Egypt, Gibraltar, Iraq, Israel, Jan Mayen, Kuwait, Lebanon, Luxembourg, Madeira, Morocco, Palestine, Oman, Saudi Arabia, Svalbard (has bred), Syria and UAE.

Oxyura jamaicensis
Ruddy Duck (V, I)

OTHER NAME North American Ruddy Duck (*O j rubida*).
TAXONOMY (Gmelin, J F, 1789). Two subspecies (one in region).
DISTRIBUTION [North America and Caribbean > S North America, Central America and Carribean. Decreasing. LC] Naturalised self-sustaining population established Britain from 1960, when escapes first bred (*rubida*); has also bred *eg* France, Iceland, Ireland, Morocco, Netherlands and Spain. Possibility of large-scale hybridisation with endangered White-headed Duck led to culling of British population from *c* 6,000 in Jan 2000 to *c* 50 by Jul 2013; elsewhere, still *eg* 41 pairs France 2011 and 10-15 pairs Netherlands 2008-09. Prior to population reduction (aimed at eventual extermination), recorded in at least 22 countries E to Cyprus and Ukraine, and S to Algeria and Israel. Single accepted record attributable to genuine transatlantic vagrancy: Azores (30-31 Oct 2009*).

Oxyura leucocephala
White-headed Duck

OTHER NAME White-headed Stifftail.
TAXONOMY (Scopoli, 1769). Monotypic.
DISTRIBUTION [Mediterranean to E Asia > NW Africa, Middle East and N India. Decreasing. EN] Breeds NW Africa, Spain, SW Russia, NW Kazakhstan, Armenia, Turkey, Syria and W Iran, possibly also Georgia; formerly much more widespread and numerous, also breeding *eg* Albania, Corsica, Egypt, Greece, Hungary, Israel, Italy, Sardinia, Serbia and probably Ukraine. Resident or dispersive, winter range extending to parts of Balkans and from N Middle East S to Israel, SE Iraq and E Iran, more rarely Azerbaijan, formerly also N Egypt. Irregular Italy. Vagrant Austria, Belgium, Bosnia-Herzegovina, France, Germany, Hungary, Jordan, Libya, Macedonia, Malta, Moldova, Netherlands, Poland, Portugal, Saudi Arabia, Serbia, Slovakia, Slovenia, Switzerland and Ukraine. Records from Britain treated as uncertain origin and placed in Cat D.

ORDER GALLIFORMES

FAMILY NUMIDIDAE

Guineafowl · one species

Numida meleagris
Helmeted Guineafowl (I)

OTHER NAMES West African (or Reichenow's/Tufted) Guineafowl.
TAXONOMY (Linnaeus, 1758). Nine subspecies (three in region, of which one now apparently extinct).
DISTRIBUTION [Sub-Saharan Africa. Stable. LC] Formerly resident Morocco (*sabyi*; not recorded since early 1970s); introduced Cape Verde Is (mainly *galeatus* stock), SW Saudi Arabia and W Yemen (*meleagris*), formerly also Canary Is (Tenerife). Domesticated semi-feral populations also in *eg* S France.

FAMILY ODONTOPHORIDAE

New World quail · two species

Callipepla californica
California Quail (I)

OTHER NAME Valley Quail.
TAXONOMY (Shaw, 1798). Five subspecies (undetermined in region). Sometimes considered conspecific with extralimital Gambel's Quail (SW North America), with which it forms a superspecies.
DISTRIBUTION [W North America. Increasing. LC] Introduced Corsica; other regional introductions (*eg* in mainland France, Germany, Italy and Spain) unsuccessful.

Colinus virginianus
Northern Bobwhite (I)

OTHER NAMES (Common) Bobwhite, Bobwhite Quail.
TAXONOMY (Linnaeus, 1758). Twenty-two subspecies (one in region).
DISTRIBUTION [E USA to E Central America. Decreasing. LC] Introduced NW Italy (*virginianus*); other populations in *eg* Croatia and France regarded as not fully self-sustaining.

FAMILY PHASIANIDAE

Grouse, partridges, francolins and pheasants · 27 species

Tetrastes bonasia
Hazel Grouse

OTHER NAMES Hazel Hen, Northern Hazelhen.
TAXONOMY (Linnaeus, 1758). Eleven subspecies (six in region).
DISTRIBUTION [W Europe to E Asia. Decreasing. LC] Resident NE France, Belgium, Luxembourg and W Germany (*rhenanus*), Alps and S Poland to Hungary (*styriacus*), Balkan Peninsula (*schiebeli*), S Germany and Czech Republic (*rupestris*), W and S Scandinavia and Poland E to Ural Mtns (*bonasia*) and N Sweden (*griseonota*).

Tetrao urogallus
Western Capercaillie

OTHER NAMES (Common or Eurasian) Capercaillie; Cantabrian Capercaillie (*T u cantabricus*); Rackelhen (male Black Grouse × Capercaillie hybrid).
TAXONOMY Linnaeus, 1758. Eight subspecies (seven in region).
DISTRIBUTION [W Europe to C Siberia. Decreasing. LC] Resident NW Spain (*cantabricus*), Pyrenees (*aquitanicus*), Scotland and Fennoscandia E to White Sea (*urogallus*), NW Russia (*kureikensis*), C and S Belarus and EC Russia (*volgensis*), Germany and Poland S through C Europe to Balkans (*crassirostris*) and SW Russia (*uralensis*); in some areas populations reintroduced either wholly (Scotland) or partly (*eg* SE France, Germany and Poland) after local extinctions and declines.

Lyrurus tetrix
Black Grouse

OTHER NAMES Eurasian (or Northern) Black Grouse; British Black Grouse (*L t britannicus*); Blackcock (male), Greyhen (female); Rackelhen (male Black Grouse × Capercaillie hybrid)
TAXONOMY (Linnaeus, 1758). Six subspecies (three in region).
DISTRIBUTION [W Europe to E Asia. Decreasing. LC] Resident W and N Britain (*britannicus*) and continental Europe from Scandinavia discontinuously S to Netherlands, Belgium, SE France, Alps, parts of C Europe and W Balkans, and E through much of E and N Europe to Ural Mtns (*tetrix*) and SW Russia (*viridanus*). Formerly more widespread, also breeding *eg* Bulgaria, Hungary, Luxembourg, Montenegro and Turkey. Vagrant Greece and Macedonia.

Lyrurus mlokosiewiczi
Caucasian Grouse (E)

OTHER NAME Caucasian Black Grouse.
TAXONOMY (Taczanowski, 1875). Monotypic.
DISTRIBUTION [Caucasus Mtns. Decreasing. NT] Endemic. Resident extreme S Russia and N Georgia (Greater Caucasus), S Georgia, Armenia and Azerbaijan (Lesser Caucasus) and discontinuously NE Turkey and extreme NW Iran.

Lagopus muta
Rock Ptarmigan

OTHER NAME Ptarmigan.
TAXONOMY (Montin, 1781). Twenty-three subspecies (at least seven in region).
DISTRIBUTION [N North America and N Eurasia. Decreasing. LC] Breeds Pyrenees (*pyrenaica*), Alps (*helvetica*), Scotland (*millaisi*), Iceland (*islandorum*), Svalbard and Franz Josef Land (*hyperborea*) and from N Fennoscandia E to Kola Peninsula (*muta*); former introduced populations Faroe Is (*islandorum* and Nearctic *rupestris*) no longer extant. Essentially resident but some Arctic populations irruptive, and *macruros* (breeding extra-

limitally NE Greenland) recorded Iceland in winter. Vagrant Jan Mayen and Bulgaria.

Lagopus lagopus
Willow Ptarmigan

OTHER NAMES Willow Grouse; Red Grouse (*L l scotica*/'*hibernans*').
TAXONOMY (Linnaeus, 1758). Sixteen subspecies (four in region). *Scotica* formerly treated as specifically distinct, differing genetically and in year-round dark plumage, while *variegata* of offshore mid-Norwegian islands has less white in winter plumage than mainland European forms; present treatment in need of review.
DISTRIBUTION [N North America and N Eurasia. Decreasing. LC] Resident Ireland and Britain (*scotica*), islands off W Norway (*variegata*), from N Fennoscandia E to N Ural Mtns (*lagopus*) and Baltic States E to S Ural Mtns (*rossica*); introduced Shetland (N Scotland). Vagrant Lithuania, Poland and Ukraine.

Tetraogallus caucasicus
Caucasian Snowcock (E)

TAXONOMY (Pallas, 1811). Monotypic.
DISTRIBUTION [Extreme SE Europe and SW Asia. Stable. LC] Endemic. Confined to Greater Caucasus Mtns in extreme S Russia, N Georgia and N Azerbaijan. Resident but some movement to lower altitudes in winter.

Tetraogallus caspius
Caspian Snowcock

TAXONOMY (Gmelin, S G, 1784). Three subspecies (all in region).
DISTRIBUTION [Extreme SE Europe and SW Asia. Decreasing. LC] Near-endemic. Resident S and E Turkey, W Armenia and SW Georgia (*tauricus*), E Armenia, Azerbaijan and N Iran (*caspius*) and W Iran (*semenowtianschanskii*). Possibly also occurs Iraq, but status requires confirmation.

Alectoris graeca
Rock Partridge (E)

OTHER NAMES Alpine Rock Partridge (*A g saxatilis*), Balkan Rock Partridge (*A g graeca*), Italian Rock Partridge (*A g orlandoi*), Sicilian (Rock) Partridge (*A g whitakeri*).
TAXONOMY (Meisner, 1804). Four subspecies. Forms superspecies with Chukar and Philby's Partridges. *Whitakeri* from Sicily well differentiated morphologically and genetically, and sometimes treated as separate species.
DISTRIBUTION [C, S and SE Europe. Decreasing. NT] Endemic resident from SE France through Alps to W Croatia (*saxatilis*), peninsular Italy (*orlandoi*), Sicily (*whitakeri*) and W, S and C Balkans, extending E in C Bulgaria (*graeca*). Released Lebanon but not considered self-sustaining.

Alectoris chukar
Chukar Partridge

OTHER NAMES Chukar, Chukor.
TAXONOMY (Gray, J E, 1830). Fourteen subspecies (six in region). Forms superspecies with Philby's and Rock Partridges.
DISTRIBUTION [SE Europe to E Asia. Stable. LC] Resident SE Balkans, N Turkey and Caucasus (*kleini*), Crete, Rhodes, Cyprus, SW and SC Turkey (*cypriotes*), SE Turkey, N Syria, N Iraq and N Iran (*kurdestanica*), Syria S to Sinai (*sinaica*), E Iraq, SW Iran, N UAE and N Oman (*werae*), and NE and E Iran (*koroviakovi*). Released elsewhere in UAE, also Italy, Malta, Montenegro and Ukraine, as well as W Europe (*eg* Britain and France) where not considered self-sustaining.

Alectoris philbyi
Philby's Partridge (E)

OTHER NAME Philby's Rock Partridge.
TAXONOMY Lowe, 1934. Monotypic. Sometimes treated as subspecies of Rock Partridge, but more generally viewed as monotypic species, forming superspecies with Rock and Chukar Partridges.
DISTRIBUTION [SW Arabia. Stable. LC] Endemic resident SW Saudi Arabia and W Yemen.

Alectoris barbara
Barbary Partridge (E)

TAXONOMY (Bonnaterre, 1792). Four subspecies.

DISTRIBUTION [N Africa. Decreasing. LC] Endemic. Resident from Canary Is and N Mauritania to S Morocco, C and S Algeria, S Tunisia and W Libya (*spatzi*), NW Morocco (*koenigi*), from NE Morocco to N and C Tunisia (*barbara*), and NE Libya, formerly also NW Egypt (*barbata*). Introduced Gibraltar (*barbara*), adjacent S Spain and (apparently) Sardinia.

Alectoris rufa
Red-legged Partridge (E)

OTHER NAME French Partridge.
TAXONOMY (Linnaeus, 1758). Three subspecies.
DISTRIBUTION [SW Europe. Decreasing. LC] Endemic. Resident W and N Iberia (*hispanica*), S and E Iberia (*intercedens*) and France, N Italy and Corsica (*rufa*), formerly also Switzerland and SW Germany. Introduced Azores, Balearic Is (*intercedens*), Britain (*rufa*), Gran Canaria, Greece (Alonnisos) and Madeira.

Alectoris melanocephala
Arabian Partridge (E)

OTHER NAMES Arabian Red-legged Partridge, Arabian Chukar.
TAXONOMY (Rüppell, 1835). Two subspecies.
DISTRIBUTION [S Arabia. Stable. LC] Endemic. Resident SW Saudi Arabia and W Yemen (*melanocephala*), and E Yemen and Oman (*guichardi*). Released UAE but not considered self-sustaining.

Ammoperdix griseogularis
See-see Partridge

OTHER NAME See-See.
TAXONOMY (von Brandt, J F, 1843). Monotypic.
DISTRIBUTION [SW and S Asia. Stable. LC] Resident SE Turkey, SW Azerbaijan, N Syria, N Iraq and Iran; introduced UAE (Sir Bani Yas) but not considered self-sustaining. Vagrant Armenia; this species or hybrids with Sand Partridge also recorded E Jordan.

Ammoperdix heyi
Sand Partridge

OTHER NAME Desert Partridge.
TAXONOMY (Temminck, 1825). Four subspecies (all in region).
DISTRIBUTION [NE Africa and Middle East. Stable. LC] Resident from C Egypt (*cholmleyi*), EC and NE Egypt (*nicolli*) through Sinai, Israel, Jordan, S Syria and widely in W Saudi Arabia (*heyi*) to Yemen and S Oman (*intermedius*). Vagrant Lebanon.

Francolinus francolinus
Black Francolin

OTHER NAME Black Partridge.
TAXONOMY (Linnaeus, 1766). Six subspecies (three in region).
DISTRIBUTION [E Mediterranean to NE India. Stable. LC] Resident Israel, Cyprus, S Turkey, S Armenia, Azerbaijan and N Iran (*francolinus*), C and S Iraq, SW Iran (*arabistanicus*) and S Iran (*bogdanovi*); formerly more widespread, and reintroduced C Italy. Records from Arabia, where has bred irregularly, relate to periodic introductions or escapes.

Francolinus pondicerianus
Grey Francolin

OTHER NAMES Grey Partridge, Indian Grey Francolin.
TAXONOMY (Gmelin, J F, 1789). Three subspecies (one in region).
DISTRIBUTION [SW and S Asia. Stable. LC] Resident S Iran (*mecranensis*); also Arabia (probably introduced) in UAE and Oman (originally *mecranensis*), and more recently in E Saudi Arabia, Qatar and Bahrain.

Pternistis bicalcaratus
Double-spurred Francolin

OTHER NAME Moroccan Double-spurred Francolin (*P b ayesha*).
TAXONOMY (Linnaeus, 1766). Three subspecies (one in region). Typically larger, darker and more rufous endemic form *ayesha* sometimes split on basis of geographical isolation

and different ecological conditions, but this treatment rarely followed.
DISTRIBUTION [NW and W Africa. Decreasing. LC] Resident Morocco (*ayesha*), where critically endangered and largely confined to small area in NW.

Pternistis erckelii
Erckel's Francolin (I)

TAXONOMY (Rüppell, 1835). Monotypic.
DISTRIBUTION [NE Africa. Stable. LC] Naturalised population established Italy (Zannone, off Lazio coast).

Perdix perdix
Grey Partridge

OTHER NAMES (Common or Hungarian) Partridge; Peat Partridge (*P p sphagnetorum*).
TAXONOMY (Linnaeus, 1758). Eight subspecies (all in region). Genetic research indicates 2.2% differentiation between western and eastern clades, suggesting long-term divergence, but geographical variation apparently largely clinal and individual variation considerable.
DISTRIBUTION [W Europe to CE Asia. Decreasing. LC] Resident NE Portugal and N Spain (*hispaniensis*), W and S France (*armoricana*), Italy (*italica*), NE Netherlands and NW Germany (*sphagnetorum*), Ireland, Britain and continental Europe from S Scandinavia and N France to SE Balkans (*perdix*), Finland E to Ural Mtns, NE Balkans and N Caucasus (*lucida*), Turkey to Caucasus and NW Iran (*canescens*), and E of lower Ural river basin and Ural Mtns (*robusta*). Introduced or reintroduced to many areas within contracting European range, *eg* parts of Britain, France, Ireland and Russia.

Coturnix coturnix
Common Quail

OTHER NAMES (Eurasian or European) Quail.
TAXONOMY (Linnaeus, 1758). Five subspecies (three in region).
DISTRIBUTION [Africa, Europe to C and S Asia > Sub-Saharan Africa and S Asia. Decreasing. LC] Breeds Azores (*conturbans*), Cape Verde Is (*inopinata*), Canary Is, Madeira, NW Africa, Europe from Mediterranean N to Ireland, Britain, S Fennoscandia and NW Russia, N Middle East and parts of E Arabia (*coturnix*). Mainly migratory, occurring widely on passage, but some overwinter in S of breeding range. Rare Channel Is (formerly bred). Vagrant Faroe Is (has bred) and Iceland.

Coturnix delegorguei
Harlequin Quail

OTHER NAME African Harlequin Quail.
TAXONOMY Delegorgue, 1847. Three subspecies (one in region).
DISTRIBUTION [C, E and S Africa, SW Arabia. Stable. LC] Breeds extreme SW Saudi Arabia and W Yemen (*arabica*). Vagrant Oman.

Syrmaticus reevesii
Reeves's Pheasant (I)

OTHER NAME White-crowned Long-tailed Pheasant.
TAXONOMY (Gray, J E, 1829). Monotypic.
DISTRIBUTION [E Asia. Decreasing. VU] Naturalised populations self-sustaining in Czech Republic (N and C Moravia) and France (mainly N and C); introductions elsewhere in Europe (*eg* Austria, Germany and Hungary) unsuccessful.

Phasianus colchicus
Common Pheasant

OTHER NAMES (Ring-necked) Pheasant; Black-necked (or Caucasus) Pheasant (*P c colchicus*). Extralimital forms potentially occurring through introduction in region: Chinese Ring-necked Pheasant (*P c torquatus*), Grey-rumped Pheasant (*P c pallasi*, *P c satschuensis*, *P c torquatus*), Kirghiz (or Mongolian Ring-necked) Pheasant (*P c mongolicus*), Tarim Pheasant (*P c tarimensis*), White-winged (or Prince of Wales) Pheasant (*P c principalis*).
TAXONOMY Linnaeus, 1758. Thirty subspecies (four native to region).
DISTRIBUTION [SE Europe to E Asia. Decreasing. LC] Resident NW Caspian and N Caucasus (*septentrionalis*), Georgia, Armenia and

N Azerbaijan (*colchicus*), SE Azerbaijan and NC Iran (*talischensis*) and NE Iran (*persicus*); historically, possibly also occurred extreme SE Europe and N Turkey. Widely introduced elsewhere across much of Europe except NE, far N and Iberia; naturalised population largely consists of intraspecific hybrids, and regularly augmented by large-scale releases, especially in Britain.

Chrysolophus pictus
Golden Pheasant (I)

TAXONOMY (Linnaeus, 1758). Monotypic.
DISTRIBUTION [E Asia. Decreasing. LC] Introduced Britain, where small established population in long-term decline.

Chrysolophus amherstiae
Lady Amherst's Pheasant (I)

TAXONOMY (Leadbeater, 1829). Monotypic.
DISTRIBUTION [E Asia. Decreasing. LC] Introduced Britain, where formerly established but latterly declining, and tiny remnant population no longer self-sustaining.

ORDER GAVIIFORMES

FAMILY GAVIIDAE

Divers · five species

Gavia stellata
Red-throated Diver

OTHER NAME Red-throated Loon (IOC).
TAXONOMY (Pontoppidan, 1763). Monotypic.
DISTRIBUTION [N North America and N Eurasia > W North America, Europe, SW and E Asia. Decreasing. LC] Breeds Iceland, Faroe Is, N and W Scotland, and across Norway, Sweden, Finland and NW Russia, also Arctic islands; has bred Estonia and Latvia. Migratory and dispersive, occurring on passage and/or in winter Atlantic and North Sea coasts from Iceland and W Norway S to Iberia (very scarce), also S Baltic, Belarus, very locally C and SE Europe, parts of N Mediterranean, Black Sea and SW Caspian. Vagrant Algeria, Armenia, Azores, Balearic Is, Corsica, Gibraltar, Iran, Israel, Jan Mayen, Luxembourg, Malta, Morocco, Slovakia and Tunisia.

Gavia arctica
Black-throated Diver

OTHER NAMES Black-throated Loon (IOC), Arctic Loon.
TAXONOMY (Linnaeus, 1758). Two subspecies (one in region). Formerly included Pacific Diver (*qv*).
DISTRIBUTION [N Eurasia and extreme NW North America > Europe, SW and E Asia. Decreasing. LC] Summer visitor N Scotland, Norway, Sweden, Finland and across NW Russia, with scattered small populations Baltic States and N Belarus (*arctica*); formerly also bred Poland, has nested Svalbard. Migratory and dispersive, occurring on passage and/or in winter Atlantic and North Sea coasts from S Baltic and SW Scandinavia S to Iberia (very scarce), also NC Mediterranean, locally C Europe and Balkans, Black Sea and S Caspian. Vagrant Algeria, Armenia, Azores, Balearic Is, Canary Is, Cyprus, Faroe Is, Gibraltar, Iran, Israel, Jordan, Latvia, Luxembourg, Morocco, Svalbard and Tunisia.

Gavia pacifica
Pacific Diver (V)

OTHER NAME Pacific Loon (IOC).
TAXONOMY (Lawrence, 1858). Monotypic. Now separated from Black-throated Diver on basis of differences in size, coloration and bill colour, and lack of hybridisation.
DISTRIBUTION [NE Asia and N North America > E Asia and W North America. Increasing. LC] Extreme vagrant (10 records, first as recently as 2007, and five involving returning wintering individuals: Britain (7: 12 Jan-4 Feb 2007; 2 Feb-20 Mar 2007, presumed same 16 Jan-11 Feb 2008 and 25-26 Feb 2009; 17 Feb-10 Mar 2007, presumed same 23-29 Nov 2007; 2 Nov and 19 Nov-9 Dec 2009, 21 Dec 2010-3 Mar 2011, 5 Dec 2012-25 Mar 2013;

18-19 and 27 Nov 2009; 16 May 2013; 10-27 Nov 2013); Channel Is (2 Jan-28 Mar 2010, presumed same 8 Jan-8 Feb 2011); Finland (10-11 Nov 2010); Ireland (12 Jan 2009, presumed same 30 Jan, 6-28 Mar and 2-23 May 2010); Spain (7 Dec 2009).

Gavia immer
Great Northern Diver

OTHER NAMES Great Northern Loon (IOC), Common (or Black-billed) Loon.
TAXONOMY (Brünnich, 1764). Monotypic.
DISTRIBUTION [N North America and Iceland > W and E North America, W Europe. Decreasing. LC] Breeds Iceland, irregularly also Bear Is. In winter numbers augmented by migrants from Greenland, Baffin Is and mainland Canada, distribution stretching from Iceland to many Atlantic and North Sea coasts from N Norway S to N Spain, rarer Baltic, with very small numbers also Mediterranean France and Spain; scarce Faroe Is, Finland and Sweden, rare but regular Austria and Switzerland, irregular Italy. Vagrant Albania, Algeria, Azores, Balearic Is, Bosnia-Herzegovina, Bulgaria, Canary Is, Croatia, Czech Republic, Estonia, Germany, Gibraltar, Greece, Hungary, Latvia, Lithuania, Luxembourg, Macedonia, Madeira, Montenegro, Morocco, Poland, Portugal, Romania, Sardinia, Slovakia, Slovenia, Turkey and Ukraine.

Gavia adamsii
White-billed Diver

OTHER NAMES Yellow-billed Loon (IOC), White-billed Loon.
TAXONOMY (Gray, G R, 1859). Monotypic.
DISTRIBUTION [N North America and arctic Russia > NW North America, NW Europe and E Asia. Decreasing. NT] Breeds Novaya Zemlya and possibly elsewhere in Russian Arctic. Winters from extreme NW Russia W along coastal Norway; scarce or very scarce Denmark, Sweden and Baltic, from where some migrate overland (mainly spring) between Finland and White Sea. Range elsewhere poorly known, but small numbers of spring migrants now known to stage annually NW Scotland (Hebrides) and NE Scotland (Aberdeenshire), presumably indicating regular wintering farther S in Atlantic/Irish Sea and North Sea respectively. Vagrant Austria, Belarus, Belgium, Bulgaria, Croatia, Czech Republic, Estonia, Faroe Is, France, Germany, Iceland, Ireland, Italy, Jan Mayen, Latvia, Lithuania, Netherlands, Poland, SE European Russia (North Ossetia), Slovakia, Slovenia, Svalbard, Switzerland and Ukraine.

ORDER PROCELLARIIFORMES

FAMILY OCEANITIDAE

Austral storm-petrels · three species

Oceanites oceanicus
Wilson's Storm-petrel

OTHER NAMES Flat-clawed (or Yellow-webbed) Storm-petrel, Wilson's Petrel.
TAXONOMY (Kühl, 1820). Three subspecies (at least one in region).
DISTRIBUTION [Antarctic and sub-Antarctic islands > N Pacific, N Atlantic and Indian Oceans. Stable. LC] Pelagic migrant NE Atlantic, mainly Mar-Apr and Sep-Oct off N Mauritania (including single record of *c* 10,000 individuals, 15-17 Jul 2005) but regularly also Jun-Sep in Azores (total *c* 650 birds*) and off SW Britain (annual mean 2000-09 of 38), Atlantic Spain (275 to 2013) and Ireland (254 to 2013, of which two confirmed as *exasperatus*); scarcer Biscay and Portugal. Also regular passage visitor off S and SE Arabia (rare Gulf). Vagrant Canary Is, Cape Verde Is, Channel Is, Egypt (Red Sea coast), France, Germany, Iceland, Iran, Israel, Italy, Jordan, Madeira, Morocco, Netherlands, Norway, Poland, Saudi Arabia, Mediterranean Spain and Svalbard.

Pelagodroma marina
White-faced Storm-petrel

OTHER NAMES (Atlantic) Frigate (or White-faced) Petrel.

TAXONOMY (Latham, 1790). Six subspecies (probably three in region).
DISTRIBUTION [Temperate and subtropical southern oceans and C Atlantic > widespread dispersal. Decreasing. LC] Breeds Salvage Is and Canary Is (*hypoleuca*) and Cape Verde Is (*eadesi*). Non-breeding range poorly known; rarely recorded in Atlantic waters around and between Azores, Madeira, at sea W of Iberia and Morocco, and may occur offshore N Mauritania. Vagrant Britain (*hypoleuca*), Netherlands, Oman and Yemen; may be more regular Arabian Sea in summer than very few records suggest (these probably being attributable to *dulciae*, breeding extralimitally S Australia).

Fregetta tropica
Black-bellied Storm-petrel (V)

OTHER NAMES Striped (or Gould's) Storm-petrel.
TAXONOMY (Gould, 1844). Two subspecies (undetermined in region). Taxonomy of South Atlantic populations of this species and extralimital White-bellied Storm-petrel *F grallaria* (subtropical and temperate southern oceans; see Appendix 4) poorly understood and possibly in need of revision.
DISTRIBUTION [Antarctic and sub-Antarctic islands > tropical and subtropical S oceans. Decreasing. LC] Extreme vagrant (seven records; old claim of regular presence Aug-Sep in C Arabian Sea requires confirmation): Canary Is (2: 10 Sep 2011; 18 Aug 2012); Madeira (8 Aug 2011*); Oman (3: Jul 1963, three; 14 Jun 1989, two; 15 Aug 2013); Yemen (1 Nov 1997). Further records of unidentified *Fregetta* storm-petrels, either this species or White-bellied Storm-petrel, from Britain (25 Nov 2009) and Cape Verde Is (17 Aug 1986).

FAMILY DIOMEDEIDAE

Albatrosses · four species

Diomedea dabbenena
Tristan Albatross (V)

TAXONOMY Mathews, 1929. Monotypic. Formerly treated as subspecies of extralimital Wandering Albatross *D exulans* (Southern Ocean), but differs in smaller size and considerably shorter bill; conflicting genetic research variously reports differentiation as low or very high, and requires clarification.
DISTRIBUTION [Tristan da Cunha and Gough Is > South Atlantic and SW Indian Ocean. Decreasing. CR] Extreme vagrant: Italy (4 Oct 1957, collected).

Thalassarche melanophris
Black-browed Albatross (V)

OTHER NAME Black-browed Mollymawk.
TAXONOMY (Temminck, 1828). Monotypic.
DISTRIBUTION [S South America, S Atlantic and S Indian Ocean > circumpolar in southern oceans. Decreasing. NT] Rare vagrant (occurrences range from at-sea sightings to long-staying individuals returning to Northern Gannet colonies over many years): Azores (2); Britain (28); Channel Is (3); Corsica (21 Feb 1991); Denmark (2); Faroe Is (2); France (9); Germany (5 Oct 1988); Iceland (2); Ireland (12); Italy (18 July 2000); Morocco (3); Norway (16); Portugal (Feb 1983); Spain (6); Svalbard (Jun 1878, collected at sea); Sweden (19-20 May 2001 and again 15 May 2004). Most reports of unidentified albatrosses in region probably also relate to this species.

Thalassarche cauta
Shy Albatross (V)

OTHER NAMES Shy Mollymawk; Tasmanian Albatross (*T c cauta*).
TAXONOMY (Gould, 1841). Two subspecies (one in region).
DISTRIBUTION [Tasmania and New Zealand > Southern Ocean. Population trend unknown. NT] Extreme vagrant (records from two countries relating to same individual, *cauta*): Egypt (20 Feb-7 Mar 1981); Israel (20-26 Feb, 2 Mar and 7 Mar 1981, found dead on last date).

Thalassarche chlororhynchos
Atlantic Yellow-nosed Albatross (V)

OTHER NAME Yellow-nosed Albatross.
TAXONOMY (Gmelin, J F, 1789). Monotypic.
DISTRIBUTION [Tristan da Cunha and Gough

Is > S Atlantic. Decreasing. EN] Extreme vagrant (five records of four birds): Britain (29 Jun and 2-3 Jul 2007, same as Sweden); Faroe Is (22 Jul 2012, dead on fishing line at sea *c* 110 nautical miles NW of islands); Norway (2: 13 Apr 1994; 28 Jun and 4-8 Jul 2007); Sweden (8 Jul 2007, same as Britain).

FAMILY HYDROBATIDAE

Northern storm-petrels · seven species

Hydrobates pelagicus
European Storm-petrel

OTHER NAMES (British) Storm Petrel; Mediterranean Storm-petrel (*H p melitensis*).
TAXONOMY (Linnaeus, 1758). Two subspecies (both in region). Has been treated as monotypic, but alternatively *melitensis* sometimes considered worthy of specific status based on lack of gene flow with nominate and differences in vocalisations, bill morphology and plumage.
DISTRIBUTION [NE Atlantic and Mediterranean > N and S Atlantic. Decreasing. LC] Breeding endemic. Breeds NE Atlantic in Canary Is, N Spain, NW France, Channel Is, W and N Britain, Ireland, Faroe Is, Iceland and Norway (*pelagicus*), and in Mediterranean in Spain, Balearic Is, Corsica, Sardinia, Sicily, mainland Italy, Malta, W Balkans, Aegean and possibly also NE Morocco; has bred Algeria and Tunisia (*melitensis*). Atlantic population moves S to winter largely beyond region, some entering North Sea (rare in S) and Mediterranean on passage; Mediterranean birds perhaps mainly winter locally, but *melitensis* chicks ringed Balearic Is, E Spain and W Sicily recovered NW Morocco, N Spain and S Portugal respectively. Large numbers recorded both passages N Mauritania. Vagrant Austria, Belgium, Cape Verde Is, Croatia, Cyprus, Czech Republic, Finland, Germany, Israel, Lebanon, Libya, Madeira, Poland, Svalbard, Sweden, Switzerland, Turkey and Ukraine.

Oceanodroma castro
Band-rumped Storm-petrel

OTHER NAMES Madeiran (or Harcourt's) Storm-petrel; Grant's Storm-petrel (*O c 'granti'*).
TAXONOMY (Harcourt, 1851). Monotypic. Formerly included Monteiro's and Cape Verde Storm-petrels (*qv*); further division between summer- and winter-breeding populations (latter provisionally being named '*granti*') also proposed but not so far adopted.
DISTRIBUTION [Tropical and subtropical Atlantic and Pacific Oceans > W Atlantic and C Pacific Oceans. Decreasing. LC] Two seasonally separated populations in Atlantic islands, breeding summer (Mar-Oct) Madeira, Desertas, Salvage Is and Canary Is, and winter (Aug-Mar, '*granti*') Azores, Madeira, Salvage Is, Canary Is and Portugal (Berlengas Is). Non-breeding distribution of this and following two species obscured by identification complexities, and thus uncertain; possibly regular late summer W Biscay, and records of up to several hundred Band-rumped Storm-petrels *sensu lato* offshore from N Mauritania perhaps involve this species and/or Cape Verde Storm-petrel. None of 21 vagrancy records (to end 2010) from Europe nor sole Middle Eastern record so far assigned specifically, and only in rare case of specimens may this prove possible through DNA sampling; undetermined extralimital records Britain, Finland, France, Ireland, Israel, Spain and Switzerland.

Oceanodroma monteiroi
Monteiro's Storm-petrel (E)

TAXONOMY Bolton *et al*, 2008. Monotypic. Formerly treated as conspecific with Band-rumped Storm-petrel *O castro*, but differs in genetics, timing of breeding, moult, diet, vocalisations and morphology.
DISTRIBUTION [Azores > EC Atlantic? Stable. VU] Endemic. Breeds Azores (Mar-Sep). Non-breeding distribution at sea unknown, but no unequivocal records away from Azores archipelago.

Oceanodroma jabejabe
Cape Verde Storm-petrel (E)

TAXONOMY (Barboza du Bocage, 1875). Monotypic. Formerly treated as conspecific with Band-rumped Storm-petrel, but genetically isolated and vocally distinct from all other populations of that species.

DISTRIBUTION [Cape Verde Is > EC Atlantic? Population trend unknown. NE] Endemic. Breeds Cape Verde Is (mainly Oct-Jun). Dispersive and pelagic, with records in Cape Verde seas every month except Jul; wider non-breeding distribution unconfirmed, but see also Band-rumped Storm-petrel.

Oceanodroma monorhis
Swinhoe's Storm-petrel (V)

OTHER NAME Swinhoe's Petrel.
TAXONOMY (Swinhoe, 1867). Monotypic.
DISTRIBUTION [NW Pacific > W Pacific and Indian Ocean. Stable. NT] Rare visitor Oman to offshore waters, usually in very small numbers, but exceptionally 99 after cyclone, 10 Nov 1999. Series of records from Atlantic since 1983, some involving returning individuals and trapped birds with brood patches, suggestive of tiny regional breeding population, perhaps on Macaronesian islands; most recently, six individuals (five females and one male) trapped on Salvage Is (Selvagem Grande) during systematic search for species between 2007 and 2013, but breeding still unconfirmed. Vagrant Azores, Balearic Is, Britain, Canary Is, France, Ireland, Israel, Italy, Madeira, Norway, Portugal, Salvage Is, Spain, UAE and Yemen.

Oceanodroma leucorhoa
Leach's Storm-petrel

OTHER NAMES (Leach's) Fork-tailed Storm-petrel, Leach's Petrel.
TAXONOMY (Vieillot, 1818). Four subspecies (one in region).
DISTRIBUTION [N Atlantic and N Pacific > tropical Atlantic and C and E Pacific. Stable. LC] Breeds W Ireland, W and N Scotland, Faroe Is, Iceland and Norway (*leucorhoa*). Moves S in autumn through NE Atlantic (W to Azores, where scarce), when numbers augmented by migrants from NE North American population; passage most obvious after NW and W gales, especially Ireland, W Britain and Biscay, some also entering North Sea and a few Mediterranean (rare Gibraltar). Small numbers winter in seas off Cape Verde Is, N Mauritania and W Morocco, less so to N off Atlantic and Biscay coasts of Iberia. Rarely observed from land on return passage in spring. Vagrant Algeria, Austria, Belgium, Channel Is, Egypt, Estonia, Germany, Israel, Italy, Latvia, Lithuania, Lebanon, Luxembourg, Madeira, Malta, Poland, Sardinia, Svalbard, Switzerland, Tunisia and UAE.

Oceanodroma matsudairae
Matsudaira's Storm-petrel (V)

OTHER NAMES Matsudaira's Petrel, Sooty Storm-petrel.
TAXONOMY Kuroda, 1922. Monotypic.
DISTRIBUTION [Indian and Pacific Oceans > C Indian and W Pacific Oceans. Population trend unknown. VU] Extreme vagrant: Oman (19 Aug 2013).

FAMILY PROCELLARIIDAE

Petrels and shearwaters · 28 species

Macronectes giganteus
Southern Giant Petrel (V)

OTHER NAMES (Antarctic) Giant Petrel, Giant Fulmar.
TAXONOMY (Gmelin, J F, 1789). Monotypic.
DISTRIBUTION [Southern Ocean and Antarctica. Increasing. LC] Extreme vagrant: Italy (2 Sep 1991, at sea). Additional record from France (2 Nov 1967, at sea) accepted as involving this species or extralimital Northern Giant Petrel *M halli* (Southern Ocean).

Fulmarus glacialis
Northern Fulmar

OTHER NAMES (Arctic) Fulmar, Fulmar Petrel; Blue Fulmar (dark morph).
TAXONOMY (Linnaeus, 1761). Three subspecies (two in region). Extralimital *rodgersii* (breeding N Pacific) is highly divergent genetically; split under consideration.
DISTRIBUTION [Arctic, N Atlantic and N Pacific Oceans. Increasing. LC] Breeds Atlantic and North Sea coasts in NW France, Chan-

nel Is, Britain, Ireland, Heligoland (Germany), W Norway, Faroe Is and Iceland (*auduboni*), and high-arctic islands in Barents Sea (*glacialis*). Outside breeding season ranges widely at sea W in Atlantic and S to about 48°N; scarce Atlantic and Biscay coasts of Spain. Vagrant Azores, Canary Is, Czech Republic, Finland, Italy, Madeira*, Montenegro, Morocco, Poland, Portugal, Slovenia and Ukraine.

Daption capense
Cape Petrel (V)

OTHER NAMES Cape Fulmar, Cape Pigeon, Pintado Petrel.
TAXONOMY (Linnaeus, 1758). Two subspecies (one in region).
DISTRIBUTION [Antarctica and subantarctic islands > Southern and SE Pacific Oceans. Stable. LC] Extreme vagrant (three records): Gibraltar (20 Jun 1979); Italy (Sep 1964, collected; under review); Norway (5 and 7 May 2010, *capense*). Other regional records considered to involve released or ship-assisted birds; hypothetical Arabian Sea.

Pterodroma incerta
Atlantic Petrel (V)

OTHER NAME Schlegel's Petrel.
TAXONOMY (Schlegel, 1863). Monotypic.
DISTRIBUTION [Tristan da Cunha and Gough Is > S Atlantic Ocean. Decreasing. EN] Extreme vagrant (two spring records): Israel (31 May 1982; 18 Apr 1989, same as Jordan); Jordan (18 Apr 1989, same as Israel).

Pterodroma mollis
Soft-plumaged Petrel (V)

OTHER NAME Soft-plumed Petrel.
TAXONOMY (Gould, 1844). Monotypic. Formerly included Zino's, Fea's and Desertas Petrels (*qv*).
DISTRIBUTION [S Atlantic, S Indian and SW Pacific Oceans. Stable. LC] Extreme vagrant (two accepted records, but see Flood and Fisher 2013 for comments about identification of both): Israel (25 Mar 1997); Norway (6 Jun 2009).

Pterodroma madeira
Zino's Petrel

OTHER NAMES Freira, Madeira Petrel.
TAXONOMY Mathews, 1934. Monotypic. Formerly considered conspecific with Soft-plumaged and Fea's Petrels; recent molecular studies indicate probably closest to Bermuda and Fea's Petrels, and that all three (as well as Black-capped) merit species status, but subsequent research suggests that Zino's, Fea's and Desertas Petrels may form a clade.
DISTRIBUTION [Madeira > E and C Atlantic Ocean. Stable. EN] Breeding endemic. Nests only on Madeira. One confirmed sight record in region away from Madeira and adjacent waters (Azores, 1 Aug 2012*, photographed), but now known from geolocator studies to range widely in NE Atlantic, N/NW of Madeira, NW to Azores, N to within several hundred miles of Ireland and S to West Africa Upwelling. See comments under Fea's Petrel regarding vagrancy within region.

Pterodroma feae
Fea's Petrel

OTHER NAMES Cape Verde Petrel, Gon-Gon.
TAXONOMY (Salvadori, 1900). Monotypic. As with Zino's Petrel, previously included within Soft-plumaged Petrel but subsequently separated at species level, with Desertas Petrel then recognised as a subspecies of Fea's. Recent molecular studies indicate that Fea's may be closely related to Zino's and Bermuda Petrels, but subsequent genetic analyses suggest that all three taxa breeding in NE Atlantic may form a clade.
DISTRIBUTION [Cape Verde Is > EC Atlantic Ocean. Increasing. NT] Breeding endemic. Nests in winter Cape Verde Is (Fogo, Santiago, Santo Antão and São Nicolau). Non-breeding range poorly understood but limited geolocator studies indicate movements in EC Atlantic that extend N beyond breeding areas of Zino's and Desertas Petrels and to S of regional boundary; single record N Mauritania, but postulated to occur there year-round offshore. Separation at sea of this, previous and following species extremely difficult, obscuring true distribution of each; except for single Zino's occurrence in Azores waters, no other regional vagrancy

sight records determined to species level (as defined here), though annual reports from NW Europe, often in Jul-Sep, perhaps more likely to involve this species rather than congeners, which are then breeding. Split of Fea's and Desertas Petrels recognised by few national committees (so far); thus three accepted records of vagrant 'Fea's Petrel' in Britain (compared with total of 54 unidentified *feae/deserta/madeira*) relate to either *feae* or *deserta*, and other regional vagrancy sight records (except above-mentioned Azores record of Zino's) typically treated as undetermined *feae/deserta/madeira*. Collectively, birds of these taxa have occurred as vagrants in Azores, Britain, Canary Is, Faroe Is, France, Ireland, Netherlands, Portugal, Spain and Sweden.

Pterodroma deserta
Desertas Petrel

OTHER NAME Bugio Petrel.
TAXONOMY Mathews, 1934. Monotypic. *Deserta* previously treated as a synonym of *feae*, but latterly recognised and now separated from that taxon on genetic evidence, morphology, vocalisations and breeding phenology, though subsequent research suggests that Desertas, Fea's and Zino's Petrels may form a clade.
DISTRIBUTION [Desertas Is > N and S Atlantic Ocean. Increasing. NT] Breeding endemic. Nests Desertas Is (Bugio and possibly Deserta Grande). Geolocator studies confirm movements to waters around Azores (where several have been trapped, suggesting possible breeding), Canary Is and Cape Verde Is, and to W fringe of region. See comments on extralimital occurrences under Fea's Petrel.

Pterodroma cahow
Bermuda Petrel (V)

OTHER NAME Cahow.
TAXONOMY (Nichols and Mowbray, 1916). Monotypic.
DISTRIBUTION [Bermuda > W N Atlantic Ocean. Increasing. EN] Extreme vagrant (one returning bird in three years): Azores (17 and 21 Nov 2002, captured in burrow, ringed and recaptured; trapped again 19 and 23 Nov 2003; also 12 and 13 Dec 2006). More recently, geolocator studies indicate new foraging areas in C Atlantic, with birds ranging NE to within *c* 125-150 miles of Ireland and Spain.

Pterodroma hasitata
Black-capped Petrel (V)

OTHER NAMES Capped (or West Indian) Petrel, Diablotin.
TAXONOMY (Kuhl, 1820). Monotypic. Distinct white-faced and black-faced types recently found also to be genetically distinct, possibly at subspecies level.
DISTRIBUTION [Caribbean > W Atlantic Ocean. Decreasing. EN] Extreme vagrant (seven records, mainly Apr-May): at sea (8 May 2010, c230 miles NE of Madeira*); Azores (3: 26 May 2007; 22 May 2009; 7 Sep 2011*); Britain (2: Mar/Apr 1850, found exhausted, then collected; 16 Dec 1984, dead); Spain (30 Apr 2002, at sea).

Pterodroma arminjoniana
Trindade Petrel (V)

OTHER NAMES South Trinidad (or Trinidade) Petrel.
TAXONOMY (Giglioli and Salvadori, 1868). Monotypic. Occurs in dark and light morphs which show no genetic differentiation and freely interbreed; possibly conspecific with extralimital Herald Petrel *P heraldica* (Pacific and Indian Oceans).
DISTRIBUTION [Trindade Is, S Atlantic Ocean, and Round Is, SW Indian Ocean > N and S Atlantic. Stable. VU] Very rare vagrant (12 records in NE Atlantic, Apr-Oct, but regular extralimitally in NW Atlantic off North Carolina, USA, and pelagic range may just extend into W oceanic extreme of region): Azores (11*); Cape Verde Is (30 Sep 2008).

Calonectris leucomelas
Streaked Shearwater (V)

OTHER NAMES Streak-headed (or White-faced/White-fronted) Shearwater.
TAXONOMY (Temminck, 1836). Monotypic.
DISTRIBUTION [NW Pacific Ocean > W

Pacific and E Indian Oceans. Decreasing. NT] Extreme vagrant (four records of at least five individuals): Israel (3: 27 Jun-18 Sep 1992, two or three, same as Jordan; early May-summer 1993, same as Jordan; 2 Feb 2003, one or two); Jordan (2: 27 Jun-18 Sep 1992, two or three, same as Israel; early May-summer 1993, same as Israel); Oman (4 Nov 1998).

Calonectris diomedea
Scopoli's Shearwater

OTHER NAMES (Mediterranean) Cory's Shearwater.
TAXONOMY (Scopoli, 1769). Monotypic. Formerly treated with Cory's and Cape Verde Shearwaters (*qv*) as a single species.
DISTRIBUTION [Mediterranean Sea > SE Atlantic Ocean. Decreasing. LC] Breeding endemic on islands in Mediterranean from Spain and Algeria E to Greece; also isolated mainland colony on Atlantic coast of France (Gironde), and occasionally breeds Salvage Is (also recorded off Madeira). Regular on passage in many non-breeding areas of Mediterranean, where some remain year-round, E to *eg* Cyprus, Egypt, Israel and Lebanon, also Atlantic coast of Morocco S to N Mauritania, where numerous in winter. Status unclear Cape Verde Is, also N Red Sea where a few apparently migrate N overland in spring but small numbers also recorded summer. Vagrant Austria, Britain, Bulgaria, Canary Is, Czech Republic, Oman, Poland, Switzerland, Syria and Ukraine; undetermined Scopoli's/Cory's vagrancy records also from Estonia and (most likely this species) Faroe Is (shot), Iran and Montenegro.

Calonectris borealis
Cory's Shearwater

OTHER NAMES Atlantic Shearwater, North Atlantic Great Shearwater.
TAXONOMY (Cory, 1881). Monotypic. Formerly considered conspecific with Scopoli's Shearwater, but differs in plumage, size and vocalisations; molecular evidence supports split in one study, but equivocal in another.
DISTRIBUTION [Macaronesian islands > C, SW and SE Atlantic. Decreasing. LC] Breeding endemic, mainly NE Atlantic on Azores, Canary Is, Madeira, Desertas, Salvage Is and Berlengas Is, Portugal, also W Mediterranean on islands off SE Spain and N Morocco. Disperses in Atlantic regularly N to Biscay/W France, Ireland and SW Britain (rare North Sea). On migration to winter quarters beyond region, recorded Morocco and N Mauritania; rare but probably regular on passage Cape Verde Is. Vagrant Belgium, Channel Is, Denmark, Germany, Italy, Netherlands, Norway, Oman, Poland, Sweden, Switzerland and UAE; see previous species for records of undetermined *borealis/diomedea*.

Calonectris edwardsii
Cape Verde Shearwater

TAXONOMY (Oustalet, 1883). Monotypic. Formerly included within Cory's Shearwater, but plumage, biometrics and vocalisations differ; most closely related genetically to Scopoli's Shearwater.
DISTRIBUTION [Cape Verde Is > C and S Atlantic. Decreasing. NT] Breeding endemic on most islands in Cape Verde group. Non-breeding range poorly understood, but only exceptionally recorded away from seas around breeding areas. Vagrant Canary Is and Salvage Is*.

Ardenna pacificus
Wedge-tailed Shearwater (V)

TAXONOMY (Gmelin, J F, 1789). Monotypic.
DISTRIBUTION [Tropical Indian and Pacific Oceans. Decreasing. LC] Regular but scarce non-breeding visitor from W Indian Ocean; uncommon Yemen, rare Oman (Mar-Nov). Vagrant Egypt, Iran and UAE.

Ardenna griseus
Sooty Shearwater

OTHER NAMES Sombre Petrel, Sombre Shearwater.
TAXONOMY (Gmelin, J F, 1789). Monotypic.
DISTRIBUTION [S South America, S Atlantic and SE Australasia > widespread most oceans. Decreasing. NT] Passage visitor NE Atlantic (mainly late Jul-Oct), especially Atlantic islands (though rare Cape Verde), Britain,

Ireland, Biscay/W France and NW Spain, also Iceland, Faroe Is, W Scandinavia, North Sea, Channel Is, Portugal, Morocco (uncommon) and N Mauritania (Mar-May and Sep-Dec); winter and spring records in W of region otherwise very rare. A few penetrate Baltic and Mediterranean (rare Gibraltar); more regular Gulf of Aqaba (especially spring but noted all months), thus also recorded Red Sea coast of Egypt, and annual UAE (rare, mainly Apr-May). Vagrant Algeria, Ceuta, Finland, Iran, Italy, Kuwait, Latvia, Malta, Oman, Poland, Svalbard, Switzerland and Tunisia.

Ardenna carneipes
Flesh-footed Shearwater

OTHER NAME Pale-footed Shearwater.
TAXONOMY Gould, 1844. Monotypic.
DISTRIBUTION [Indian and Pacific Oceans. Stable. LC] Regular SE Iran and Oman, probably also Gulf of Aden. Vagrant Israel, Jordan and UAE (possibly regular).

Ardenna gravis
Great Shearwater

OTHER NAME Greater Shearwater.
TAXONOMY (O'Reilly, 1818) Monotypic.
DISTRIBUTION [S Atlantic islands > S and N Atlantic. Stable. LC] Passage visitor NE Atlantic (mainly Aug-Oct, later farther S), where regularly recorded Atlantic islands and also reaching NW Spain, Biscay/W France, SW Britain and Ireland, less commonly N Mauritania (year-round but mainly spring and autumn), Morocco and Portugal; volume of passage and extent of movements may vary significantly year to year. Very scarce Faroe Is, North Sea coasts of Britain and S Spain/Gibraltar. Vagrant Algeria, Balearic Is, Belgium, Ceuta, Channel Is, Denmark, Germany, Greece, Iceland, Israel, Italy, Lebanon, Libya, Netherlands, Norway, Sardinia, Mediterranean Spain and Sweden.

Puffinus puffinus
Manx Shearwater

TAXONOMY (Brünnich, 1764). Monotypic. Formerly included Balearic and Yelkouan Shearwaters (*qv*).
DISTRIBUTION [E North America, NW Europe and Atlantic islands > C and S Atlantic. Decreasing. LC] Breeds Azores, Canary Is, Madeira, Brittany, Channel Is, W and N Britain, Ireland, Faroe Is and Iceland. Disperses as far as N Norway and into North Sea, with major autumn passage (mainly Sep-Oct) S to winter quarters off Atlantic coasts S to N Mauritania; spring passage much less apparent from land. Very small numbers winter in SW of region. Vagrant Bulgaria, Finland, Mediterranean France, Malta, Poland, Romania, Mediterranean Spain, Svalbard and Switzerland.

Puffinus yelkouan
Yelkouan Shearwater (E)

OTHER NAMES Levantine Shearwater; Mediterranean Shearwater (prior to split from *P mauretanicus*).
TAXONOMY (Acerbi, 1827). Monotypic. Has been considered conspecific with both Manx and Balearic Shearwaters, but morphological, behavioural and ecological analyses, as well as genetic differences, indicate all three best regarded as separate species.
DISTRIBUTION [Mediterranean and Black Sea. Decreasing. VU] Endemic. Breeds Mediterranean islands W to S France and Menorca (where population perhaps intermediate with Balearic Shearwater) and E to Aegean, a few also Algerian and Tunisian coasts; has bred Bulgaria. Disperses from Mediterranean mainly NE into Black Sea, some also W to Strait of Gibraltar (where formerly much more numerous); exceptional in NE Atlantic (*eg* unrecorded Portugal), but several documented occurrences from NW Spain and France. Vagrant Austria, Britain, E Egypt (Red Sea), S Israel (Gulf of Aqaba), Switzerland and Syria.

Puffinus mauretanicus
Balearic Shearwater (E)

OTHER NAME Mediterranean Shearwater (prior to split with *P yelkouan*).
TAXONOMY Lowe, 1921. Monotypic. Formerly considered conspecific with Manx Shearwater,

then as polytypic species with Yelkouan Shearwater (*qv*).
DISTRIBUTION [Mediterranean and NE Atlantic. Decreasing. CR] Endemic. Breeds Balearic Is (but not Menorca, where replaced by Yelkouan Shearwater or intermediate population), possibly also NE Spain. Disperses widely in W and C Mediterranean and especially W into NE Atlantic, regularly S to Morocco and N to Biscay, S Ireland, SW Britain and English Channel, less frequently N Ireland, Irish Sea, Scotland (especially SW) and North Sea, including Netherlands. Vagrant Azores, Belgium, Canary Is, Denmark, Estonia, Germany, Greece, Israel, Libya, Madeira, Malta, Poland, Norway and Sweden.

Puffinus persicus
Persian Shearwater

OTHER NAME Arabian Shearwater.
TAXONOMY Hume, 1872. Two subspecies (one in region). Formerly treated as a subspecies of both Tropical Shearwater and extralimital Audubon's Shearwater *P lherminieri* (W Atlantic Ocean and Caribbean Sea), but differs in plumage, measurements and biology.
DISTRIBUTION [Arabian Sea and Indian Ocean. Population trend unknown. LC] Breeds Oman (*persicus*). Common non-breeding visitor Iran, UAE and Yemen, but rare Persian Gulf; status uncertain SW Saudi Arabia but may be scarce non-breeding visitor. Vagrant Israel and Kuwait.

Puffinus bailloni
Tropical Shearwater (V)

OTHER NAMES Baillon's Shearwater; Mascarene Shearwater (when erroneously proposed as new species, *P atrodorsalis*).
TAXONOMY Bonaparte, 1857. Five subspecies (undetermined in region). Variously treated as conspecific with other small shearwaters, including Persian (*qv*).
DISTRIBUTION [Indian and Pacific Oceans. Population trend and status not assessed] Extreme vagrant (two records, first of which originally believed to involve previously undescribed species, '*Puffinus atrodorsalis*'): Israel (18-21 Jun 1992; 15 May 1999).

Puffinus baroli
Barolo Shearwater (E)

OTHER NAMES (North) Atlantic (or Canarian/Madeiran) Little Shearwater, Barolo's (or Macaronesian) Shearwater.
TAXONOMY (Bonaparte, 1857). Monotypic. Forms a monophyletic group with Boyd's Shearwater (*qv*) and extralimital Audubon's Shearwater *P lherminieri* (W Atlantic Ocean and Caribbean Sea); former name of Little Shearwater misleading as not closely related to extralimital *P assimilis* (widespread in southern oceans), formerly also known as Little Shearwater (now Subantarctic Shearwater).
DISTRIBUTION [NE Atlantic. Population trend unknown. NE] Endemic. Breeds Azores, Madeira, Desertas, Salvage Is and Canary Is. Disperses into NE Atlantic waters, regularly occurring in small numbers N to Biscay and Atlantic coast of Iberia; may also be more regular in NW African offshore waters than records suggest. Winter range poorly known, but many remain near breeding areas. Vagrant Belgium, Britain, Channel Is, France, Germany, Gibraltar, Ireland, Israel, Italy, N Mauritania, Morocco and Sardinia.

Puffinus boydi
Boyd's Shearwater (E)

OTHER NAMES Cape Verde Little (or Macaronesian) Shearwater.
TAXONOMY Mathews, 1912. Monotypic. Formerly considered conspecific with Barolo Shearwater, extralimital Audubon's Shearwater *P lherminieri* (W Atlantic Ocean and Caribbean Sea) and Subantarctic Shearwater *P assimilis* (widespread in southern oceans), but differs in plumage, proportions, vocalisations and to limited extent mtDNA.
DISTRIBUTION [Cape Verde Is. Population trend not assessed. NE] Provisionally classified as endemic (has been recorded Senegal, where status uncertain but may possibly occur on passage). Breeds Cape Verde Is. Apparently mainly sedentary and not reliably recorded in region away from Macaronesian and Mauritanian waters, but may move N and NE into Atlantic. Vagrant Canary Is.

Bulweria bulwerii
Bulwer's Petrel

TAXONOMY (Jardine and Selby, 1828). Monotypic.
DISTRIBUTION [Tropical oceans. Stable. LC] Breeds in summer Azores, Madeira, Desertas, Salvage Is and Canary Is, and in winter Cape Verde Is. Outside breeding season disperses into Atlantic, apparently wintering in SW of region, and possibly more regular in offshore African and W Iberian waters than records suggest. Vagrant France, Ireland, Italy, Morocco, Portugal, Spain and at sea in Mediterranean between Corsica and NW Italy. Record from Netherlands (21 Aug 1995) now rejected after review.

Bulweria fallax
Jouanin's Petrel

TAXONOMY Jouanin, 1955. Monotypic.
DISTRIBUTION [NW Indian Ocean. Increasing. NT] Regular visitor from breeding grounds just S of region to Gulf of Aden and Arabian Sea coasts of Yemen and S Oman (where may possibly also breed). Formerly rare vagrant UAE on Gulf of Oman coast, but hundreds encountered on pelagic trips Oct-Dec 2012, max 609 on 21 Dec. Vagrant Saudi Arabia. Old report from Italy (2 Nov 1953, three, one of which collected) now considered dubious.

ORDER PODICIPEDIFORMES

FAMILY PODICIPEDIDAE

Grebes · six species

Tachybaptus ruficollis
Little Grebe

OTHER NAMES Common (or Red-throated) (Little) Grebe, Dabchick.
TAXONOMY (Pallas, 1764). Seven subspecies (four in region).
DISTRIBUTION [Africa, Europe, SW to SE Asia. Decreasing. LC] Breeds N Africa W to N Mauritania, across much of Europe from Mediterranean N to S Fennoscandia, W to Ireland and E to SE European Russia, also W Turkey S to Israel (*ruficollis*), Caucasus, NW and SE Iran (*albescens*), SW Iran and SE Iraq (*iraquensis*), and E, C, CS and W Arabia (*capensis*), this last subspecies also being known from Western Desert oases of Egypt. Mainly resident in W and S of region but largely migratory C and E Europe N to Baltic, wintering across much of Middle East and also N Africa. Vagrant Azores, Canary Is, Faroe Is, Gibraltar, Iceland and Madeira.

Podilymbus podiceps
Pied-billed Grebe (V)

TAXONOMY (Linnaeus, 1758). Three subspecies (undetermined in region, but presumably migratory *podiceps*).
DISTRIBUTION [N and E Canada to S South America > USA to S South America. Stable. LC] Vagrant (mainly autumn and winter, with some long-staying and returning birds): Azores (34 to 2011*); Belgium (5-10 Jan 2009); Britain (42; has hybridised with Little Grebe); Canary Is (13-20 Jan 1991); France (10); Germany (6 Jul-18 Aug 2002); Iceland (3); Ireland (12); Netherlands (2); Norway (5); Spain (6).

Podiceps grisegena
Red-necked Grebe

OTHER NAMES Grey-cheeked Grebe; Holbøll's (or American Red-necked) Grebe (*P g holbollii*).
TAXONOMY (Boddaert, 1783). Two subspecies (both in region).
DISTRIBUTION [N North America, N and E Europe to NE Asia > NW and NE North America, W and S Europe, SW and E Asia. Decreasing. LC] Breeds from S Fennoscandia, Netherlands (rare), Germany, E Europe and Balkans E to Ural River, W Caspian and Turkey (*grisegena*). Migratory, occurring on passage through Baltic to winter Atlantic and North Sea coasts of NW Europe from W Norway S to N Spain, and from C Europe, N

Italy, coasts of W and S Balkans to Black Sea, S Caspian and E Iran. Vagrant Algeria, Azores, Balearic Is, Cyprus, Egypt, Faroe Is, Iceland, Israel, Kuwait, Lebanon, Libya, Luxembourg, Moldova, Saudi Arabia, Sardinia, Svalbard, Syria and Tunisia; *holbollii* (breeding N North America) vagrant to Britain, France, Iceland and Spain.

Podiceps cristatus
Great Crested Grebe

TAXONOMY (Linnaeus, 1758). Three subspecies (one in region).
DISTRIBUTION [N, NE, E and S Africa, Eurasia and Australia > N, NE, E and S Africa, W and S Europe, SW to SE Asia and Australia. Population trend unknown. LC] Breeds parts of N Africa, across much of Europe from Mediterranean N to Fennoscandia, W to Ireland and E to SE European Russia, also Caucasus, locally in N Middle East and NE Arabia (*cristatus*). Mainly resident in W and S of region but largely migratory C and E Europe N to Baltic, wintering through non-breeding areas of Europe, locally in N Africa and widely in N Middle East. Vagrant Azores, Canary Is, Faroe Is, Iceland, Oman and UAE.

Podiceps auritus
Horned Grebe

OTHER NAME Slavonian Grebe.
TAXONOMY (Linnaeus, 1758). Two subspecies (one in region).
DISTRIBUTION [N North America and N Eurasia > W and E North America, W and S Europe, SW and E Asia. Decreasing. VU] Summer visitor Iceland, N Scotland and from Fennoscandia and Baltic States E to Ural Mtns; has bred Faroe Is (*auritus*). Migrates W to winter on Atlantic and North Sea coasts of NW Europe from W Norway S to W France, and SW across E and C Europe to winter N Italy, C Mediterranean, Adriatic, Black Sea and S Caspian. Vagrant Armenia, Azores, Corsica, Cyprus, Czech Republic, Gibraltar, Greece, Israel, Jan Mayen, Kuwait, Lebanon, Libya, Luxembourg, Madeira, Moldova, Morocco, Portugal, Romania, Sardinia, Slovakia, Spain, Svalbard and Tunisia.

Podiceps nigricollis
Black-necked Grebe

OTHER NAME Eared Grebe.
TAXONOMY Brehm, C L, 1831. Three subspecies (one in region).
DISTRIBUTION [N North America, S and E Africa, Eurasia > W and S North America, N, S and E Africa, SW to E Asia. Population trend unknown. LC] Breeds N Morocco, N Tunisia, very locally in W Europe N to S Scandinavia and W to Britain and Ireland, and more widely from E Europe and Baltic States to Black Sea, N Caspian and Caucasus, with scattered populations N Middle East; occasionally breeds Arabia (*nigricollis*). Mainly migratory N of Mediterranean, becoming more widespread (and often coastal) W and S Europe and Mediterranean in winter, also N Africa (irregular Canary Is), Black and S Caspian Seas and widely in Middle East. Vagrant Azores, Finland, Liechtenstein, Madeira and Norway.

FAMILY PHOENICOPTERIDAE

Flamingos · two species

Phoenicopterus ruber
Greater Flamingo

OTHER NAMES (Eurasian or Rosy) Flamingo.
TAXONOMY Pallas, 1811. Monotypic.
DISTRIBUTION [Africa, S Europe, SW, S and C Asia. Increasing. LC] Breeds N Mauritania, N Tunisia, S and E Iberia, S France, Sardinia, NE Italy, Greece, Turkey, Syria, NW and S Iran, and Kuwait; formerly also Cape Verde Is, and has bred UAE and W Yemen. Some populations resident but others dispersive and partly migratory, occurring widely from Mediterranean basin W to N Mauritania and E through N Middle East and Arabia (including wintering migrants from C Asia); sometimes oversummers far from breeding areas. Scarce visitor Malta. Vagrant Austria, Belarus, Bosnia-Herzegovina, Bulgaria, Canary Is, Czech Republic, Hungary, Lithuania, Macedonia, Netherlands, Poland, Romania, Slovenia, Switzerland and Ukraine; vagrancy pattern in W and N Europe complicated by presence of feral birds and frequent escapes

from captivity, and wild individuals may occur more widely than recognised (*eg* Cat D in Britain, Estonia, Germany, Ireland, Latvia and Norway).

Phoeniconaias minor
Lesser Flamingo

TAXONOMY (Geoffroy Saint-Hilaire, É, 1798). Monotypic.
DISTRIBUTION [W, S and E Africa, S Asia. Decreasing. NT] Regular non-breeding visitor N Mauritania (all months); more erratic visitor Yemen (has built nests), ranging N to SW Saudi Arabia and E to S Oman (where vagrant, but one of five records involved flock of up to 270 birds present for 4+ years). Also vagrant Algeria, Canary Is, Egypt, Iran, Kuwait, Morocco, Spain (has bred), Tunisia, Turkey and UAE; true occurrence pattern obscured by birds of unknown origin (*eg* Cat D France and Portugal) and escapes (*eg* Balearic Is, Germany and Netherlands).

ORDER PHAETHONTIFORMES

FAMILY PHAETHONTIDAE

Tropicbirds · two species

Phaethon aethereus
Red-billed Tropicbird

OTHER NAME Arabian Red-billed Tropicbird (*P a indicus*).
TAXONOMY Linnaeus, 1758. Three subspecies (two in region).
DISTRIBUTION [Tropical oceans. Decreasing. LC] Summer visitor Cape Verde Is, more or less regularly also reaching N Mauritanian waters; increasingly frequent vagrant and irregular breeder Canary Is (*mesonauta*). Resident but dispersive E Egypt, SW, S and SE Arabia and S Iran (*indicus*); individuals from Red Sea population occasionally wander N to S Israel. Vagrant Azores (has bred), Britain, Ireland, Jordan, Kuwait, Madeira and Portugal; unidentified tropicbird, probably this species, also recorded Morocco.

Phaethon lepturus
White-tailed Tropicbird (V)

OTHER NAME Yellow-billed Tropicbird.
TAXONOMY Daudin, 1802. Six subspecies (undetermined in region).
DISTRIBUTION [Tropical oceans. Decreasing. LC] Extreme vagrant (six records): Azores (3: 14-27 Oct 2011; 15 Sep 2012; 2-12 Sep and 7 Oct 2013*); Cape Verde Is (3: 20 Feb 1999; 3 May 2011; 16 Mar 2013*). Other records from waters around Azores and Cape Verde archipelagos have occurred outside regional boundary. Tideline corpse found Britain (6 Jan 2013) not admitted to national list.

ORDER CICONIIFORMES

FAMILY CICONIIDAE

Storks · seven species

Mycteria ibis
Yellow-billed Stork

OTHER NAMES (African or Yellow-billed) Wood Stork, Wood Ibis.
TAXONOMY (Linnaeus, 1766). Monotypic.
DISTRIBUTION [Sub-Saharan Africa. Decreasing. LC] Regular non-breeding summer visitor S Egypt; occasional farther N along Nile Valley. Vagrant Bulgaria, Israel, Jordan, N Mauritania (several large flocks), Morocco, Tunisia and Turkey.

Anastomus lamelligerus
African Openbill (V)

OTHER NAMES African Open-billed Stork, Openbill Stork.
TAXONOMY Temminck, 1823. Two subspecies (one in region).

DISTRIBUTION [Sub-Saharan Africa. Decreasing. LC] Extreme vagrant (two records of three individuals, *lamelligerus*): Egypt (2: 26 May 2009; 31 May 2013, two*).

Ciconia nigra
Black Stork

TAXONOMY (Linnaeus, 1758). Monotypic.
DISTRIBUTION [S Africa, W Europe to E Asia > sub-Saharan Africa, SW Europe, S and SE Asia. Population trend unknown. LC] Summer visitor Iberia, W France E to Germany, chiefly E Europe N to Baltic States, S to Balkans and E to Ural Mtns, also Caucasus, Asia Minor and N Iran; formerly more widespread, also breeding *eg* Denmark, Finland and S Sweden. On migration regular N and W to Baltic Fennoscandia and widely S of breeding range, passage channelled especially through Bosphorus, Caucasus and W Middle East/E Egypt, also Strait of Gibraltar, fewer broad front over Mediterranean (very scarce but annual Malta); some winter Iberia, small numbers also SW Arabia. Vagrant Azores, Bahrain, Britain, Canary Is, Cape Verde Is, Channel Is, Iceland, Ireland, Kuwait, Libya, Madeira, Norway, Oman, Qatar and UAE.

Ciconia abdimii
Abdim's Stork

OTHER NAME White-bellied Stork.
TAXONOMY Lichtenstein, M H C, 1823. Monotypic.
DISTRIBUTION [W to NE Africa, SW Arabia > E to S and SW Africa. Decreasing. LC] Breeds SW Saudi Arabia and W Yemen. Mainly migratory, most wintering Africa, but some (mainly younger birds) remain in winter. Uncommon and irregular visitor Oman, where occasional influxes recorded between autumn and spring.

Ciconia episcopus
Woolly-necked Stork (V)

OTHER NAMES White-necked (or Bishop) Stork.
TAXONOMY (Boddaert, 1783). Three subspecies (undetermined in region).
DISTRIBUTION [Sub-Saharan Africa, S and SE Asia. Decreasing. LC] Extreme vagrant: Iran (19 Mar 1901, three).

Ciconia ciconia
White Stork

OTHER NAMES European Stork; Eastern White Stork (*C c asiatica*), Western White Stork (*C c ciconia*).
TAXONOMY (Linnaeus, 1758). Two subspecies (both in region).
DISTRIBUTION [NW Africa and SW Europe to C Asia > SW Europe, sub-Saharan Africa and S Asia. Increasing. LC] Breeds N Africa, Iberia, locally elsewhere in W and S continental Europe, more commonly from C Europe N to Denmark and Baltic States (recently also Finland) and E to Balkans, Black Sea and Ural Mtns, also S Caucasus and N Middle East (*ciconia*); formerly more widespread in NW Europe. Mainly migratory, with passage concentrated at Strait of Gibraltar and Bosphorus, also Israel and Suez; regular on passage W to Canary Is, scarce *eg* Britain, Finland and Malta. Some overwinter Iberia and S Middle East; *asiatica* (breeding C Asia) winters S Iran, and individuals matching measurements of this taxon also recorded Israel. Vagrant Azores, Cape Verde Is, Channel Is, Iceland, Ireland, Madeira and Norway.

Leptoptilos crumenifer
Marabou Stork (V)

OTHER NAME Marabou.
TAXONOMY (Lesson, 1831). Monotypic.
DISTRIBUTION [Sub-Saharan Africa. Increasing. LC] Rare vagrant (most records concentrated late summer-autumn): Hungary (25 May-1 Jun 2002); Israel (2: May 1951, collected; Mar-May 1957, two, one collected 31 Mar); N Mauritania (29 Mar 1966); Spain (27 records of 29 individuals). Further records from France placed in Cat E and Portugal in Cat D, but temporal pattern of occurrences in both countries broadly matches that in Spain.

ORDER PELECANIFORMES

FAMILY THRESKIORNITHIDAE

Ibises and spoonbills · five species

Threskiornis aethiopicus
African Sacred Ibis

OTHER NAME Sacred Ibis.
TAXONOMY (Latham, 1790). Monotypic.
DISTRIBUTION [Sub-Saharan Africa and Middle East. Decreasing. LC] Breeds W Yemen, SE Iraq and possibly SW Iran, where some of Iraq population winters; historical breeding range included Egypt and apparently Azerbaijan, but disappeared from both in first half of 19th century. Naturalised populations also established Bahrain, Canary Is, France, Italy and UAE, and has bred *eg* Belgium and Germany. Vagrant Azerbaijan, Kuwait, Oman and Saudi Arabia; wandering individuals from naturalised populations recorded *eg* Channel Is, Netherlands and Spain, probably also Britain (though records there and elsewhere in W Europe generally attributed to escapes from captivity).

Geronticus eremita
Northern Bald Ibis

OTHER NAMES Bald (or Hermit) Ibis, Waldrapp (Ibis).
TAXONOMY (Linnaeus, 1758). Monotypic.
DISTRIBUTION [NW Africa and Middle East > NW and NE Africa. Decreasing. CR] Breeding endemic. Resident W Morocco, with relict migrant population Syria (formerly also SE Turkey) now virtually extirpated, last breeding successfully 2011 and single individual returning 2013; latter population formerly migrated through Jordan, Saudi Arabia and Yemen. Historical range much more extensive, occurring as migrant breeder C Europe and possibly Balkans until 17th century, more recently also breeding Algeria erratically until early 1980s. Reintroduction projects established with varying success S Germany, Italy, Spain and Turkey. Vagrant Azores, Cape Verde Is, Croatia, Iraq, Israel and N Mauritania.

Plegadis falcinellus
Glossy Ibis

TAXONOMY (Linnaeus, 1766). Monotypic.
DISTRIBUTION [S North America, N South America, Africa, S Eurasia and Australia. Decreasing. LC] Breeds S Europe from SW Spain to Hungary, Balkans, Black and Caspian Seas, also Caucasus, Asia Minor, N Israel, SE Iraq and W and S Iran; formerly more widespread W Europe and NW Africa, and recently breeding again Portugal, Morocco and Tunisia. Largely migratory, occurring on passage widely Mediterranean basin and Middle East, some wintering N Africa and Middle East; present year-round S Iberia but expanding population also dispersive, leading to increasing records N to *eg* Britain and Netherlands. Vagrant Austria, Azores, Belarus, Belgium, Canary Is, Cape Verde Is, Channel Is, Czech Republic, Denmark, Estonia, Faroe Is, Finland, Germany, Gibraltar, Iceland, Ireland, Latvia, Lithuania, Luxembourg, Madeira, Norway, Poland, Slovakia, Sweden and Switzerland.

Platalea leucorodia
Eurasian Spoonbill

OTHER NAMES (Common/European or White) Spoonbill; Mauritanian Spoonbill (*P l balsaci*), Red Sea Spoonbill (*P l archeri*).
TAXONOMY Linnaeus, 1758. Three subspecies (all in region).
DISTRIBUTION [S Eurasia, W and NE Africa > S Eurasia, W, N and NE Africa, Sahel. Population trend unknown. LC] Breeds N Mauritania (*balsaci*), W Europe from Iberia N to Denmark (recently also Britain), S and EC Europe from NE Italy, Hungary and Slovakia to Balkans, Black and Caspian Seas, also Caucasus and from N Middle East discontinuously to Gulf (*leucorodia*), W Arabia and E Egypt (*archeri*). Partially migratory, occurring more widely continental Europe to winter S Europe, N Africa S to N Mauritania, Canary Is and Cape Verde Is, and parts of Middle East, where Gulf and W Arabian populations largely resident. Vagrant Azores, Belarus, Estonia, Faroe Is, Finland, Gibraltar, Iceland, Ireland, Latvia, Lithuania, Luxembourg, Madeira, Norway, Poland and Sweden.

Platalea alba
African Spoonbill

TAXONOMY Scopoli, 1786. Monotypic.
DISTRIBUTION [Sub-Saharan Africa. Stable. LC] Occasional Yemen, where breeding attempted. Vagrant Oman.

FAMILY ARDEIDAE

Bitterns, herons and egrets · 30 species

Botaurus stellaris
Eurasian Bittern

OTHER NAMES (Common or Great) Bittern.
TAXONOMY (Linnaeus, 1758). Two subspecies (one in region).
DISTRIBUTION [S Africa, Europe, SW to C and E Asia > Sahel, S Africa, W and S Europe, SW to SE and E Asia. Decreasing. LC] Breeds Europe widely from Mediterranean N to S Fennoscandia, W to Britain and E to Ural River and NW Kazakhstan, also Caucasus, Turkey, N and E Iran and SE Iraq (*stellaris*); range heavily fragmented in W and S, and formerly also bred NW Africa. N and E continental European populations migratory, wintering to W and S Europe, N Middle East, Morocco and Algeria (rare), and Nile Delta, some also moving S to Afrotropics; rare Libya and Arabia. Vagrant Azores, Canary Is, Cape Verde Is, Faroe Is, Iceland, Ireland, Liechtenstein, Norway and Tunisia (has bred).

Botaurus lentiginosus
American Bittern (V)

OTHER NAME North American Bittern.
TAXONOMY (Rackett, 1813). Monotypic.
DISTRIBUTION [N and C North America > W, S and E USA to West Indies and N Central America. Decreasing. LC] Rare vagrant (most records Sep-Mar, especially Oct-Nov): Azores (20*); Britain (38); Canary Is (10 Dec 2002); Channel Is (27 Oct 1870, shot); Denmark (2, including territorial male each spring 2001-2004); Faroe Is (3); Iceland (6); Ireland (21); Norway (26 Aug 1956); Spain (4); Sweden (24 Apr 2009).

Ixobrychus exilis
Least Bittern (V)

OTHER NAME Chory's Bittern (erythristic individuals of subspecies *exilis* and *erythromelas*).
TAXONOMY (Gmelin, J F, 1789). Six subspecies (undetermined in region).
DISTRIBUTION [E and S North America, West Indies, N and C South America > S USA to N and C South America. Stable. LC] Extreme vagrant (eight records, Sep-Nov): Azores (7: undated, collected*; 7 Sep 1951, collected*; 27 Nov 1951, collected*; 11 Sep 1952, collected*; 8 Oct 1964, collected*; 5 Nov 2001, taken into care and released; 28 Sep 2007, trapped); Iceland (17 Sep 1970, collected).

Ixobrychus minutus
Little Bittern

TAXONOMY (Linnaeus, 1766). Three subspecies (two in region).
DISTRIBUTION [W Europe to C Asia, N and sub-Saharan Africa > N Egypt and sub-Saharan Africa. Decreasing. LC] Summer visitor NW and N Africa, continental Europe from Iberia discontinuously N to Germany and S Baltic States, E to Balkans, Black and Caspian Seas, Ural River, also Caucasus and N Middle East; resident N Egypt and (mainly) C and E Arabia (*minutus*), possibly also S Egypt (*payesii*). Occurs widely S of breeding range on migration, W to Canary Is (has bred). Vagrant Azores, Britain (has bred), Cape Verde Is, Denmark, Estonia, Faroe Is, Finland, Gibraltar, Iceland, Ireland, Liechtenstein, Madeira, N Mauritania, Norway and Sweden; *payesii* vagrant Canary Is.

Ixobrychus sinensis
Yellow Bittern

OTHER NAMES Chinese Little (or Long-nosed) Bittern, Little Yellow Bittern.
TAXONOMY (Gmelin, J F, 1789). Monotypic.
DISTRIBUTION [SW to E Asia > S and SE Asia, Melanesia. Population trend unknown. LC] Breeds S Oman, where present mainly late Apr-late Nov but also recorded winter. In 2012 discovered breeding Red Sea coast of Egypt, where apparently established.

Ixobrychus eurhythmus
Von Schrenck's Bittern (V)

OTHER NAMES Schrenck's (Little) Bittern.
TAXONOMY (Swinhoe, 1873). Monotypic.
DISTRIBUTION [SE Siberia to E Asia > SE Asia. Decreasing. LC] Extreme vagrant: Italy (12 Nov 1912, collected). Another record involving specimen from Germany not accepted because of doubts about location and date of collection.

Ixobrychus cinnamomeus
Cinnamon Bittern (V)

OTHER NAME Chestnut Bittern.
TAXONOMY (Gmelin, J F, 1789). Monotypic.
DISTRIBUTION [S, SE and E Asia > S and SE Asia. Stable. LC] Extreme vagrant: UAE (20 Nov 2000).

Ixobrychus sturmii
Dwarf Bittern (V)

OTHER NAMES African Dwarf Bittern, Rail Heron.
TAXONOMY (Wagler, 1827). Monotypic.
DISTRIBUTION [Sub-Saharan Africa. Population trend unknown. LC] Extreme vagrant (five records): Canary Is (3: Oct 1886, collected; 21-30 Jan 2000; 23 Aug 2002-10 May 2003); Cape Verde Is (12 Jun and 17 Jul 2011); Oman (1 Nov 2013).

Nycticorax nycticorax
Black-crowned Night Heron

OTHER NAMES (Black-capped) Night Heron.
TAXONOMY (Linnaeus, 1758). Four subspecies (two in region).
DISTRIBUTION [Widespread North and South America, Africa and Eurasia. Decreasing. LC] Breeds NW Africa (probably also N Mauritania), Europe from Mediterranean discontinuously N to N France (irregularly to Netherlands), Poland, Ukraine and NW Kazakhstan, also Caucasus, N Middle East, C, E and S Arabia and N Egypt (*nycticorax*); introduced Mallorca. Chiefly migratory, bulk of population wintering S of region, and migrants regular N to Britain (scarce) and W to Canary Is (has bred); small numbers overwinter *eg* Morocco, Iberia, Italy, Levant, Egypt and increasingly E Arabia. Vagrant Azores, Cape Verde Is, Channel Is, Denmark, Estonia, Faroe Is, Finland, Jan Mayen, Latvia, Lithuania, Iceland, Ireland, Liechtenstein, Luxembourg, Madeira, Norway and Sweden; *hoactli* (breeding North and South America) also vagrant Iceland.

Nyctanassa violacea
Yellow-crowned Night Heron (V)

TAXONOMY (Linnaeus, 1758). Five subspecies (undetermined in region).
DISTRIBUTION [C and E North America to E South America > S USA to E South America. Stable. LC] Extreme vagrant (six records, most involving birds discovered autumn and winter, but following long stays now recorded all months): Azores (5: 25 Oct 2009*, photographed, presumed same 28 Jul 2010-18 Apr 2011*, and again 8 Dec 2011; 18-28 Oct 2010; 28 Jun 2011; 26 Sep-12 Oct 2011*; 31 Dec 2013-Jan 2014*); Madeira (end Jan-22 Dec 2011*, presumed same nearby 22 Jan 2012).

Butorides virescens
Green Heron (V)

OTHER NAMES Green-backed (or Little Green) Heron.
TAXONOMY (Linnaeus, 1758). Four subspecies (undetermined in region). Fomerly treated as conspecific with Striated Heron, but no mixed pairings known from zones of contact.
DISTRIBUTION [SW and SE Canada, USA to Central America and West Indies > S USA to Central America and West Indies. Population trend unknown. LC] Very rare vagrant (mostly mid-Aug to Nov): Azores (10*); Britain (7); Canary Is (7-12 May 2006); Channel Is (17-21 Aug and 13 Sep-early Oct 1992); France (2, second of which wintered three consecutive seasons from 2006-07 and summered in Netherlands); Iceland (2); Ireland (11-13 Oct 2005, later seen in Britain); Madeira (1, date unknown, collected and specimen preserved*); Netherlands (25 Apr-24 Jun and

14-30 Sep 2006, returning in 2007 and 2009, same as France).

Butorides striata
Striated Heron

OTHER NAME Mangrove Heron.
TAXONOMY (Linnaeus, 1758). Twenty subspecies (at least two in region). Formerly also included Green Heron (*qv*).
DISTRIBUTION [South America, sub-Saharan Africa, SW to E Asia, Australasia. Decreasing. LC] Breeds Nile Valley (presumed *atricapilla*) and from Red Sea coasts (*brevipes*) E to S and E Arabia (mainly coastal) and S Iran (subspecies undetermined, either *brevipes* or possibly *chloriceps*). Rare visitor S Israel, mainly in winter but has bred. Vagrant Jordan and Kuwait.

Ardeola ralloides
Squacco Heron

OTHER NAMES Common Squacco (or Squacco Pond) Heron.
TAXONOMY (Scopoli, 1769). Monotypic.
DISTRIBUTION [N and sub-Saharan Africa, SW Europe to C Asia > sub-Saharan Africa. Decreasing. LC] Breeds NW Africa, S Europe from Iberia and W France to Hungary, Balkans, Black Sea and NW Caspian, also Caucasus, locally N Middle East, C and E Arabia. Largely migratory, occurring regularly WC Europe and throughout Mediterranean, N Africa (W to Canary Is and N Mauritania) and Middle East en route to main wintering grounds S of region; some remain Mediterranean basin and Middle East in winter. Vagrant Azores, Belarus, Belgium, Britain, Cape Verde Is, Channel Is, Czech Republic, Denmark, Finland, Germany, Gibraltar, Iceland, Ireland, Luxembourg, Madeira, Netherlands, Poland, Slovakia and Sweden.

Ardeola grayii
Indian Pond Heron

OTHER NAMES (Indian) Paddybird, (Indian) Pondbird.
TAXONOMY (Sykes, 1832). Monotypic.
DISTRIBUTION [SW to SE Asia. Population trend unknown. LC] Resident S Iran; common passage migrant and winter visitor Oman and UAE. Vagrant Kuwait.

Ardeola bacchus
Chinese Pond Heron (V)

OTHER NAME Paddybird.
TAXONOMY (Bonaparte, 1855). Monotypic.
DISTRIBUTION [E Asia > E and SE Asia. Stable. LC] Extreme vagrant (two fully accepted records and five further occurrences, all mid Jul-mid Nov): Oman (2: 27 Aug 2010; 20 Sep 2012). Five additional occurrences: one in Hungary (14-23 Aug 2000) apparently wild but "origination from a non-wild population cannot be excluded with absolute certainty" and record placed in Cat D_A; another from Norway (autumn 1973, shot) considered to involve bird of uncertain origin and placed in Cat D; three others, from Britain (31 Oct and 13 Nov 2004) and Finland (2: 17-19 Jul 2007; 4 Aug 2012), considered of captive origin and placed in Cat E.

Bubulcus ibis
Western Cattle Egret

OTHER NAMES Buff-backed Heron, Cattle Egret.
TAXONOMY (Linnaeus, 1758). Monotypic. Formerly included Eastern Cattle Egret (*qv*).
DISTRIBUTION [Widespread North and South America, Africa, S Europe and SW Asia. Increasing. LC] Breeds Cape Verde Is, NW and N Africa, Iberia, France, Italy, Volga Delta, S Caucasus and widely separated areas of Middle East; introduced Mallorca. Mainly resident but more migratory in E of range; also dispersive, expanding range steadily, and regularly wanders to *eg* Britain (scarce, has bred), Channel Is, Belgium, Netherlands, Switzerland and Croatia. Very rare on passage Malta. Winters more widely NW Africa (W to Canary Is), SW Europe and throughout Middle East. Vagrant Albania, Austria, Azores, Belarus, Bosnia-Herzegovina, Bulgaria, Czech Republic, Denmark, Faroe Is, Finland, Germany, Greece, Hungary, Iceland, Ireland, Latvia, Liechtenstein, Lithuania, Luxembourg, Macedonia, Madeira, Moldova, Norway, Poland, Romania, Serbia, Slovakia, Slovenia, Sweden and Ukraine.

Bubulcus coromandus
Eastern Cattle Egret (V)

OTHER NAMES Buff-backed Heron, (Indian) Cattle Egret.
TAXONOMY (Boddaert, 1783). Monotypic. Formerly considered conspecific with Western Cattle Egret but differs constantly in adult breeding plumage, proportions and vocalisations.
DISTRIBUTION [S to E Asia, Australasia. Population trend not assessed. NE] Extreme or very rare vagrant (though historically also reported breeding E Iran): Oman (6); UAE (6* records, all of which possibly involving same returning individual, and one also with second bird present).

Ardea cinerea
Grey Heron

OTHER NAMES (Common or European) Heron; Mauritanian (or Pallid/White) Heron (*A c monicae*).
TAXONOMY Linnaeus, 1758. Four subspecies (two in region). Mauritanian *monicae* sometimes separated at species level because of constantly paler plumage and geographical isolation from nominate *cinerea*.
DISTRIBUTION [Widespread Africa and Europe > Africa, W and S Europe, SW to E Asia. Population trend unknown. LC] Breeds N Mauritania (*monicae*), N Tunisia and Egypt, much of Europe W to Ireland, N to W Norway and S Finland and E to Ural Mtns and NW Caspian Sea, also Caucasus, Asia Minor, N and E Iran, Kuwait and very locally E, C and W Arabia (*cinerea*). Mainly resident W Europe but C and E populations largely migratory, occurring widely to S on passage and in winter, when range extends to Iceland, Faroe Is, Atlantic islands (has bred Azores and Cape Verde Is), W and N Africa and across Middle East. Vagrant Jan Mayen and Svalbard; *monicae* vagrant Morocco.

Ardea herodias
Great Blue Heron (V)

TAXONOMY Linnaeus, 1758. Five subspecies (undetermined in region).
DISTRIBUTION [Widespread North America > W, S and E North America to N South America. Increasing. LC] Very rare vagrant: Azores (32*); Britain (7 Dec 2007); Canary Is (5-30 Dec 1998); Cape Verde Is (4-5 Mar 2002); France (11-27 Apr 1996); Iceland (2: 5 Nov 1983, shot; 6 Apr 1984, taken into captivity injured and later died). Several further records involving ship-assisted vagrancy in Atlantic, also within *eg* Belgian and British territorial limits, not admitted to national lists as birds were fed during transit.

Ardea melanocephala
Black-headed Heron

TAXONOMY Children and Vigors, 1826. Monotypic.
DISTRIBUTION [Sub-Saharan Africa and SW Arabia. Increasing. LC] Winter visitor Yemen and, since 2002, presumed colonising breeder. Vagrant Cape Verde Is, France, Israel, Jordan, Oman and Saudi Arabia.

Ardea goliath
Goliath Heron

TAXONOMY Cretzschmar, 1829. Monotypic.
DISTRIBUTION [Sub-Saharan Africa and SW Asia. Stable. LC] Breeds SE Egypt (rare), SW Saudi Arabia, W Yemen, SE Iraq and SW Iran. Mainly resident, but some local movements on Red Sea coasts. Vagrant Israel, Jordan, Oman and Syria.

Ardea purpurea
Purple Heron

OTHER NAMES Cape Verde (Purple) or Bourne's Heron (*A p bournei*).
TAXONOMY Linnaeus, 1766. Four subspecies (two in region). *Bournei* sometimes elevated to species level on basis of diagnosably distinct plumage and different breeding ecology.
DISTRIBUTION [Sub-Saharan Africa, W Europe to E Asia > sub-Saharan Africa, Egypt, S to E Asia. Decreasing. LC] Resident Cape Verde Is on Santiago (*bournei*); summer visitor NW Africa, from Iberia discontinuously N to Netherlands, E through C Europe to Balkans, Black and Caspian Seas, also Caucasus, patchily in N Middle East S to Levant and Kuwait, and W and C Arabia (*purpurea*). Occurs widely to S of breeding range on migration, W to Britain

(scarce, has bred) and Canary Is, and irregularly N to Poland (has bred); some winter Egypt (has bred) and Middle East. Vagrant Azores, Belarus, Cape Verde Is (*purpurea*), Denmark, Estonia, Faroe Is, Finland, Iceland, Ireland, Liechtenstein, Lithuania, Madeira, Norway and Sweden; sole 'extralimital' record of Cape Verdean *bournei* concerns one on Boa Vista in same archipelago.

Ardea alba
Great Egret

OTHER NAMES Great White (or Large) Egret, Great White Heron, Western Great Egret; (Great) American Egret (*A a egretta*), Eastern Great Egret (*A a modesta*).
TAXONOMY Linnaeus, 1758. Four subspecies (all in region). Genetically differentiated American form *egretta*, which also has distinctive soft-part coloration in peak breeding condition, sometimes separated at species level; eastern *modesta* also occasionally split, but division not supported by genetic data and apparently hybridises with *alba* where ranges overlap.
DISTRIBUTION [Cosmopolitan. Population trend unknown. LC] Breeds locally S and E Iberia, France and NE Italy, more widely from Hungary through E and SE Europe to NW Kazakhstan and N Caucasus, also C and W Turkey and Syria (*alba*) and SE Iran (*modesta*); range expanding at least in W, where recently also breeding N to *eg* S Britain, Netherlands and parts of S Baltic, and now regular non-breeding visitor S Scandinavia. Mainly resident but dispersive in W and S; migratory in E and SE Europe, occurring more widely on passage and in winter Middle East (including *modesta* in SE Arabia), S Europe (rare but increasing Malta), N Africa, and N Mauritania (*melanorhynchos*, breeding sub-Saharan Africa). Vagrant Azores, Canary Is, Cape Verde Is, Channel Is, Faroe Is, Iceland, Ireland, Norway and Svalbard; *egretta* (North and South America) vagrant Azores. Note: *modesta*-type birds documented C Europe may be variant *alba* individuals.

Ardea intermedia
Intermediate Egret

OTHER NAMES Yellow-billed (or Median/Smaller) Egret.
TAXONOMY (Wagler, 1829). Three subspecies (two in region).
DISTRIBUTION [Sub-Saharan Africa, S, SE and E Asia, Australia. Decreasing. LC] Fairly common passage migrant and winter visitor Oman (*intermedia* believed to occur). Vagrant Cape Verde Is (*brachyrhyncha*), Israel, Italy, Jordan, Qatar, UAE and Yemen.

Egretta ardesiaca
Black Heron (V)

OTHER NAME Black Egret.
TAXONOMY (Wagler, 1827). Monotypic.
DISTRIBUTION [Sub-Saharan Africa. Stable. LC] Very rare vagrant: Cape Verde Islands (8 records of at least 9 individuals); Greece (3-12 Jun 2012); Israel (19-20 Oct 1982); Oman (9 Sep-25 Oct 1994); Yemen (2: 30-31 Mar and 26 Apr 1996; 23 Jan 1997, possibly same as in 1996).

Egretta tricolor
Tricolored Heron (V)

OTHER NAME Louisiana Heron.
TAXONOMY (Statius Müller, 1776). Two subspecies (one in region).
DISTRIBUTION [S North America, Central America and N South America > S North America, West Indies, Central America and N South America. Stable. LC] Extreme vagrant (four records but perhaps involving only three individuals, *ruficollis*): Azores (3: 22-24 Oct 1985; 2 Oct 2007; 20 Oct 2012*); Canary Is (15 Nov 2007 intermittently to 6 Jun 2008, possibly same individual as Azores, Oct 2007).

Egretta caerulea
Little Blue Heron (V)

TAXONOMY (Linnaeus, 1758). Monotypic.
DISTRIBUTION [S and E North America, West Indies, N South America > SW to E North America, Central America, West Indies and N South America. Decreasing. LC] Extreme vagrant (six records, mainly Sep-Oct): Azores (5: 28 Nov 1964*, ringed New Jersey, USA; 18 Sep-9 Oct 1997; 4-7 Oct 1998; 24 Jun-15 Jul 2012; 21 Oct 2012*); Ireland (24 Sep-22 Oct 2008).

Egretta thula
Snowy Egret (V)

TAXONOMY (Molina, 1782). Two subspecies (undetermined in region).
DISTRIBUTION [N USA to S South America > S USA to S South America. Increasing. LC] Very rare vagrant: Azores (15*); Britain (5 Nov 2001 intermittently to 17 Sep 2002); Cape Verde Is (1-3 Nov 2005); Germany (26 Oct 1937); Iceland (4).

Egretta garzetta
Little Egret

TAXONOMY (Linnaeus, 1766). Two subspecies (one in region). Sometimes considered to include Western Reef Egret (*qv*).
DISTRIBUTION [Africa, W, C and S Europe, SW to E Asia, Australasia. Increasing. LC] Breeds Cape Verde Is, Canary Is, N Mauritania, NW Africa, Europe from Mediterranean discontinuously N to *eg* Ireland, Britain, Slovakia, Ukraine and NW Kazakhstan, also Caucasus, N Middle East and Nile Valley (*garzetta*). Mainly resident in W and S but dispersive and expanding range N towards S Baltic (has bred Poland); largely migratory in NE and E, occurring widely on passage S through Middle East and N Africa, regularly W to Azores and Madeira. Vagrant Belarus, Czech Republic, Iceland, Faroe Is, Finland, Latvia, Lithuania, Norway and Sweden.

Egretta gularis
Western Reef Egret

OTHER NAMES Western Reef Heron (IOC); Arabian (or Indian) Reef Egret (*E g schistacea*), West African Reef Egret (*E g gularis*).
TAXONOMY (Bosc, 1792). Two subspecies (both in region). Has been treated as conspecific with Little Egret, but interbreeding limited and habitat preferences differ. Dimorphic; white plumage rare in *gularis* but present in up to 50 per cent of *schistacea*, which further differs in longer bill and shorter tarsus.
DISTRIBUTION [W and E Africa, SW and S Asia. Stable. LC] Breeds coastally N Mauritania (*gularis*), E Egypt, Arabia (less widely in S) and S Iran (*schistacea*). Mainly resident but some post-breeding dispersal, when regular visitor Cape Verde Is, and Red Sea population regular N in Gulf of Aqaba to S Israel and S Jordan. Vagrant Azores, Bulgaria, Canary Is, Corsica, Croatia, France, Greece, Hungary, Italy, Malta, Montenegro, Morocco, Portugal, Sardinia, Spain and Tunisia.

FAMILY SCOPIDAE

Hamerkop · one species

Scopus umbretta
Hamerkop

OTHER NAMES Hammerhead, Hammer-headed Stork.
TAXONOMY Gmelin, J F, 1789. Two subspecies (one in region).
DISTRIBUTION [Sub-Saharan Africa and SW Arabia. Stable. LC] Resident SW Saudi Arabia and W Yemen (*umbretta*).

FAMILY PELECANIDAE

Pelicans · three species

Pelecanus onocrotalus
Great White Pelican

OTHER NAMES (Eastern or European) White Pelican, Rosy Pelican.
TAXONOMY Linnaeus, 1758. Monotypic.
DISTRIBUTION [Sub-Saharan Africa, SE Europe to C and S Asia > sub-Saharan Africa to S Asia. Population trend unknown. LC] Breeds N Mauritania, Greece, Danube Delta, Ukraine, SE European Russia, Caucasus, Turkey and NW Iran; formerly more widespread C and SE Europe, and also bred SE Iraq and Kuwait. Mauritanian birds move S after breeding; those from Europe migrate through Bosphorus and (with NC Asian birds) Caucasus. Some winter N Israel, SE Iraq, S Iran and SW Arabia, but most continue S through W Middle East and Egypt to winter S of region. Irregular Italy. Vagrant Algeria, Austria,

Bahrain, Belarus, Bosnia-Herzegovina, Cape Verde Is, Corsica, Croatia, Czech Republic, Finland, France, Germany, Hungary, Latvia, Libya, Lithuania, Malta, Moldova, Montenegro, Morocco, Netherlands, Oman, Poland, Qatar, Sardinia, Slovakia, Slovenia, Spain, Switzerland, Tunisia and UAE. Extralimital status complicated by possibility of (or known) escapes from captivity, many individuals thus being placed in Cat D, but overall pattern of such occurrences established to support genuine vagrancy to N and W Europe of this and following two species.

Pelecanus rufescens
Pink-backed Pelican

TAXONOMY Gmelin, J F, 1789. Monotypic.
DISTRIBUTION [Sub-Saharan Africa and SW Arabia. Stable. LC] Breeds SW Saudi Arabia and W Yemen; regular non-breeding summer visitor S Egypt. Vagrant Balearic Is, Israel, Jordan, Kuwait, Lebanon, Oman, Spain and Turkey; see comments under Great White Pelican, also applicable to this species, on genuine vagrancy to N and W Europe (though records generally retained in Cat D, *eg* in Austria, France, Germany, Netherlands, Poland and Portugal).

Pelecanus crispus
Dalmatian Pelican

TAXONOMY Bruch, 1832. Monotypic.
DISTRIBUTION [SE Europe, SW to C Asia > SE Europe, SW to SE Asia. Decreasing. VU] Breeds S and E Balkans, Danube Delta, SE European Russia, NW Kazakhstan, Caucasus, Turkey and SW Iran; formerly more widespread, also breeding *eg* Hungary and Ukraine. N populations migratory, occurring on passage Black Sea and W Caspian coasts, wintering locally parts of S Balkans, W Turkey, Levant, NW and S Iran, rarely to SE Iraq, sometimes farther N in milder winters. Vagrant Algeria, Austria, Croatia, Cyprus, Denmark, Egypt, Hungary (formerly bred), Israel, Italy, Kuwait, Latvia, Lithuania, Moldova, Netherlands, Oman, Poland, Slovakia, Slovenia, Syria and UAE; see comments under Great White Pelican, also applicable to this species, on likelihood of genuine vagrancy to N and W Europe.

ORDER SULIFORMES

FAMILY FREGATIDAE

Frigatebirds · four species

Fregata aquila
Ascension Frigatebird (V)

OTHER NAME Ascension Island Frigatebird.
TAXONOMY (Linnaeus, 1758). Monotypic.
DISTRIBUTION [Ascension Is > Atlantic Ocean. Population trend unknown. VU] Extreme vagrant: Britain (2: 9 or 10 July 1953, caught, later died and skin preserved; 5 July 2013).

Fregata magnificens
Magnificent Frigatebird

TAXONOMY Mathews, 1914. Monotypic.
DISTRIBUTION [E Pacific Ocean, Caribbean and E Atlantic Ocean. Increasing. LC] Resident Cape Verde Is, where tiny relict E Atlantic population (three individuals in 2014) no longer breeds; formerly more numerous and widespread in archipelago. Vagrant Azores*, Britain, Denmark, France, Portugal and Spain. Most vagrant frigatebirds in region are not identified to species, but away from Middle East (*eg* those recorded Belgium, Canary Is, Ireland, Madeira, Morocco, Norway and Sweden), records also likely to involve this species; on statistical grounds, origin of such vagrants clearly more likely to be W Atlantic or Caribbean than Cape Verde Is.

Fregata minor
Great Frigatebird (V)

OTHER NAMES Greater (or Pacific) Frigatebird.
TAXONOMY (Gmelin, J F, 1789). Five subspecies (undetermined in region).
DISTRIBUTION [Tropical oceans. Decreasing. LC] Extreme vagrant: Oman (11-18 Jun 1982).

Fregata ariel
Lesser Frigatebird (V)

OTHER NAME Least Frigatebird.
TAXONOMY (Gray, G R, 1845). Three subspecies (undetermined in region).
DISTRIBUTION [Tropical oceans. Decreasing. LC] Extreme vagrant: Israel (2: 1 Dec 1997, same as Jordan; 6 May 1999); Jordan (1 Dec 1997, same as Israel); Kuwait (1 Apr 2008); Oman (5: 9 Aug 1986, found dead on 17 Aug; 5-8 Jul 1993; 30 Oct 1997; 22 Sep 2002; 11 Oct 2002); Yemen (29 Oct 2008).

FAMILY SULIDAE

Gannets and boobies · five species

Morus bassanus
Northern Gannet

OTHER NAMES (North) Atlantic Gannet, Gannet.
TAXONOMY (Linnaeus, 1758). Monotypic.
DISTRIBUTION [N Atlantic Ocean. Increasing. LC] Breeds NW France, Channel Is, N and W Britain, Ireland, Faroe Is, Iceland, Norway and NW Germany. Pelagic outside breeding season, some remaining in breeding areas but mainly younger birds dispersing widely in Arctic waters, North Sea and NE Atlantic, S to N Macaronesian islands and N Mauritania; also penetrates Mediterranean, where much less numerous in E. Vagrant Austria, Belarus, Bulgaria, Cape Verde Is, Croatia, Czech Republic, Estonia, Finland, Jordan, Latvia, Lithuania, Luxembourg, Montenegro, Poland, Slovenia, Switzerland and Syria.

Morus capensis
Cape Gannet (V)

OTHER NAMES (South) African Gannet.
TAXONOMY (Lichtenstein, M H C, 1823). Monotypic.
DISTRIBUTION [S Africa > W, S and E Africa. Decreasing. VU] Extreme vagrant: Oman (14 Mar 2004). Formerly accepted record of ringed individual supposedly recovered at sea off S Morocco found to be erroneous; further record from Spain (28 Jan 1985) judged to have involved either Cape Gannet or possibly extralimital Australian Gannet *M serrator*, and treated as of uncertain origin. All other claims from Atlantic and Mediterranean now considered not to eliminate fully Northern Gannet.

Sula dactylatra
Masked Booby

OTHER NAMES Blue-faced (or White) Booby.
TAXONOMY Lesson, 1831. Four subspecies (one, probably two, in region).
DISTRIBUTION [Tropical Oceans. Decreasing. LC] Breeds on islands off coasts of W Yemen and S Oman (*melanops*); some non-breeding birds wander N to Gulf of Oman. Vagrant Azores, Cape Verde Is, France, Iran, Israel, Morocco, Saudi Arabia, Spain and UAE; records in Atlantic sector most likely involve *dactylatra* (breeding Caribbean and SW Atlantic).

Sula sula
Red-footed Booby (V)

TAXONOMY (Linnaeus, 1766). Three subspecies (at least one in region).
DISTRIBUTION [Tropical Oceans. Decreasing. LC] Very rare vagrant (most records in second half of year, those in Atlantic sector probably involving *sula*): Canary Is (23 Sep 2012); Cape Verde Is (8); France (2-14 July 2011); Iran (28 Aug 1979*, on ship, same as first Oman/UAE record); Oman (3, *rubripes*); Spain (11 Aug and 3-15 Dec 2010, taken into care but later died); UAE (2, *rubripes*).

Sula leucogaster
Brown Booby

OTHER NAMES White-bellied Booby; Forster's (Brown) Booby (*S l plotus*).
TAXONOMY (Boddaert, 1783). Four subspecies (two in region).
DISTRIBUTION [Tropical Oceans. Decreasing. LC] Breeds Cape Verde Is (*leucogaster*) and on islands in Red Sea off Egypt, Saudi Arabia and Yemen, also Gulf of Aden (*plotus*).

Regularly wanders N along Gulf of Aqaba to Israel and Jordan, also off S Arabia E to Oman (uncommon). Vagrant Azores, Canary Is, Italy, Madeira, N Mauritania, Morocco, Portugal, Spain and UAE.

FAMILY PHALACROCORACIDAE

Cormorants and shags · seven species

Microcarbo africanus
Reed Cormorant

OTHER NAME Long-tailed Cormorant.
TAXONOMY (Gmelin, J F, 1789). Two subspecies (one in region).
DISTRIBUTION [Sub-Saharan Africa. Decreasing. LC] Breeds coastal N Mauritania (*africanus*); present year-round, though at least some of breeding population moves S in winter. Vagrant Egypt (formerly bred) and Morocco; record involving reidentified specimen from Spain placed in Cat D.

Microcarbo pygmeus
Pygmy Cormorant

OTHER NAME Little Cormorant.
TAXONOMY (Pallas, 1773). Monotypic.
DISTRIBUTION [SE Europe to C Asia > SE Europe and SW Asia. Increasing. LC] Breeds from NE Italy, Austria, Hungary and Balkans E to S Ukraine, SW Russia and NW Kazakhstan, also Caucasus, Turkey, Levant, N Israel, SE Iraq and N Iran; formerly bred Algeria until *c* 1857. Mainly migratory in N of range, occurring on passage and in winter N to Slovakia (has bred) and S through breeding range, formerly as far S as Jordan. Vagrant Belarus, Belgium, Cyprus (formerly wintered), Czech Republic, Egypt*, France, Germany, Kuwait, Latvia, Lithuania, Malta, Netherlands, Poland, Sardinia, Sweden, Switzerland and Tunisia.

Phalacrocorax auritus
Double-crested Cormorant (V)

OTHER NAME White-crested Cormorant.
TAXONOMY (Lesson, 1831). Five subspecies (undetermined in region).
DISTRIBUTION [North America > W and S North America. Increasing. LC] Rare vagrant (mainly autumn and winter): Azores (50+*); Britain (11 Jan-21 Apr 1989); Canary Is (22 Oct 2012); France (5-8 Oct 2000); Ireland (18 Nov 1995-6 Jan 1996).

Phalacrocorax aristotelis
European Shag (E)

OTHER NAMES (Common or Green) Shag, Green Cormorant; Mediterranean Shag (*P a desmarestii*), Moroccan Shag (*P a riggenbachi*).
TAXONOMY (Linnaeus, 1761). Three subspecies.
DISTRIBUTION [NW and N Africa, Europe. Decreasing. LC] Endemic. Breeds Atlantic coasts of Morocco (*riggenbachi*), Iberia, NW France, Channel Is, W and N Britain, Ireland, Iceland, Norway and extreme NW Russia (*aristotelis*), and on scattered Mediterranean and Black Sea coastal locations and islands (*desmarestii*). Migratory in N Norway and NW Russia, when more likely *eg* Sweden and S North Sea in winter, but otherwise generally sedentary or moving relatively short distances from colonies. Vagrant Austria, Czech Republic, Denmark, Egypt, Finland, Georgia, Germany, Iraq, Israel, Lithuania, Macedonia, Madeira, Malta, Montenegro, Poland, Romania, Slovakia, Switzerland and Syria.

Phalacrocorax carbo
Great Cormorant

OTHER NAMES (Common) Cormorant, Large (Black) Cormorant; Atlantic (Great) Cormorant (*P c carbo*), Continental (Great) Cormorant (*P c sinensis*), Moroccan Cormorant (*P c maroccanus*).
TAXONOMY (Linnaeus, 1758). Five subspecies (three in region). Formerly included White-breasted Cormorant (*qv*).
DISTRIBUTION [NE North America, NW Africa, Eurasia and Australasia. Increasing. LC] Breeds Atlantic coasts of Morocco (*maroccanus*), Iceland, Faroe Is, Norway, Brit-

ain and Ireland (*carbo*), and scattered areas of interior and coastal Europe from Britain, France and Iberia E to Black and Caspian Seas and Asia Minor, N to Baltic and S to N Mediterranean (*sinensis*). Present year-round in most of W and S of range, numbers augmented by migrants from N and E Europe; also winters throughout Mediterranean basin and Middle East (has bred Bahrain). Vagrant Azores, Madeira, N Mauritania and Svalbard; *sinensis* vagrant Ireland; undetermined *maroccanus* or White-breasted Cormorant vagrant Malta.

Phalacrocorax lucidus
White-breasted Cormorant

OTHER NAME White-necked Cormorant.
TAXONOMY (Lichtenstein, M H C, 1823). Monotypic. Sometimes included within Great Cormorant but often considered a separate species, with significant differences in plumage and subtler distinctions in measurements.
DISTRIBUTION [W, S and E Africa. Population trend unknown. NE] Breeds coastal N Mauritania, probably intergrading with *maroccanus* Great Cormorant in S Morocco. Vagrant Belgium*, Cape Verde Is (possibly formerly resident), Egypt*, Israel, Saudi Arabia and Yemen; this species or *maroccanus* Great Cormorant also vagrant Malta.

Phalacrocorax nigrogularis
Socotra Cormorant

OTHER NAMES Socotra Shag.
TAXONOMY Ogilvie-Grant and Forbes, H O, 1899. Monotypic.
DISTRIBUTION [Persian Gulf and S Arabia > Persian Gulf to NE Africa. Decreasing. VU] Breeding endemic. Nests chiefly on islands in Persian Gulf off coasts of Bahrain, Qatar, Saudi Arabia and UAE, with isolated colonies on S coasts of Oman and Yemen; scarce resident S Iran. More widespread outside breeding season, dispersing N to Kuwait (uncommon spring-autumn; has bred) and more widely along coasts of SE and S Arabia, though rare S Red Sea coast of Yemen and essentially vagrant to W coast of Saudi Arabia.

FAMILY ANHINGIDAE

Anhingas and darters · one species

Anhinga rufa
African Darter

OTHER NAME Darter.
TAXONOMY (Daudin, 1802). Three subspecies (two in region).
DISTRIBUTION [Sub-Saharan Africa, Madagascar and Middle East. LC] Breeds SE Iraq (*chantrei*); rare visitor Iran, where several recent summer records in SW. Also bred Turkey until early 1960s, when wetland breeding area drained; birds from this population formerly wintered N Israel and N Jordan until late 1950s, after which no occurrences except single record Israel, 31 May 2004. Vagrant Egypt (*rufa*), Kuwait and Morocco (presumed *rufa*).

ORDER ACCIPITRIFORMES

FAMILY PANDIONIDAE

Ospreys · one species

Pandion haliaetus
Western Osprey

OTHER NAMES Osprey; American Osprey (*P h carolinensis*).
TAXONOMY (Linnaeus, 1758). Three subspecies (two in region). Previously included extralimital SE Asian and Australasian *P cristatus* as a subspecies.
DISTRIBUTION [North America, N Africa and Eurasia > S North America and South America, Africa and S Eurasia. Increasing. LC] Summer visitor NW France, Britain (including reintroduced population EC England), Fennoscandia and S Baltic region E to Ural Mtns and NW Caspian, also Danube Delta; resident Cape Verde Is, Canary Is, coastal Morocco and Algeria (formerly also Tunisia), S Portugal, Balearic Is, Corsica, N Iran and coasts of Arabian Gulf, Arabian Sea and Red Sea (*haliae-*

tus). Occurs on passage across much of region. Vagrant Azores, Faroe Is, Iceland and Madeira; *carolinensis* (breeding North America) vagrant Azores (13 Oct-4 Nov 2011), Faroe Is (16 Sep 2013*) and Iceland (22 Sep 2008).

FAMILY ACCIPITRIDAE

Kites, hawks and eagles · 56 species

Elanus caeruleus
Black-winged Kite

OTHER NAMES (Common) Black-shouldered Kite.
TAXONOMY (Desfontaines, 1789). Three subspecies (two in region).
DISTRIBUTION [Sub-Saharan and N Africa, SW Europe, SW to SE Asia. Stable. LC] Breeds NW Africa, Iberia, S France and Nile Valley (*caeruleus*), also W Yemen, and apparently colonising Iraq, Iran, Israel and possibly Turkey (*vociferus*). Mainly sedentary in region but range expanding, and some movement evidenced by vagrancy N to Scandinavia. Vagrant Austria, Belgium, Bulgaria, Cyprus, Czech Republic, Denmark, Germany, Gibraltar, Greece, Hungary, Israel, Italy, Jordan, Kuwait, Lebanon, Libya, N Mauritania, Netherlands, Oman, Qatar, Saudi Arabia, Serbia, Slovakia, Sweden, Switzerland, Turkey and UAE.

Chelictinia riocourii
Scissor-tailed Kite (V)

OTHER NAMES (African) Swallow-tailed (or Fork-tailed) Kite.
TAXONOMY (Temminck, 1821). Monotypic.
DISTRIBUTION [W to NE and E Africa. Decreasing. LC] Extreme vagrant: Yemen (Jun or Jul 1960).

Gypaetus barbatus
Bearded Vulture

OTHER NAME Lammergeier.
TAXONOMY (Linnaeus, 1758). Three subspecies (all in region).
DISTRIBUTION [S Europe, NW, E and S Africa, SW to CE Asia. Decreasing. LC] Breeds Morocco and perhaps still N Algeria (*barbatus*), Pyrenees, Alps (reintroduced), Corsica, Crete, S Balkans, Caucasus, Turkey, N Iraq and N Iran (*aureus*), and E Egypt, SW Saudi Arabia and W Yemen (*meridionalis*). Formerly more widespread, including *eg* NE Portugal, C and S Spain (where reintroduction under way), WC Europe, NW Balkans, Levant, NW Saudi Arabia, Sinai and Tunisia. Mainly resident with limited dispersal, especially of younger birds; wandering individuals from reintroduced C European population responsible for most recent records NW Europe. Vagrant Belgium, Bulgaria, Croatia, Cyprus, Czech Republic, N Germany, Gibraltar, Lebanon, Liechtenstein, Poland, Romania, Slovenia and Ukraine.

Neophron percnopterus
Egyptian Vulture

OTHER NAMES Indian (or White) Scavenger Vulture.
TAXONOMY (Linnaeus, 1758). Three subspecies (two in region).
DISTRIBUTION [NW, NE and E Africa, S Europe to C and S Asia > SW, W to NE and E Africa, SW and S Asia. Decreasing. EN] Breeds Canary Is (*majorensis*), Cape Verde Is, NW Africa discontinuously E to Egypt, much of S Europe from Iberia E to Crimea and Caucasus, and widely in N Middle East, Levant and Arabia (*percnopterus*). Resident Atlantic islands, S Morocco and Arabia but otherwise mainly migratory, European population largely moving S via Gibraltar and Bosphorus-Levant routes and, as many N African breeders, crossing Sahara; small numbers remain Mediterranean basin and S Iran. Vagrant Austria, Azores, Belgium, Britain, Czech Republic, Denmark, Estonia, Finland, Germany, Hungary, Luxembourg, Madeira, Malta, Netherlands, Norway, Poland, Qatar, Romania, Sardinia, Slovakia, Slovenia, Sweden, Switzerland and Ukraine.

Pernis apivorus
European Honey Buzzard

OTHER NAMES (Eurasian or Western) Honey Buzzard.
TAXONOMY (Linnaeus, 1758). Monotypic.

DISTRIBUTION [W Europe to S Siberia and SW Asia > sub-Saharan Africa. Stable. LC] Summer visitor Britain (very scarce) and much of continental Europe from C Iberia N to Fennoscandia and E to Ural Mtns, Caucasus and Balkans, also Asia Minor and N Iran. Migration focused at key 'bottlenecks' such as S Sweden/N Denmark, Strait of Gibraltar, W Caucasus and S Israel, some also via Tunisia-Sicily and Libya-Greece, but routes vary between spring and autumn, and according to weather; significant spring passage also noted N Mauritania, but rare on migration E Arabia. Exceptional in winter. Vagrant Canary Is, Faroe Is, Iceland, Ireland and Madeira.

Pernis ptilorhyncus
Crested Honey Buzzard

OTHER NAMES Oriental (or Asian/Asiatic) Honey Buzzard, Eastern (or Siberian) Honey Buzzard.
TAXONOMY (Temminck, 1821). Six subspecies (two in region).
DISTRIBUTION [S Siberia to NE, SE and S Asia > S and SE Asia. Stable. LC] Uncommon but regular migrant/winter visitor S Iran, Saudi Arabia, UAE and Oman (where apparently mainly *ruficollis* but *orientalis* also documented); perhaps under-recorded because of confusion with European Honey Buzzard. Elsewhere in E of region, mainly detected on passage (*orientalis*) in large movements of congener, especially at migration 'bottlenecks' in S Israel (ratio approx 1:1,000, with typical annual counts of 15-20 in spring and 5-12 autumn) and Georgia; some hybrids also recorded. Rare Kuwait. Vagrant Bahrain, Cyprus, Egypt, Italy, Jordan, Lebanon, Qatar, Turkey and Yemen.

Elanoides forficatus
Swallow-tailed Kite (V)

OTHER NAME American Swallow-tailed Kite.
TAXONOMY (Linnaeus, 1758). Two subspecies (undetermined in region).
DISTRIBUTION [SE USA, Middle America and South America > South America. Increasing. LC] Extreme vagrant (two records): Azores (24 Aug-7 Sep 2008); Canary Is (19-23 Mar 1993). Further published historical claim from Malta not submitted for assessment.

Necrosyrtes monachus
Hooded Vulture (V)

TAXONOMY (Temminck, 1823). Monotypic.
DISTRIBUTION [Sub-Saharan Africa. Decreasing. CR] Extreme vagrant (at least two records of three individuals): N Mauritania (Dec 1960); Morocco (7 Jun 1955, two). May possibly also occur on S fringe of region in N Niger. Further record from N Mauritania (Dec 1960) relates to unspecified area of Banc d'Arguin, while another Banc d'Arguin record (8-9 Apr 2001) relates to mainland site just outside regional boundary; exact location, date and validity of additional published N Mauritanian record from Adrar is unclear. *Contra* Fenech 2010, published historical claim from Malta has not been accepted, species unrecorded from Algeria and Tunisia, and recent Spanish records considered to be escapes.

Gyps africanus
White-backed Vulture (V)

OTHER NAMES African White-backed Vulture, White-backed Griffon Vulture.
TAXONOMY Salvadori, 1865. Monotypic.
DISTRIBUTION [Sub-Saharan Africa. Decreasing. CR] Extreme vagrant (three records): Spain (3: 7 Sep 2008; 25 Jun 2009, dead; 19 Sep 2011). A record from Portugal (14 Oct 2006) currently in Cat D may deserve reconsideration in view of subsequent series of Spanish records; old specimen record from Italy (Jun 1933) also placed in Cat D.

Gyps bengalensis
White-rumped Vulture (V)

OTHER NAMES Indian (or Asian/Oriental) White-rumped (or White-backed) Vulture.
TAXONOMY (Gmelin, J F, 1788). Monotypic.
DISTRIBUTION [S and SE Asia. Decreasing. CR] Formerly resident SE Iran; current status unclear, but no confirmed records since 1972.

Gyps rueppelli
Rüppell's Vulture

OTHER NAMES Rueppell's Vulture, Rüppell's Griffon (Vulture).

TAXONOMY (Brehm, A E, 1852). Two subspecies (one in region).
DISTRIBUTION [Sub-Saharan Africa. Decreasing. CR] Rare but increasingly regular nonbreeding visitor to Iberia (*rueppelli*); mostly occurs late spring and especially autumn, but records in all months. First recorded Spain 1992 and Portugal 1993; subsequently, 77 records of 92 birds by 2013 in Spain and 15 records by 2010 in Portugal, some involving long-staying individuals. Has apparently hybridised with Griffon Vulture in Spain. Vagrant Egypt, Gibraltar, N Mauritania and Morocco; two French records considered to involve bird(s) not of wild origin.

Gyps himalayensis
Himalayan Vulture (V)

OTHER NAMES Himalayan Griffon (Vulture).
TAXONOMY Hume, 1869. Monotypic.
DISTRIBUTION [Himalayas, SC to CE Asia. Stable. LC] Extreme vagrant: UAE (13-16 Oct 2012).

Gyps fulvus
Griffon Vulture

OTHER NAMES Eurasian (or Common/European) Griffon.
TAXONOMY (Hablizl, 1783). Two subspecies (one in region).
DISTRIBUTION [NW Africa and S Europe to C Asia > Sahel, NE Africa, SW to S Asia. Increasing. LC] Breeds NW Africa, S Europe from Iberia, S France and Mediterranean islands (non-breeders also summering Alps) to Italy (reintroduced), Balkans, Crimea and Caucasus, and in parts of Middle East from Turkey, Iran and Levant S to C, W and SW Arabia (*fulvus*); formerly more widespread. Mainly resident, but younger birds dispersive or migratory. Much of Turkish population moves S in winter, when recorded more widely through Middle East (though rare Arabia away from breeding areas) and W to E Egypt; also significant passage of younger birds Strait of Gibraltar, with limited migration/wintering Morocco and N Mauritania. Vagrant Belarus, Belgium, Channel Is, Corsica, Czech Republic, Denmark, Estonia, Finland, Germany, Hungary, Ireland, Latvia, Libya, Lithuania, Malta, Moldova, Netherlands, Poland, Romania, Slovakia, Sweden and UAE.

Aegypius monachus
Cinereous Vulture

OTHER NAMES (Eurasian or European) Black Vulture, Monk Vulture.
TAXONOMY (Linnaeus, 1766). Monotypic.
DISTRIBUTION [SW and SE Europe, SW, C and E Asia. Decreasing. NT] Breeds Spain, Portugal (rare), Mallorca, S France (reintroduced), E Balkans, S Ukraine, Turkey, Caucasus and Iran; formerly more widespread, also breeding *eg* Austria, Cyprus, Israel, Italy, Moldova, Montenegro, Romania and Slovenia. Generally resident, but some dispersal of mainly young birds, including from reintroduced population; occasional migrant Gibraltar, and winter movement noted especially in E of range S to Egypt (rare), Israel, Palestine, N Arabia and S Iran. Vagrant Algeria, Belarus, Bosnia-Herzegovina, Ceuta, Cyprus, Czech Republic, Estonia, Finland, Germany, Hungary, Jordan, Latvia, Lebanon, Macedonia, Moldova, Morocco, Netherlands, Oman, Poland, Romania, Slovakia, Slovenia, Sweden*, Switzerland, Serbia, Syria, Tunisia and UAE.

Torgos tracheliotos
Lappet-faced Vulture

OTHER NAME Nubian Vulture.
TAXONOMY (Forster, J R, 1796). Two subspecies (both in region).
DISTRIBUTION [Sub-Saharan Africa and Arabia. Decreasing. EN] Breeds SE Egypt (*tracheliotos*), W and C Saudi Arabia, E Yemen (rare) and Oman, formerly also UAE (*negevensis*). Range formerly much larger, also encompassing NW Africa, but apparently now extirpated N Mauritania, Morocco, Algeria and Tunisia, with no records since 1970s; farther E, range also extended over much larger area of Egypt and N to Israel and Palestine. Mainly resident, but young birds may disperse over long distances. Vagrant France, Israel, Jordan, Kuwait, Libya and Spain.

Circaetus gallicus
Short-toed Snake Eagle

OTHER NAME Short-toed Eagle.
TAXONOMY (Gmelin, J F, 1788). Monotypic.
DISTRIBUTION [NW Africa, SW Europe to C Asia, also S, CE and SE Asia > Sahel, S, CE and SE Asia. Stable. LC] Summer visitor NW Africa, S Europe, E Europe N to Baltic and E to S Ural Mtns, Caucasus, N Middle East, Levant and very locally C and S Arabia. Migrants now regular N to *eg* Netherlands and occur widely S of breeding areas on passage, but migration into/out of Europe concentrated at Strait of Gibraltar and Bosphorus, with smaller numbers using Sicily-Tunisia route (very scarce Malta); a few winter Arabia, less predictably Spain and N Africa. Vagrant Belgium, Britain, Canary Is, Channel Is, Czech Republic, Denmark, Finland, Germany (formerly bred), Luxembourg, Norway and Sweden.

Circaetus cinereus
Brown Snake Eagle (V)

OTHER NAME Brown Harrier-eagle.
TAXONOMY Vieillot, 1818. Monotypic.
DISTRIBUTION [Sub-Saharan Africa. Stable. LC] Extreme vagrant: N Mauritania (14 Apr 2003).

Terathopius ecaudatus
Bateleur (V)

OTHER NAME Bateleur Eagle.
TAXONOMY (Daudin, 1800). Monotypic.
DISTRIBUTION [Sub-Saharan Africa and SW Arabia. Decreasing. NT] Probably scarce former breeding visitor SW Saudi Arabia and W Yemen, but not recorded from presumed historical range since 1999; current status requires confirmation. Vagrant Cyprus, Egypt, Iraq, Israel, N Mauritania, Palestine, Spain and Tunisia.

Clanga nipalensis
Lesser Spotted Eagle

TAXONOMY (Brehm, C L, 1831). Monotypic.
DISTRIBUTION [C Europe to SW Asia > S and E Africa. Population trend unknown. LC] Breeding endemic. Summer visitor to C, E and SE Europe, Turkey, Caucasus and N Iran; has also bred Algeria. On passage wanders N to Finland and migrates to winter quarters chiefly through NE Egypt, Israel and Levant; scarce Cyprus and annual Malta, with a few crossing Mediterranean farther W via N Tunisia and Sicily. Rare on passage and in winter Oman. Vagrant Belgium, Corsica, Denmark, France, Gibraltar, Iraq, Kuwait, Libya, Liechtenstein, Malta, Morocco, Netherlands, Portugal, Sardinia, Saudi Arabia, Slovenia, Spain, Sweden, Switzerland, UAE and Yemen.

Clanga clanga
Greater Spotted Eagle

OTHER NAME Spotted Eagle.
TAXONOMY (Pallas, 1811). Monotypic.
DISTRIBUTION [E Europe to E Asia > S Europe and Sahel to SE Asia. Decreasing. VU] Summer visitor E Europe N to E Finland and NW Russia, E to Ural Mtns and S to SE Balkans. Wanders W on passage to C Europe but mainly migrates S to winter *eg* Turkey, N Israel, Syria and Iraq, N Persian Gulf, Oman, SW Saudi Arabia and W Yemen, also locally *eg* S Balkans, NE and SE Italy, and NE and S France; formerly more widespread and numerous in winter, and some modern records relate to returning individuals in successive years. Vagrant Algeria, Belgium, Britain, Cyprus, Czech Republic, Denmark, Germany (has bred), Gibraltar, Ireland, Jordan, Libya, Liechtenstein, Luxembourg, Morocco, Netherlands, Norway, Portugal, Sardinia, Spain, Sweden and Tunisia.

Hieraaetus wahlbergi
Wahlberg's Eagle (V)

TAXONOMY (Sundevall, 1850). Monotypic.
DISTRIBUTION [Sahel, E and S Africa > W to E Africa. Stable. LC] Extreme vagrant: Egypt (3 May 2013*).

Hieraaetus pennatus
Booted Eagle

TAXONOMY (Gmelin, J F, 1788). Monotypic.

DISTRIBUTION [S Africa, NW Africa and SW Europe to CE Asia > Mediterranean, sub-Saharan Africa and S Asia. Decreasing. LC] Summer visitor NW Africa, Iberia and France, SE and E Europe E to W European Russia, Caucasus and N Middle East; resident Balearic Is, has bred Hungary. Occurs widely S of breeding areas and in EC and S Europe on passage, with significant numbers at Strait of Gibraltar but movement otherwise less confined to main migration 'bottlenecks', even extending W to Canary Is; small numbers winter Mediterranean basin, Levant and Arabia. Vagrant Belgium, Cape Verde Is, Channel Is, Corsica, Czech Republic, Denmark, Germany, Estonia, Finland, Iceland, Latvia, Lithuania, Luxembourg, Madeira, Malta, Netherlands, Norway*, Qatar, Slovakia, Slovenia, Sweden and Switzerland; Cat D records from Britain and Ireland, involving at least one but possibly two or three individuals, may well relate to genuine vagrant(s).

Aquila rapax
Tawny Eagle

OTHER NAME Indian Tawny Eagle (*A r vindhiana*).
TAXONOMY (Temminck, 1828). Three subspecies (two in region). *Vindhiana* sometimes treated as full species, but limited differences do not support separation. Formerly also included Steppe Eagle (*qv*).
DISTRIBUTION [NW and sub-Saharan Africa, SW Arabia and S Asia. Stable. LC] Breeds SW Morocco and Algeria (very rare), formerly Tunisia (last recorded 1988), perhaps N Chad, also SW Saudi Arabia and W Yemen (*belisarius*); may breed Iran (*vindhiana*). Mainly resident but some nomadic and seasonal movement. Vagrant Egypt, Israel, Italy, N Mauritania and Sardinia (*belisarius*), and Oman (probably *vindhiana)*.

Aquila nipalensis
Steppe Eagle

OTHER NAMES Eastern Steppe Eagle (*A n nipalensis*), Western Steppe Eagle (*A n orientalis*).
TAXONOMY Hodgson, 1833. Two subspecies (at least one in region). Previously regarded as conspecific with Tawny Eagle, but differs in morphology, behaviour and ecology, and treatment as separate species supported by molecular analysis.
DISTRIBUTION [SE Europe to E Asia > E and S Africa, SW to SE Asia. Decreasing. EN] Summer visitor Caucasus (uncommon), SW Russia and NW Kazakhstan, formerly also Ukraine, Moldova and possibly Romania (*orientalis*); has bred Turkey. Migrates to winter quarters through Caucasus and Middle East, with largest numbers S Israel, Sinai and Suez, to lesser extent SW Yemen; some *nipalensis* (C and E Asia) may use same route and individuals showing characters of this taxon observed *eg* Israel, but confirmation required. Some overwinter Middle East, notably Oman. Vagrant Austria, Belarus, Belgium, Bulgaria, Croatia, Czech Republic, Denmark, Estonia, Finland, France, Germany, Greece, Hungary, Italy, Latvia, Lithuania, Macedonia, Moldova, Netherlands, Norway, Poland, Romania, Sardinia, Serbia, Slovakia, Spain, Sweden and Tunisia.

Aquila adalberti
Spanish Imperial Eagle (E)

OTHER NAMES Spanish (or Adalbert's) Eagle.
TAXONOMY Brehm, C L, 1861. Monotypic. Formerly treated as conspecific with Eastern Imperial Eagle, but differs in adult and especially juvenile plumages, and in ecology; genetic divergence considered relatively recent.
DISTRIBUTION [Iberia. Increasing. VU] Endemic. Breeds C and SW Spain and E Portugal. Mainly resident but some dispersal (mostly young birds) in winter, a few as far S as Morocco, where formerly bred and nesting again confirmed in 1995; also bred Algeria in 19th century. Vagrant France, Gibraltar, Libya, N Mauritania and Netherlands.

Aquila heliaca
Eastern Imperial Eagle

OTHER NAMES (Asian) Imperial Eagle.
TAXONOMY Savigny, 1809. Monotypic. Formerly included Spanish Imperial Eagle (*qv*).
DISTRIBUTION [C Europe to S Siberia > E

Africa to E Asia. Decreasing. VU] Breeds EC and SE Europe E to S Ural Mtns and NW Kazakhstan, also Caucasus, Asia Minor and NW Iran. Resident and partial migrant, occurring widely but generally uncommonly on passage S through Middle East and E Egypt; some remain in breeding range in winter, when more widespread Turkey and parts of Middle East S to Egypt and Oman, and E to Iran. Vagrant Belarus, Cyprus (formerly bred), Czech Republic, Denmark, Finland, France, Germany, Italy, Latvia, Libya, Lithuania, Montenegro, Netherlands, Norway, Poland, Slovenia, Spain, Sweden and Tunisia.

Aquila chrysaetos
Golden Eagle

OTHER NAME Homeyer's Golden Eagle (*A c homeyeri*).
TAXONOMY (Linnaeus, 1758). Six subspecies (three in region).
DISTRIBUTION [Widespread North America, Eurasia and N Africa. Stable. LC] Breeds from N Britain and Scandinavia E to Ural Mtns, S through E and C Europe to Corsica, Sardinia, Italy and Balkans (*chrysaetos*), from Iberia, parts of N Africa and Middle East N to Crete, Turkey and Caucasus, E to Oman and W and N Iran (*homeyeri*), and in E Iran (*daphanea*); formerly bred Ireland (reintroduction programme under way). Mainly resident, but some dispersal and NW Russian population migratory, occurring through Baltic to winter mainly *eg* S Scandinavia and E Europe (few even reaching N Africa). Vagrant Belgium, Channel Is, Cyprus, Czech Republic, Gibraltar, Kuwait, Malta and Netherlands.

Aquila verreauxii
Verreaux's Eagle

OTHER NAMES (African) Black Eagle.
TAXONOMY Lesson, 1831. Monotypic.
DISTRIBUTION [S, E, NC and NE Africa, Arabia. Stable. LC] Breeds SE Egypt, Sinai, SW Jordan (rare), W Saudi Arabia, W Yemen and S Oman; occasional visitor Israel (has bred). Vagrant Algeria, NW Chad (Tibesti), Lebanon and N Mauritania.

Aquila fasciata
Bonelli's Eagle

TAXONOMY Vieillot, 1822. Two subspecies (one in region).
DISTRIBUTION [N Africa, S Europe, SW to E Asia. Decreasing. LC] Breeds NW Africa (irregularly also Libya), Iberia and S France, range becoming very fragmented E through parts of S Europe and widely scattered areas of Middle East (*fasciata*). Resident but dispersive, mainly involving movement S by young birds. Vagrant Austria, Balearic Is, Belgium, Canary Is, Czech Republic, Denmark, Germany, Hungary, Kuwait, Netherlands, Qatar, Romania, Slovakia and Slovenia.

Micronisus gabar
Gabar Goshawk

TAXONOMY (Daudin, 1800). Three subspecies (one in region).
DISTRIBUTION [Sub-Saharan Africa and SW Arabia. Stable. LC] Resident SW Saudi Arabia and W Yemen (*niger*). Vagrant Egypt.

Melierax metabates
Dark Chanting Goshawk

OTHER NAME Chanting Goshawk.
TAXONOMY Heuglin, 1861. Four subspecies (three in region).
DISTRIBUTION [NW and sub-Saharan Africa, SW Arabia. Stable. LC] Resident SW Saudi Arabia and Yemen (*ignoscens*) and formerly SW Morocco (*theresae*), where very few records by late 1990s and endemic taxon now possibly extinct. Perhaps also occurs on S fringe of region in N Niger (*metabates*). Vagrant Algeria, Israel (*ignoscens*) and N Mauritania (*metabates*).

Accipiter badius
Shikra

OTHER NAMES Indian Sparrowhawk, Little Banded Goshawk; Northern (or Asian) Shikra (*A b cenchroides*), Southern (or African/ Tihama) Shikra (*A c sphenurus*).
TAXONOMY (Gmelin, J F, 1788). Six subspe-

cies (at least two in region). African subspecies *sphenurus* and extralimital *polyzonoides* are closely related to Levant Sparrowhawk, and similar sequence divergences supportive of equivalent taxonomic rank treatment; any taxonomic revision of *badius/brevipes* superspecies should include Asian subspecies of former.
DISTRIBUTION [Sub-Saharan Africa, SW to C and SE Asia. Stable. LC] Resident SW Saudi Arabia and W Yemen (*sphenurus*) and UAE (presumed *cenchroides* or possibly introduced *badius*); summer visitor N Iran and probably also breeds regularly SE Azerbaijan (*cenchroides*), though few records, and perhaps now Armenia, where first fully documented in 2009 when two pairs bred. Winter visitor Kuwait and NE Saudi Arabia (*cenchroides*). Vagrant Israel, Oman and Turkey.

Accipiter brevipes
Levant Sparrowhawk

TAXONOMY (Severtzov, 1850). Monotypic. See comments under Shikra.
DISTRIBUTION [SE Europe to SW and C Asia > NE Africa. Stable. LC] Near-endemic breeding species, ranging from Balkans and Hungary (rare) E through SE Europe and W and N Turkey to NW Kazakhstan, Caucasus, N Iraq and W Iran. Migratory, entire population funnelling through W Middle East, especially Israel, on both passages; rare S or W of Sinai, but occasional Yemen on migration. Vagrant Austria, Cyprus, Czech Republic, Italy, Poland, Qatar, Slovakia, Tunisia and UAE.

Accipiter nisus
Eurasian Sparrowhawk

OTHER NAMES (European or Northern) Sparrowhawk.
TAXONOMY (Linnaeus, 1758). Seven subspecies (four in region).
DISTRIBUTION [NW Africa and Eurasia > N and E Africa, Europe, SW to E Asia. Stable. LC] Breeds Canary Is and Madeira (*granti*), NW Africa and Libya (*punicus*), Corsica and Sardinia (*wolterstorffi*) and across much of Europe and parts of N Middle East (*nisus*). Mainly resident but migratory in C and N Fennoscandia and NW Russia, passage and winter range extending S to non-breeding areas in S Europe and especially through Middle East, more locally also N Africa. Rare N Mauritania. Vagrant Faroe Is, Iceland, Jan Mayen and Svalbard.

Accipiter gentilis
Northern Goshawk

OTHER NAMES Goshawk; American Goshawk (*A g atricapillus*).
TAXONOMY (Linnaeus, 1758). Ten subspecies (five in region).
DISTRIBUTION [Widespread North America and Eurasia. Stable. LC] Breeds N Morocco, W, C, E and much of N Europe (*gentilis*), N Fennoscandia and NW Russia (*buteoides*), Corsica and Sardinia (*arrigonii*), and Italy E through SE Europe and N Middle East to Caucasus and N Iran (*marginatus*). Re-established British population largely or wholly derived from escaped falconers' birds, hence mainly *gentilis* with influence from other subspecies. Largely resident but migratory in far N and NE of region, with some movement into S and SE Europe and N Middle East in winter. Vagrant Algeria (possibly breeding), Canary Is, Egypt, Faroe Is, Iraq, Ireland, Libya, Oman, Tunisia and UAE; *atricapillus* (breeding North America) vagrant Britain (28 Dec 1935) and Ireland (6), though all records under review.

Circus aeruginosus
Western Marsh Harrier

OTHER NAMES (Eurasian or Northern) Marsh Harrier.
TAXONOMY (Linnaeus, 1758). Two subspecies (both in region). Sometimes treated as including Eastern Marsh Harrier (*qv*).
DISTRIBUTION [NW Africa and Europe to C Asia > W and S Europe, sub-Saharan Africa, SW to S Asia. Increasing. LC] Breeds NW Africa (*harterti*), much of Europe W to Britain, N to S Scandinavia and NW Russia, E to Ural Mtns, Caucasus and Black Sea and discontinuously S to Mediterranean, also parts of N Middle East (*aeruginosus*). Migratory in continental European range (less so in W and

S), regularly W to Canary Is, wintering mainly S and W Europe and more locally N Africa (S to Cape Verde Is) and Middle East. Vagrant Azores, Faroe Is, Iceland and Madeira.

Circus spilonotus
Eastern Marsh Harrier (V)

OTHER NAME Spotted Marsh Harrier.
TAXONOMY Kaup, 1847. Monotypic. Sometimes treated as conspecific with Western Marsh Harrier, but hybridisation zone limited.
DISTRIBUTION [NE and E Asia > E and SE Asia. Stable. LC] Extreme vagrant (both old specimen records): Iran (2: 5 Oct 1898, collected; 11 Nov 1900, collected).

Circus cyaneus
Hen Harrier

TAXONOMY (Linnaeus, 1766). Monotypic. Formerly included Northern Harrier (*qv*).
DISTRIBUTION [W and N Eurasia > N Africa, W and S Eurasia. Decreasing. LC] Breeds widely across W, N and E Europe from Iberia, Ireland and Britain E to Ural Mtns and N to N Scandinavia, but absent from much of C and S Europe. Migratory in N and E, occurring widely on passage and in winter S to Mediterranean, N Middle East and also N Africa, where generally scarce/rare (irregular Canary Is). Vagrant Azores, Bahrain, Faroe Is, Iceland, Madeira, N Mauritania, Svalbard, UAE and Yemen.

Circus hudsonius
Northern Harrier (V)

OTHER NAMES (North) American Harrier, Marsh Hawk.
TAXONOMY (Linnaeus, 1766). Monotypic. Previously considered conspecific with Hen Harrier, but differs genetically and in plumage.
DISTRIBUTION [N and C North America > USA to extreme NW South America. Decreasing. LC] Very rare vagrant (mostly autumn, also winter; records increasing with greater awareness of ID criteria, and numerous other claims currently under assessment): Azores (several records accepted/documented with photos but most of 13 *cyaneus/hudsonius* reports probably involve latter taxon); Britain (6); Faroe Is (17 Oct 1989, dead); Iceland (15 Oct 1957, collected*); Ireland (7).

Circus macrourus
Pallid Harrier

TAXONOMY (Gmelin, S G, 1770). Monotypic.
DISTRIBUTION [E Europe to C Asia > sub-Saharan Africa, SW and S Asia. Decreasing. NT] Chiefly summer visitor from Ukraine E through SE European Russia and NW Kazakhstan; may also breed Turkey and Iran. Breeding range expanding; now nests annually Finland in very small numbers. Widespread on passage mainly through Balkans, Caucasus and Middle East in autumn, but increasingly frequent if very scarce N Europe, including *eg* Denmark, Netherlands and Sweden, also Hungary; return passage in spring includes Mediterranean crossing farther W via Algeria, Tunisia and Italy, also Libya and Malta. Some remain to winter Tunisia, N Egypt and Middle East, occasionally elsewhere. Formerly very rare W Europe, but marked recent increase in records far outside normal range, despite overall population trend. Vagrant Albania, Balearic Is, Belarus (has bred), Belgium, Bosnia-Herzegovina, Britain, Canary Is, Cape Verde Is, Channel Is, Corsica, Croatia, Czech Republic, Denmark, Estonia, France, Germany, Gibraltar, Iceland, Ireland, Latvia, Liechtenstein, Lithuania, Luxembourg, Macedonia, N Mauritania, Morocco, Norway, Poland, Portugal, Slovakia, Slovenia, Spain, Svalbard and Switzerland.

Circus pygargus
Montagu's Harrier

TAXONOMY (Linnaeus, 1758). Monotypic.
DISTRIBUTION [W Europe to C Asia > sub-Saharan Africa, SW and S Asia. Decreasing. LC] Summer visitor NW Africa, much of continental Europe N to S Scandinavia and S Finland (though range heavily fragmented in C and S), S Britain (very rare), parts of Turkey and W Iran. More widespread on migration, especially N Africa (W to Canary Is) and Middle East, some wintering S Arabia. Vagrant Azores, Bahrain, Cape Verde Is, Iceland, Ireland (has bred), Madeira and Norway.

Milvus milvus
Red Kite (E)

OTHER NAMES Kite; Cape Verde Kite (*M m fasciicauda*).
TAXONOMY (Linnaeus, 1758). Two subspecies (both in region). *Fasciicauda* previously considered a full species by some authorities, or hybrid (with Black Kite) by others, but DNA analysis established that taxon is nested within Red Kite clade.
DISTRIBUTION [NW Africa and Europe > NW Africa, W, SW and S Europe. Decreasing. NT] Endemic. Breeds Morocco, W Mediterranean islands, from Iberia N to S Scandinavia, Wales (also reintroduced England and Scotland), and parts of S, C and E Europe E to W Ukraine (*milvus*); formerly much more widespread, including Cape Verde Is (*fasciicauda*), where now believed extinct. Mainly migratory N and E Europe, wintering France S to parts of N Africa, very locally/irregularly E to Turkey (formerly bred). Vagrant Armenia, Azerbaijan, Azores, Canary Is, Cyprus, Egypt, Finland, Iceland, Iraq, Iran, Israel, Jordan, Lebanon, Libya, Madeira, Malta, N Mauritania and Norway; bred in some of these countries before undergoing significant range retraction.

Milvus migrans
Black Kite

OTHER NAMES Black-eared (or Eared/Large Indian) Kite (*M m lineatus*).
TAXONOMY (Boddaert, 1783). Five subspecies (two in region). Formerly included Yellow-billed Kite (*qv*); *lineatus* (E Asia) sometimes also separated at species level.
DISTRIBUTION [Africa, Eurasia and Australia > sub-Saharan Africa, SW to E Asia, Australia. Population trend unknown. LC] Breeds Cape Verde Is (rare), NW Africa, much of continental Europe except NW, and N Middle East (*migrans*). Chiefly summer visitor, occurring through non-breeding areas on passage W to Britain (scarce) and Canary Is (has bred and overwintered), and N to Denmark and Sweden; small numbers winter in S, and in E Arabia and Iran wintering population largely comprises *lineatus*. Vagrant Azores, Faroe Is, Iceland, Ireland, Madeira and Norway; *lineatus* has occurred W to *eg* Croatia, Cyprus, Iraq and Kuwait.

Milvus aegyptius
Yellow-billed Kite

OTHER NAMES African Black Kite (*M a parasitus*), Arabian Black Kite (*M a aegyptius*).
TAXONOMY (Gmelin, J F, 1788). Two subspecies (both in region). Formerly treated as conspecific with Black Kite, but Yellow-billed is as genetically distinct from that species as Black Kite is from Red Kite; the two Yellow-billed subspecies are further similarly distinct from each other. More geographical sampling needed to help clarify sister relationships among major groups in genus.
DISTRIBUTION [Egypt, sub-Saharan Africa and S Arabia. Population trend unknown. NE] Breeds Nile Valley, SW Saudi Arabia, Yemen and S Oman (rare), in 19th century apparently also Israel, where last recorded late 1960s (*aegyptius*); also known to have occurred N Mauritania (*parasitus*), but frequency and range limits there require confirmation. Vagrant Morocco.

Haliastur indus
Brahminy Kite (V)

OTHER NAMES Red-backed (or Rufous-backed) Kite, White-headed Kite, White-headed Sea-eagle.
TAXONOMY (Boddaert, 1783). Four subspecies (undetermined in region).
DISTRIBUTION [S and SE Asia, Australasia. Decreasing. LC] Extreme vagrant: Oman (20 Nov 2015). Further records from UAE (12 Apr and 21 Oct 1986-30 Jan 1987, with two present 20-25 Dec 1986) currently placed in Cat D.

Haliaeetus vocifer
African Fish Eagle (V)

OTHER NAMES Fish (or River) Eagle.
TAXONOMY (Daudin, 1800). Monotypic.
DISTRIBUTION [Sub-Saharan Africa. Stable. LC] Extreme vagrant: Egypt (1 Nov 1947, collected).

Haliaeetus leucoryphus
Pallas's Fish Eagle (V)

OTHER NAMES Pallas's Sea Eagle, Band-tailed (or Ring-tailed) Fishing-Eagle.
TAXONOMY (Pallas, 1771). Monotypic.
DISTRIBUTION [C, S, SE and E Asia > SC to SE

Asia. Decreasing. VU] Formerly bred (last in 1947), or at least oversummered, between Volga and Ural rivers in extreme SE Russia and NW Kazakhstan, where last recorded late Jun 1990. Previously also rare or uncommon winter visitor Iraq, but no records since 1940s. Vagrant Azerbaijan, Finland, Iran, Norway, Poland, Oman, Saudi Arabia, UAE and Ukraine.

Haliaeetus albicilla
White-tailed Eagle

OTHER NAMES (White-tailed) Sea (or Fish) Eagle.
TAXONOMY (Linnaeus, 1758). Two subspecies (one in region).
DISTRIBUTION [Greenland and N Eurasia > Greenland, W Europe to SW, N and E Asia. Increasing. LC] Breeds Iceland and from W Scandinavia E across much of N Europe to Ural Mtns, range extending S through Germany and E Europe discontinuously to parts of Balkans, Turkey and N Iran; formerly more widespread, also breeding Britain, Ireland and Israel (where now reintroduced) and *eg* Algeria, Corsica, Egypt, Faroe Is, Iraq, Montenegro, Sardinia and Syria (*albicilla*). Population in NE of region migratory and younger birds throughout range more dispersive; winters locally S to France and parts of C, S and SE Europe and N Middle East. Vagrant Balearic Is, Belgium, Channel Is, Cyprus, Jan Mayen, Kuwait, Lebanon, Luxembourg, Malta, Morocco, Portugal, Spain, Saudi Arabia, Svalbard, Switzerland and Tunisia.

Haliaeetus leucocephalus
Bald Eagle (V)

TAXONOMY (Linnaeus, 1766). Two subspecies (undetermined in region).
DISTRIBUTION [N, E and W North America > S Canada to N Mexico. Increasing. LC] Extreme vagrant (two late autumn/winter records): Ireland (2: 11 Jan 1973, shot; 17 Nov 1987, caught exhausted, returned to USA).

Butastur rufipennis
Grasshopper Buzzard (V)

OTHER NAMES Grasshopper Buzzard-Eagle, Grasshopper Buzzard-Hawk.
TAXONOMY (Sundevall, 1850). Monotypic.
DISTRIBUTION [W, NC, NE and E Africa. Decreasing. LC] Extreme vagrant: N Mauritania (7-8 May 2004).

Butastur teesa
White-eyed Buzzard

OTHER NAMES White-eyed Buzzard-Eagle, White-eyed Buzzard-Hawk.
TAXONOMY (Franklin, 1831). Monotypic.
DISTRIBUTION [S and SE Asia. Stable. LC] Scarce resident SE Iran, where records somewhat irregular but species known to have bred. Vagrant Oman (22 Mar 2004); two older records from UAE (8-9 Dec 1997; 22 Sep 1998-30 Apr 2002, perhaps same bird) placed in Cat D but possibly worthy of review.

Buteo lagopus
Rough-legged Buzzard

OTHER NAMES Rough-leg; Rough-legged Hawk (or American Rough-legged Buzzard) (*B l sanctijohannis*).
TAXONOMY (Pontoppidan, 1763). Four subspecies (two in region).
DISTRIBUTION [Holarctic > S Canada, N USA and N Eurasia. Stable. LC] Breeds from W Scandinavia to NW Russia (*lagopus*). Moves S to winter Britain (scarce) and continental Europe variably from Low Countries E through S Scandinavia, C and E Europe, Italy, Balkans and NW Turkey to Ural Mtns, more rarely N Iran, exact range depending on food supply. Vagrant Azores, Channel Is, Cyprus, Faroe Is, France, Greece, Iceland, Ireland, Israel, Jan Mayen, Jordan, Lebanon, Liechtenstein, Luxembourg, Madeira, Malta, Sardinia, Spain, Svalbard, Switzerland and Tunisia; *sanctijohannis* (breeding N North America) vagrant Azores, Faroe Is, Iceland and Ireland.

Buteo hemilasius
Upland Buzzard (V)

OTHER NAME Mongolian Buzzard.
TAXONOMY Temminck and Schlegel, 1844. Monotypic.
DISTRIBUTION [EC Asia to NE Asia > N India,

SC and E Asia. Stable. LC] Extreme vagrant: Iran (30 Nov 1900, collected).

Buteo rufinus
Long-legged Buzzard

OTHER NAMES Atlas (or North African) Long-legged Buzzard (*B r cirtensis*); Gibraltar Buzzard (hybrid *cirtensis* Long-legged × *buteo* Common Buzzard).
TAXONOMY (Cretzschmar, 1829). Two subspecies (both in region). Sometimes considered conspecific with Upland Buzzard. Subspecies *cirtensis* may merit treatment as separate species; more research required.
DISTRIBUTION [N Africa, C Europe to C and SW Asia. Stable. LC] Resident from N Mauritania through NW Africa discontinuously E to N Libya, N Egypt and Arabia (*cirtensis*), and from EC Europe, Balkans and N Middle East E to Iran and S to Israel, possibly also Sinai (*rufinus*). European population largely migratory, and regular on passage Italy (first instance of breeding also recently reported*), but otherwise mainly resident, though with local movements within range (including breeders from Israel dispersing N as far as Ukraine); numbers wintering in E Mediterranean and Middle East augmented by immigrants from C Asia. Vagrant Belarus, Belgium, Canary Is, Croatia, Czech Republic, Denmark, Finland, France, Germany, Gibraltar, Lithuania, Malta, Netherlands, Norway, Poland, Portugal*, Sardinia, Slovakia, Slovenia, Spain (has bred), Sweden and Switzerland.

Buteo bannermani
Cape Verde Buzzard (E)

TAXONOMY Swann, 1919. Monotypic. Formerly considered a subspecies of Common Buzzard, but genetic studies found present species to share common ancestry with Long-legged Buzzard, and to have identical nucleotide sequences to extralimital Socotran Buzzard *B socotrae*, now also elevated to species level.
DISTRIBUTION [Cape Verde Is. Population trend not assessed. NE] Endemic. Resident Cape Verde Is on Santiago and Santo Antão, perhaps also Fogo, and formerly São Nicolau and Boa Vista; single records also from Brava and São Vicente.

Buteo buteo
Common Buzzard

OTHER NAMES (Eurasian) Buzzard; Azores (or Rothschild's) Buzzard (*B b rothschildi*), Canary Islands (or Canarian Common) Buzzard (*B b insularum*), Caucasian Buzzard (*B b menetriesi*), Steppe Buzzard (*B b vulpinus*); Gibraltar Buzzard (hybrid *cirtensis* Long-legged × *buteo* Common Buzzard).
TAXONOMY (Linnaeus, 1758). Seven subspecies (all in region); six listed by IOC, but long-recognised Azorean *rothschildi* presumably omitted in error. Formerly also included Cape Verde Buzzard (*qv*).
DISTRIBUTION [Widespread Europe, Asia and Africa. Increasing. LC] Mainly resident across much of N, W, C and parts of S Europe (*buteo*), with sedentary populations also Azores (*rothschildi*), Canary Is (*insularum*), Madeira (*harterti*), Corsica, Sardinia and C and S Italy (*pojana*). Widespread in summer NE, E and SE Europe (*vulpinus*) and E Turkey (partly resident) through Caucasus to N Iran (*menetriesi*). *Buteo* leaves most of S Scandinavia in winter, when also moves S into *eg* NW Africa (rare N Mauritania) and Balkans, but unrecorded Arabia; vagrant Faroe Is, Iceland and apparently Algeria. *Vulpinus* migrates to winter S of Sahara, and regular on passage W to Tunisia; in Europe rare W to *eg* Czech Republic and Hungary, vagrant W to *eg* Belgium, Croatia, Netherlands and Spain.

FAMILY OTIDIDAE

Bustards · six species

Otis tarda
Great Bustard

TAXONOMY Linnaeus, 1758. Two subspecies (one in region).
DISTRIBUTION [NW Africa and SW Europe to E Asia. Decreasing. VU] Breeds N Morocco (rare) and Iberia, C Europe N to Germany and Poland, in Turkey and Iran, and from N Black

Sea E through Ukraine and S Russia (*tarda*); bred Britain until 19th century (reintroduction in progress). E populations migratory, wintering S to parts of Balkans and N Middle East N to Azerbaijan (rare, formerly bred); irregular Italy. Vagrant Algeria, Armenia, Belgium, Bosnia-Herzegovina, Britain, Croatia, Cyprus, Czech Republic, Denmark, Egypt, Estonia, Finland, France, Gibraltar, Greece, Ireland, Israel, Latvia, Lebanon, Lithuania, Luxembourg, Macedonia, Malta, Montenegro, Netherlands, Norway, Sardinia, Saudi Arabia, Slovenia, Sweden, Switzerland and Tunisia.

Ardeotis arabs
Arabian Bustard

TAXONOMY (Linnaeus, 1758). Four subspecies (two or three in region).
DISTRIBUTION [NW Africa, Sahel and SW Arabia. Decreasing. NT] Rare resident SW Saudi Arabia and W Yemen (*arabs*). Formerly resident W Morocco but probably now extinct (*lynesi*); rare local resident or winter visitor S Morocco (probably *stieberi*). Vagrant Algeria.

Chlamydotis undulata
Houbara Bustard

OTHER NAMES (African) Houbara.
TAXONOMY (Jacquin, 1784). Two subspecies (both in region). Formerly included Macqueen's Bustard (*qv*).
DISTRIBUTION [N Africa. Decreasing. VU] Near-endemic. Resident E Canary Is on Lanzarote and Fuerteventura, with records in past also Gran Canaria (*fuertaventurae*), and from Morocco discontinuously E through N Sahara to Nile Valley (*undulata*); formerly also known from N Mauritania. Generally sedentary, with limited dispersal recorded away from breeding areas. Vagrant Italy, Malta, Romania and Switzerland; majority of pre-split records of 'Houbara Bustard' *sensu lato* likely to refer to Macqueen's Bustard (*qv*).

Chlamydotis macqueenii
Macqueen's Bustard

OTHER NAMES Asian Houbara (Bustard).
TAXONOMY (Gray, J E, 1832). Monotypic. Formerly treated as conspecific with Houbara Bustard, but separated on differences in plumage, sexual display and genetic divergence.
DISTRIBUTION [Middle East to E Asia. Decreasing. VU] Breeds NE Egypt, Israel, Saudi Arabia, Yemen, Oman and Iran; reintroductions have taken place Jordan, Saudi Arabia and UAE. Formerly bred Armenia and Syria, apparently Turkey and perhaps Azerbaijan (no recent records); has also bred Kuwait. Middle Eastern population augmented in winter by migrants from Asia, and before sharp decline vagrant 'Houbaras' *sensu lato* occurred across Europe (very few records in recent decades), majority involving this species. Vagrant Austria, Belgium, Britain, Cyprus, Czech Republic, Denmark, Finland, France, Germany, Italy, Latvia, Lebanon, Lithuania, Netherlands, Poland, Qatar, Romania, Slovenia, Spain, Sweden, Switzerland and Ukraine.

Neotis nuba
Nubian Bustard (V)

TAXONOMY (Cretzschmar, 1826). Monotypic.
DISTRIBUTION [Sahel. Decreasing. NT] Extreme vagrant: N Chad (Jun 1961, collected; species then said to be not uncommon near Zouarké, Tibesti). Perhaps also recorded around regional boundary in N Mauritania (several imprecisely documented records from Banc d'Arguin NP, part of which lies within region), N Mali and possibly NW Niger (known to occur in Ténéré desert which extends N just into region from Aïr Massif).

Tetrax tetrax
Little Bustard

TAXONOMY (Linnaeus, 1758). Monotypic. Eastern population sometimes separated as *orientalis*.
DISTRIBUTION [NW Africa and S Europe to C Asia. Decreasing. NT] Breeds N Morocco, Iberia, France, Sardinia, Turkey (rare), Ukraine, N Caspian, parts of Caucasus and N Iran; formerly also bred *eg* Austria, Azerbaijan, Greece, Hungary, mainland Italy, Macedonia, Serbia, Syria and Tunisia. N populations migratory, wintering *eg* N Africa, Crimea, Turkey, Iraq

(rare) and Israel. Vagrant Albania, Armenia, Austria, Belgium, Bosnia-Herzegovina, Britain, Canary Is, Channel Is, Corsica, Croatia, Czech Republic, Cyprus, Denmark, Egypt, Estonia, Finland, Germany, Greece, Hungary, Iraq, Ireland, Latvia, Lebanon, Libya, Luxembourg, Madeira, Malta, Moldova, Netherlands, Norway, Oman, Poland, Romania, Serbia, Slovenia, Sweden, Switzerland and Tunisia.

ORDER GRUIFORMES

FAMILY RALLIDAE

Rails, crakes and coots · 21 species

Rallus aquaticus
Water Rail

OTHER NAME European Water Rail.
TAXONOMY Linnaeus, 1758. Two subspecies (both in region).
DISTRIBUTION [N Africa, Europe and W Asia. Decreasing. LC] Breeds widely in Europe from Iberia and Ireland N to S Fennoscandia, E to S Ural Mtns and S to Mediterranean, also N Africa, Turkey (*aquaticus*) and S and E Iran (*korejewi*); formerly also bred Iceland, where endemic taxon (*hibernans*) now extinct. Mainly resident in W and S of range but otherwise migratory, occurring widely on passage N to Faroe Is and S to N Africa and across Arabia. Vagrant Azores, Canary Is, Iceland, Madeira, N Mauritania, Jan Mayen and Svalbard.

Crex egregia
African Crake (V)

TAXONOMY (Peters, W, 1854). Monotypic.
DISTRIBUTION [Sub-Saharan Africa. Stable. LC] Extreme vagrant (seven records): Canary Is (5, all taken into care, four of which later died: 23 Nov 2001; 15 Nov 2006; 16 Nov 2006; 12 Jan 2007; 5 Jan 2011); Cape Verde Is (4 Feb 2004, dead); Morocco (11 Dec 2009, present for about a week). Further published record (29 Nov 2011) from Banc d'Arguin, N Mauritania, relates to mainland location marginally outside regional boundary; earlier individual reported from same location (late Jan 2007) lacked documentation.

Crex crex
Corn Crake

OTHER NAME Corncrake.
TAXONOMY (Linnaeus, 1758). Monotypic.
DISTRIBUTION [Europe to C Siberia > E and SE Africa. Stable. LC] Declining summer visitor to Ireland, NW Scotland and from C Scandinavia S to France and E across Europe N of Italy and S Balkans, also Caucasus, Turkey and N Iran; reintroduced EC England. On passage S of breeding range, more widespread Egypt and parts of Middle East, scarcer/rare in W. Vagrant Azores, Bahrain, Canary Is, Channel Is (formerly bred), Corsica, Faroe Is (formerly bred), Iceland, Libya, Madeira, Morocco, Portugal and Spain.

Amaurornis phoenicurus
White-breasted Waterhen

OTHER NAME White-breasted Swamphen.
TAXONOMY (Pennant, 1769). Three subspecies (one in region).
DISTRIBUTION [S to E Asia > S and SE Asia. Population trend unknown. LC] Rare to uncommon passage migrant and winter visitor Oman; rare in winter UAE (presumed *phoenicurus*). Vagrant Iran, Qatar, Saudi Arabia and Yemen.

Porzana parva
Little Crake

TAXONOMY (Scopoli, 1769). Monotypic.
DISTRIBUTION [W Europe to NW China. Decreasing. LC] Summer visitor mainly to C and E Europe, extending E to S Ural Mtns and Caucasus, locally with smaller populations W to France, N to Estonia and S to Greece, also very locally S Turkey and N Arabia; has bred Iraq. On passage more widespread S Europe, Middle East and N Africa, W to Canary Is; rare Algeria (formerly bred), Finland, Spain (has bred) and Sweden. Occasionally winters Mediterranean and Middle East. Vagrant

Azores*, Balearic Is, Belgium, Britain, Channel Is, Denmark, Ireland, Liechtenstein, Luxembourg, Madeira, Morocco, Netherlands, Norway, Portugal, Qatar, Sweden and UAE.

Porzana pusilla
Baillon's Crake

OTHER NAMES Marsh (or Tiny) Crake.
TAXONOMY (Pallas, 1776). Six subspecies (two in region).
DISTRIBUTION [Widespread sub-Saharan Africa, Eurasia and Australasia. Population trend unknown. LC] Summer visitor mainly to E Europe, especially Ukraine and SW Russia, with outlying and often irregular populations *eg* Balkans, Egypt, Hungary, Jordan, Morocco, Netherlands and Spain (formerly also Portugal), occasionally Austria, Cyprus, France, Israel, Switzerland and probably Turkey (*intermedia*); also breeds Iran (*pusilla*). More widespread on migration SE Europe to Middle East and N Africa; a few winter Mediterranean and S Arabia. Vagrant Algeria (formerly bred), Armenia, Azerbaijan (has bred), Azores*, Balearic Is, Belarus, Belgium, Britain (has bred), Canary Is, Corsica, Czech Republic, Denmark, Estonia, Finland, Germany (has bred), Gibraltar, Ireland, Latvia, Libya, Liechtenstein, Lithuania, Luxembourg*, Madeira, Malta, Morocco, Poland, Qatar, Slovenia, Sweden, Switzerland (has bred), Tunisia and UAE.

Porzana porzana
Spotted Crake

OTHER NAME Eurasian Spotted Crake.
TAXONOMY (Linnaeus, 1766). Monotypic.
DISTRIBUTION [W Europe to C Asia > E and S Africa, S Asia. Stable. LC] Generally declining summer visitor, with main breeding range in Russia, Belarus, Romania and Ukraine, but also breeds very locally Britain and from C Scandinavia (also Finland) patchily S to C Iberia, C Italy and N Balkans, as well as Caucasus and Iran. More widespread on passage elsewhere in Europe and Middle East, and in N Africa (W to Canary Is); a few winter Iberia and N Africa. Vagrant Azores, Cape Verde Is, Faroe Is, Iceland and Madeira.

Porzana carolina
Sora (V)

OTHER NAMES Sora (or Carolina) Crake, Sora Rail.
TAXONOMY (Linnaeus, 1758). Monotypic.
DISTRIBUTION [N North America > S North America, West Indies and N South America. Increasing. LC] Very rare vagrant: Azores (3); Britain (18); France (2); Iceland (25-27 Apr 2011); Ireland (11 Apr 1920, dead); Madeira (4 Dec 2006, again on 29-30 Jan 2007); Morocco (29 Mar 1989); Spain (2); Sweden (3).

Porzana fusca
Ruddy-breasted Crake (V)

OTHER NAME Ruddy Crake.
TAXONOMY (Linnaeus, 1766). Four subspecies (undetermined in region).
DISTRIBUTION [S, SE and E Asia > S and SE Asia. Decreasing. LC] Extreme vagrant: Oman (23 Nov-4 Dec 2012).

Aenigmatolimnas marginalis
Striped Crake (V)

TAXONOMY (Hartlaub, 1857). Monotypic.
DISTRIBUTION [Sub-Saharan Africa. Decreasing. LC] Extreme vagrant (five records): Algeria (Jan 1867, collected); Italy (4 Jan 1997, taken into care, later died); Malta (2: 29 Mar 1981, collected; Apr 2004, collected); Spain (12 Dec 2010).

Gallicrex cinerea
Watercock (V)

TAXONOMY (Gmelin, J F, 1789). Monotypic.
DISTRIBUTION [S, SE and E Asia > S and SE Asia. Decreasing. LC] Extreme vagrant (four records between autumn and spring): Oman (4: 18 Apr-8 May 1992; 28 Dec 2008-28 Jan 2009; 24-25 Nov 2011; 6 Jan 2013).

Porphyrio porphyrio
Western Swamphen

OTHER NAMES Purple Gallinule, Purple Swamphen.

TAXONOMY (Linnaeus, 1758). Monotypic. Formerly also included African and Grey-headed Swamphens (*qv*).
DISTRIBUTION [NW Africa and SW Europe. Population trend not assessed. NE] Mainly resident parts of coastal NW Africa, S Portugal, Spain, Balearic Is, Sardinia and Sicily. Iberian population augmented by reintroductions in Portugal and NE Spain, leading to colonisation of S France; further small naturalised population established mainland Italy. Vagrant Corsica, Gibraltar and Malta; there are further records of vagrant undetermined swamphens from Austria, Croatia, Cyprus, Germany, Greece, Hungary and Madeira*, often prior to 1950 and long pre-dating split of *P porphyrio* into multiple species.

Porphyrio madagascariensis
African Swamphen

OTHER NAMES Green-backed Gallinule, (African) Purple Swamphen.
TAXONOMY (Latham, 1801). Monotypic. Formerly treated as conspecific with Western and Grey-headed Swamphens and other extralimital taxa, but now separated on basis of genetic and morphological differences.
DISTRIBUTION [NE and sub-Saharan Africa, Madagascar. Population trend not assessed. NE] Resident Egypt, mainly along Nile from delta S to Aswan, also Suez Canal area, Lake Qarun and Wadi el Rayan. Released into wild in UAE, where has bred, but apparently not established. Vagrant Canary Is, Israel, N Mauritania and Oman; for details of undetermined vagrant swamphens, see Western Swamphen.

Porphyrio poliocephalus
Grey-headed Swamphen

TAXONOMY (Latham, 1801). Three subspecies (two in region). As with African Swamphen, now separated from Western Swamphen (and other extralimital taxa) on basis of genetic divergence and morphology.
DISTRIBUTION [SW to SE Asia. Population trend not assessed. NE] Mainly resident scattered areas of N Middle East from SC Turkey to S Georgia and E to Caspian Sea and NW Iran (*caspius*), and from Iraq and Kuwait to E Iran (*seistanicus*). May be colonising E Saudi Arabia, where previously irregular; formerly also scarce winter visitor N Israel but unrecorded since 1950s. Vagrant Jordan, Lebanon, Norway, Oman, Qatar, UAE (has bred) and Ukraine; for details of undetermined vagrant swamphens, see Western Swamphen.

Porphyrio alleni
Allen's Gallinule (V)

OTHER NAME Lesser Gallinule.
TAXONOMY Thomson, 1842. Monotypic.
DISTRIBUTION [Sub-Saharan Africa. Decreasing. LC] Rare vagrant: Algeria (occasional 19th century records); Azores (4); Balearic Is (2); Britain (2); Canary Is (11); Ceuta (Dec 2000); Cyprus (2); Denmark (30 Dec 1929); Finland (10 May 1979); France (4); Germany (2); Gibraltar (13 Dec 2001); Greece (18 Jan 1996); Italy (8); Madeira (2); Malta (4); N Mauritania (4 Jan 1980, dead); Melilla (21 Nov 2008); Morocco (8); Oman (8); Portugal (3); Spain (10); Tunisia (Dec 1902, killed); Turkey (26-27 Apr 2013*).

Porphyrio martinica
Purple Gallinule (V)

OTHER NAMES American Purple (or Yellow-legged) Gallinule.
TAXONOMY (Linnaeus, 1766). Monotypic.
DISTRIBUTION [SE North America to C South America > Gulf of Mexico to C South America. Decreasing. LC] Very rare vagrant: Azores (8*); Britain (3); Canary Is (2); Cape Verde Is (2); Iceland (2); Italy (2); Madeira (3*); Malta (autumn 1977); Norway (autumn 1883); Portugal (8 Nov 2013*).

Gallinula chloropus
Common Moorhen

OTHER NAMES (Eurasian) Moorhen.
TAXONOMY (Linnaeus, 1758). Five subspecies (two in region). Formerly included extralimital subspecies *cachinnans*, now separated as Common Gallinule and potential vagrant from North America.
DISTRIBUTION [Widespread Eurasia and Africa. Population trend unknown. LC] Breeds across

much of Europe, W to Atlantic islands, E to Ural Mtns, N to C Scandinavia and S to Mediterranean, also N Africa, W and N Middle East and parts of coastal Arabia (*chloropus*), as well as N Mauritania (*meridionalis*). Migratory in N and E of continental range, occurring widely on passage S to parts of N Africa and Middle East. Vagrant Cape Verde Is (formerly bred), Gibraltar, Iceland, Jan Mayen and Svalbard.

Paragallinula angulata
Lesser Moorhen (V)

TAXONOMY Sundevall, 1850. Monotypic.
DISTRIBUTION [Sub-Saharan Africa. Population trend unknown. LC] Extreme vagrant (four records of five individuals): Madeira (26 Jan 1895, collected; formerly accepted as Black Crake *Amaurornis flavirostris*, which is otherwise unrecorded in region); Oman (2: 7-9 Oct 1991, two; 1 Oct 2008); Spain (10 Mar 2003). Previously accepted record from Egypt (6 May 1997) now considered more likely to relate to Common Moorhen.

Fulica cristata
Red-knobbed Coot

OTHER NAME Crested Coot.
TAXONOMY Gmelin, J F, 1789. Monotypic.
DISTRIBUTION [SW Europe, NW, E and S Africa. Decreasing. LC] Breeds SW Spain and N Morocco, and reintroduced E Spain and Mallorca; historically also bred N Algeria, N Tunisia and Portugal. Vagrant France, Italy, Malta, Oman, Portugal, Sardinia and UAE (where hybridised with Eurasian Coot).

Fulica atra
Eurasian Coot

OTHER NAMES Common (or Black/European) Coot.
TAXONOMY Linnaeus, 1758. Four subspecies (one in region).
DISTRIBUTION [Eurasia, N Africa and Australasia. Decreasing. LC] Widespread across much of Europe from Atlantic islands in W to Ural Mtns in E, N to C Scandinavia and S to Mediterranean, also parts of N Africa, N Middle East and coastal Arabia (*atra*). In NE and E of range migratory, occurring widely on passage S to N Africa and Middle East. Formerly wintered Gibraltar (none since 1949). Vagrant Cape Verde Is, Iceland (has bred), Jan Mayen and Svalbard.

Fulica americana
American Coot (V)

TAXONOMY Gmelin, J F, 1789. Two subspecies (one in region).
DISTRIBUTION [N North America to N South America, Caribbean > S and W North America to N South America, Caribbean. Decreasing. LC] Very rare vagrant (presumed *americana*): Azores (25*); Britain (6); Faroe Is (1 Nov 1985, dead); Iceland (4); Ireland (3); Madeira (20 Jan 2012*), Portugal (17 Sep 1992); Spain (3).

FAMILY GRUIDAE

Cranes · four species

Grus leucogeranus
Siberian Crane

OTHER NAMES Siberian (or Asiatic/Great) White Crane.
TAXONOMY Pallas, 1773. Monotypic.
DISTRIBUTION [W Siberia and EC Siberia > SW, S and E Asia. Decreasing. CR] Formerly scarce migrant Volga Delta, SW Russia, and presumably also NW Kazakhstan, and winter visitor S Caspian Sea in N Iran, but now reduced to single wintering individual of near-extirpated western population. Previously more numerous (*eg* 13 Volga Delta 1996, up to 300 in 19th century) and historically also wintered E Azerbaijan; status uncertain Turkey (perhaps formerly vagrant or scarce but regular passage migrant). Vagrant Azerbaijan, Jordan and Ukraine.

Grus canadensis
Sandhill Crane (V)

OTHER NAMES Greater Sandhill Crane (*G c tabida* and sometimes other subspecies except *canadensis*); Lesser Sandhill Crane (*G c canadensis*).

TAXONOMY (Linnaeus, 1758). Five subspecies (at least one in region).
DISTRIBUTION [NE Siberia and N North America > S North America. Increasing. LC] Very rare vagrant (many of the recent records in continental Europe are thought to relate to just one or two individuals): Azores (26 Jun-3 Jul 2000); Britain (4, of which first was *canadensis*); Denmark (27-28 May 2013); Estonia (6-9 Sep 2011); Faroe Is (14 Oct 1980, collected, *canadensis*); Finland (2); France (7-12 Oct 2009); Germany (2); Ireland (*c* 11-14 Sep 1905, collected, *canadensis*); Netherlands (28-30 Sep 1991, previously seen Britain); Norway (31 May-5 June 2012); Spain (1: 4 Nov 2011, 2 Jan 2012 and 24-26 Feb 2013).

Grus virgo
Demoiselle Crane

TAXONOMY (Linnaeus, 1758). Monotypic.
DISTRIBUTION [SE Europe to NE China > E Sahel, S Asia. Increasing. LC] Summer visitor SE Ukraine and S Russia, also NE Turkey (now very rare); formerly bred NW Africa, where last recorded Morocco in late 1980s, Italy (Lampedusa) and Romania. More widespread Turkey on passage, when also regular through Caucasus, Cyprus and W Arabia; rarer Israel, Lebanon and (also in winter) Iraq and Oman. Vagrant Balearic Is, Bulgaria, Denmark, Estonia, Finland, France, Germany, Greece, Hungary, Iran, Italy, Jordan, Kuwait, Malta, Montenegro, Netherlands, Norway, Poland, Portugal, Romania, Serbia, Slovakia, Spain, Sweden, Syria and UAE.

Grus grus
Common Crane

OTHER NAMES (Eurasian) Crane.
TAXONOMY (Linnaeus, 1758). Monotypic.
DISTRIBUTION [N Eurasia > S Europe, N and NE Africa, Middle East, S and E Asia. Population trend unknown. LC] Mainly summer visitor, breeding from Scandinavia and Germany E to Ural Mtns and N Ukraine, also Turkey and Armenia; small resident population Britain, where partly reintroduced. Widespread on passage, some wintering *eg* France, Spain, from Morocco to Libya, N Israel and Iran. Vagrant Azores, Bahrain, Canary Is, Faroe Is, Iceland (has bred), Ireland, Jan Mayen, Kuwait, Madeira, N Mauritania, Qatar, Svalbard and UAE.

ORDER CHARADRIIFORMES

FAMILY TURNICIDAE

Buttonquail · one species

Turnix sylvaticus
Common Buttonquail

OTHER NAMES Small (or Little/Striped) Buttonquail; Andalusian Buttonquail or Andalusian Hemipode (*T s sylvaticus*).
TAXONOMY (Desfontaines, 1789). Nine subspecies (three in region).
DISTRIBUTION [NW and sub-Saharan Africa, SW Arabia, S and SE Asia. Decreasing. LC] Resident N Morocco (*sylvaticus*); status uncertain SW Saudi Arabia and W Yemen (*lepurana*), where very few records but probably breeding resident, and E Iran (*dussumier*), where may be scarce breeder. Modern range of relict W Mediterranean population poorly known; may still extend N to S Spain, where last recorded 2002, and E of Morocco in coastal NW Africa, where last recorded NW Algeria 1994. Formerly more widespread but range now greatly contracted and last recorded Tunisia 1984, Portugal 1973 and Sicily 1913; may have also occurred W Libya and, doubtfully, Montenegro. Vagrant Oman.

FAMILY BURHINIDAE

Stone-curlews and thick-knees · four species

Burhinus oedicnemus
Eurasian Stone-curlew

OTHER NAMES (Eurasian, Common or Northern) Thick-knee, (Northern or European) Stone Curlew.
TAXONOMY (Linnaeus, 1758). Five subspecies

(all in region). IOC subspecies followed here *pro tem* but traditional subspecific limits not supported by recent genetic assessment; more research needed into taxonomic status of genetically divergent and possibly specifically distinct Canary Is population. Formerly also included extralimital Indian Stone-curlew *B indicus*, now considered separate species.
DISTRIBUTION [W Europe and N Africa to C Asia > SW Europe, N Africa and SW Asia. Decreasing. LC] Breeds W Canary Is (*distinctus*), E Canary Is (*insularum*), N Mauritania, NW and N Africa, and Mediterranean islands E through N Middle East and Asia Minor to Iran (*saharae*), parts of W, C, S and E Europe E to Balkans, Ukraine and Caucasus (*oedicnemus*) and NW Kazakhstan (*harterti*). Resident Canary Is, W and N Africa, much of Iberia and most Mediterranean islands, but otherwise mainly summer visitor. Vagrant Azores, Belgium, Denmark, Estonia, Finland, Iceland, Ireland, Lithuania, Luxembourg, Madeira, Norway, Slovenia and Sweden.

Burhinus senegalensis
Senegal Thick-knee

OTHER NAME Senegal Stone-curlew.
TAXONOMY (Swainson, 1837). Monotypic.
DISTRIBUTION [W, C and E Africa N to Egypt. Population trend unknown. LC] Resident Egypt, mainly in Nile Delta and Valley. Vagrant Saudi Arabia.

Burhinus capensis
Spotted Thick-knee

OTHER NAMES Spotted (or Cape) Dikkop, Cape Thick-knee.
TAXONOMY (Lichtenstein, M H C, 1823). Four subspecies (one in region).
DISTRIBUTION [Sub-Saharan Africa and S Arabia. Stable. LC] Resident SW Saudi Arabia, Yemen and S, C and E Oman (*dodsoni*).

Esacus recurvirostris
Great Stone-curlew

OTHER NAMES Great Stone Plover, Great(-billed) Thick-knee.
TAXONOMY (Cuvier, 1829). Monotypic.
DISTRIBUTION [SW to SE Asia. Decreasing. NT] Breeds SE Iran. Winters N Oman (scarce). Vagrant UAE.

FAMILY HAEMATOPODIDAE

Oystercatchers · one species

Haematopus ostralegus
Eurasian Oystercatcher

OTHER NAMES (Common or European/Northern) Oystercatcher, Pied Oystercatcher.
TAXONOMY Linnaeus, 1758. Four subspecies (three in region).
DISTRIBUTION [W and N Europe to E Asia > W Europe, W, N and E Africa, SW to SE Asia. Decreasing. NT] Breeds primarily on coasts from N Iberia discontinuously N in W Europe to Iceland, Scandinavia and Finland, also *eg* Balkans (*ostralegus*), and extensively inland from NW Russia S to Ukraine and Turkey (*longipes*) and NW Kazakhstan (*buturlini*). Mainly resident in W, but E and most N populations migratory, occurring widely on passage, and wintering *eg* W Iberia, parts of coastal N Africa (S to N Mauritania) and Middle East. Vagrant Algeria, Azores, Cape Verde Is, Czech Republic, Madeira, Moldova, Slovakia and Svalbard.

FAMILY DROMADIDAE

Crab-plover · one species

Dromas ardeola
Crab-plover

OTHER NAME Crab Plover.
TAXONOMY Paykull, 1805. Monotypic.
DISTRIBUTION [Indian Ocean basin. Stable. LC] Breeds Iraq, Iran, Kuwait, Oman, Saudi Arabia, UAE and Yemen; small population also recently discovered SE Egypt. Partially migratory, when regular on passage Bahrain and Qatar; some winter S Red Sea coast of

Egypt but others leave Red Sea and Gulf to winter outside region. Vagrant N Egypt, Israel, Jordan and Turkey.

FAMILY RECURVIROSTRIDAE

Stilts and avocets · two species

Himantopus himantopus
Black-winged Stilt

OTHER NAME Common Stilt.
TAXONOMY (Linnaeus, 1758). Monotypic.
DISTRIBUTION [Africa, S Europe to CE Asia > Africa, SW Europe, SW to SE Asia. Increasing. LC] Resident parts of SW Iberia, N Africa and Middle East, otherwise summer visitor S Europe N to *eg* NW France, Austria, Slovakia and Ukraine, E to SW Russia and NW Kazakhstan, and Asia Minor. Occurs widely on passage S through region and frequently overshoots N in spring, when occasionally breeds N of normal range. Vagrant Azores, Belarus (has bred), Britain (has bred), Channel Is, Denmark, Estonia, Finland, Gibraltar, Iceland, Ireland, Latvia, Lithuania, Madeira, Norway and Sweden.

Recurvirostra avosetta
Pied Avocet

OTHER NAMES (Eurasian) Avocet.
TAXONOMY Linnaeus, 1758. Monotypic.
DISTRIBUTION [S and E Africa, W Europe to CE Asia > Africa, W and S Europe, SW Asia. Population trend unknown. LC] Fragmented breeding distribution from NW Africa and parts Middle East discontinuously N to Britain, S Scandinavia, Baltic States, Ukraine, SW Russia and NW Kazakhstan, but absent large areas of interior Europe. Populations in W Europe and Mediterranean often present year round, but those in Scandinavia, E Europe, Caucasus and Asia Minor mainly migratory, occurring S of breeding areas widely on passage and in winter. Vagrant Azores, Belarus, Faroe Is, Iceland, Ireland (has bred), Latvia, Liechtenstein and Madeira.

FAMILY CHARADRIIDAE

Lapwings and plovers · 23 species

Vanellus vanellus
Northern Lapwing

OTHER NAMES (Common or Eurasian) Lapwing, Green Plover, Peewit.
TAXONOMY (Linnaeus, 1758). Monotypic.
DISTRIBUTION [Europe to E Asia > W and S Europe, N Africa to SE Asia. Decreasing. NT] Breeds Britain, Ireland and much of continental Europe (less widespread in S and extreme N), also N Morocco, Asia Minor, Caucasus and NW Iran. Mainly resident in W and S of range but otherwise largely migratory, occurring widely on passage and in winter S to coastal N Africa (perhaps irregularly S to N Mauritania) and much of Middle East except SW Arabia. Scarce or irregular Azores, Canary Is, Iceland (has bred), Madeira, N Mauritania and Svalbard. Vagrant Cape Verde Is, Jan Mayen and Yemen.

Vanellus spinosus
Spur-winged Lapwing

OTHER NAME Spur-winged Plover.
TAXONOMY (Linnaeus, 1758). Monotypic.
DISTRIBUTION [W to E Africa, Nile Valley, SW Asia. Increasing. LC] Breeds SE Greece, Turkey, Cyprus and Levant S to Nile Valley and Yemen, and in Iraq and C Saudi Arabia. N populations migratory; occasional NW Iran. Vagrant Armenia, Bahrain, Bulgaria, Cape Verde Is, Croatia, Czech Republic, Hungary, Italy, Kuwait, Libya, Malta, Montenegro, Oman, Qatar, Romania, Serbia, UAE and Ukraine.

Vanellus tectus
Black-headed Lapwing (V)

OTHER NAMES Black-headed (or Blackhead) Plover, Crested Wattled Plover.
TAXONOMY (Boddaert, 1783). Two subspecies (one in region).
DISTRIBUTION [Sahel and E Africa. Population trend unknown. LC] Extreme vagrant: Israel/

Jordan (1869, shot in area of modern-day border, *latifrons*).

Vanellus cinereus
Grey-headed Lapwing (V)

OTHER NAME Grey-headed Plover.
TAXONOMY (Blyth, 1842). Monotypic.
DISTRIBUTION [E Asia > S and SE Asia. Decreasing. LC] Extreme vagrant: Oman (3 Jan 2012).

Vanellus indicus
Red-wattled Lapwing

OTHER NAMES Indian Decorated Lapwing, Red-wattled Plover.
TAXONOMY (Boddaert, 1783). Four subspecies (one in region).
DISTRIBUTION [SW to SE Asia. Population trend unknown. LC] Breeds SE Turkey and possibly NE Syria, and from Iraq E through S Iran and S to Kuwait, Saudi Arabia, Qatar (rare), UAE and Oman (*aigneri*). Vagrant Azerbaijan, Bahrain, Israel and Jordan.

Vanellus gregarius
Sociable Lapwing

OTHER NAME Sociable Plover.
TAXONOMY (Pallas, 1771). Monotypic.
DISTRIBUTION [Extreme SE Europe to C Asia > NE Africa to S Asia. Decreasing. CR] Summer visitor extreme SE European Russia and NW Kazakhstan; breeding range formerly extended W to Ukraine. Migrates SSW through Volga-Ural and Caspian regions, staging *eg* S Russia, SE Turkey and Syria, and NW and SE Arabia; rare/irregular Azerbaijan, Iraq and Iran. Winters Israel (scarce), Oman and UAE (rare), possibly also Syria. Vagrant Armenia, Austria, Bahrain, Balearic Is, Belarus, Belgium, Britain, Bulgaria, Canary Is, Czech Republic, Cyprus, Denmark, Egypt, Estonia, Finland, France, Georgia, Germany, Greece, Hungary, Ireland, Italy, Jordan, Kuwait, Lebanon, Luxembourg, Malta, Morocco, Netherlands, Poland, Portugal, Qatar, Romania, Slovakia, Slovenia, Spain, Sweden, Switzerland, Tunisia and Yemen.

Vanellus leucurus
White-tailed Lapwing

OTHER NAME White-tailed Plover.
TAXONOMY (Lichtenstein, M H C, 1823). Monotypic.
DISTRIBUTION [SW to C Asia > NE Africa to S Asia. Population trend unknown. LC] Fragmented breeding distribution in C Turkey, Caucasus (rare), extreme SE European Russia, NW Kazakhstan, Iran, Iraq and Syria, also locally in Jordan and in Gulf states E to UAE; has bred Kuwait, Romania and Ukraine. Mainly migratory in N of range and more widespread in Middle East on passage W to Cyprus and Egypt, and in winter. Vagrant Algeria, Austria, Belgium, Britain, Bulgaria, Canary Is, Croatia, Cyprus, Czech Republic, Denmark, Estonia, Finland, France, Germany, Greece, Hungary, Italy, Libya, Lithuania, Malta, Morocco, Netherlands, Norway, Poland, Romania, Spain, Sweden and Tunisia.

Pluvialis apricaria
European Golden Plover

OTHER NAMES (Eurasian or Greater) Golden Plover.
TAXONOMY (Linnaeus, 1758). Monotypic.
DISTRIBUTION [N Europe and NW Asia > Europe, N Africa and SW Asia. Decreasing. LC] Breeds Iceland, Ireland, N Britain, Scandinavia, N Germany and Baltic States E through far NW Russia. Mainly migratory, N populations occurring widely on passage to S, and wintering S to Canary Is (irregular), Mediterranean basin, W Middle East, S Caspian and UAE. Vagrant Azores, Bahrain, Cape Verde Is, Gibraltar, Iraq, Jordan, Kuwait, Madeira, N Mauritania, Oman and Saudi Arabia.

Pluvialis fulva
Pacific Golden Plover

OTHER NAMES Eastern (or Asian/Asiatic) Golden Plover; Lesser Golden Plover (prior to split from American Golden Plover).
TAXONOMY (Gmelin, J F, 1789). Monotypic. Formerly treated as conspecific with American Golden Plover, but differs in plumage and

structure, and no evidence of interbreeding where ranges overlap.
DISTRIBUTION [N Siberia to W Alaska > NE Africa, SW to SE Asia, Australasia and Pacific islands. Decreasing. LC] Regular on passage and in winter Gulf from Kuwait and Iraq (rare) E to SE Iran and Oman, also SW Saudi Arabia and W Yemen. Vagrant Albania, Algeria, Austria, Azerbaijan, Azores, Belgium, Britain, Cyprus, Denmark, Estonia, Finland, France, Georgia, Germany, Greece, Hungary, Iceland, Ireland, Israel, Italy, Jordan, Latvia, Lebanon, Lithuania, Madeira, Malta, Netherlands, Norway, Poland, Portugal, Romania, Slovenia, Spain, Svalbard (has bred), Sweden, Switzerland, Tunisia, Turkey and Ukraine; undetermined *fulva/dominica* also recorded Jan Mayen.

Pluvialis dominica
American Golden Plover

OTHER NAME Lesser Golden Plover (prior to split from *P fulva*).
TAXONOMY (Statius Müller, 1776). Monotypic. Formerly included Pacific Golden Plover (*qv*).
DISTRIBUTION [Arctic North America > C South America. Decreasing. LC] Scarce migrant Britain (annual mean 2000-09 of 17), especially autumn but occasional spring; rarer but regular Azores, where largest flock in region (10) recorded 7 Oct 2012. Vagrant Austria*, Balearic Is, Belgium, Canary Is, Cape Verde Is, Channel Is, Czech Republic, Denmark, Finland, France, Germany, Hungary, Iceland, Ireland, Israel, Italy, Lithuania, Morocco, Netherlands, Norway, Oman, Poland, Portugal, Spain, Svalbard, Sweden, Tunisia and Turkey; undetermined *dominica/fulva* also recorded Jan Mayen.

Pluvialis squatarola
Grey Plover

OTHER NAMES Black-bellied (or Silver) Plover.
TAXONOMY (Linnaeus, 1758). Three subspecies (one in region).
DISTRIBUTION [Holarctic > near cosmopolitan. Decreasing. LC] Breeds extreme NW Russia (*squatarola*). Occurs across much of region on migration, but generally scarcer or rare inland. Winters widely on coasts from North Sea, Atlantic and Mediterranean to Red Sea, Arabian Sea and Persian Gulf. Vagrant Armenia, Faroe Is, Gibraltar, Iceland, Luxembourg, Madeira, Moldova and Svalbard.

Charadrius hiaticula
Common Ringed Plover

OTHER NAMES (Greater) Ringed Plover; Greenland Ringed Plover (*C h psammodromus*), Tundra Ringed Plover (*C h tundrae*).
TAXONOMY Linnaeus, 1758. Three subspecies (all in region).
DISTRIBUTION [NE Canada, Greenland and N Eurasia > W and SW Europe, SW Asia and Africa. Decreasing. LC] Breeds Iceland (*psammodromus*), Ireland, Britain and W France N to S Scandinavia and Baltic States (*hiaticula*) and N Scandinavia, Finland and far NW Russia (*tundrae*). Occurs widely across region on migration, but generally less numerous inland. Winters mainly on coasts in North Sea and Atlantic, parts of W and E Mediterranean, S Caspian, and Red Sea, Arabian Sea and Persian Gulf. Vagrant Gibraltar and Moldova.

Charadrius semipalmatus
Semipalmated Plover

OTHER NAME Semipalmated Ringed Plover.
TAXONOMY Bonaparte, 1825. Monotypic.
DISTRIBUTION [N North America > USA to S South America. Stable. LC] Scarce migrant Azores, where regular in autumn (average 23* individuals annually 2005-2011); has overwintered, and latterly recorded all months. Vagrant Britain, Canary Is, Cape Verde Is, France, Iceland, Ireland, Madeira, Norway and Spain.

Charadrius dubius
Little Ringed Plover

OTHER NAME Little Plover.
TAXONOMY Scopoli, 1786. Three subspecies (one in region).
DISTRIBUTION [NW Africa and Eurasia > W to E Africa, SW to SE Asia. Stable. LC] Summer visitor NW Africa, much of Europe N to Britain, S Fennoscandia and NW Russia, and

more patchily in Middle East N to Asia Minor and Caucasus, S through Levant to NE Egypt, and in E Arabia and much of Iran (*curonicus*). Occurs widely across region on migration; a few winter S Europe, N Africa (resident Canary Is) and S Middle East. Vagrant Azores, Ireland and Madeira.

Charadrius vociferus
Killdeer (V)

TAXONOMY Linnaeus, 1758. Three subspecies (one in region).
DISTRIBUTION [North America and NW South America > SW Canada, W and S USA to N South America. Decreasing. LC] Vagrant, mainly to W Europe in late autumn and winter, and two breeding records (*vociferus*): Azores (29*: single pair bred 2010 and 2011, with minimum of nine individuals recorded in latter year); Britain (56); Channel Is (2); Faroe Is (Sep 1939); France (6 records, including 1 of flock of 5); Hungary (2); Iceland (3); Ireland (22); Madeira (17 Feb 2004); Netherlands (3-5 Apr 2005); Norway (14 May 1913); Portugal (28 Feb 1998); Spain (4); Switzerland (2).

Charadrius pecuarius
Kittlitz's Plover

OTHER NAME Kittlitz's Sand Plover.
TAXONOMY Temminck, 1823. Monotypic.
DISTRIBUTION [Egypt and sub-Saharan Africa. Population trend unknown. LC] Resident Egypt (Nile Delta S to Fayoum and Suez, Lake Nasser). Vagrant Bahrain, Cyprus, France, Greece, Israel, Italy, Morocco, Norway, Saudi Arabia and UAE.

Charadrius tricollaris
Three-banded Plover

OTHER NAMES African Three-banded Plover, Collared (Sand) Plover, Treble-banded (or Tricollared) (Sand) Plover.
TAXONOMY Vieillot, 1818. Two subspecies (one in region).
DISTRIBUTION [NE, E, C and S Africa. Population trend unknown. LC] Recorded only in Egypt and formerly considered extreme vagrant, with four records 1993-2003 (*tricollaris*). Since 2006, recorded annually in S and breeding first confirmed 2009, since when has nested annually.

Charadrius alexandrinus
Kentish Plover

TAXONOMY Linnaeus, 1758. Four subspecies (one in region). Formerly also included extralimital Snowy Plover *C nivosus* (North and South America).
DISTRIBUTION [Widespread Europe, Africa and Asia. Decreasing. LC] Breeds mainly on coasts from Atlantic islands and N Africa N in continental W Europe to Denmark, E through much of Mediterranean to Black and Caspian Seas, and SE in Red Sea, Arabian Sea and Persian Gulf (*alexandrinus*); in some areas, *eg* NW Africa, SC Spain, Hungary, Ukraine and SW Russia, significant populations also breed inland. NW European and NE regional populations mainly migratory; occurs more widely on passage, especially in Middle East, and a few reach N to *eg* Channel Is and S Britain (scarce; formerly bred), Sweden and Switzerland. Vagrant Czech Republic, Finland, Gibraltar (has bred), Ireland, Latvia, Lithuania, Luxembourg, Malta, Moldova, Norway, Poland (has bred) and Slovakia (has bred).

Charadrius mongolus
Lesser Sand Plover

OTHER NAMES Tibetan (or Spot-fronted) Sand Plover (*C m atrifrons*), Mongolian (or Black-fronted) (Sand) Plover (*C m mongolus/stegmanni*).
TAXONOMY Pallas, 1776. Five subspecies (three in region). Populations in EC Asia (*pamirensis*, *atrifrons* and *schaeferi*, '*atrifrons* group') and NE Asia (*mongolus* and *stegmanni*, '*mongolus* group') are morphologically distinct and geographically isolated from each other, and may merit treatment as two separate species.
DISTRIBUTION [EC and NE Asia > E and S Africa to Australia. Population trend unknown. LC] Occurs on passage and in winter coastal Arabia and S Iran; rare Red Sea coast of Egypt (*pamirensis/atrifrons*). Vagrant elsewhere (probably mainly involving '*atrifrons* group',

but minority of records determined as '*mongolus* group' and some individuals not identified to subspecies): Belgium, Britain (4 of 6 records '*mongolus* group'), Bulgaria, Canary Is ('*mongolus* group'), Denmark, Finland, France, Ireland (sole record '*mongolus* group'), Israel, Italy, Libya, Norway ('*mongolus* group'), Poland, Portugal ('*mongolus* group'), Spain, Sweden ('*mongolus* group', possibly also '*atrifrons* group'), Syria and Turkey*.

Charadrius leschenaultii
Greater Sand Plover

OTHER NAMES Geoffroy's (or Large Sand) Plover; Anatolian Sand Plover (*C l columbinus*), Transcaspian Greater Sand Plover (*C l scythicus*).
TAXONOMY Lesson, 1826. Three subspecies (two in region). For nomenclatural reasons long-established subspecies name *crassirostris* now replaced by *scythicus*.
DISTRIBUTION [Middle East to EC Asia > E and S Africa to Australia. Population trend unknown. LC] Breeds Turkey, Syria and Jordan (*columbinus*) and from N Iran just into Armenia (rare) and Azerbaijan (rare) (*scythicus*). Migrates through Middle East and E/SE Mediterranean (W to Libya), wintering Cyprus and extensively SE Mediterranean, Red Sea and Arabian coasts. Vagrant Austria, Belgium, Britain, Bulgaria, Denmark, Estonia, Finland, France, Georgia, Germany, Greece, Hungary, Iceland, Italy, Latvia, Malta, Morocco, Netherlands, Norway, Poland, Romania, Spain, Sweden, Tunisia and Ukraine.

Charadrius asiaticus
Caspian Plover

OTHER NAMES Caspian Sand Plover, Lesser Oriental Plover.
TAXONOMY Pallas, 1773. Monotypic.
DISTRIBUTION [C Asia > E and S Africa. Decreasing. LC] Summer visitor SE European Russia and NW Kazakhstan. Migrates through Caspian region and Middle East, but generally scarce or rare on passage, when recorded E Egypt, N Iran, Israel, Kuwait, Oman, Qatar, Turkey and UAE; formerly regular Azerbaijan (last in 1985). Vagrant Armenia, Azores, Britain, Bulgaria, Cyprus, Finland, France, Georgia, Germany, Greece, Hungary, Italy, Lebanon, Malta, Netherlands, Norway, Sweden, Syria, Ukraine and Yemen.

Charadrius veredus
Oriental Plover (V)

OTHER NAMES Asiatic (or Eastern) Sand Plover, Greater Oriental Plover, Oriental Dotterel.
TAXONOMY Gould, 1848. Monotypic.
DISTRIBUTION [S Siberia to CE Asia > Australasia. Population trend unknown. LC] Extreme vagrant: Finland (25 May 2003).

Charadrius morinellus
Eurasian Dotterel

OTHER NAMES (Mountain) Dotterel.
TAXONOMY Linnaeus, 1758. Monotypic.
DISTRIBUTION [N Eurasia and W Alaska > N Africa and SW Asia. Decreasing. LC] Summer visitor mainly to upland N Britain, W and N Fennoscandia and NW Russia, with tiny outlying populations *eg* Pyrenees, Alps, Apennines, Carpathians and Urals; formerly also bred Netherlands at sea level. On migration occurs S across region but generally scarce or rare where present (though larger flocks recorded *eg* Hungary and Turkey). Winters from Canary Is (irregular) and NW Africa (mainly Algeria and Tunisia) E to N Middle East and Iran. Vagrant Armenia, Azores, Bahrain, Belarus, Bosnia-Herzegovina, Corsica, Croatia, Faroe Is, Iceland, Ireland, Latvia, Lithuania, Luxembourg, N Mauritania, Moldova, Oman, Slovakia, Slovenia, Svalbard and UAE.

FAMILY PLUVIANIDAE

Egyptian Plover · one species

Pluvianus aegyptius
Egyptian Plover (V)

OTHER NAMES Crocodile-bird, Crocodile-plover, Egyptian Courser.
TAXONOMY (Linnaeus, 1758). Monotypic.
DISTRIBUTION [W, C, E and NE Africa. Decreas-

ing. LC] Formerly resident Egypt in Nile Valley, but no confirmed records since 1937. Vagrant Libya and Spain; other European occurrences (including *eg* France and Poland) considered to relate to escapes from captivity.

FAMILY ROSTRATULIDAE

Painted-snipes · one species

Rostratula benghalensis
Greater Painted-snipe

OTHER NAME Painted Snipe.
TAXONOMY (Linnaeus, 1758). Monotypic. Formerly included extralimital Australian Painted-snipe *R australis*.
DISTRIBUTION [Egypt and sub-Saharan Africa, S and SE Asia. Decreasing. LC] Resident Egypt, mainly in Nile Delta and Valley; very rare visitor Israel, especially spring and summer (has bred). Vagrant Iran, Jordan, Morocco, Oman and UAE.

FAMILY JACANIDAE

Jacanas · one species

Hydrophasianus chirurgus
Pheasant-tailed Jacana

OTHER NAMES (Chinese) Water-pheasant.
TAXONOMY (Scopoli, 1786). Monotypic.
DISTRIBUTION [SE Arabia, S and SE Asia. Decreasing. LC] Winter visitor Oman; a few oversummer (two breeding records). Vagrant Qatar, Saudi Arabia, UAE and Yemen.

FAMILY SCOLOPACIDAE

Sandpipers and snipes · 63 species

Scolopax rusticola
Eurasian Woodcock

OTHER NAMES (European) Woodcock.
TAXONOMY Linnaeus, 1758. Monotypic.
DISTRIBUTION [W Europe to E Asia > W and S Europe to SE Asia. Stable. LC] Breeds mainly from N Iberia N through W, C and E Europe to N Scandinavia and NW Russia, range extending W to Iceland (recent colonist), Britain, Ireland and Azores, Canary Is and Madeira, with scattered outlying populations in *eg* N Italy, Balkans and Caucasus. Resident in W of range but largely migratory elsewhere, occurring widely on passage and in winter as far S as N Africa and N Middle East. Vagrant Bahrain, Jan Mayen, Jordan, Oman, Saudi Arabia, Svalbard and UAE.

Scolopax minor
American Woodcock (V)

TAXONOMY Gmelin, J F, 1789. Monotypic.
DISTRIBUTION [SE Canada and E USA > SE USA. Decreasing. LC] Extreme vagrant: France (28 Oct 2006, shot).

Lymnocryptes minimus
Jack Snipe

TAXONOMY (Brünnich, 1764). Monotypic.
DISTRIBUTION [N Eurasia > W and S Europe, N and W to E Africa, SW to SE Asia. Stable. LC] Summer visitor from Scandinavia E through Baltic and NW Russia to Ural Mtns. Migrates mainly SW, occurring across much of region on passage and wintering (generally locally) from Denmark and Faroe Is S through NW Europe to Iberia and Morocco (irregular Canary Is), also widely in Mediterranean basin and Middle East. Vagrant Azores, Cape Verde Is, Gibraltar, Iceland, Jan Mayen, Madeira, Moldova and Svalbard.

Gallinago solitaria
Solitary Snipe (V)

OTHER NAMES Hermit (or Tibet) Snipe.
TAXONOMY Hodgson, 1831. Two subspecies (one in region).
DISTRIBUTION [C, E and NE Asia. Stable. LC] Formerly occurred on passage and in winter Iran (*solitaria*), but no confirmed records since 1956 and status now unknown. Vagrant Saudi Arabia.

Gallinago stenura
Pin-tailed Snipe

OTHER NAMES Pintail (or Asiatic) Snipe.
TAXONOMY (Bonaparte, 1831). Monotypic.
DISTRIBUTION [W Siberia to NE Asia > SW to SE Asia. Population trend unknown. LC] Summer visitor N Ural Mtns. On passage and in winter (mainly Sep-Apr) occurs Bahrain, Oman, UAE and Yemen. Vagrant Iran (formerly possibly regular), Israel, Italy, Jordan, Saudi Arabia and Tunisia; *cf* Swinhoe's Snipe.

Gallinago megala
Swinhoe's Snipe (V)

OTHER NAMES Chinese (or Forest/Marsh) Snipe.
TAXONOMY Swinhoe, 1861. Monotypic.
DISTRIBUTION [S Siberia, NC and NE Asia > S and SE Asia, N Australia. Population trend unknown. LC] Extreme vagrant (five summer records; those in Russia, all from same area of Ural Mtns, may indicate that species' Asian range extends W to edge of region): Finland (13 Jun-6 Jul 2008); Russia (4: 23-24 Jun 2002; 19-20 Jun 2010, pair; Jun 2011, 2+ displaying males*; Jun 2012, 2+ displaying males*). This species or Pin-tailed Snipe also recorded Egypt (22 Oct 2008).

Gallinago media
Great Snipe

TAXONOMY (Latham, 1787). Monotypic.
DISTRIBUTION [Scandinavia to C Siberia > W, C and E Africa. Decreasing. NT] Summer visitor to C and N Scandinavia, Finland (rare), Baltic States, Poland and Belarus E through NW Russia and N Ukraine to Ural Mtns; formerly much more numerous and widespread. Occurs on migration from S Fennoscandia, C, S and SE Europe to N Africa (mainly spring, rare W to Morocco) and Turkey, Levant, Israel (very rare), Iraq and Kuwait E to Iran, also Oman (very rare), but some individuals known to make non-stop flights in autumn to African wintering grounds. Rare winter records mostly from Mediterranean countries. Vagrant Balearic Is, Belgium, Britain, Cape Verde Is, Channel Is, Corsica, Czech Republic, France, Germany (formerly bred), Jordan, Luxembourg, Madeira, Malta, Moldova, Netherlands, Portugal, Romania, Slovenia, Spain, Svalbard, Switzerland, UAE and Yemen.

Gallinago gallinago
Common Snipe

OTHER NAMES (Eurasian or Fantail) Snipe; Faroe Snipe (*G g faeroeensis*).
TAXONOMY (Linnaeus, 1758). Two subspecies (both in region). Formerly included Wilson's Snipe (*qv*).
DISTRIBUTION [N Eurasia > W and S Europe, N, W, C and E Africa, SW to SE Asia. Decreasing. LC] Breeds Iceland, Faroe Is, Shetland and Orkney (*faeroeensis*), and from Ireland and much of Britain E across N continental Europe and S to *eg* C France, C Europe, N Ukraine and S Ural Mtns, with outlying populations Azores and NW Iberia; may breed Turkey (*gallinago*). Mainly resident in W but migratory in N and E, occurring widely to S across region on passage, and wintering S Europe, N Africa (W to Atlantic islands) and parts of Middle East. Rare N Mauritania. Vagrant Gibraltar and Svalbard.

Gallinago delicata
Wilson's Snipe

TAXONOMY (Ord, 1825). Monotypic. Previously considered conspecific with Common Snipe, but differs in winnowing display sounds and morphology.
DISTRIBUTION [N North America > W and S North America to N South America. Decreasing. LC] Very scarce but apparently regular autumn migrant and occasional winterer Azores; displaying individuals also recorded there (6 Jun 2010*), Finland (each summer 2008-2010) and Iceland (14-16 Jun 2010). Vagrant Britain, Canary Is, Cape Verde Is, France, Ireland, Madeira and Portugal.

Limnodromus griseus
Short-billed Dowitcher (V)

OTHER NAME Common Dowitcher.
TAXONOMY (Gmelin, J F, 1789). Three subspecies (undetermined in region).

DISTRIBUTION [N North America > S North America to N South America. Decreasing. LC] Very rare vagrant (mainly Sep-Oct, fewer Jul-Aug; several long-stayers, but otherwise exceptional in winter and spring): Azores (24*); Britain (4); France (2: 31 Aug-10 Oct 1998; 22-27 Oct 2012); Germany (2*); Iceland (30 Jun and 18-20 July 2011); Ireland (3); Norway (6 Jul 1971); Spain (9-25 Mar 2013); Sweden (7 Sep 2012). See Long-billed Dowitcher for comment on *Limnodromus* records not assigned to species.

Limnodromus scolopaceus
Long-billed Dowitcher (V)

TAXONOMY (Say, 1822). Monotypic.
DISTRIBUTION [NE Asia and NW North America > Japan, S USA to West Indies and Central America. Population trend unknown. LC] Vagrant (mainly Sep-Oct, with smaller peak Apr-May; occasionally overwinters): Azores (10*); Balearic Is (9-30 Oct 2002); Belgium (7); Britain (243); Canary Is (2); Channel Is (2); Denmark (15); Estonia (2); Finland (10); France (44); Germany (6); Greece (2); Hungary (30 Apr-1 May 1995); Iceland (10); Ireland (123); Israel (21 Jul-25 Aug 1984); Italy (7); Malta (5-25 Oct 2012); Morocco (9); Netherlands (39); Norway (16); Oman (12); Poland (2); Portugal (7+); Spain (17); Sweden (19); UAE (25 Jan 2013). Many other records of dowitchers not assigned to species level, mainly from Britain (78 to 2002) and Ireland (42), mostly prior to 1980s when ID critieria poorly understood; majority probably relate to this species.

Limnodromus semipalmatus
Asian Dowitcher (V)

OTHER NAMES Asiatic (or Oriental/Snipe-billed) Dowitcher, Semi-palmated Snipe, Snipe-billed Godwit.
TAXONOMY (Blyth, 1848). Monotypic.
DISTRIBUTION [W Siberia and C Asia to E Asia > S and SE Asia. Decreasing. NT] Extreme vagrant: Oman (3 Nov 2015-5 Jan 2016). Old record from Yemen (22-23 Aug 1958) previously discounted but now under review as potential first regional record.

Limosa limosa
Black-tailed Godwit

OTHER NAMES Continental Black-tailed Godwit (*L l limosa*), Eastern (or Asiatic) Black-tailed Godwit (*L l melanuroides*), Icelandic Black-tailed Godwit (*L l islandica*).
TAXONOMY (Linnaeus, 1758). Three subspecies (two in region). Sometimes considered as three separate species, each having unique haplotypes lacking gene flow from the others, and thus indicative of long-standing reproductive isolation; all three also morphologically distinct in breeding plumage, and *melanuroides* notably smaller.
DISTRIBUTION [N Eurasia > W Europe and W Africa to Australia. Decreasing. NT] Summer visitor Iceland, Faroe Is, N Scotland (rare) and N Norway (*islandica*) and E England (rare), Low Countries N to S Scandinavia and E through C and E Europe to Ural Mtns, with isolated minor populations to S in *eg* France and N Italy (*limosa*). Widespread on migration S of breeding range, wintering from Britain S to Morocco (*islandica*) and parts of Mediterranean basin and Middle East (*limosa*). Vagrant Cape Verde Is, Gibraltar, Jan Mayen, Liechtenstein, Madeira and Svalbard.

Limosa haemastica
Hudsonian Godwit (V)

TAXONOMY (Linnaeus, 1758). Monotypic.
DISTRIBUTION [N North America > S South America. Decreasing. LC] Extreme vagrant (six records, including two returning individuals): Azores (25 Jul 2007); Britain (2: 10 Sep-3 Oct 1981, presumed same 22 Nov 1981-14 Jan 1982 and 26 Apr-6 May 1983; 26 Sep 1988); Denmark (6 Sep 1986); Norway (24 Jun 2008); Sweden (14-20 Jun 2003, presumed same 27-30 Jun 2006).

Limosa lapponica
Bar-tailed Godwit

OTHER NAMES Lapland Bar-tailed Godwit (*L l lapponica*), Siberian Bar-tailed Godwit (*L l baueri*).
TAXONOMY (Linnaeus, 1758). Four subspecies (two in region). Easternmost subspecies *baueri* recently proposed as separate species.

DISTRIBUTION [N Eurasia to W Alaska > W Europe and W Africa to S Asia and Australasia. Decreasing. NT] Summer visitor from N Scandinavia to NW Russia (*lapponica*). Primarily coastal on migration, especially Baltic, North Sea and Atlantic coasts, and exclusively so in winter, chiefly from W Europe S to N Mauritania; also stages and/or winters parts of Mediterranean basin (mainly in W) and in Middle East, where Siberian *baueri* occurs alongside *lapponica* in Oman (and other coasts locally?). Rare/irregular Atlantic islands. Vagrant Cyprus, Faroe Is, Gibraltar, Iceland, Jordan, Lebanon, Libya, Liechtenstein, Luxembourg, Malta, Moldova, Romania, Serbia and Svalbard.

Numenius minutus
Little Curlew (V)

OTHER NAME Little Whimbrel.
TAXONOMY Gould, 1841. Monotypic.
DISTRIBUTION [N and NE Asia > SE Asia to N Australia. Stable. LC] Extreme vagrant (seven records): Belgium (18 Sep 2010); Britain (2: 30 Aug-6 Sep 1982; 24 Aug-3 Sep 1985); Finland (1-2 Oct 1996); Kuwait (12-29 Dec 2007); Norway (14 Jul 1969); Sweden (12-13 Sep, 24-30 Sep and 8 Oct-8 Nov 2005).

Numenius borealis
Eskimo Curlew (V)

TAXONOMY (Forster, J R, 1772). Monotypic.
DISTRIBUTION [NW Canada and possibly N Alaska > SE South America. Population trend unknown. CR-PE] Former extreme vagrant now considered possibly extinct (five records Sep-Oct, most or all following storms or hurricanes off NE USA): Britain (4, all collected: 6 Sep 1855; 28 Sep 1878; 21 Sep 1880; 10 Sep 1887); Ireland (21 Oct 1870, collected). No certain records of this species worldwide since 1962 (Texas, USA) and 1963 (Barbados).

Numenius phaeopus
Eurasian Whimbrel

OTHER NAMES Whimbrel (IOC); Siberian Whimbrel (*N p variegatus*), Steppe Whimbrel (*N p alboaxillaris*).
TAXONOMY (Linnaeus, 1758). Three subspecies (two in region). Formerly included Hudsonian Whimbrel (*qv*).
DISTRIBUTION [N Eurasia > sub-Saharan Africa to Australasia. Decreasing. LC] Summer visitor Iceland, Faroe Is, N Scotland and N Scandinavia to NW Russia (*phaeopus*), and SE European Russia and NW Kazakhstan (*alboaxillaris*). Occurs widely on migration S of breeding range, notably on Baltic, North Sea and Atlantic coasts but also staging inland *eg* Low Countries and Hungary, and regular migrant Balkans. Winters mainly N Mauritania, with smaller numbers N to Iberia and Atlantic islands (*phaeopus*), and on coasts of Arabia and Iran (*phaeopus* and *alboaxillaris*). Uncommon Israel, rare Jan Mayen, Lebanon, Malta and Svalbard. Vagrant Armenia (*phaeopus* and *alboaxillaris*), Jordan, Liechtenstein, Moldova and Syria.

Numenius hudsonicus
Hudsonian Whimbrel (V)

TAXONOMY Latham, 1790. Monotypic. Traditionally considered conspecific with Eurasian Whimbrel (and still so by IOC), but exhibits diagnostic differences in plumage and mean morphometric measurements, and marked divergence in mtDNA sequences from Eurasian Whimbrels from both Scandinavia (*phaeopus*) and E Asia (*variegatus*). Split accepted by BOURC and IRBC.
DISTRIBUTION [N and NW North America > S North America and South America. Population trend unknown. NE] Rare vagrant (most frequent in autumn but occasional in winter, more rarely spring and summer): Azores (44*); Britain (8); Cape Verde Is (2); Ireland (3); Madeira (4*); Norway (7 Jun-15 Jul 1989); Spain (20 Jan-22 Dec 2009, but apparently absent May-Jun).

Numenius tenuirostris
Slender-billed Curlew

TAXONOMY Vieillot, 1817. Monotypic.
DISTRIBUTION [SW Siberia and possibly NC Asia > S Europe, N Africa and Middle East.

Decreasing. CR] Former migrant and winter visitor, occurring regularly in 19th and early 20th centuries on passage *eg* Balkans, Hungary, Italy, Malta, Spain and Ukraine, and wintering Greece, Italy, Morocco, Algeria and Tunisia; also regular but rare on passage Azerbaijan, with further records from *eg* Nile Delta and Turkey perhaps indicative of regular passage/winter occurrence in Middle East. Dramatic decline during 20th century, with very few recent confirmed occurrences; last recorded Morocco (18 Sep 1994-23 Feb 1995, up to two, and 11 Feb 1998), Croatia (24 Mar 1996), Sicily (31 Mar 1996) and Hungary (15 Apr 2001). Widely recorded prior to 1980s as rare migrant/vagrant: Albania, Azores*, Austria, Bosnia-Herzegovina, Belgium, Bulgaria, Croatia, France, Georgia, Germany, Greece, Hungary, Iran, Iraq, Israel, Latvia, Libya, Macedonia, Netherlands, Poland, Portugal, Romania, Serbia, Slovakia, Spain, Switzerland and Yemen. Formerly accepted record from Britain (4-7 May 1998) now rejected after review, and all records from Cyprus (3) and Oman (5) currently under review (see Kirwan *et al* 2015 for recent analysis of Middle Eastern records). Lack of any confirmed records worldwide since 2001 possibly indicative of extinction.

Numenius arquata
Eurasian Curlew

OTHER NAMES Common (or European/Western) Curlew.
TAXONOMY (Linnaeus, 1758). Three subspecies (all in region).
DISTRIBUTION [N Eurasia > W and S Europe, Africa, SW to E Asia. Decreasing. NT] Breeds from Ireland, Britain and Faroe Is E across much of N Europe to Ural Mtns and S to *eg* Pyrenees, C Europe and N Black Sea (*arquata*), and NNW Kazakhstan (*suschkini*). Partly resident Ireland, Britain and Netherlands but otherwise largely migratory, occurring widely on migration S of breeding range, wintering W Europe from Iceland (has bred) S to N Mauritania, also parts of Mediterranean basin and Middle East; passage and wintering populations include some *orientalis* (breeding Asia) as far W as *eg* Hungary and Iberia. Vagrant Azores, Cape Verde Is, Jan Mayen, Madeira and Svalbard.

Numenius madagascariensis
Far Eastern Curlew (V)

OTHER NAMES Eastern (or Australian) Curlew, Long-billed (or Red-rumped) Curlew.
TAXONOMY (Linnaeus, 1766). Monotypic.
DISTRIBUTION [E and NE Asia > E and SE Asia to Australasia. Decreasing. EN] Extreme vagrant (two autumn records): Iran (2 Oct 1898, collected); Oman (23 Oct 1997).

Bartramia longicauda
Upland Sandpiper (V)

OTHER NAMES Bartram's Sandpiper, Upland Plover.
TAXONOMY (Bechstein, 1812). Monotypic.
DISTRIBUTION [N North America > C and SE South America. Increasing. LC] Vagrant (generally Sep-Nov, but occasional in spring): Azores (15*); Britain (46); Canary Is (20-22 Sep 2002); Channel Is (2); Croatia (31 Oct 1936); Denmark (3 Nov 1920); France (8); Germany (early 19th century); Greece (23 Nov 2004); Hungary (18-21 Nov 2007); Iceland (3); Ireland (12); Italy (10); Madeira (2); Malta (2); N Mauritania (11 May 1986); Morocco (11 Oct 2013); Norway (3); Portugal (4); Spain (3); Sweden (2).

Tringa erythropus
Spotted Redshank

OTHER NAME Dusky Redshank.
TAXONOMY (Pallas, 1764). Monotypic.
DISTRIBUTION [N Eurasia > W and S Europe, W Africa to SE Asia. Stable. LC] Summer visitor N Fennoscandia and NW Russia. Migrates S across much of region but often local or uncommon, occurring in numbers mainly at favoured staging areas in *eg* NW Africa, Spain and Hungary. Winters mainly on freshwater coastal wetlands from W Europe, Morocco and parts of Mediterranean basin E to Arabia and Iran. Vagrant Azores, Cape Verde Is,

Faroe Is, Iceland, Liechtenstein, Madeira and Svalbard.

Tringa totanus
Common Redshank

OTHER NAMES (European) Redshank (*T t totanus*), Icelandic Redshank (*T t robusta*).
TAXONOMY (Linnaeus, 1758). Six subspecies (two in region). See also Hale *et al* 2005 for postulated influence of extralimital *ussuriensis* (S Siberia) intergrading with European breeding population as far W as Italy.
DISTRIBUTION [Eurasia > W and S Europe, W Africa to Australia. Population trend unknown. LC] Breeds Iceland and Faroe Is (*robusta*) and from Ireland and Britain E across much of N continental Europe (except most of N European Russia), S to *eg* NW France, C Europe, Black Sea and W Iran, with scattered populations S Europe, Tunisia and Turkey (*totanus*). Mainly migratory in N and E, occurring widely to S on migration. Winters Atlantic coasts of W Europe S to N Mauritania, also Canary Is and Cape Verde Is, and E through parts of Mediterranean basin and Middle East. Vagrant Azores, Gibraltar, Liechtenstein and Madeira.

Tringa stagnatilis
Marsh Sandpiper

TAXONOMY (Bechstein, 1803). Monotypic.
DISTRIBUTION [E Europe to CE Asia > SW Europe, Africa, SW to SE Asia and Australia. Decreasing. LC] Summer visitor mainly N Ukraine and S and C European Russia E to Ural Mtns, also Romania (Black Sea coast), with some recent range expansion N to *eg* Poland (rare), Belarus, Finland and NW European Russia. Occurs widely to S and SW on migration, reaching W to *eg* Germany (has bred), Hungary (uncommon; formerly bred), Italy, Netherlands (very scarce), E and S Spain (scarce, mainly spring), Switzerland (rare) and Morocco (uncommon). Winters mainly S Middle East, with smaller numbers *eg* N Africa W to N Mauritania (rare Morocco), and very locally in S Spain. Vagrant Albania, Azores, Canary Is, Cape Verde Is, Belgium, Britain, Denmark, Ireland, Luxembourg, Moldova, Norway, Portugal and Sweden.

Tringa nebularia
Common Greenshank

OTHER NAMES (Eurasian or Greater) Greenshank.
TAXONOMY (Gunnerus, 1767). Monotypic.
DISTRIBUTION [N Eurasia > W and S Europe, Africa, SW to SE Asia and Australasia. Stable. LC] Summer visitor N Scotland and from C and N Scandinavia E to NW Russia, also S to Estonia, Belarus and N Ukraine. Occurs widely across region to S on migration, but less numerous Atlantic islands. Winters mainly coastally from Britain and Ireland S along Atlantic coasts of region, and locally in Mediterranean and S Middle East. Vagrant Faroe Is, Iceland and Svalbard.

Tringa melanoleuca
Greater Yellowlegs (V)

TAXONOMY (Gmelin, J F, 1789). Monotypic.
DISTRIBUTION [N North America > S North America to S South America. Stable. LC] Vagrant (mainly Aug-Oct but recorded throughout year, largely in W Europe): Azores (19*); Balearic Is (16 Apr-29 Jun 1995); Belgium (27 Nov-2 Dec 1994); Britain (31); Canary Is (18-19 Nov 1989); Cape Verde Is (2); Croatia (13 Jul 2010); Czech Republic (26 Aug 1964, shot); France (6); Iceland (2); Ireland (12); Morocco (16 Nov 2005); Netherlands (4); Norway (8-9 May 1993); Poland (2); Portugal (2); Spain (10*); Sweden (3*).

Tringa flavipes
Lesser Yellowlegs (V)

OTHER NAME Yellowshank.
TAXONOMY (Gmelin, J F, 1789). Monotypic.
DISTRIBUTION [N North America > S North America to S South America. Decreasing. LC] Vagrant (occurs mainly in autumn, as Greater Yellowlegs, but outnumbers it approx 8.5:1, with records extending much farther E across region): Austria (10 May 1997); Azores (131); Balearic Is; Belgium (4); Britain (347); Canary Is (15); Cape Verde Is (23); Denmark (7); Finland (5); France (56); Germany (14); Greece (9 Aug 1986); Hungary (3); Iceland (21*); Ireland (146); Israel (2); Italy (10); Morocco (6); Neth-

erlands (31); Norway (18); Oman (13-14 Oct 2003); Poland (26 Oct-9 Nov 2013); Portugal (18+); Slovenia (23 Apr 1993); Spain (60); Sweden (19); Turkey (27 Nov 2006); UAE (28 Jan 2007-23 Apr 2008).

Tringa ochropus
Green Sandpiper

TAXONOMY Linnaeus, 1758. Monotypic.
DISTRIBUTION [N Eurasia > W and S Europe, Africa, SW to E Asia. Stable. LC] Summer visitor N Europe, breeding mainly from Poland and Scandinavia E to Ural Mtns, also irregularly or in very small numbers W to Germany and S to *eg* Czech Republic, Slovenia and Bulgaria. Occurs widely across region on migration, wintering from Britain, Ireland and Netherlands S through parts of W, C and S Europe, N Africa and Middle East. Vagrant Azores, Cape Verde Is, Faroe Is, Gibraltar, Iceland, Jan Mayen, Madeira and Svalbard.

Tringa solitaria
Solitary Sandpiper (V)

TAXONOMY Wilson, A, 1813. Two subspecies (unconfirmed in region). Substantial differences in certain DNA sequences between subspecies, indicating long period of reproductive isolation and suggesting that they may be better treated as separate species.
DISTRIBUTION [N North America > S North America to C South America. Stable. LC] Rare vagrant (all records from W seaboard of region, especially late Sep-early Oct): Azores (19*); Britain (37, including one apparent *solitaria*); Cape Verde Is (2); Channel Is (2); France (6); Iceland (6); Ireland (5); Madeira (2 Sep 2010); Netherlands (14-17 May 2006); Portugal (26 June 1989); Spain (2); Svalbard (14-18 July 2010); Sweden (25 May 1987).

Tringa glareola
Wood Sandpiper

TAXONOMY Linnaeus, 1758. Monotypic.
DISTRIBUTION [N Eurasia > SW Europe, Africa, SW to E Asia. Stable. LC] Summer visitor N Scotland and Scandinavia E across N Europe to Ural Mtns, S to Belarus and N Ukraine. Widespread across region on migration, but scarcer *eg* Britain, Ireland and Iberia. Winters locally parts of Iberia, Canary Is (irregular), Cape Verde Is, coastal N Africa and S Arabia. Vagrant Azores, Faroe Is, Iceland (has bred), Madeira and Svalbard.

Tringa brevipes
Grey-tailed Tattler (V)

OTHER NAMES Grey-rumped (or Polynesian/Siberian) Tattler, Grey-tailed Sandpiper.
TAXONOMY (Vieillot, 1816). Monotypic.
DISTRIBUTION [NC Siberia and NE Asia > SE Asia to Australia. Stable. LC] Extreme vagrant (four records falling equally in July and autumn-early winter): Britain (2: 13 Oct-17 Nov 1981; 27 Nov-27 Dec 1994); Netherlands (28 July 2010); Sweden (12 July 2003). Another unidentified tattler, either this species or extralimital Wandering Tattler *T incana* (NE Asia and NW North America), reported Azores (7 Sep 2008*).

Tringa semipalmata
Willet (V)

OTHER NAMES Eastern Willet (*T s semipalmata*), Western Willet (*C s inornata*).
TAXONOMY (Gmelin, J F, 1789). Two subspecies (both in region). Differences in structure, plumage, moult and vocalisations, as well as allopatric breeding, suggest that the two subspecies are well separated from each other.
DISTRIBUTION [S Canada and N USA > W and S North America to N South America. Stable. LC] Very rare vagrant (recorded all months except May and Aug, but especially Sep): Azores (7*); Finland (21 Sep 1983); France (2: 1867, collected; 12-13 Sep 1998, *inornata*); Italy (2 Feb 2008); Norway (14 Oct 1992-23 Mar 1993, *semipalmata*); Portugal (29 Apr 2009, *inornata*).

Xenus cinereus
Terek Sandpiper

TAXONOMY (Güldenstädt, 1775). Monotypic.
DISTRIBUTION [N Eurasia > S and E Africa,

SW to SE Asia, Australasia. Stable. LC] Summer visitor from NW Russia E to Ural Mtns, with outlying populations Finland, Latvia, Poland, Belarus and Ukraine. Most migrate through E of region, W to *eg* Ukraine and E Egypt; rare Balkans, Hungary and Israel but small numbers regular Italy and Tunisia. Winters mainly Arabian coasts, S Iraq and S Iran. Vagrant Austria, Azores*, Bear Is, Belgium, Britain, Canary Is, Corsica, Croatia, Cyprus, Czech Republic, Denmark, Estonia, France, Germany, Ireland, Jordan, Libya, Lebanon, Lithuania, Malta, N Mauritania, Moldova, Morocco, Netherlands, Norway, Portugal, Romania, Sardinia, Slovakia, Slovenia, Spain, Sweden, Switzerland and Syria.

Actitis hypoleucos
Common Sandpiper

OTHER NAME Eurasian Sandpiper.
TAXONOMY (Linnaeus, 1758). Monotypic.
DISTRIBUTION [Widespread Eurasia > W and S Europe, Africa, SW to SE Asia, Australia. Decreasing. LC] Summer visitor across much of Europe (though local or absent parts of Iberia N through W Europe to S Britain and Denmark), and also E into S Caucasus, E Turkey and N and W Iran. Widespread elsewhere on passage, a few reaching NW to Faroe Is, where has bred. Small numbers winter from *eg* S Britain and W France to Iberia, Morocco, Canary Is, Cape Verde Is, parts of Mediterranean basin and N Middle East. Vagrant Azores, Iceland, Jan Mayen, Madeira and Svalbard.

Actitis macularius
Spotted Sandpiper

TAXONOMY (Linnaeus, 1766). Monotypic.
DISTRIBUTION [North America > S North America to C South America. Decreasing. LC] Scarce visitor Azores, mainly on passage but also occasional in winter (average 20* annually 2006-11). Vagrant Austria, Belgium, Britain, Bulgaria, Canary Is, Cape Verde Is, Channel Is, Denmark, Finland, France, Germany, Greece, Iceland, Ireland, Italy, Lithuania, Madeira, Morocco, Netherlands, Norway, Poland, Portugal, Spain, Sweden, Switzerland, Svalbard, Tunisia and Turkey.

Arenaria interpres
Ruddy Turnstone

OTHER NAME Turnstone.
TAXONOMY (Linnaeus, 1758). Two subspecies (both in region).
DISTRIBUTION [Holarctic > S North America, South America, W Europe, Africa, SW to E Asia, Australasia. Decreasing. LC] Summer visitor Scandinavia, Baltic, NW Russia and Svalbard (*interpres*). Primarily coastal on migration, especially Baltic, North Sea and Atlantic coasts NW to Iceland, and exclusively so in winter, chiefly from W Europe S to N Mauritania; also stages and/or winters parts of Mediterranean basin and Middle East. Passage and winter populations include birds breeding N Europe and NE Canada; recovery in NW Canada of bird ringed N Mauritania indicates occurrence of Nearctic *morinella* in region, but exact status requires clarification. Vagrant Albania, Luxembourg, Slovenia and Syria*.

Calidris tenuirostris
Great Knot

OTHER NAMES Eastern (or Greater) Knot.
TAXONOMY (Horsfield, 1821). Monotypic.
DISTRIBUTION [NE Asia > SW to SE Asia, Australia. Decreasing. EN] Winters locally in Gulf, from S Iran and Kuwait (rare) discontinuously S and E to UAE and Oman. Vagrant Bahrain, Britain, Germany, Ireland, Israel, Morocco, Netherlands, Norway, Poland, Qatar, Saudi Arabia, Spain, Sweden and Yemen.

Calidris canutus
Red Knot

OTHER NAMES (European or Lesser) Knot; Greenland Knot (*C c islandica*), Siberian Knot (*C c canutus*).
TAXONOMY (Linnaeus, 1758). Six subspecies (two in region).
DISTRIBUTION [Holarctic > W and S Europe, Africa, S Asia, Australia, S North America and South America. Decreasing. NT] Occurs primarily on passage and in winter on N and W coasts of region, *islandica* (breeding N Canada and Greenland) migrating via Iceland to winter chiefly Britain and Ireland (some S to Iberia),

and *canutus* (breeding NC Siberia, occasionally W to Svalbard) staging Wadden Sea and wintering Morocco and especially N Mauritania. Some *canutus* also migrate via Black Sea to Mediterranean basin, where regular *eg* Italy, Tunisia and Turkey, or directly overland (uncommon/rare C Europe). Regular but uncommon Azores, Canary Is and Madeira; rare Israel and Egypt. Vagrant Azerbaijan, Belarus, Cape Verde Is, Cyprus, Czech Republic, Gibraltar, Iran, Iraq, Jordan, Kuwait, Lebanon, Libya, Liechtenstein, Malta, Oman, Romania, Saudi Arabia, Slovakia, Slovenia, UAE and Yemen.

Calidris alba
Sanderling

TAXONOMY (Pallas, 1764). Two subspecies (one in region).
DISTRIBUTION [Holarctic > W and S Europe, Africa, SW to E Asia, Australasia, S North America and South America. Population trend unknown. LC] Summer visitor Svalbard (*alba*). Greenland and Siberian populations occur on passage on coasts widely to S, more rarely inland, wintering (mainly Greenland birds) chiefly from Britain, Ireland and NW continental Europe S to N Mauritania, also parts of Mediterranean basin and Middle East. Vagrant Armenia, Jordan, Lebanon, Liechtenstein and Luxembourg.

Calidris pusilla
Semipalmated Sandpiper

TAXONOMY (Linnaeus, 1766). Monotypic.
DISTRIBUTION [N North America > West Indies and Central America to N South America. Decreasing. NT] Scarce migrant Azores, mainly Sep-Oct (*eg* average 25* annually 2006-11). Vagrant Albania, Austria, Britain, Canary Is, Cape Verde Is, Croatia, Denmark, Faroe Is, France, Germany, Hungary, Iceland, Ireland, Israel, Madeira, Morocco, Netherlands, Norway, Poland, Portugal, Spain and Sweden.

Calidris mauri
Western Sandpiper (V)

TAXONOMY (Cabanis, 1857). Monotypic.
DISTRIBUTION [Extreme NE Asia, W and N Alaska > S North America to N South America. Decreasing. LC] Very rare vagrant (most records Sep-Oct): Azores (15*); Britain (8); France (5); Iceland (2); Ireland (4); Madeira (1 Nov 1979, collected); Norway (12-13 Jul 2008); Spain (3); Sweden (2).

Calidris ruficollis
Red-necked Stint (V)

OTHER NAMES Rufous-necked (or Eastern Little) Stint.
TAXONOMY (Pallas, 1776). Monotypic.
DISTRIBUTION [N and NE Asia > SE Asia to Australia. Population trend unknown. NT] Rare vagrant (almost all records relating to breeding-plumaged adults): Belgium (16 Aug 1988); Britain (7); Czech Republic (18 Sep 1988); Denmark (2); Finland (4); France (2); Germany (Jul 1979); Iran (2 records of three birds); Ireland (4); Israel (15-22 Apr 2003); Italy (2); Netherlands (6); Norway (4); Spain (9 Jul 2008); Sweden (11); Tunisia (16 Feb 1994); UAE (5).

Calidris minuta
Little Stint

OTHER NAME Lesser Stint.
TAXONOMY (Leisler, 1812). Monotypic.
DISTRIBUTION [N Eurasia > W and S Europe, Africa, SW and S Asia. Decreasing. LC] Summer visitor far N Norway and NW Russia; has bred N Finland and N Sweden. Occurs widely on passage throughout region, wintering S Europe from Biscay, Iberian and Mediterranean coasts S to N Africa (scarce/irregular Azores, Canary Is and Cape Verde Is) and Middle East (mainly in S). Vagrant Bear Is, Faroe Is, Gibraltar, Iceland, Liechtenstein, Madeira, Moldova and Svalbard.

Calidris temminckii
Temminck's Stint

TAXONOMY (Leisler, 1812). Monotypic.
DISTRIBUTION [N Eurasia > S Europe, N and W to E Africa, SW to SE Asia. Population trend unknown. LC] Summer visitor N Scotland (rare) and from WC and N Scandinavia E to far NW

Russia. Occurs widely on passage throughout region but least numerous in W maritime countries. Some winter very locally parts of Mediterranean basin, Arabia and S Iran. Vagrant Armenia, Azores, Canary Is, Cape Verde Is, Iceland, Ireland, Madeira and Svalbard.

Calidris subminuta
Long-toed Stint

TAXONOMY (Middendorff, 1853). Monotypic.
DISTRIBUTION [N and NE Asia > SW to SE Asia, Australia. Population trend unknown. LC] Uncommon passage migrant and winter visitor Oman (most Sep-Dec, fewer Jan-Apr), rare UAE. Vagrant Bahrain, Britain, Finland, France, Germany, Iran, Ireland, Israel, Netherlands, Saudi Arabia, Sweden and Yemen.

Calidris minutilla
Least Sandpiper (V)

OTHER NAMES American (or Least) Stint.
TAXONOMY (Vieillot, 1819). Monotypic.
DISTRIBUTION [N North America > S North America to N South America. Decreasing. LC] Vagrant (most records Jul-Oct, but has overwintered and some, *eg* all from Cape Verde Is, in spring): Azores (82*+); Belgium (3-14 Aug 1996); Britain (36); Canary Is (6-8 Aug 2009); Cape Verde Is (6); Finland (5 Jun 1847); France (14); Germany (20-21 Sep 1985); Iceland (4); Ireland (13); Italy (12 February 1994, dead); Poland (9 Aug 2005); Spain (3).

Calidris fuscicollis
White-rumped Sandpiper

OTHER NAME Bonaparte's Sandpiper.
TAXONOMY (Vieillot, 1819). Monotypic.
DISTRIBUTION [N North America > SE South America. Decreasing. LC] Regular autumn migrant Azores, mainly late Aug-Oct (*c* 25 to 100 or more annually, exceptionally several hundred), and Britain (annual mean 2000-09 of 18, all autumn). Vagrant Austria, Balearic Is, Belgium, Canary Is, Cape Verde Is, Czech Republic, Denmark, Finland, France, Germany, Greece, Hungary, Iceland, Ireland, Israel, Italy, Jan Mayen, Madeira, Morocco, Netherlands, Norway (including one instance of pair displaying in Arctic), Poland, Portugal, Spain, Svalbard, Sweden, Switzerland, Tunisia, Turkey and UAE.

Calidris bairdii
Baird's Sandpiper (V)

TAXONOMY (Coues, 1861). Monotypic.
DISTRIBUTION [Extreme NE Asia and N North America > W and S South America. Stable. LC] Vagrant (mainly Sep-Oct): Azores (27*); Austria (7); Balearic Is (8-12 Oct 2011); Belgium (2); Britain (255); Canary Is (5); Cape Verde Is (1-3 Nov 2012); Cyprus (2-3 Dec 2009); Czech Republic (2); Denmark (3); Finland (6); France (53); Germany (9); Greece (2); Hungary (3); Iceland (6*), Ireland (148); Israel (17 Oct 1998); Italy (2); Netherlands (9); Norway (5); Oman (6 Nov 1987); Poland (4); Portugal (14 Dec 2004); Spain (10); Svalbard (20-22 May 2005); Sweden (7); Turkey (23 May 2011, two).

Calidris melanotos
Pectoral Sandpiper

OTHER NAME American Pectoral Sandpiper.
TAXONOMY (Vieillot, 1819). Monotypic.
DISTRIBUTION [NE Asia, N North America > N and C South America, Pacific islands, SE Australasia. Stable. LC] Breeding range in Russian Arctic may extend W (at least irregularly) into region, and probably breeds annually Svalbard; has also bred N Scotland. Primarily scarce migrant parts of W Europe in autumn (mainly Jul-Oct, much rarer spring), mostly Azores (annual mean 2006-11 of 24*), Britain (annual mean 2000-09 of 121), France (409 to 2005) and Ireland (566 to 2001). Has occurred widely elsewhere across region as vagrant; most numerous in N and W (*eg* 372 Sweden, 287 records of 359 birds Spain, 223 Germany to 2009, 246 Norway to 2009), but records extend across Europe and multiple occurrences also documented S to Malta, N Africa and Middle East.

Calidris acuminata
Sharp-tailed Sandpiper (V)

OTHER NAME Siberian Pectoral Sandpiper.

TAXONOMY (Horsfield, 1821). Monotypic.
DISTRIBUTION [NE Asia > Australasia and Polynesia. Stable. LC] Rare vagrant (mainly Aug-Sep): Austria (6-8 Aug 1983); Azores (23 Aug 2000); Belgium (3-8 Sep 1989); Britain (32); Bulgaria; Denmark (2); Estonia (2); Finland (2); Germany (2); France (7); Hungary (2); Ireland (7); Jan Mayen (27 Jun-1 Jul 2012); Madeira (30 Aug-4 Sep 2010); Netherlands (7); Norway (7); Oman (11 Feb 2000); Portugal (21 March 2007); Spain (3); Svalbard; Sweden (8); Yemen (*c* 12 Dec 1982).

Calidris ferruginea
Curlew Sandpiper

TAXONOMY (Pontoppidan, 1763). Monotypic.
DISTRIBUTION [N Siberia > Africa, SW to SE Asia, Australia. Increasing. NT] Passage migrant across much of region, migrating to and from Arctic breeding grounds mainly on coasts in W Europe but overland in E; has bred extreme NE European Russia. Winters mainly N Mauritania, Morocco and Tunisia, also SW Iberia and coastal Arabia. Rare but regular on passage and in winter Azores; irregular Cape Verde Is. Vagrant Albania, Bear Is, Faroe Is, Gibraltar, Iceland, Madeira, Moldova and Svalbard.

Calidris maritima
Purple Sandpiper

TAXONOMY (Brünnich, 1764). Monotypic.
DISTRIBUTION [Holarctic > E North America, N and W Europe. Decreasing. LC] Breeds Iceland, Faroe Is, N Scotland (rare), C and N Scandinavia, Kola Peninsula, Jan Mayen, Bear Is, Svalbard, Franz Josef Land and Novaya Zemlya. Strictly coastal in winter, some remaining near breeding grounds but others (including entire high-arctic population and some from NE Canada) moving S to Scandinavian, N Baltic, North Sea and Atlantic coasts as far S as W (more rarely S) Iberia. Irregular/rare Mediterranean coasts of W Europe. Vagrant Austria, Azores, Belarus, Bulgaria, Canary Is, Croatia, Czech Republic, Greece, Gibraltar, Hungary, Latvia, Lithuania, Madeira, Malta, N Mauritania, Morocco, Poland, Sardinia, Slovakia, Switzerland and Ukraine.

Calidris alpina
Dunlin

OTHER NAMES Red-backed Sandpiper; Greenland (or Schiøler's) Dunlin (*C a arctica*), Hudsonian Dunlin (*C a hudsonia*), Northern Dunlin (*C a alpina*), Southern Dunlin (*C a schinzii*).
TAXONOMY (Linnaeus, 1758). Ten subspecies (six in region). Taxonomy unsettled, some authorities recognising as few as five subspecies.
DISTRIBUTION [N North America and N Eurasia > W and S North America, W Europe and W Africa to E Asia. Decreasing. LC] Summer visitor Iceland, Faroe Is, Ireland, N Britain, most of Scandinavia and Baltic (*schinzii*) and N Norway and far NW Russia (*alpina*). Widespread across region on migration, when numbers augmented on W coasts by *arctica* (breeding NE Greenland) and along E overland routes by *alpina* (breeding NW Siberia); regular on passage Atlantic islands (especially Canary Is). Winters North Sea and Atlantic coasts of NW Europe (mainly *alpina*) discontinuously S through Iberia (*alpina* and *schinzii*) to Morocco and N Mauritania (mainly *schinzii*, plus *arctica*), also parts of Mediterranean basin and Middle East (*alpina*). Multiple ringing recoveries of birds from breeding range of *centralis* (NC to NE Siberia), and single Norwegian specimen of *sakhalina* (E Russia); one of these Siberian forms, probably former, postulated to occur Israel, and latter possibly Spain. Vagrant Gibraltar and Moldova; *hudsonia* (breeding C Canada) vagrant Azores*.

Calidris himantopus
Stilt Sandpiper (V)

TAXONOMY (Bonaparte, 1826). Monotypic.
DISTRIBUTION [N North America > S North America to C South America. Increasing. LC] Rare vagrant (mainly Aug-Sep, occasional in spring): Azores (7 Oct 2007); Balearic Is (6-7 May 1983); Britain (31), Denmark (23-24 Apr 1998); Finland (2); France (9); Germany (2); Hungary (13-15 Jul 2006); Iceland (17 Jul 1985, collected); Ireland (14); Morocco (28 Mar-3 Apr 1996); Netherlands (3); Norway (4); Sweden (3); Tunisia (29 Dec 1999).

Eurynorhynchus pygmeus Spoon-billed Sandpiper (V)

OTHER NAME Spoonbill Sandpiper.
TAXONOMY (Linnaeus, 1758). Monotypic.
DISTRIBUTION [NE Siberia > SE Asia. Decreasing. CR] Extreme vagrant: Ukraine (20 Aug 1952, shot).

Limicola falcinellus Broad-billed Sandpiper

TAXONOMY (Pontoppidan, 1763). Two subspecies (one in region).
DISTRIBUTION [N Europe and NE Asia > SW to SE Asia and Australia. Decreasing. LC] Summer visitor C and N Scandinavia, N Finland and far NW Russia (*falcinellus*). Migrates S and SE to winter quarters across C and E Europe to Middle East, wintering mainly coastal Arabia. Rare on passage and/or in winter Armenia, Azerbaijan, Egypt, Netherlands and Tunisia. Vagrant Algeria, Bear Is, Belgium, Britain, Corsica, Croatia, Czech Republic, France, Iceland, Ireland, Libya, Malta, Morocco, Portugal, Slovenia, Spain, Svalbard, Switzerland and Syria.

Tryngites subruficollis Buff-breasted Sandpiper

TAXONOMY (Vieillot, 1819). Monotypic.
DISTRIBUTION [Extreme NE Asia and N North America > C and SE South America. Decreasing. NT] Scarce autumn migrant (far rarer spring), occasionally in small groups, to Britain (annual mean 2000-09 of 32, mainly Sep) and Ireland (annual mean 2006-11 of 36, mainly Aug-Oct). Vagrant Austria, Azores, Balearic Is, Belgium, Bulgaria, Canary Is, Channel Is, Czech Republic, Denmark, Estonia, Faroe Is, Finland, France, Germany, Greece, Hungary, Iceland, Italy, Madeira, Malta, Morocco, Netherlands, Norway, Oman, Poland, Portugal, Romania, Saudi Arabia, Spain, Svalbard, Sweden, Switzerland, Tunisia and Ukraine.

Philomachus pugnax Ruff

OTHER NAME Reeve (female).
TAXONOMY (Linnaeus, 1758). Monotypic.
DISTRIBUTION [N Eurasia > W and S Europe, Africa, SW to SE Asia. Decreasing. LC] Breeds from NW France (rare), Britain (rare) and Netherlands discontinuously N to N Scandinavia, and E across large areas of N Europe to Ural Mtns, Ukraine and NW Kazakhstan. Widespread on passage across much of region, wintering mainly from S Britain S to Iberia, Morocco, parts of Mediterranean basin and Middle East. Scarce/very scarce on migration Azores, Faroe Is and Madeira. Vagrant Bear Is, Iceland, Jan Mayen, Liechtenstein and Svalbard.

Phalaropus tricolor Wilson's Phalarope (V)

TAXONOMY (Vieillot, 1819). Monotypic.
DISTRIBUTION [N North America > W, C and S South America. Increasing. LC] Vagrant (mainly mid-Aug to late Sep): Azores (11*); Balearic Is; Britain (242); Belgium (13); Bulgaria; Canary Is (5); Denmark (9); Estonia (2); Finland (5); France (76); Germany (16); Greece (3); Hungary (2); Iceland (6*); Ireland (89); Italy (5); Madeira (2); Morocco (5); Netherlands (24); Norway (6); Oman (2); Portugal (4+); Romania (24-26 May 2012); Spain (20); Sweden (10); Turkey (2); UAE (3-11 Jan 2010).

Phalaropus lobatus Red-necked Phalarope

OTHER NAME Northern Phalarope.
TAXONOMY (Linnaeus, 1758). Monotypic.
DISTRIBUTION [Holarctic > seas off NW South America, S Arabia and SE Asia. Decreasing. LC] Summer visitor Iceland, Faroe Is, N Scotland (rare), C and N Scandinavia, Svalbard, Finland and far NW Russia; has bred Ireland. Migrates mainly across N and E of region, varying in abundance/rarity locally; scarce/rare much of W and C Europe, also Tunisia, generally becoming increasingly frequent farther E. Winters in huge flocks at sea off S Arabia. Rare Armenia. Vagrant Algeria, Azores, Canary Is, Cape Verde Is, Croatia, Czech Republic, Gibraltar, Jan Mayen, Lebanon, Luxembourg, Madeira, Malta, N Mauritania, Moldova, Morocco, Portugal, Slovakia, Slovenia, Spain and Switzerland.

Phalaropus fulicarius
Red Phalarope

OTHER NAME Grey Phalarope.
TAXONOMY (Linnaeus, 1758). Monotypic.
DISTRIBUTION [Holarctic > seas off W South America and W and SW Africa. Decreasing. LC] Summer visitor Iceland (rare), Bear Is, Svalbard and Novaya Zemlya. Highly pelagic on migration, occurring inshore or inland usually only in adverse weather. Some Nearctic breeders also involved in passage movements, which range E to arctic Scandinavia and NW Russia, and S through E Atlantic sector (far fewer in North Sea) off W Europe and NW Africa, including Canary Is. Scarce Azores, where small flocks occasionally recorded on passage. Common winter visitor to offshore waters of Saharan Atlantic Morocco and N Mauritania, but rarely recorded Cape Verde seas. Irregular migrant and winter visitor Italy. Vagrant Austria, Azerbaijan, Belarus*, Bosnia-Herzegovina, Bulgaria, Croatia, Cyprus, Czech Republic, Egypt, Estonia, Faroe Is, Finland, Germany, Gibraltar, Greece, Hungary, Iran, Iraq, Israel, Jan Mayen, Jordan, Kuwait, Latvia, Lithuania, Luxembourg, Madeira, Malta, Poland, Oman, Romania, Saudi Arabia, Sardinia, Serbia, Slovakia, Slovenia, Sweden, Switzerland, Tunisia, Turkey, UAE and Ukraine.

FAMILY GLAREOLIDAE

Coursers and pratincoles · six species

Cursorius cursor
Cream-coloured Courser

OTHER NAME Desert Courser.
TAXONOMY (Latham, 1787). Three subspecies (all in region).
DISTRIBUTION [Sahel, W and N Africa, SW Asia. Stable. LC] Breeds Cape Verde Is (*exsul*), Canary Is, NW and N Africa E to parts of Arabia, and N to Iraq (*cursor*), SE Turkey and Iran (*bogolubovi*). Mainly resident, but partially migratory in NW Africa and N Middle East, occurring to S on passage and in winter. Regular migrant Cyprus; irregular/rare Italy and Malta. Vagrant Austria, Azerbaijan, Belgium, Britain, Channel Is, Croatia, Czech Republic, Denmark, Finland, France, Germany, Greece, Ireland, Madeira, Netherlands, Norway, Portugal, Sardinia, Serbia, Slovakia, Slovenia, Spain (has bred), Sweden, Switzerland and Ukraine.

Cursorius temminckii
Temminck's Courser

TAXONOMY Swainson, 1822. Three subspecies (one in region).
DISTRIBUTION [Sub-Saharan Africa. Stable. LC] Winter visitor (probably rare) N Mauritania (*temminckii*).

Glareola pratincola
Collared Pratincole

OTHER NAMES (Common or Red-winged) Pratincole.
TAXONOMY (Linnaeus, 1766). Two subspecies (one in region).
DISTRIBUTION [N, C and S Africa, S Europe to SW and C Asia > W to NE, E and SE Africa. Decreasing. LC] Summer visitor NW Africa and Iberia discontinuously E through Mediterranean basin (N to Hungary and irregularly Switzerland), Balkans, Black Sea, Caucasus and parts of Middle East (*pratincola*). Occurs widely to S of breeding areas on migration, W to Canary Is and Cape Verde Is; a few winter S Arabia. Vagrant Belgium, Britain, Cape Verde Is, Czech Republic, Denmark, Estonia, Finland, Germany, Gibraltar, Iceland, Ireland, Latvia, Madeira, Malta, Netherlands, Norway, Poland, Slovakia, Slovenia and Sweden.

Glareola maldivarum
Oriental Pratincole (V)

OTHER NAMES Eastern (Collared) Pratincole, Large Indian Pratincole.
TAXONOMY Forster, J R, 1795. Monotypic.
DISTRIBUTION [S to E Asia > S and SE Asia to Australia. Decreasing. LC] Very rare vagrant: Britain (7); Cyprus (23 Apr 1993); Denmark (26 May 2010); Iran (18 Jan 2004); Israel (2-3 Oct 1999); Kuwait (25 Mar 2008, two); Netherlands (1-5 Aug 1997); Norway (22-23 Aug 2009); Sweden (2 Jul-6 Oct 2001); UAE (4 records of 5 birds).

Glareola nordmanni
Black-winged Pratincole

TAXONOMY Fischer von Waldheim, 1842. Monotypic.
DISTRIBUTION [SE Europe to NC Asia > Sahel, W and S Africa. Decreasing. NT] Summer visitor Ukraine, SE European Russia and NW Kazakhstan, very locally also Armenia; status unclear Azerbaijan, where formerly bred, and E Turkey, where probably has done so. Uncommon or rare on passage Balkans (mainly E, formerly breeding Bulgaria and Romania), Caucasus and E Turkey to Levant, Israel and Egypt, also S Gulf from Iraq E to E Saudi Arabia. Vagrant Austria, Balearic Is, Belgium, Bahrain, Britain, Croatia, Cyprus, Denmark, Estonia, Finland, France, Germany, Hungary (has bred), Iceland, Iran, Ireland, Italy, Jordan, Latvia, Libya, Lithuania, Macedonia, Morocco, Netherlands, Norway, Oman, Poland, Romania, Slovakia, Spain, Sweden, Switzerland, Tunisia, UAE and Yemen; record from Czech Republic under review.

Glareola lactea
Small Pratincole

OTHER NAMES Little (or Milky/Small Indian) Pratincole.
TAXONOMY Temminck, 1820. Monotypic.
DISTRIBUTION [S and SE Asia > SW to SE Asia. Population trend unknown. LC] Uncommon winter visitor Oman (early Nov-late Feb). Vagrant Bahrain, Iran, Qatar, UAE and Yemen.

FAMILY LARIDAE

Gulls, terns and skimmers · 60 species

Anous stolidus
Brown Noddy

OTHER NAMES Common Noddy, Noddy Tern.
TAXONOMY (Linnaeus, 1758). Four subspecies (at least one in region).
DISTRIBUTION [Tropical and subtropical Atlantic, Indian and Pacific Oceans. Stable. LC] Breeds Oman, Saudi Arabia and Yemen (*pileatus*); winters off S Arabia. Formerly vagrant to UAE but now considered uncommon visitor; recorded annually since 2006, with up to 17 birds at one site 31 Jul-24 Oct 2008. Vagrant Germany and Iran. Two other records of unidentified noddies *Anous sp* perhaps involve this species (Norway, Aug 1974) or extralimital Black Noddy *A minutus* (N Mauritania, 7 Dec 1984).

Anous tenuirostris
Lesser Noddy

OTHER NAME Sooty Noddy.
TAXONOMY (Temminck, 1823). Two subspecies (one in region). Forms superspecies with extralimital Black Noddy *A minutus* (tropical oceans); further study may show the two to be conspecific.
DISTRIBUTION [Tropical Indian Ocean. Stable. LC] Rare non-breeding summer visitor Oman (*tenuirostris*). Vagrant UAE and Yemen.

Rynchops flavirostris
African Skimmer

OTHER NAME Scissorbill.
TAXONOMY Vieillot, 1816. Monotypic.
DISTRIBUTION [Sub-Saharan Africa. Decreasing. NT] Scarce non-breeding summer visitor to Upper Nile Valley, Egypt, exceptionally also Red Sea coast. Vagrant Israel, Morocco and Yemen.

Rynchops albicollis
Indian Skimmer (V)

OTHER NAME Scissorbill.
TAXONOMY Swainson, 1838. Monotypic.
DISTRIBUTION [S Asia. Decreasing. VU] Extreme vagrant: Oman (13-15 Aug 1979, three). Claimed early 20th-century status as rare vagrant in Iran now discounted because of lack of traceable evidence.

Rissa tridactyla
Black-legged Kittiwake

OTHER NAME Kittiwake.
TAXONOMY (Linnaeus, 1758). Two subspecies (one in region).

DISTRIBUTION [N North America and N Eurasia > N Atlantic and N Pacific. Decreasing. LC] Breeds discontinuously NW Spain, Brittany, Germany (Heligoland), Britain, Ireland, Faroe Is, Iceland, W and N Scandinavia, Jan Mayen, Svalbard, Franz Josef Land and NW Russia (*tridactyla*); has bred Netherlands, formerly also Channel Is (Sark) and N Portugal. Winters North Sea and in Atlantic S of breeding range to *c* 35°N, irregularly to Atlantic islands, small numbers reaching farther S to N Mauritania and also into W Mediterranean; scarcer Baltic, E Mediterranean and Turkey, rare Poland and S Israel. Largely maritime/pelagic outside breeding season, but occasional flocks recorded *eg* C Europe. Vagrant Azerbaijan, Belarus, Bosnia-Herzegovina, Corsica, Cyprus, Czech Republic, Hungary, Iran, Jordan, Kuwait, Latvia, Lebanon, Libya, Lithuania, Luxembourg, Macedonia, Montenegro, Oman, Romania, Serbia, Syria, UAE and Ukraine.

Pagophila eburnea
Ivory Gull

TAXONOMY (Phipps, 1774). Monotypic.
DISTRIBUTION [Circumpolar from N Canada to NC Siberia > Arctic Ocean. Decreasing. NT] Breeds Svalbard, Franz Josef Land and Novaya Zemlya; rare but annual Iceland (278 records to 2013), where most frequent in winters when Arctic pack ice extends S. Vagrant Britain, Channel Is, Denmark, Estonia, Faroe Is, Finland, France, Germany, Ireland, Italy, Jan Mayen, Netherlands, Norway, Poland, Sweden and Switzerland.

Xema sabini
Sabine's Gull

TAXONOMY (Sabine, 1819). Monotypic.
DISTRIBUTION [Circumpolar in Arctic > S America and S Africa. Stable. LC] Breeds Svalbard (uncommon) and regular Jan Mayen (has bred). Mainly pelagic on autumn migration but regular (if generally scarce) NE Atlantic sector, especially W coasts of Britain, Ireland, France (notably Bay of Biscay) and Spain during and after NW/W gales; less numerous W Scandinavia, Baltic, North Sea and Portugal. Passage off N Mauritania apparently regular but some distance from coast. In spring, migrates to breeding grounds direct through mid-Atlantic and rarely encountered on regional coasts. Vagrant Algeria, Austria, Azores, Bulgaria, Canary Is, Cape Verde Is, Channel Is, Corsica, Czech Republic, Estonia, Faroe Is, Finland, Gibraltar, Hungary, Iceland, Israel, Italy, Jordan, Latvia, Lithuania, Luxembourg, Madeira, N Mauritania, Montenegro, Morocco, Oman, Poland, Romania, Slovakia, Mediterranean Spain, Switzerland, Tunisia and UAE.

Chroicocephalus genei
Slender-billed Gull

TAXONOMY Brème, 1839. Monotypic.
DISTRIBUTION [S Europe, N Africa and W Asia. Increasing. LC] Breeds N Mauritania, Morocco (scarce), Tunisia and NE Egypt, in Mediterranean Europe discontinuously from S and E Spain, S France, Sardinia and Italy to Greece, N Black Sea and Caspian Sea, and in parts of N Middle East from Turkey E to Gulf and SE Iran. Non-breeding distribution primarily coastal and more widespread within same broad areas, but many Spanish and W Mediterranean birds winter NW Africa and Black and Caspian Sea populations move S to Mediterranean and Gulf. Scarce visitor S Portugal. Vagrant Austria, Belarus, Belgium, Britain, Canary Is, Cape Verde Is, Croatia, Denmark, Finland, Germany, Gibraltar, Hungary, Lithuania, Netherlands, Montenegro, Poland, Slovenia, Sweden and Switzerland.

Chroicocephalus philadelphia
Bonaparte's Gull (V)

TAXONOMY Ord, 1815. Monotypic.
DISTRIBUTION [N North America > S North America and Caribbean. Increasing. LC] Annual vagrant in small numbers, mainly in autumn but also winter and spring, with occasional summer records: Azores (31 to 2012*); Belgium (4); Britain (211); Canary Is (27 Nov 2000); France (26); Czech Republic (24 Apr 1988); Denmark (5); Germany (4); Iceland (31*); Ireland (71); Madeira (2); Morocco (3); Netherlands (5); Norway (14); Portugal (6, including one wintering annually 1990-1999); Spain (32); Svalbard (9-11 Jun 2008); Sweden (4).

Chroicocephalus brunnicephalus
Brown-headed Gull (V)

OTHER NAME Indian Black-headed Gull.
TAXONOMY (Jerdon, 1840). Monotypic.
DISTRIBUTION [C Asia to C China > S and SE Asia. Stable. LC] Very rare vagrant (most records Dec-Mar): Iran (4); Israel (12 May 1985); Oman (6 Jan 1988, at sea); UAE (7).

Chroicocephalus ridibundus
Black-headed Gull

OTHER NAMES Common (or Northern) Black-headed Gull.
TAXONOMY (Linnaeus, 1766). Monotypic.
DISTRIBUTION [Eurasia, N Africa and NE North America. Decreasing. LC] Breeds across much of Europe from Iceland and Ireland in W to Ural Mtns in E, N to N Scandinavia and S to parts of Iberia, Italy, N Balkans, Black Sea and Caspian Sea, also Caucasus, locally Asia Minor, N Iraq and N Morocco. On migration and in winter even more widespread, moving W and S into Atlantic (uncommon Cape Verde Is and N Mauritania), Mediterranean, N Africa and Middle East, where perhaps least common Red Sea. Vagrant Svalbard (has bred).

Chroicocephalus cirrocephalus
Grey-headed Gull

OTHER NAME Grey-hooded Gull.
TAXONOMY (Vieillot, 1818). Two subspecies (one in region).
DISTRIBUTION [S America and sub-Saharan Africa. Stable. LC] Breeds Mauritania on islands in Banc d'Arguin, where now uncommon (*poiocephalus*); small numbers also present in winter. Vagrant Algeria, Canary Is, Egypt, Gibraltar, Israel, Italy, Jordan, Morocco, Saudi Arabia, Spain, Tunisia and Yemen.

Hydrocoloeus minutus
Little Gull

TAXONOMY (Pallas, 1776). Monotypic.
DISTRIBUTION [E North America and N Eurasia > E North America, Europe, N Africa, SW and E Asia. Increasing. LC] Breeds across Fennoscandia (more local in W), Baltic States, Belarus and European Russia, also Netherlands, Poland and N Ukraine, formerly also *eg* Denmark, Germany and C Europe; has bred Britain. Widespread W Europe on passage, less so C Europe, moving overland and especially along Baltic and North Sea coasts W and SW to main wintering grounds in Atlantic (N to Brittany and S to Morocco, where uncommon) and W Mediterranean, also E Mediterranean, Black and Caspian Seas. Rare N Mauritania. Vagrant Azores, Bahrain, Canary Is, Faroe Is, Iceland (has bred), Iraq, Jordan, Kuwait, Liechtenstein, Madeira, Saudi Arabia, Svalbard and UAE.

Rhodostethia rosea
Ross's Gull

OTHER NAME Rosy Gull.
TAXONOMY (MacGillivray, 1824). Monotypic.
DISTRIBUTION [Arctic Siberia, Greenland and Canada > Arctic Ocean and Beaufort, N Labrador, Okhotsk and Bering Seas. Population trend unknown. LC] Regular visitor Svalbard (has bred) and Franz Josef Land. Vagrant Britain, Denmark, Estonia, Faroe Is, Finland, France, Germany, Iceland, Ireland, Italy, Latvia, Lithuania, Macedonia, Netherlands, Norway, Poland, Sardinia, Spain and Sweden.

Leucophaeus atricilla
Laughing Gull (V)

TAXONOMY (Linnaeus, 1758). Two subspecies (one in region).
DISTRIBUTION [E and S North America and Caribbean > S North America to N South America. Increasing. LC] Vagrant (*megalopterus*), chiefly to W Europe in autumn and winter, with major influx (*eg* over 40 per cent of all Iberian records) attributable to aftermath of Hurricane Wilma in Nov 2005: Azores (50); Belgium (28-29 Oct 1994); Bulgaria (3-7 Jun 2004); Britain (196); Canary Is (7); Denmark (5); Finland (1-8 May and 11 May-22 Jul 1997); France (41); Gibraltar (3); Germany (11); Greece (15 Aug 1984); Iceland (19*); Ireland (42); Italy (5); Lithuania (24 Jul 2011); Madeira (3); N Mauritania (15 Feb 1980); Morocco (6); Netherlands (3); Norway (4);

Poland (5-20 Jul and 22 Aug 2007); Portugal (5); Spain (41+); Sweden (6); Switzerland (2).

Leucophaeus pipixcan
Franklin's Gull (V)

TAXONOMY (Wagler, 1831). Monotypic.
DISTRIBUTION [C North America > W South America. Increasing. LC] Vagrant, some probably arriving direct from winter quarters, hence broader seasonal and geographical distribution of records than Laughing Gull: Azores (6); Belgium (3); Britain (67); Canary Is (2); Croatia (2-3 May 2006); Cyprus (30 Jul-5 Aug 2006); Denmark (27 Nov 2005); Egypt (12 Mar 2011); Faroe Is (2); Finland (2); France (33); Germany (6); Hungary (2); Iceland (4); Ireland (11); Israel (2); Italy (3); Jordan (1 Jul 2006*, same as second Israel record); Kuwait (5 Jun 2012); Libya (1 Feb 2011); Madeira (1); Morocco (7); Netherlands (8); Norway (9); Portugal (4); Spain (16); Svalbard (26 Jun 2011); Sweden (7); Switzerland (14 Jan-13 Feb 2006); UAE (17-21 May 2011).

Ichthyaetus relictus
Relict Gull (V)

OTHER NAMES Central Asian (or Lonnberg's) Gull.
TAXONOMY (Lönnberg, 1931). Monotypic.
DISTRIBUTION [C Asia to NC China > E Asia. Decreasing. VU] Extreme vagrant (two records): Bulgaria (25 Mar 1978, ring recovery of bird ringed Lake Alakol', E Kazakhstan, 21 Jun 1975; ring not submitted and no specimen); Russia/NW Kazakhstan (7 May 2000, flock of 14, Malyy Uzen River along border between countries). Further ring recovery from Turkey (30 Mar 1990, involving leg of bird also ringed Lake Alakol', E Kazakhstan, on 22 Jun 1979) not admitted to national list, as neither ring nor remains retained for full verification.

Ichthyaetus audouinii
Audouin's Gull

TAXONOMY (Payraudeau, 1826). Monotypic.
DISTRIBUTION [Mediterranean > W Africa. Stable. LC] Breeding endemic, distributed from N Morocco, S Portugal (rare) and Spain discontinuously E through Mediterranean and Aegean to Cyprus (formerly also Lebanon); two-thirds of world population breeds in two main areas, in Spain (Ebro Delta and environs) and off N Morocco (Spanish-controlled Islas Charafinas), Spain holding almost 90 per cent of world population. More widespread Mediterranean outside breeding season, when many also move W and SW to winter Atlantic on NW African coast and beyond. Vagrant Azores, Belgium, Britain, Bulgaria, Canary Is, Cape Verde Is, Czech Republic, Denmark, Egypt, Finland, Atlantic coast of France, Georgia, Germany, Hungary, Israel, Jordan, Madeira, Montenegro, Netherlands, Slovenia and Switzerland.

Ichthyaetus melanocephalus
Mediterranean Gull

OTHER NAMES Mediterranean Black-headed Gull.
TAXONOMY (Temminck, 1820). Monotypic.
DISTRIBUTION [Europe > Europe, N Africa and SW Asia. Stable. LC] Breeding endemic. Main range historically centred on Black Sea in Ukraine, also SW Russia and W and C Turkey, but has expanded across Europe W to Spain, NW to S Britain, Belgium and Netherlands, and N to S Baltic and Belarus. Disperses after breeding, when widespread Atlantic and North Sea coasts N to S Sweden and S to NW Africa, Mediterranean basin and S Black Sea. Vagrant Azores, Canary Is, Estonia, Finland, Iceland*, Jordan, Kuwait, Latvia, Madeira, Norway, Saudi Arabia and UAE.

Ichthyaetus ichthyaetus
Pallas's Gull

OTHER NAME Great Black-headed Gull.
TAXONOMY (Pallas, 1773). Monotypic.
DISTRIBUTION [SE Europe to C Asia > Middle East, E Africa and S Asia. Decreasing. LC] Breeds SE Ukraine, S Russia, NW Kazakhstan and NW Iran. Some winter Black and Caspian Seas, but many move S to E Mediterranean (scarce Cyprus), Armenia, Azerbaijan (for-

merly bred), N Middle East (rare Jordan), N Egypt and Red Sea, Arabian coasts and S Iran. Recorded regularly W to parts of Balkans and C Europe, especially Hungary, and now regular Sicily and Libya in winter. Vagrant Albania, Austria, Belarus, Belgium, Britain, Corsica, Czech Republic, Denmark, France, Germany, Greece, Latvia, Macedonia, Madeira, Malta, Moldova, Montenegro, Netherlands, Poland, Sardinia, Slovenia, Sweden and Tunisia.

Ichthyaetus leucophthalmus White-eyed Gull

TAXONOMY (Temminck, 1825). Monotypic.
DISTRIBUTION [NE Africa and W Middle East. Stable. NT] Breeds Red Sea in E Egypt, W Saudi Arabia and W Yemen. Mainly resident, but frequent non-breeding visitor Israel and Jordan in Gulf of Aqaba, also three records from Mediterranean coast of N Israel; rare passage migrant and winter visitor Oman. Vagrant Greece, Iran, Turkey and UAE.

Ichthyaetus hemprichii Sooty Gull

OTHER NAMES Aden (or Hemprich's) Gull.
TAXONOMY (Bruch, 1853). Monotypic.
DISTRIBUTION [E Africa and Arabia.] Breeds coastal Saudi Arabia, Yemen, Oman and UAE. Present year-round Arabian Sea, Gulf of Aden and in Red Sea, where most numerous in S but reaches NE Egypt; also disperses into Indian Ocean. In Persian Gulf occurs N to Qatar. Vagrant Bahrain, Israel, Jordan and Lebanon.

Larus canus Common Gull

OTHER NAMES Mew Gull (IOC); Kamchatka Gull (*L c camtschatchensis*), Russian Common Gull (*L c heinei*), Short-billed (or Mew) Gull (*L c brachyrhynchus*).
TAXONOMY Linnaeus, 1758. Four subspecies (three in region). Differences in plumage and mtDNA between Eurasian and Nearctic taxa have led some authorities to recognise North American *brachyrhynchus* as a full species, Mew (or Short-billed) Gull. Taxonomic status of E Asian *camtschatchensis* also merits investigation.
DISTRIBUTION [NW North America and N Eurasia > W North America, Europe, SW and E Asia. Population trend unknown. LC] Breeds from Iceland, Ireland, Britain and France E across N Europe to NW Russia (*canus*) and NE Russia (*heinei*). More widespread in winter, including C and E Mediterranean, Black Sea, Caspian Sea, N Middle East and (rarely) Gulf, when numbers in E of region augmented by Siberia-breeding *heinei*; rare S Iberia, N Africa and Malta. Vagrant Azores, Bahrain, Liechtenstein, Madeira, Canary Is, Saudi Arabia and UAE. One accepted record of *brachyrhynchus* (North America): Azores (18 Feb-24 Mar 2003).

Larus delawarensis Ring-billed Gull

TAXONOMY Ord, 1815. Monotypic.
DISTRIBUTION [N and W North America > S and W North America, West Indies. Increasing. LC] Scarce but regular winter visitor Azores (mainly Terceira, where flocks of up to 50 recorded), Britain (annual mean 54, 2000-09; total 1,821, 1973-2010) and Ireland (*c* 70 annually, 2000-08; total *c* 1,770, 1979-2008); records in last two countries include regular long-staying individuals returning in successive winters, and hybrid pairing with Common Gull in Ireland. Accidental elsewhere, with occurrences across Europe but bulk of records from countries on Atlantic seaboard. Vagrant Austria, Belgium, Bulgaria, Canary Is, Cape Verde Is, Channel Is, Croatia, Czech Republic, Denmark, Faroe Is, France, Germany, Gibraltar, Hungary, Iceland, Italy, Jan Mayen, Madeira, Morocco, Netherlands, Norway, Poland, Portugal, Slovakia, Spain, Svalbard and Sweden.

Larus marinus Great Black-backed Gull

OTHER NAME Greater Black-backed Gull.
TAXONOMY Linnaeus, 1758. Monotypic.
DISTRIBUTION [NE North America and N Europe > E North America and NW Europe. Increasing. LC] Breeds W France, Channel Is, Britain, Ireland, Faroe Is, Iceland, Scandi-

navia, Baltic, far NW Russia and Svalbard; a few pairs recently also Spain, Portugal and S Morocco. Resident and partial migrant; mainly coastal on passage but some move S/SE and regular in small numbers C Europe, Croatia and perhaps Italy, but rare Montenegro. Winters mainly N and W Europe S to Azores and Atlantic coast of Iberia, but increasingly scarce towards S of range; irregular Canary Is. Vagrant Albania, Algeria, Balearic Is, Belarus, Bulgaria, Cape Verde Is, Croatia, Cyprus, Czech Republic, Gibraltar, Greece, Iran, Israel, Jordan, Lebanon, Libya, Luxembourg, Madeira, Malta, N Mauritania, Morocco, Romania, Sardinia, Serbia, Slovakia, Slovenia, Syria, Tunisia, Turkey and Ukraine.

Larus dominicanus
Kelp Gull (V)

OTHER NAMES Dominican (or Southern Black-backed) Gull; Cape Gull (*L d vetula*).
TAXONOMY Lichtenstein, M H C, 1823. Five subspecies (one in region). Mainland African form *vetula* sometimes given specific status.
DISTRIBUTION [S South America, Antarctica, Southern Ocean, W and S Africa, Australia and New Zealand. Increasing. LC] Very rare vagrant but potential coloniser (*vetula*): Canary Is (15 Apr 2001); France (Jan 1995); N Mauritania (23 Apr 1997 to at least 21 Dec 2007, paired with Yellow-legged Gull); Morocco (9 records, first on 28 Aug 2006, followed by multiple individuals at one site and two breeding pairs among up to 10 adults in spring 2009); Portugal (2: 22-23 May 2013*; 5 Jun 2013*).

Larus glaucescens
Glaucous-winged Gull (V)

TAXONOMY Naumann, J F, 1840. Monotypic.
DISTRIBUTION [NW and W North America, NE Asia. Increasing. LC] Extreme vagrant (six records): Britain (2: 15 Dec 2006 intermittently to 18 Apr 2007; 31 Dec 2008-10 Jan 2009); Canary Is (7-10 Feb 1992); Denmark (27 Nov 2009 intermittently to 27 Feb 2010, returning 27 Aug intermittently to 28 Dec 2010 and 24 Jan-11 Feb 2012); Morocco (31 Jan 1995); Norway (4 and 14-27 July 2011).

Larus hyperboreus
Glaucous Gull

TAXONOMY Gunnerus, 1767. Four subspecies (two in region).
DISTRIBUTION [Holarctic > N North America, N Europe and NE Asia. Stable. LC] Breeds Iceland (*leuceretes*), Bear Is, Svalbard, Franz Josef Land and NW Russia (*hyperboreus*). Winters S to Scandinavia, Baltic, North Sea coasts, Britain, Ireland and NW France. Vagrant Austria, Azerbaijan, Azores, Bulgaria, Canary Is, Channel Is, Czech Republic, Hungary, Israel, Italy, Jordan, Latvia, Lithuania, Madeira, Malta, Morocco, Portugal, Romania, Sardinia, Slovakia, Slovenia, Spain, Tunisia, Turkey and Ukraine.

Larus glaucoides
Iceland Gull

OTHER NAME Kumlien's Gull (*L g kumlieni*).
TAXONOMY Meyer, B, 1822. Two subspecies (both in region). Status of variable *kumlieni* unclear; sometimes regarded as hybrid population originating from interbreeding Iceland and Thayer's Gulls, but evidence of extensive hybridisation weak and taxon may be stable, separate and self-sustaining. More research needed.
DISTRIBUTION [N North America > NE North America and N Europe. Stable. LC] May have bred Jan Mayen 19th century, possibly also Novaya Zemlya and NW Russia. Winter visitor principally to Iceland, with smaller numbers Faroe Is, Britain, Ireland and Scandinavia (*glaucoides*); rarer *kumlieni* also regular Iceland, some reaching Faroe Is and Britain (very scarce) in variably low numbers. Vagrant Azores, Belgium, Canary Is, Channel Is, Czech Republic, Estonia, Finland, France, Greece, Hungary, Italy, Latvia, Lithuania, Morocco, Poland, Portugal, Spain; *kumlieni* vagrant Azores, Belgium, Channel Is, France, Ireland, Madeira*, Netherlands, Norway, Portugal, Spain and Sweden.

Larus thayeri
Thayer's Gull (V)

TAXONOMY Brooks, W S, 1915. Monotypic. Variable taxon sometimes regarded as subspecies of Iceland Gull, but despite limited

hybridisation retains constant wing pattern and distinctive plumages.
DISTRIBUTION [Arctic North America > C and W North America. Increasing. LC] Very rare vagrant: Britain (2); Denmark (2); Iceland (3+*); Ireland (9); Norway (3 Jan 2000); Spain (Feb 2008, returning 7 Dec 2009, 20 Mar-8 Apr 2010, 13-23 Mar 2011 and 10 Mar-17 Apr 2013*).

Larus argentatus European Herring Gull (E)

OTHER NAMES Herring Gull; British Herring Gull (*L a argenteus*), Marsh Gull (*L a 'omissus'*), Scandinavian Herring Gull (*L a argentatus*).
TAXONOMY Pontoppidan, 1763. Two subspecies (both in region). Formerly considered to include Yellow-legged, Armenian, Caspian and American Herring Gulls (*qv*), all of which now separated at species level.
DISTRIBUTION [N and W Europe. Decreasing. LC] Endemic. Breeds W and N France, Channel Is, Britain, Ireland, Faroe Is and Iceland (*argenteus*), and from NW Germany through Scandinavia and Baltic to Belarus, NW Russia, Svalbard and Jan Mayen (*argentatus*); subspecies intergrade where ranges meet on North Sea coast of continental Europe. *Argentatus* winters S to Britain and France, but otherwise mainly sedentary or short-distance migrant, very small numbers reaching Spain, C and S Europe, N Mediterranean, N Balkans and W Black Sea in winter. Vagrant Azerbaijan, Azores, Balearic Is, Canary Is, Cyprus, Gibraltar, Greece, Israel, Luxembourg, Madeira, Malta, Morocco, Portugal, Romania, Serbia, Slovakia and Tunisia.

Larus smithsonianus American Herring Gull (V)

OTHER NAMES (Arctic) Herring Gull, Smithsonian Gull.
TAXONOMY Coues, 1862. Monotypic. Now widely considered a separate species from European Herring Gull. Taxonomic treatment complicated by position of extralimital Vega Gull *L vegae* (NE Asia), sometimes treated as subspecies of present species; IOC treatment of *L smithsonianus* as monotypic followed here.
DISTRIBUTION [N North America > W, S and NE North America. Decreasing. LC] Vagrant: Azores (72 to 2011*); Britain (28); France (12); Iceland (up to 20 records under review); Ireland (92); Portugal (2); Spain (6).

Larus cachinnans Caspian Gull

OTHER NAME Pontic Gull.
TAXONOMY Pallas, 1811. Monotypic. Formerly considered subspecies of Herring Gull, then as nominate eastern form of Yellow-legged Gull before recognition as full species, initially with subspecies *barabensis* and extralimital *mongolicus* (EC Asia); following further genetic studies, former now assigned to Lesser Black-backed Gull, latter to extralimital Vega Gull *L vegae* (NE Asia).
DISTRIBUTION [C Europe to W Asia > W Europe to SW Asia. Stable. LC] Breeds from Germany and C Europe discontinuously E to Caspian Sea. Winters mainly Black and S Caspian Seas, E Mediterranean (W to Libya) and S through Middle East; scarcer W of breeding range, dispersing and/or wintering locally N to S Scandinavia and S Baltic, W to *eg* Britain, France and Low Countries, and S to Italy. Very scarce Finland and Spain. Vagrant Estonia, Ireland, Luxembourg, Malta*, Norway, Portugal, Sardinia, Slovenia and Tunisia.

Larus michahellis Yellow-legged Gull

OTHER NAMES Yellow-legged Herring Gull; Azores (or Atlantic/Azorean Yellow-legged) Gull (*L m atlantis*).
TAXONOMY Naumann, J F, 1840. Two subspecies (both in region). Previously treated as form of European Herring Gull, but now separated on basis of sympatry, behaviour and morphology. Taxonomic status of morphologically distinct and geographically isolated *atlantis* in Azores may merit investigation; further study of somewhat intermediate Canary Is and Madeira populations, here included within *atlantis*, also needed.
DISTRIBUTION [Atlantic islands, N Africa and S Europe > Atlantic islands, W and N Africa, W and S Europe. Increasing. LC] Near-endemic. Breeds Azores, Canary Is and Madeira (*atlantis*), and from W France, W Iberia, W Morocco and

Mediterranean basin E to Black Sea, also locally N to C and E Europe (*michahellis*). Nominate population disperses N to *eg* Ireland (scarce), S Britain, S Scandinavia and S Baltic; winter range extends SW to N Mauritania (uncommon) and E to Jordan. *Atlantis* more sedentary, but regular non-breeding visitor Cape Verde Is. Vagrant Estonia, Finland, Iceland, Latvia, Luxembourg, Norway and Saudi Arabia; *atlantis* vagrant *eg* Britain*, Ireland and mainland Portugal*.

Larus armenicus
Armenian Gull (E)

TAXONOMY Buturlin, 1934. Monotypic. Historically included within the 'yellow-legged Herring Gulls', but differs from congeners in *eg* morphology and moult.
DISTRIBUTION [SW Asia > Middle East. Decreasing. NT] Endemic. Breeds Armenia, Georgia, NW Iran and Turkey; status uncertain Azerbaijan and Russian Caucasus. Some remain SE Black Sea but others move S to winter E Mediterranean, especially N Israel but also Cyprus (scarce), Egypt, Palestine, Syria and Turkey, as well as N Red Sea, Bahrain, Iraq, Kuwait (scarce) and W and NE coasts of Saudi Arabia (scarce). Vagrant Greece, Italy and Lebanon. Unrecorded Oman and UAE, where previous reports now known to refer to *barabensis* Lesser Black-backed Gull.

Larus schistisagus
Slaty-backed Gull (V)

OTHER NAMES Kamchatka (or Pacific) Gull.
TAXONOMY Stejneger, 1884. Monotypic.
DISTRIBUTION [NE Asia > E Asia. Population trend unknown. LC] Extreme vagrant (five individuals in six countries): Belarus (26 Dec 2012-4 Jan 2013); Britain (13 Jan intermittently to 26 Feb 2011); Finland (3-4 Nov 2012); Iceland (14 May 2012*); Latvia (13-21 Apr 2009, same as Lithuania); Lithuania (17-20 Nov 2008, same as Latvia).

Larus fuscus
Lesser Black-backed Gull

OTHER NAMES Baltic (or Scandinavian) Lesser Black-backed Gull (*L f fuscus*), British (or Western) Lesser Black-backed Gull (*L f graellsii*), Continental (or Intermediate) Lesser Black-backed Gull (*L f intermedius*), Heuglin's (or Siberian/Tundra) Gull (*L f heuglini*), Steppe (or Baraba) Gull (*L f barabensis*).
TAXONOMY Linnaeus, 1758. Five subspecies (all in region). Traditional species arrangement included only *graellsii* and *intermedius* with nominate *fuscus*, but genetic similarity now established in allopatric *barabensis* (breeding extralimitally C Asia) and sympatric *heuglini*. Latter taxon occupies different breeding habitat and exhibits little evidence of gene flow with *fuscus*, but IOC treatment provisionally followed here pending further research.
DISTRIBUTION [E North America, N and W Europe, NW and W Asia > E North America, W and S Europe, SW Asia, E and W Africa. Increasing. LC] Breeds W Iberia, France, Channel Is, Britain, Ireland, Faroe Is and Iceland (*graellsii*), E Spain, Belgium and Netherlands N along North Sea coast to W and N Scandinavia (*intermedius*), Baltic coasts and N Norway (*fuscus*), and Russian tundra (*heuglini*); breeding ranges of subspecies overlap in some areas (*eg graellsii* and *intermedius* intergrade Belgium and Netherlands, *intermedius* and *fuscus* both breed N Norway). N populations migratory, *graellsii* and *intermedius* moving SW in W Europe to winter mainly from S Britain (some farther N) S to NW Africa and Atlantic islands, and *fuscus* and *heuglini* moving S/SE in E Europe to winter E Mediterranean and Middle East (where *fuscus* scarcer in E). *Barabensis* is winter visitor chiefly to Iran, Gulf coasts and Oman, also S Red Sea, but rare Israel (mainly Mediterranean coast); recently recorded also extreme SE European Russia in summer. Vagrancy of all taxa complicated by identification issues (and in some countries confirmed only through ringing), but *graellsii* vagrant E to *eg* Finland and Hungary, *intermedius* also *eg* Finland and Latvia, *fuscus* W to *eg* Belgium, Britain, Netherlands and Spain and S to Armenia and Azerbaijan, and *heuglini* occurs Finland, probably scarce migrant Baltic and vagrant SW to *eg* Germany, Austria, Hungary and Ukraine.

Gelochelidon nilotica
Gull-billed Tern

OTHER NAME Common Gull-billed Tern.

TAXONOMY (Gmelin, J F, 1789). Six subspecies (one in region). Extralimital *macrotarsa* (Australia) now separated by some authorities as Australian Gull-billed Tern.
DISTRIBUTION [Cosmopolitan. Decreasing. LC] Mainly summer visitor, breeding discontinuously from N Mauritania to W Libya, Iberia and France E to N Caspian, Caucasus and N Middle East (*nilotica*); formerly also bred *eg* Austria, Denmark, Morocco and Netherlands. More widespread on passage S Europe, N Africa (W to Canary Is) and Middle East, including Atlantic coast from W France S to Morocco. Present year-round N and W Persian Gulf and SE Iraq; elsewhere in winter mainly Iran and S Arabia. Vagrant Azores, Belgium, Britain, Cape Verde Is, Czech Republic, Estonia, Finland, Germany, Iceland, Ireland, Luxembourg, Madeira, Montenegro, Norway, Poland, Slovakia, Slovenia and Sweden.

Hydroprogne caspia
Caspian Tern

TAXONOMY (Pallas, 1770). Monotypic.
DISTRIBUTION [Subcosmopolitan. Increasing. LC] Breeds N Mauritania, in Sweden, Finland and Estonia (formerly also Denmark), in SE Europe from N Black Sea to Caspian Sea, and in Turkey, Iran, Persian Gulf and Red Sea; may breed Iraq and Syria, has also bred *eg* Azerbaijan, France, Latvia, Spain and Tunisia. Baltic breeders mainly migrate S overland, fewer along continental North Sea and Atlantic coasts, with passage via Mediterranean and Moroccan coasts to main sub-Saharan wintering areas or to Arabia. Some winter N Mauritania and on Arabian coasts, few also *eg* Morocco, S Spain, Tunisia, Libya and Red Sea. Vagrant Armenia, Azores, Balearic Is, Britain, Canary Is, Cape Verde Is, Cyprus, Faroe Is, Ireland, Luxembourg, Madeira and Norway (has bred).

Thalasseus maximus
Royal Tern

TAXONOMY (Boddaert, 1783). Two subspecies (both in region).
DISTRIBUTION [S North America, Caribbean, W and E South America and W Africa. Stable. LC] Breeds N Mauritania, with limited dispersal and wintering N to S Morocco, less frequently N Morocco (*albididorsalis*). Vagrant Britain, Cape Verde Is, France, Gibraltar, Ireland, Norway, Portugal and Spain; *maximus* (North and South America) vagrant Azores, Britain and Spain.

Thalasseus bergii
Greater Crested Tern

OTHER NAMES (Great) Crested Tern, Swift Tern.
TAXONOMY (Lichtenstein, M H C, 1823). Four subspecies (one in region).
DISTRIBUTION [E and S Africa, SW to SE Asia and Australasia. Stable. LC] Breeds Red Sea in Egypt, Saudi Arabia and Yemen, Gulf coasts in Bahrain, Iran, Iraq, Kuwait, Qatar, Saudi Arabia and UAE, and in Oman (*velox*). Disperses widely S to Gulf of Aden and Arabian Sea. Vagrant Israel and Jordan.

Thalasseus elegans
Elegant Tern (V)

TAXONOMY (Gambel, 1849). Monotypic.
DISTRIBUTION [SW North America > W Central and W South America. Stable. NT] Very rare vagrant: Denmark (30 May-12 Jun 2000); France (9 individuals 1974-2012, often returning birds); Germany (26 Sep-5 Oct 2004); Ireland (5); Spain (24-30 Apr 1993). Several other records (*eg* Britain and Ireland) under consideration, and three colour-ringed individuals (two breeding France and one breeding both France and Spain, mainly with Sandwich Terns and producing hybrid chicks) now confimed as pure Elegant Terns through genetic sampling.

Thalasseus bengalensis
Lesser Crested Tern

TAXONOMY (Lesson, 1831). Three subspecies (at least two in region).
DISTRIBUTION [N Africa to N Australia > W, S and E Africa to Australia. Stable. LC] Mainly summer visitor, breeding Libya (*emigrata*), widely on Gulf coasts (*emigrata* or sometimes considered to be *torresii*) and on Red

Sea coasts (*bengalensis*); formerly bred Lebanon, and has also nested (and/or paired with Sandwich Tern) Britain, France, Italy, Spain and Tunisia. Regular on passage NW African coasts; very scarce S and E Spain, occasional Gibraltar. Most leave Mediterranean, Red Sea and Gulf in winter, but a few remain N Mauritania. Vagrant Austria, Britain, Canary Is*, Corsica, France, Germany, Greece, Ireland, Italy, Malta, Portugal, Sardinia, Slovenia, Switzerland and Turkey.

Thalasseus sandvicensis
Sandwich Tern

OTHER NAME Eurasian Sandwich Tern.
TAXONOMY (Latham, 1787). Monotypic. Formerly included Cabot's Tern (*qv*).
DISTRIBUTION [Europe and SW Asia > S Europe, Africa and Middle East. Stable. LC] Breeds Britain, Ireland, and North Sea/S Baltic coasts of continental Europe to Biscay, and from E Spain sparsely through Mediterranean to Black and Caspian Seas and W Turkey; has also bred Saudi Arabia. Widespread along regional coasts on passage but rarer inland, *eg* WC Europe. Winters S Europe, N Africa and Middle East; irregular in winter Canary Is, Cape Verde Is and Madeira. Vagrant Azores, Belarus, Czech Republic, Faroe Is, Hungary, Iceland, Jan Mayen, Luxembourg, Slovakia and Svalbard.

Thalasseus acuflavidus
Cabot's Tern (V)

OTHER NAMES American Sandwich Tern; Cayenne Tern (*S a eurygnatha*).
TAXONOMY (Cabot, S, 1847). Two subspecies (one in region). Molecular analyses indicate closer relationship to Elegant Tern than to morphologically similar Sandwich Tern, with which formerly considered conspecific; also subtle plumage differences.
DISTRIBUTION [SE USA to Caribbean and C America > N South America. Population trend not assessed. NE] Extreme vagrant (two recoveries of nominate *acuflavidus* ringed as chicks in North Carolina, USA): Britain (28 Nov 1984, dead); Netherlands (23 Dec 1978, dead).

Sternula albifrons
Little Tern

TAXONOMY (Pallas, 1764). Three subspecies (two in region). Forms subcosmopolitan superspecies with four, possibly five, other small terns.
DISTRIBUTION [Europe, W Africa, SW and E Asia to Australasia. Decreasing. LC] Resident N Mauritania (*guineae*) but otherwise mainly summer visitor, breeding widely but locally Europe (absent far N, rare C), Asia Minor, Iraq, Iran, Persian Gulf, Red Sea and N African coast (*albifrons*). More widespread on passage, W European populations wintering Africa and E European populations probably moving to N and E Middle East, where non-breeding range partly obscured through confusion with Saunders's Tern. Scarce on passage Canary Is. Vagrant Azores, Cape Verde Is, Czech Republic, Faroe Is, Iceland, Lebanon (formerly bred), Luxembourg, Madeira, Malta, Norway and Yemen.

Sternula saundersi
Saunders's Tern

OTHER NAMES Black-shafted (or Saunders's Little) Tern.
TAXONOMY (Hume, 1877). Monotypic. Sometimes considered conspecific with Little Tern.
DISTRIBUTION [SW and S Asia > SW to SE Asia. Decreasing. LC] Fragmented breeding range on islands and coasts of Arabia N to Kuwait, where rare, and in S Iran; since 2012 also known to breed Egyptian Red Sea. Some remain to winter around Arabian Peninsula, Gulf of Oman and Gulf coasts. Known Iraq in 19th century but more recent status uncertain. Vagrant Israel.

Sternula antillarum
Least Tern (V)

TAXONOMY Lesson, 1847. Three subspecies (undetermined in region). Sometimes considered conspecific with Little Tern.
DISTRIBUTION [W, C, E and S USA to Central America and Caribbean > Central America and Caribbean to NE South America. Decreasing. LC] Extreme vagrant (one record of returning summering bird): Britain (*c* 8 Jun-5 Jul 1983, returning to same location on single or mul-

tiple dates each subsequent year until 1992, when last recorded 12 Jul; in 1990 also present at different site 31 May-1 Jul).

Onychoprion aleuticus
Aleutian Tern (V)

OTHER NAME Kamchatka Tern.
TAXONOMY (Baird, S F, 1869). Monotypic.
DISTRIBUTION [NE Asia and Alaska > SE Asia. Stable. LC] Extreme vagrant: Britain (28-29 May 1979).

Onychoprion anaethetus
Bridled Tern

OTHER NAME Brown-winged Tern.
TAXONOMY (Scopoli, 1786). Four subspecies (two in region).
DISTRIBUTION [Tropical and subtropical oceans. Population trend unknown. LC] Very rare summer visitor N Mauritania and S Morocco (*melanopterus*); breeds in much larger numbers along coasts of E Egypt, Arabia and S Iran (*antarcticus*). NW African population moves S to tropical waters after breeding; in Middle East, occurs widely on passage Red Sea (rarely N to Israel, where exceptionally also recorded Mediterranean coast), Arabian Sea, Gulf of Oman and Gulf coasts, some also wintering Arabian Sea. Vagrant Azores (has bred), Britain, Denmark, France, Germany, Greece, Libya (has bred), Netherlands, Norway, Spain and Sweden.

Onychoprion fuscatus
Sooty Tern

OTHER NAME Wideawake (Tern).
TAXONOMY (Linnaeus, 1766). Seven subspecies (two presumed in region).
DISTRIBUTION [Tropical and subtropical oceans. Population trend unknown. LC] Breeds Oman, but rare (*nubilosus*); irregular in summer Azores, where nesting occasionally attempted (presumed *fuscatus*). Omani population disperses into Gulf of Oman and N Arabian Sea, with occasional records throughout year except late winter; scarce and irregular visitor S Iran. Vagrant Bahrain, Britain, Canary Is, France, Germany, Iceland, Ireland, Israel, Italy, Madeira (has bred Salvage Is), N Mauritania, Morocco, Netherlands, Portugal, Saudi Arabia, Spain, Sweden, Tunisia, UAE and Yemen.

Sterna aurantia
River Tern (V)

OTHER NAME Indian River Tern.
TAXONOMY Gray, J E, 1831. Monotypic.
DISTRIBUTION [S and SE Asia. Decreasing. NT] Extreme vagrant: Iran (18 Jan 2005, two).

Sterna dougallii
Roseate Tern

TAXONOMY Montagu, 1813. Five subspecies (two in region).
DISTRIBUTION [Temperate and tropical oceans. Population trend unknown. LC] Summer visitor Azores, Britain, Ireland and NW France, irregularly also Madeira (*dougallii*) and Oman (*arideensis*). Maritime on passage, occurring from Britain and Ireland S in W Atlantic to Spain (rare, has bred), Morocco (rare) and N Mauritania, and in Gulf of Oman and N Arabian Sea. Claimed status as migrant in Libya doubtful, as no other recent N African records E of Morocco. Vagrant Austria, Bahrain, Belgium, Canary Is, Cape Verde Is, Channel Is, Denmark, Germany, Gibraltar, Iceland*, Israel, Italy, Malta, Netherlands, Norway, Poland, Portugal, Sardinia, E Saudi Arabia, Sweden, Switzerland, Tunisia (old breeding record) and UAE.

Sterna hirundo
Common Tern

OTHER NAME Eastern Common Tern (*S h longipennis*).
TAXONOMY Linnaeus, 1758. Four subspecies (all in region).
DISTRIBUTION [North America and Eurasia > tropical and temperate southern oceans. Decreasing. LC] Summer visitor Europe and N Middle East (*hirundo*); widely distributed in N but more fragmented or absent large areas in S, where breeding range extends to some Mediterranean and Atlantic islands, N

Mauritania, Tunisia, Libya, Israel, Iraq and Iran. Widespread on coasts on migration, with many inland and E breeding populations filtering W to North Sea and Atlantic seaboard, then S to main winter quarters beyond region. Some winter *eg* S Arabia and N Mauritania, less frequently S European waters. Vagrant Cape Verde Is, Faroe Is (has bred), Iceland, Liechtenstein, Malta and Svalbard. Birds showing characters of *longipennis* (breeding NE Asia) reportedly uncommon on migration Israel, vagrant Turkey and also provisionally documented *eg* Britain and Belgium; *minussensis* (breeding C Siberia to N Mongolia) vagrant European Russia; *tibetana* (breeding Himalayas to W Mongolia) vagrant Iran.

Sterna repressa
White-cheeked Tern

TAXONOMY Hartert, 1916. Monotypic.
DISTRIBUTION [NE and E Africa, SW and S Asia > Indian Ocean basin. Decreasing. LC] Summer visitor to islands off E Egypt, Arabia and S Iran; status uncertain Iraq, where has bred but no recent records. Regularly occurs in small numbers N to Israel and Jordan, and occasionally N to Suez Canal; exceptionally recorded Mediterranean coast of N Israel. Migrates S to Indian Ocean in autumn; scarce S Arabian waters in winter.

Sterna paradisaea
Arctic Tern

TAXONOMY Pontoppidan, 1763. Monotypic.
DISTRIBUTION [N North America and N Eurasia > Southern Ocean. Decreasing. LC] Summer visitor Britain, Ireland and North Sea coasts of continental Europe N to Faroe Is, Iceland, Scandinavia, Baltic, NW Russia and Arctic islands. Coastal on migration, moving through NW European waters and S along Atlantic seaboard, usually some distance from shore; regular on passage Canary Is and Madeira. Vagrant Algeria, Austria, Azores, Belarus, Bulgaria, Cape Verde Is, Cyprus, Czech Republic, Gibraltar, Greece, Hungary, Israel, Italy, Jan Mayen, Jordan, Kuwait, Malta, Oman, Romania, Serbia, Slovakia, Switzerland, Turkey, UAE and Ukraine.

Sterna forsteri
Forster's Tern (V)

TAXONOMY Nuttall, 1834. Monotypic.
DISTRIBUTION [S Canada and USA > S USA, West Indies and Central America. Increasing. LC] Rare vagrant (mainly in winter, with some individuals returning in successive years): Azores (6*); Belgium (17 May-12 Jun 2003); Britain (20); France (6); Iceland (22 Oct 1959); Ireland (37); Netherlands (4); Portugal (2); Spain (6); Sweden (26 Apr 1993).

Chlidonias hybrida
Whiskered Tern

TAXONOMY (Pallas, 1811). Three subspecies (one in region).
DISTRIBUTION [S Eurasia and S Africa > Africa, SW to SE Asia, Australasia. Stable. LC] Breeds mainly S and E Europe, less widely N Middle East, with highly fragmented range extending N to NW France, Poland, Lithuania, Belarus, W Russia and NW Kazakhstan, E to Iran, also extreme N Morocco (*hybrida*). Largely migratory, occurring regularly N to Benelux countries and widely S of breeding areas on passage as far W as Canary Is. Small numbers winter *eg* Sicily, coastal N Africa and S Middle East, and present year-round SW Iberia and N Morocco. Vagrant Azores, Britain, Cape Verde Is, Channel Is, Corsica, Czech Republic, Denmark, Gibraltar, Estonia, Finland, Iceland, Ireland, Latvia, Luxembourg, Madeira, Malta, Norway and Sweden.

Chlidonias leucopterus
White-winged Tern

OTHER NAME White-winged Black Tern.
TAXONOMY (Temminck, 1815). Monotypic.
DISTRIBUTION [E Europe to C and E Asia > Africa, S and SE Asia, Australasia. Stable. LC] Summer visitor E Europe, breeding N to Baltic States, W to Germany, S to Black Sea and Caucasus, and E to Ural Mtns, with outlying populations NW Italy, Iraq and recently Kuwait (rare); has bred E Saudi Arabia and Turkey, and may do so Iran. Occurs widely on passage S of breeding range, regularly (though often scarce or rare) N to Finland and W to *eg* Britain, France, Morocco, Netherlands and Spain. Scarce in win-

ter S Arabia. Vagrant Albania, Azores, Belgium, Cape Verde Is, Czech Republic, Denmark, Faroe Is, Gibraltar, Iceland, Ireland, Luxembourg, Madeira, Norway, Portugal and Sweden.

Chlidonias niger
Black Tern

OTHER NAME American Black Tern (*C n surinamensis*).
TAXONOMY (Linnaeus, 1758). Two subspecies (both in region).
DISTRIBUTION [North America, W Europe to C Asia > Central and N South America, W and SW Africa. Decreasing. LC] Summer visitor, breeding mainly N and E Europe from Benelux countries and S Scandinavia E to Balkans, Black Sea, N Caucasus and S Ural Mtns, but with scattered small populations in SW Europe, Tunisia and W Turkey (*niger*); has bred Britain. Widespread on migration, most travelling W to Baltic, North Sea and Atlantic, and many staging in Netherlands; passage also through Mediterranean basin and NW Middle East (scarce Iran), again mostly W, few moving S via Egypt. Major passage on Atlantic seaboard, W to Canary Is, to wintering areas off N Mauritania; a few also remain Black and Caspian Seas. Vagrant Azores, Bahrain, Faroe Is, Iceland (has bred), Iraq, Kuwait, Lebanon, Liechtenstein, Madeira, Oman, Saudi Arabia, Svalbard, UAE and Yemen; *surinamensis* (breeding North America) vagrant Azores, Britain, Denmark, Iceland and Ireland.

FAMILY STERCORARIIDAE

Skuas · six species

Stercorarius maccormicki
South Polar Skua (V)

OTHER NAME McCormick's Skua.
TAXONOMY Saunders, H, 1893. Monotypic.
DISTRIBUTION [Antarctica > N Pacific and NW Atlantic. Stable. LC] Very rare vagrant, but known from satellite-tracking studies to move regularly through C and E areas of N Atlantic, thus certainly under-recorded in vast pelagic regional range: Azores (5*); Canary Is (2 records of 4 birds: 6 Oct 2005*, accepted as South Polar or Brown Skua but photos show former species; 11 Sep 2011, three); Faroe Is (24 Sep 1889, collected at sea); Israel (2: 3-6 Jun 1983, same as Jordan; 28 Jun 1992); Jordan (3-6 Jun 1983, same as Israel); Madeira (13 Jun 2012*); Oman (4); Spain (3 Oct 2013). Other well-supported claims await formal assessment. See also Brown and Great Skuas.

Stercorarius antarcticus
Brown Skua (V)

OTHER NAMES Antarctic Skua; Falkland (or Southern) Skua (*S a antarcticus*), Subantarctic (or Lonnberg's) Skua (*S a lonnbergi*), Tristan Skua (*S a hamiltoni*).
TAXONOMY (Lesson, 1831). Three subspecies (undetermined in region). Extralimital *antarcticus* (Falkland Is and SE Argentina) sometimes separated but retained here following IOC and pending further studies.
DISTRIBUTION [Antarctica and Southern Ocean > Tropic of Capricorn. Stable. LC] Extreme vagrant (three records): Oman (3: 25 Oct 1991; 18 Nov 2002; 5 Sep 2003). Vagrant large skuas in Iran also considered to be this species, and record from UAE involved either Brown or South Polar Skua; similarly, two unidentified skuas taken into care and DNA-tested Britain (Oct 2001 and Feb 2002), and potentially also one sight record (27 Jan-4 Feb 1996*), also either this species or South Polar Skua.

Stercorarius skua
Great Skua

OTHER NAMES Bonxie, Northern Skua.
TAXONOMY (Brünnich, 1764). Monotypic.
DISTRIBUTION [N Europe > C Atlantic. Stable. LC] Breeding endemic, with main populations Britain, Faroe Is and Iceland, more recently also Bear Is, Jan Mayen, Svalbard, Norway and Kola Peninsula (Russia). Widespread N and E Atlantic (including most Atlantic islands) and North Sea on passage, some also in winter; rarer W Baltic, irregular Poland and W Black Sea. Winters mainly from Biscay S to C Atlantic, NW Africa and Mediterranean E to Malta (very scarce). Vagrant Albania, Algeria, Austria, Belarus, Cape Verde Is, Corsica, Croatia, Czech Republic, Estonia, Finland, Greece,

Hungary, Israel, Latvia, Libya, Lithuania, Luxembourg, Slovenia, Switzerland, Turkey and Ukraine; record from Qatar originally claimed as this species may refer to South Polar Skua.

Stercorarius pomarinus
Pomarine Skua

OTHER NAMES Pomarine Jaeger (IOC), Pomatorhine Skua.
TAXONOMY (Temminck, 1815). Monotypic.
DISTRIBUTION [Circumpolar in Arctic > mainly at sea Tropic of Cancer to Equator, Australia. Stable. LC] Breeds extreme NW Russia; has also summered Bear Is and Svalbard but breeding not confirmed. Widespread on migration much of Baltic (irregular Poland), North Sea and NE Atlantic coasts of NW and W Europe, some Atlantic islands, some W Mediterranean coasts, W Black Sea, N Red Sea, Arabian Sea and Gulf. Main regional wintering area off S Morocco and N Mauritania, irregular Cape Verde Is. Vagrant Albania, Algeria, Armenia, Austria, Azerbaijan, Balearic Is, Belarus, Bosnia-Herzegovina, Cape Verde Is, Corsica, Croatia, Cyprus, Czech Republic, Greece, Hungary, Jordan, Latvia, Lithuania, Luxembourg, Malta, Romania, Sardinia, Serbia, Montenegro, Slovakia, Slovenia, Switzerland and Tunisia.

Stercorarius parasiticus
Arctic Skua

OTHER NAMES Parasitic Jaeger (IOC), Parasitic (or Richardson's) Skua.
TAXONOMY (Linnaeus, 1758). Monotypic.
DISTRIBUTION [Circumpolar in Arctic and subarctic > S Atlantic, SE Pacific and Southern Oceans. LC] Breeds Iceland, Faroe Is, Scotland, Norway, Sweden, Finland, Jan Mayen, Svalbard, Bear Is, Franz Josef Land and NW Russia. Widespread on migration North Sea and NE Atlantic coasts of NW and W Europe, also Atlantic islands, Baltic and parts of Mediterranean, less so Black and S Caspian Seas, and in Red Sea, Gulf and Arabian Sea; rare but regular inland continental Europe. Some winter off S Morocco, N Mauritania and S Arabia. Vagrant Albania, Armenia, Azerbaijan, Balearic Is, Belarus, Bosnia-Herzegovina, Corsica, Croatia, Cyprus, Czech Republic, Iraq, Libya, Malta, Montenegro, Serbia, Slovakia, Slovenia and Switzerland.

Stercorarius longicaudus
Long-tailed Skua

OTHER NAMES Long-tailed Jaeger (IOC), Buffon's Skua.
TAXONOMY Vieillot, 1819. Two subspecies (both in region).
DISTRIBUTION [Circumpolar in Arctic and subarctic > S Atlantic, SE Pacific and Southern Oceans. Stable. LC] Breeds Iceland (rare), Norway, Sweden, Finland and NW Russia (*longicaudus*), and Svalbard (*pallescens*, though population perhaps intermediate with *longicaudus*). Widespread on migration but oceanic, so generally scarce/rare along NE Atlantic and North Sea coasts of NW and W Europe, also Baltic (irregular Poland); may occur inshore in numbers in adverse weather conditions. Similarly, rare but variably regular migrant N Mauritania (where some remain in winter), Spain, Italy, W Black Sea and Gulf of Aqaba. Vagrant Albania, Armenia, Austria, Azores, Balearic Is, Belarus, Bosnia-Herzegovina, Canary Is, Channel Is, Corsica, Croatia, Czech Republic, Georgia, Greece, Hungary, Iran, Iraq, Jordan, Kuwait, Lithuania, Luxembourg, Madeira, Malta, Morocco, Oman, Portugal, Qatar, Romania, Serbia, Slovakia, Slovenia, Switzerland, Turkey, UAE and Ukraine.

FAMILY ALCIDAE

Auks · 11 species

Alle alle
Little Auk

OTHER NAMES Dovekie; Franz Josef Land Little Auk (*A a polaris*).
TAXONOMY (Linnaeus, 1758). Two subspecies (both in region).
DISTRIBUTION [Arctic North America and Eurasia > N Atlantic. Decreasing. LC] Breeds Bear Is, Svalbard, Franz Josef Land (*polaris*) and Novaya Zemlya (*alle*); formerly also bred Iceland. Winters arctic waters S to North Sea and far NE Atlantic (rare Ireland); less predictable E into Baltic and S to Channel Is, Biscay and

rarely Atlantic coasts of Spain, where mostly recorded far offshore. Vagrant Austria, Azores, Canary Is, Corsica, Czech Republic, Estonia, Italy, Latvia, Lithuania, Madeira, Malta, Morocco, Poland, Portugal and Ukraine.

Uria lomvia
Brünnich's Guillemot

OTHER NAMES Thick-billed Murre (IOC), Brünnich's Murre.
TAXONOMY (Linnaeus, 1758). Four subspecies (one in region).
DISTRIBUTION [N North America and N Eurasia > N Pacific and N Atlantic Oceans. Increasing. LC] Breeds Iceland, N Norway, Svalbard, Franz Josef Land and NW Russia (*lomvia*). Winters Barents Sea and extreme NE Atlantic. Vagrant Austria, Azores, Belgium, Britain, Denmark, Finland, France, Germany, Ireland, Netherlands, Poland, Spain and Sweden; two historical records from Latvia placed in Cat D.

Uria aalge
Common Guillemot

OTHER NAMES Common Murre (IOC), Thin-billed Murre.
TAXONOMY (Pontoppidan, 1763). Five subspecies (three in region).
DISTRIBUTION [NW and NE North America, NW and NE Eurasia > N Pacific and N Atlantic Oceans. Increasing. LC] Breeds NW Spain, NW France, Channel Is, Heligoland, Britain N to S Scotland and Ireland (*albionis*), Scotland, Faroe Is, Iceland, S Norway and Baltic Sea (*aalge*) and N Norway, Bear Is, Franz Josef Land, Svalbard, NW Russia and Novaya Zemlya (*hyperborea*). Winters arctic and subarctic waters S to North Sea, Baltic, Biscay and W Iberia, rarely W Mediterranean. Vagrant Austria, Balearic Is, Bulgaria, Canary Is, Czech Republic, Gibraltar, Italy, N Mauritania, Morocco and Sardinia.

Alca torda
Razorbill

OTHER NAME Razor-billed Auk.
TAXONOMY Linnaeus, 1758. Two subspecies (both in region).
DISTRIBUTION [NE North America and NW Europe. Decreasing. NT] Breeds NW France, Channel Is, Ireland, Britain, Heligoland, Faroe Is and Iceland (*islandica*) and Bear Is, Scandinavia, Baltic and NW Russia (*torda*). Widespread Baltic, North Sea and NE Atlantic in winter, S to N Morocco (uncommon) and W Mediterranean. Vagrant Algeria, Austria, Azores, Croatia, Czech Republic, Hungary, Malta, N Mauritania, Montenegro, Serbia and Tunisia.

Cepphus grylle
Black Guillemot

OTHER NAMES Tystie; Mandt's Guillemot (*C g mandtii*).
TAXONOMY (Linnaeus, 1758). Five subspecies (all in region).
DISTRIBUTION [NE North America, N Europe and NW Asia. Increasing. LC] Breeds Iceland (*islandicus*), Faroe Is (*faeroeensis*), Svalbard and other high-arctic islands (*mandtii*), Baltic Sea (*grylle*) and White Sea, S Scandinavia, N Britain and Ireland (*arcticus*). Limited dispersal outside breeding season, more marked in high latitudes, but rare *eg* North Sea and from S Britain southwards. Vagrant Belgium, Croatia, France, Netherlands, Slovenia and Spain; record from Czech Republic under review.

Brachyramphus perdix
Long-billed Murrelet (V)

TAXONOMY (Pallas, 1811). Monotypic. Formerly treated as a subspecies of extralimital Marbled Murrelet *B marmoratus* (NW North America).
DISTRIBUTION [NE Asia > NW Pacific Ocean. Decreasing. NT] Extreme vagrant (three records, all Nov-Dec): Britain (7-14 Nov 2006); Romania (21-23 Dec 2006); Switzerland (between 15 and 18 Dec 1997, found dead).

Synthliboramphus antiquus
Ancient Murrelet (V)

TAXONOMY (Gmelin, J F, 1789). Two subspecies (undetermined in region).
DISTRIBUTION [NE Asia, NW and W North America > Pacific Ocean. Decreasing. LC] Extreme vagrant (one record of returning indi-

vidual): Britain (27 May-26 Jun 1990, 14 Apr-20 Jun 1991 and 30 Mar-29 Apr 1992).

Aethia psittacula
Parakeet Auklet (V)

OTHER NAME Paroquet Auklet.
TAXONOMY (Pallas, 1769). Monotypic.
DISTRIBUTION [NE Asia and NW North America > Pacific Ocean. Decreasing. LC] Extreme vagrant: Sweden (mid-Dec 1860, found dead).

Aethia cristatella
Crested Auklet (V)

TAXONOMY (Pallas, 1769). Monotypic.
DISTRIBUTION [NE Asia and NW North America > Pacific Ocean. Decreasing. LC] Extreme vagrant: Iceland (between 12 and 20 Aug 1912, at sea, collected).

Fratercula arctica
Atlantic Puffin

OTHER NAMES (Common) Puffin.
TAXONOMY (Linnaeus, 1758). Monotypic.
DISTRIBUTION [NE North America and NW Europe > N Atlantic and W Mediterranean. Decreasing. VU] Breeds from NW France, Channel Is, Britain, Ireland, Faroe Is and Iceland to Bear Is, Norway, Svalbard, NW Russia and Novaya Zemlya; formerly also NW Germany (Heligoland) and SW Sweden. Passage occurs S to Strait of Gibraltar; pelagic in winter, range extending from NE Atlantic and North Sea to W Mediterranean; regular Italy, scarce Morocco. Vagrant Algeria, Austria, Azores, Belgium, Canary Is, Croatia, Finland, Germany, Hungary, Lithuania, Madeira, Malta, Poland, Slovenia and Tunisia.

Fratercula cirrhata
Tufted Puffin (V)

TAXONOMY (Pallas, 1769). Monotypic.
DISTRIBUTION [NE Asia, NW and W North America. Decreasing. LC] Extreme vagrant (two records): Britain (16 Sep 2009); Sweden (1 and 8 Jun 1994).

FAMILY PTEROCLIDAE

Sandgrouse · seven species

Syrrhaptes paradoxus
Pallas's Sandgrouse

TAXONOMY (Pallas, 1773). Monotypic.
DISTRIBUTION [W, C and E Asia. Stable. LC] Presumed scarce visitor NW Kazakhstan, small flocks occurring Volga-Ural region during dispersal from breeding areas to E of regional boundary. Otherwise very rare vagrant outside infrequent irruption years, when recorded in 30 countries across region W to Faroe Is and Ireland; many records relate to influxes in late 19th and early 20th centuries (notably in 1863, 1888 and 1908), after which extralimital breeding widely recorded in region. Few vagrancy occurrences after 1908 irruption, and very few since 1950: last recorded Finland (29 Jul-2 Aug 2010).

Pterocles alchata
Pin-tailed Sandgrouse

OTHER NAMES Large (or White-bellied) Sandgrouse.
TAXONOMY (Linnaeus, 1766). Two subspecies (both in region).
DISTRIBUTION [N Africa, SW Europe, Middle East and SW Asia. Stable. LC] Resident or mainly so Iberia and SE France (*alchata*) and from Morocco to NW Libya, Israel and through parts of N Middle East N to SE Turkey and E to SW Iran, also C Arabia (*caudacutus*); introduced UAE. In winter also S Caspian region, N Arabia and more widespread NW Africa. Vagrant Armenia, Azerbaijan (may have bred in past), Cyprus, Egypt, Gibraltar, Greece, Italy, Lebanon, Malta, Oman (records may relate to UAE releases) and Sardinia.

Pterocles exustus
Chestnut-bellied Sandgrouse

OTHER NAMES Lesser (or Small) Pin-tailed Sandgrouse.
TAXONOMY Temminck, 1825. Six subspecies (three in region).
DISTRIBUTION [Sahel, E Africa, S Arabia and S Asia. Stable. LC] Resident Egypt (where pre-

sumed extinct subspecies *floweri* rediscovered in 2012), W, S and E Arabia (*erlangeri*, possibly *hindustan* in E), and SE Iran (*hindustan*). Both wild and escaped/released birds have bred UAE; species also breeds in captivity in Qatar, where high probability of escaped or released birds. Vagrant Hungary, Israel, Jordan (probably bred 19th century) and Kuwait (where first of three records may relate to birds introduced for hunting).

Pterocles senegallus
Spotted Sandgrouse

OTHER NAMES Northern Spotted Sandgrouse, Saharan Sandgrouse.
TAXONOMY (Linnaeus, 1771). Monotypic.
DISTRIBUTION [N Africa, S Middle East and SW Asia. Stable. LC] Breeds from N Mauritania and Morocco E across N Africa, in Israel, Jordan and Syria, Yemen (rare), Oman and E Saudi Arabia, and from Iraq and Kuwait through S Iran. Vagrant Italy, Turkey and UAE.

Pterocles orientalis
Black-bellied Sandgrouse

OTHER NAMES Imperial (or Large) Sandgrouse.
TAXONOMY (Linnaeus, 1758). Two subspecies (both in region).
DISTRIBUTION [SW Europe, NW Africa, Middle East and SW Asia. Decreasing. LC] Mainly resident Iberia, Canary Is (Fuerteventura; also recorded Lanzarote and La Graciosa), from Morocco to NW Libya, Israel, from Turkey E through N Middle East to W Iran (*orientalis*) and S Iran (*arenarius*); also in summer from NW Kazakhstan (*arenarius*) sporadically to Transcaucasia (*orientalis*), passage/wintering range extending S in Middle East to C Arabia (rare). Vagrant Bahrain, Belgium, Cyprus, Germany, Greece, Lebanon and Malta.

Pterocles coronatus
Crowned Sandgrouse

OTHER NAMES Coronetted (or Yellow-throated) Sandgrouse.
TAXONOMY Lichtenstein, M H C, 1823. Five subspecies (four in region).
DISTRIBUTION [N Africa, S Middle East and SW Asia. Stable. LC] Breeds W and C Sahara from N Mauritania discontinuously to E Egypt (*coronatus*), from Sinai to Israel and probably Jordan (*vastitas*), and W Saudi Arabia, Yemen, Oman and S Iran (*atratus*); *saturatus* also described from N Oman (1927) but not definitely collected or recorded since. Has also bred ferally UAE following introduction in 1990s.

Pterocles lichtensteinii
Lichtenstein's Sandgrouse

OTHER NAMES Abyssinian (or Close-barred) Sandgrouse.
TAXONOMY Temminck, 1825. Five subspecies (four in region).
DISTRIBUTION [N and E Africa, S Middle East and SW Asia. Stable. LC] Breeds from N Mauritania and S Morocco E to N Chad (*targius*), E Egypt, Israel and Jordan (*lichtensteinii*), E Yemen (*ingramsi*) and W Saudi Arabia, W Yemen, Oman and UAE to S Iran (*arabicus*). Probably rare winter Iraq in 1920s but not recorded since.

FAMILY COLUMBIDAE

Pigeons and doves · 23 species

Columba livia
Rock Dove

OTHER NAMES Blue Rock Dove, Rock Pigeon; Feral (or Common) Pigeon, Feral Rock Dove (feral/introduced populations).
TAXONOMY Gmelin, J F, 1789. Nine subspecies (seven in region). Extensive variation known in feral populations, often through considerable influence of domestic varieties and because darker males outcompete paler individuals for mates, but some birds of mixed ancestry closely resemble pure *C livia*.
DISTRIBUTION [Europe, N Africa and S Asia. Decreasing. LC] Wild populations resident N Mauritania (*gymnocycla*), S Algeria, S Libya, N Niger and N Chad (*targia*), W Egypt (*dakhlae*), E Egypt (*schimperi*), NW Egypt to Morocco, Atlantic islands, Iberia, Ireland, NW Scotland, Faroe Is and from Alps and Mediterranean E through S Europe to SE Russia, N Caucasus and C Turkey (*livia*), E Turkey, S Caucasus and

Iran (*gaddi*), and from Sinai to Syria and much of Arabia (*palaestinae*). Feral populations widely established throughout region, in some areas also among wild Rock Doves.

Columba guinea
Speckled Pigeon (V)

OTHER NAMES Red-eyed Pigeon, (African or Speckled) Rock Pigeon.
TAXONOMY Linnaeus, 1758. Two subspecies (one in region).
DISTRIBUTION [Sub-Saharan Africa. Stable. LC] Extreme vagrant: Yemen (18 Dec 1997, presumed *guinea*).

Columba oenas
Stock Dove

OTHER NAMES Eurasian (or Western) Stock Dove, Stock Pigeon.
TAXONOMY Linnaeus, 1758. Two subspecies (one in region).
DISTRIBUTION [Europe to C Asia, NW Africa. Stable. LC] Breeds across much of Europe except far N, also Morocco, Turkey, Caucasus and N Iran (*oenas*). Migratory across C and E of range; winter visitor N Middle East S to Syria and Israel. Very rare Malta; status uncertain Tunisia. Vagrant Algeria, Andorra, Bahrain, Egypt, Faroe Is, Gibraltar, Iceland, Kuwait, Lebanon, Libya, Madeira*, Oman and UAE.

Columba eversmanni
Yellow-eyed Pigeon

OTHER NAMES Eastern (or Eversmann's) Stock Dove/Pigeon, Pale-backed Pigeon, Yellow-eyed (Stock) Dove.
TAXONOMY Bonaparte, 1856. Monotypic.
DISTRIBUTION [C Asia > S Asia. Decreasing. VU] Resident NE Iran; rare winter visitor SE Iran. Vagrant Russia.

Columba palumbus
Common Wood Pigeon

OTHER NAMES (European) Woodpigeon, Ring Dove/Pigeon.
TAXONOMY Linnaeus, 1758. Five subspecies (all in region; sixth subspecies, *maderensis* from Madeira, apparently now extinct).
DISTRIBUTION [Europe to C Asia, NW Africa. Increasing. LC] Widespread across Europe except extreme N, also NW Africa and parts of N Middle East (*palumbus*), with isolated populations N Oman (*iranica* and possibly *casiotis*), SW and N Iran (*iranica*), SE Iran (*casiotis*) and Azores (*azorica*); has bred Faroe Is and Iceland. Migratory across much of C and E of range; winter distribution extends to non-breeding areas of NW Africa, Malta (very scarce) and in Middle East S to Israel. Vagrant Egypt, Gibraltar, Jan Mayen, Jordan, Svalbard and UAE.

Columba trocaz
Trocaz Pigeon (E)

OTHER NAMES Long-toed (or Madeira Laurel) Pigeon.
TAXONOMY Heineken, 1829. Monotypic. Formerly sometimes considered conspecific with Bolle's Pigeon.
DISTRIBUTION [Madeira. Increasing. LC] Endemic resident Madeira, formerly also Porto Santo.

Columba bollii
Bolle's Pigeon (E)

OTHER NAMES Bolle's (or Dark-tailed) Laurel Pigeon.
TAXONOMY Godman, 1872. Monotypic. Formerly sometimes treated as conspecific with Trocaz Pigeon.
DISTRIBUTION [Canary Is. Increasing. LC] Endemic resident W Canary Is on La Palma, El Hierro, La Gomera and Tenerife, possibly formerly also Gran Canaria.

Columba junoniae
Laurel Pigeon (E)

OTHER NAME White-tailed Laurel Pigeon.
TAXONOMY Hartert, 1916. Monotypic.
DISTRIBUTION [Canary Is. Increasing. NT] Endemic resident W Canary Is on La Palma, El Hierro, La Gomera and Tenerife, formerly also Gran Canaria.

Columba arquatrix
African Olive Pigeon

OTHER NAMES Rameron (or Yellow-eyed) Pigeon, Speckled Wood Pigeon.
TAXONOMY Temminck, 1808. Monotypic.
DISTRIBUTION [E and S Africa, SW Arabia. Decreasing. LC] Presumed rare breeder SW Saudi Arabia and W Yemen; probably resident but possibly migratory (most records Mar-Nov).

Streptopelia turtur
European Turtle Dove

OTHER NAMES (Common or Eurasian) Turtle Dove; Isabelline Turtle Dove (desert forms).
TAXONOMY (Linnaeus, 1758). Four subspecies (all in region). Male *rufescens* (Egypt/Libya) may be dimorphic.
DISTRIBUTION [Europe to C Asia, N Africa > Sahel. Decreasing. VU] Summer visitor across much of Europe except N (*turtur*), also Canary Is, N Africa, N Middle East and few disjunct areas of Arabia (*arenicola*); resident forms present in C Sahara mtns (*hoggara*) and Nile Valley and Western/Libyan Desert (*rufescens*). Widespread on passage S of breeding range, when also regular Ireland (scarce) and Sweden. Vagrant Andorra, Azores, Cape Verde Is, Iceland, Jan Mayen, Madeira, Norway and Svalbard; intriguing record of *rufescens* from Faroe Is (3 Nov 1987, taken into care and later died) placed in Cat D.

Streptopelia lugens
Dusky Turtle Dove

OTHER NAMES Black (or Pink-breasted) Dove.
TAXONOMY (Rüppell, 1837). Monotypic. Arabian birds have previously been described as *bishahensis*.
DISTRIBUTION [E Africa, SW Arabia. Stable. LC] Breeds SW Saudi Arabia and W Yemen; possibly migratory in N of range.

Streptopelia orientalis
Oriental Turtle Dove

OTHER NAMES Eastern (or Rufous) Turtle Dove; Eastern Oriental Turtle Dove (*S o orientalis*), Western Oriental (or Mountain/Turkestan) Turtle Dove (*S o meena*).
TAXONOMY (Latham, 1790). Six subspecies (two in region). W Asian form *meena* may merit species status on basis of morphological and vocal differences.
DISTRIBUTION [Asia > S and SE Asia. Stable. LC] Summer visitor Ural Mtns, possibly also extreme SE European Russia (*meena*). Uncommon passage migrant N Caspian region, and in autumn (*meena*) also Oman and UAE. Vagrant Austria, Azerbaijan, Britain, Denmark, Egypt, Estonia, Faroe Is, France, Germany, Greece, Hungary, Iran, Iraq, Israel, Italy, Kuwait, Latvia, Netherlands, Norway, Poland, Saudi Arabia, Serbia, Spain, Sweden, Turkey and Ukraine; both *meena* (mainly Oct-Nov) and *orientalis* (Oct-Apr) occur as vagrants but subspecies undetermined in many extralimital records.

Streptopelia decaocto
Eurasian Collared Dove

OTHER NAME Collared Dove.
TAXONOMY (Frivaldszky, 1838). Two subspecies (one in region). Sometimes considered conspecific with African Collared Dove.
DISTRIBUTION [Eurasia and North America. Increasing. LC] Widespread Europe except far N (has bred Iceland), also Middle East (where at least partly introduced and range expanding), small disjunct areas of N Africa W to N Mauritania, and Canary Is (*decaocto*). Apparently colonising Azores and Cape Verde Is. Vagrant Madeira, Svalbard and Yemen.

Streptopelia roseogrisea
African Collared Dove

OTHER NAMES Pink-headed Dove, Rose-grey Turtle Dove; Barbary (or Ring-necked) Dove (domesticated '*risoria*' form).
TAXONOMY (Sundevall, 1857). Two subspecies (both in region). Sometimes treated as conspecific with Eurasian Collared Dove.
DISTRIBUTION [Sahel to SW Arabia. Stable. LC] Resident W Saudi Arabia and Yemen (*arabica*); regular extreme S Egypt (breeding status uncertain), N Mauritania and perhaps

N Niger (*roseogrisea*). Rare disperser Kuwait (where recent breeding may relate to escaped birds), and rare resident or migrant Oman. Probably migratory in N of range. Vagrant Bahrain, Israel and Jordan. Feral populations of domestic form Barbary Dove occur *eg* Balearic Is, Canary Is, Malta, Spain and UAE.

Streptopelia decipiens
Mourning Collared Dove (V)

OTHER NAME African Mourning Dove.
TAXONOMY (Hartlaub and Finsch, 1870). Six subspecies (at least one in region).
DISTRIBUTION [Sub-Saharan Africa. Stable. LC] Extreme vagrant (two records of three individuals): Egypt (displaying pair, 28 Dec 2010-11 May 2012 at least; one individual to 1 Aug 2013); N Mauritania (27 March 2003, *shelleyi*).

Streptopelia semitorquata
Red-eyed Dove

OTHER NAMES Black (or Half-collared) Dove, Black Pigeon.
TAXONOMY (Rüppell, 1837). Monotypic.
DISTRIBUTION [Sub-Saharan Africa and SW Arabia. Increasing. LC] Resident SW Saudi Arabia and W Yemen.

Streptopelia tranquebarica
Red Turtle Dove (V)

OTHER NAMES Red Collared (or Red Ring) Dove.
TAXONOMY (Hermann, 1804). Two subspecies (undetermined in region).
DISTRIBUTION [S, SE and E Asia. Stable. LC] Extreme vagrant (five records): Oman (4: 7 and 19-24 Oct 1975; 5-11 Jan 2002; 4 Oct 2006; 1 Feb-27 Mar 2008); UAE (21 Oct 1997). Additional records from Iran (29 Jan 1975, two) and Lebanon (2 Oct 1999) may refer to escapes.

Spilopelia senegalensis
Laughing Dove

OTHER NAMES Egyptian Turtle Dove, Palm (or Senegal) Dove.
TAXONOMY (Linnaeus, 1766). Six subspecies (four in region).
DISTRIBUTION [Africa, Middle East and S Asia. Stable. LC] Resident N Mauritania and N Mali (*senegalensis*), Canary Is and Morocco (*phoenicophila* and/or *senegalensis*), Algeria to NW Libya (*phoenicophila*), Nile Valley (*aegyptiaca*), N Middle East (*phoenicophila/senegalensis*) N to S Caucasus and E to W Arabia (*senegalensis*), E Arabia and E Iran (*cambayensis*). Some E populations perhaps partly introduced. Vagrant Balearic Is, Bulgaria, Cyprus (has bred), Greece, Italy, Madeira, Malta, Sardinia and Spain; further European occurrences placed in Cat D (Portugal) and Cat E (Finland, Germany and Norway).

Turtur abyssinicus
Black-billed Wood Dove (V)

OTHER NAMES Abyssinian (or Black-billed Blue-spotted) Wood Dove.
TAXONOMY (Sharpe, 1902). Monotypic.
DISTRIBUTION [Sahel. Stable. LC] Extreme vagrant: Yemen (14 Nov 2004, two).

Oena capensis
Namaqua Dove

OTHER NAMES Cape Dove, Long-tailed (or Masked) Dove.
TAXONOMY (Linnaeus, 1766). Two subspecies (one in region).
DISTRIBUTION [Sub-Saharan Africa and Middle East. Increasing. LC] Breeds N Mauritania, N Mali, SE Egypt, Israel, Jordan and widely across parts of Arabia N to *eg* Kuwait (*capensis*); also now regular Iraq (has bred) and recently recorded Syria, where possibly resident. Highly dispersive. Vagrant Cape Verde Is, Cyprus, Iran, Iraq, Lebanon and Turkey.

Geopelia striata
Zebra Dove (I)

OTHER NAME Barred (Ground) Dove, Zebra Ground Dove.
TAXONOMY (Linnaeus, 1766). Monotypic.

DISTRIBUTION [SE Asia. Stable. LC] Naturalised introduction with self-sustaining population in Qatar.

Zenaida macroura
Mourning Dove (V)

OTHER NAMES American Mourning (or Carolina) Dove.
TAXONOMY (Linnaeus, 1758). Five subspecies (undetermined in region).
DISTRIBUTION [North America, Central America and West Indies. Increasing. LC] Very rare vagrant: Azores (5*); Britain (4); Denmark (19-21 May 2008); Germany (4 May 2008); Iceland (2: 19 Oct 1995; 12 Sep 2012, caught by cat, died next day*); Ireland (2: 2-15 Nov 2007; 25 Oct 2009). Danish and German records are believed to involve same bird as first Irish record.

Treron waalia
Bruce's Green Pigeon

OTHER NAME Yellow-bellied Green Pigeon.
TAXONOMY (Meyer, F A A, 1793). Monotypic.
DISTRIBUTION [Sahel to NE Africa and S Arabia. Decreasing. LC] Breeds in summer SW Saudi Arabia, Yemen and SW Oman. Partial migrant, some movement recorded S Red Sea and Gulf of Aden. Vagrant Egypt.

ORDER CUCULIFORMES

FAMILY CUCULIDAE

Cuckoos · 13 species

Centropus senegalensis
Senegal Coucal

TAXONOMY (Linnaeus, 1766). Three subspecies (one in region).
DISTRIBUTION [Sub-Saharan Africa and Egypt. Stable. LC] Resident Egypt in Nile Delta and discontinuously S along Nile to about 28°N (*aegyptius*); recorded three times S to 24°N.

Centropus superciliosus
White-browed Coucal

TAXONOMY Hemprich & Ehrenberg, 1829. Three subspecies (one in region).
DISTRIBUTION [E and S Africa, SW Arabia. Stable. LC] Resident SW Saudi Arabia and W Yemen (*sokotrae*).

Clamator glandarius
Great Spotted Cuckoo

TAXONOMY (Linnaeus, 1758). Monotypic.
DISTRIBUTION [S Europe, Middle East and Africa. Stable. LC] Summer visitor Iberia, S France and NW Italy, also parts of Balkans (mainly SE), Cyprus, Turkey, Iraq, W Iran and N Middle East; resident N Morocco, NW Algeria and Nile Valley, formerly also N Tunisia, and has bred N Mauritania. More widespread on passage, including NW Africa and N and W Arabia, but very rare Malta. Vagrant Armenia, Austria, Bahrain, Belgium, Britain, Canary Is, Cape Verde Is, Channel Is, Corsica, Denmark, Finland, Germany, Georgia, Hungary, Ireland, Libya, Madeira, Moldova, Netherlands, Norway, Oman, Qatar, Romania, Slovakia, Slovenia, Sweden, Switzerland, UAE, Ukraine and Yemen.

Clamator jacobinus
Jacobin Cuckoo

OTHER NAMES Pied (or Black-and-white) Cuckoo, (Black/Pied) Crested Cuckoo.
TAXONOMY (Boddaert, 1783). Three subspecies (at least two in region). Subspecific identity of SW Arabian population requires clarification.
DISTRIBUTION [Sub-Saharan Africa, SW and S Asia > W to E Africa, S Asia. Stable. LC] Breeds W Yemen and possibly SW Saudi Arabia (either *pica* or undescribed taxon). Uncommon passage migrant Oman (*jacobinus*) and non-breeding visitor N Mauritania (*pica*). Vagrant N Chad (presumed *pica*),

Iran and UAE; further record from Finland (11 Sep 1976) currently in Cat E, but review expected.

Coccyzus americanus
Yellow-billed Cuckoo (V)

TAXONOMY (Linnaeus, 1758). Monotypic.
DISTRIBUTION [E, C and S North America, West Indies > N and C South America. Decreasing. LC] Vagrant (chiefly Oct): Azores (39*); Balearic Is (28 Oct 1994); Belgium (2); Britain (63); Denmark (23 Oct 1936); France (5-6 Oct 2005); Iceland (3); Ireland (10); Italy (7); Morocco (25 Oct 1977); Norway (2).

Coccyzus erythropthalmus
Black-billed Cuckoo (V)

TAXONOMY (Wilson, A, 1811). Monotypic.
DISTRIBUTION [E and C North America > NW South America. Decreasing. LC] Very rare vagrant (mainly Oct): Azores (4*); Britain (13); Denmark (16 Oct 1970, shot); France (15 Nov 2013); Germany (8 Oct 1952); Iceland (2); Ireland (25 Sep 1871); Italy (1858).

Eudynamys scolopaceus
Asian Koel

OTHER NAMES Common (or Indian/Western) Koel.
TAXONOMY (Linnaeus, 1758). Five subspecies (one in region).
DISTRIBUTION [S and E Asia. Stable. LC] Fairly common but irregular non-breeding visitor Oman (probably *scolopaceus* on range). Vagrant Iran, Kuwait, Qatar and UAE; two further records from Bahrain under review.

Chrysococcyx caprius
Diederik Cuckoo

OTHER NAMES Dideric (or Didric) Cuckoo.
TAXONOMY (Boddaert, 1783). Monotypic.
DISTRIBUTION [Sub-Saharan Africa and S Arabia. Stable. LC] Breeding summer visitor to SW Oman and E Yemen, and in W Yemen and (marginally) SW Saudi Arabia. Vagrant Cyprus, Israel and N Mauritania.

Chrysococcyx klaas
Klaas's Cuckoo

OTHER NAME Klass's Cuckoo.
TAXONOMY (Stephens, 1815). Monotypic.
DISTRIBUTION [Sub-Saharan Africa and SW Arabia. Stable. LC] Breeding summer visitor to SW Saudi Arabia and W Yemen.

Cacomantis passerinus
Grey-bellied Cuckoo (V)

OTHER NAMES (Indian) Plaintive Cuckoo.
TAXONOMY (Vahl, 1797). Monotypic. Now considered specifically distinct from extralimital Plaintive Cuckoo *C merulinus* (SE Asia), based chiefly on grey belly and different vocalisations.
DISTRIBUTION [S Asia. Stable. LC] Extreme vagrant (four records, mainly late autumn): Oman (4: 7 Nov 1988; 3 Nov 2002; 18 Oct-3 Nov 2006; 1 Apr 2010).

Hierococcyx varius
Common Hawk-Cuckoo (V)

TAXONOMY (Vahl, 1797). Two subspecies (one in region).
DISTRIBUTION [S Asia. Stable. LC] Extreme vagrant (two records): Oman (2: 18 Nov 1988; 13 Jan 2010).

Cuculus optatus
Oriental Cuckoo

OTHER NAMES Horsfield's (or Northern Oriental) Cuckoo.
TAXONOMY Gould, 1845. Monotypic. Now separated from extralimital Himalayan *C saturatus* and Sunda *C lepidus* Cuckoos (SE Asia), based primarily on distinct song types and measurements.
DISTRIBUTION [N and E Asia > SE Asia to N and E Australia. Stable. LC] Summer visitor

Russia to Ural Mtns, occasionally W to St Petersburg region. Vagrant Iran and Israel.

Cuculus canorus
Common Cuckoo

OTHER NAMES (Eurasian or European) Cuckoo, Grey Cuckoo.
TAXONOMY Linnaeus, 1758. Four subspecies (three in region). Recent evidence also suggests assortative breeding within 'host-specific races', in which male calls differ from each other and females parasitise different host species according to habitat type.
DISTRIBUTION [Eurasia > sub-Saharan Africa and SE Asia. Decreasing. LC] Summer visitor NW Africa, Iberia and Balearic Is (*bangsi*), Ireland, Britain and across Europe N to N Scandinavia and E to Ural Mtns and Israel (*canorus*) and N Iran (*subtelephonus*). May breed elsewhere in Middle East, and widespread there and parts of N Africa on passage. Vagrant Azores, Canary Is, Cape Verde Is, Faroe Is, Iceland, Jan Mayen, Madeira and Svalbard.

ORDER STRIGIFORMES

FAMILY TYTONIDAE

Barn owls · one species

Tyto alba
Western Barn Owl

OTHER NAMES Common (or Pale Western) Barn Owl; African Barn Owl (*T a affinis*), Canary Islands (or Slender-billed) Barn Owl (*T a gracilirostris*), Cape Verde (Barn) Owl (*T a detorta*), Dark-breasted Barn Owl (*T a guttata*), Madeira Barn Owl (*T a schmitzi*), Pale Barn Owl (*T a alba*).
TAXONOMY (Scopoli, 1769). Twenty-seven subspecies (eight in region). Taxonomy not yet fully clarified; often treated as a single cosmopolitan entity, but also variably separated into New World, Old World and island species, in region notably with forms on E Canary Is, Cape Verde Is and Madeira sometimes separated at species level.
DISTRIBUTION [Europe, S Asia and Africa. Stable. LC] Resident Cape Verde Is (*detorta*), E Canary Is (*gracilirostris*), Madeira (*schmitzi*), N Mauritania (*affinis*), N Africa, W and S Europe E to Balkans (*alba*), C and E Europe to Ukraine (*guttata*), Corsica and Sardinia (*ernesti*), and Crete and Cyprus E through N Middle East to SW Iran, locally also Arabia (*erlangeri*). Vagrant Azerbaijan, Estonia, Finland, Jan Mayen and Norway.

FAMILY STRIGIDAE

Typical owls · 21 species

Otus brucei
Pallid Scops Owl

OTHER NAMES Bruce's (or Striated) Scops Owl.
TAXONOMY (Hume, 1872). Four subspecies (two in region). Has often been considered a subspecies of Eurasian Scops Owl or extralimital Oriental Scops Owl *O sunia* (S, SE and E Asia), but differs significantly from both in voice and morphology, with no hybridisation in areas of range overlap. Arabian Scops Owl (*qv*) has occasionally been included within present species.
DISTRIBUTION [SW and C Asia. Stable. LC] Breeds SE Turkey, N Iran and possibly Syria (*obsoletus*), and C Iraq, E UAE, N and E Oman, and S Iran (*exiguus*). Eastern populations resident but otherwise mostly migratory, wintering S and W to Jordan, Israel, NE Egypt and W Saudi Arabia. Vagrant Bahrain, Kuwait, Qatar and Yemen.

Otus pamelae
Arabian Scops Owl (E)

TAXONOMY Bates, 1937. Monotypic. Formerly treated as conspecific with extralimital African Scops Owl *O senegalensis* (sub-Saharan Africa), but well differentiated genetically, phylogenetically, morphologically and vocally.
DISTRIBUTION [Arabia. Stable. LC] Endemic resident SW Saudi Arabia, W and S Yemen, and W Oman.

Otus scops
Eurasian Scops Owl

OTHER NAMES (Common or European) Scops Owl; Cyprus Scops Owl (*O s cyprius*).
TAXONOMY (Linnaeus, 1758). Six subspecies (all in region). Despite limited genetic differences, *cyprius* differs in plumage, voice and largely non-migratory behaviour, and has recently been proposed as specifically distinct.
DISTRIBUTION [NW Africa, S and E Europe, SW and C Asia > W, C and E Africa. Decreasing. LC] Summer visitor to NW Africa, Iberia and Balearic Is (*mallorcae*), from France E through S and E Europe to N Turkey and S Caucasus (*scops*), N Caucasus (*pulchellus*), S Greece and Crete to S Turkey, Syria and Jordan (*cycladum*), and Iraq and Iran (*turanicus*); mainly resident Cyprus (*cyprius*). Widespread on migration S of breeding range, but some winter Mediterranean; rarely occurs also far N of normal range, especially in spring, when occasionally remains on territory. Vagrant Belgium, Britain, Canary Is, Cape Verde Is, Denmark, Faroe Is, Finland, Germany (has bred), Iceland, Ireland, Latvia, Liechtenstein, Luxembourg, Madeira, Netherlands, Norway and Sweden.

Bubo scandiacus
Snowy Owl

TAXONOMY (Linnaeus, 1758). Monotypic.
DISTRIBUTION [Holarctic > N North America and N Eurasia. Decreasing. LC] Breeds arctic and subarctic latitudes, principally Svalbard, Jan Mayen, Norway and NW Russia, variably also Iceland, Finland and Sweden, and has bred N Britain; range linked to food availability. Moves S in winter into N and NE Europe, mainly from S Scandinavia E to N Russia, rarely in irruptive movements farther S. Some W European records known to involve ship-assisted individuals from E North America. Vagrant Albania, Austria, Azores, Belarus, Belgium, Channel Is, Czech Republic, Denmark, Estonia, France, Germany, Iran, Ireland, Latvia, Luxembourg, Netherlands, Serbia, Slovakia, Slovenia and Ukraine.

Bubo bubo
Eurasian Eagle-Owl

OTHER NAMES Eagle Owl; Byzantine Eagle-Owl (*B b interpositus*).
TAXONOMY (Linnaeus, 1758). Sixteen subspecies (eight in region). Status of genetically distinct *interpositus* worthy of investigation, as is extent of reported interbreeding in Middle East with now-split Pharaoh Eagle-Owl (*qv*).
DISTRIBUTION [Widespread Eurasia. Decreasing. LC] Breeds Iberia (*hispanus*), S, C and E Europe to Scandinavia, with isolated populations Benelux countries and N Germany (*bubo*), C, E and S European Russia (*ruthenus*), Ural Mtns (*sibiricus*), N Caspian region (*turcomanus*), Caucasus, S Ukraine, SE Balkans to N Middle East (*interpositus*), and N, E and S Iraq and much of Iran (*nikolskii*) except NE Iran (*omissus*). Status uncertain Algeria, where two specimens determined as close to Iberian *hispanus* and postulated as locally breeding; similarly Morocco, where *hispanus* suspected but not proven. Status also undetermined Britain, where apparently present in fossil record and birds of unknown (but probably mainly captive) origin have bred in wild since 1987 (vagrancy never formally accepted). Vagrant Corsica, Gibraltar and Sardinia.

Bubo ascalaphus
Pharaoh Eagle-Owl

OTHER NAMES Desert (or Savigny's) Eagle-Owl.
TAXONOMY Savigny, 1809. Monotypic. Formerly considered a subspecies of Eurasian Eagle-Owl but vocalisations, morphology and DNA evidence indicate that *ascalaphus* is specifically distinct; possibly formerly sympatric Algeria and apparently largely still so in areas of range overlap in N Middle East, though some interbreeding reported Palestine and Syria. Paler S Saharan and Arabian populations sometimes separated as *desertorum*.
DISTRIBUTION [N Africa and Arabia. Stable. LC] Breeds from N Mauritania and Morocco E across parts of N Africa to E Egypt, and from Sinai, Israel and Palestine through N Middle East to W Iraq, and S through much of Arabia (especially N, C and E); scarce S Oman, rare Yemen. Vagrant Bahrain.

Bubo africanus
Spotted Eagle-Owl

OTHER NAME Arabian (Spotted) Eagle Owl (*B a milesi*).
TAXONOMY (Temminck, 1821). Two subspecies (one in region). Arabian population is smaller and differs in plumage and vocalisations from geographically separated nominate African form, and may be specifically distinct.
DISTRIBUTION [E and S Africa, S Arabia. Stable. LC] Resident SW Saudi Arabia, Yemen and Oman (*milesi*).

Ketupa zeylonensis
Brown Fish Owl

OTHER NAMES Brown Fishing-owl; Turkish (or Western Brown) Fish Owl (*K z semenowi*).
TAXONOMY (Gmelin, J F, 1788). Four subspecies (one in region). Genetic sampling of feathers from Turkey revealed 2 per cent divergence from other Brown Fish Owl subspecies and from extralimital Buffy Fish Owl *K ketupu* (S and SE Asia), possibly indicating specific status for isolated W population; current IOC taxonomy followed here pending further research.
DISTRIBUTION [N Middle East to SE Asia. Decreasing. LC] Recently rediscovered S Turkey and SE Iran (*semenowi*). Formerly also bred or occurred Iraq, Israel, Jordan, Lebanon and Syria, but very few historic records and present status and distribution in region little known.

Strix aluco
Tawny Owl

OTHER NAMES Eurasian Tawny (or Brown) Owl, Tawny (or Eurasian) Wood-Owl; Maghreb Wood Owl (*S a mauritanica*).
TAXONOMY Linnaeus, 1758. Eight subspecies (six in region). Populations in Iberia, Asia Minor and Middle East, and those from SW Russia and from S Caspian, sometimes further subspecifically separated.
DISTRIBUTION [Europe, N Africa, SW and C Asia. Stable. LC] Resident NW Africa (*mauritanica*), W and S Europe to W Turkey and S to N Israel (*sylvatica*), N, C and SE Europe E to Ukraine and W Russia (*aluco*), Ural Mtns (*siberiae*), NE Turkey and Caucasus to NW Iran (*willkonskii*), and W Iran, possibly also NE Iraq (*sanctinicolai*); absent Atlantic islands and Balearic Is, Crete, Cyprus, Faroe Is, Iceland and Ireland. Vagrant Sardinia.

Strix butleri
Omani Owl (E)

OTHER NAMES Hume's (Tawny) Owl.
TAXONOMY (Hume, 1878). Monotypic. Only one *Strix* species, Hume's Owl *S butleri*, was known to occur in desert habitats of Middle East until discovery in 2013 of unknown owl in N Oman which differed significantly in vocalisations and plumage, and was thus proposed as new species, Omani Owl *S omanensis*. Subsequent genetic research and examination of specimens confirmed *Strix* from Oman to be same species as type specimen of *S butleri* (which scientific name therefore has priority), but that all other specimens originally identified as latter belonged to separate, hitherto unrecognised species now described as Desert Owl (*qv*).
DISTRIBUTION [Middle East. Population trend unknown. NE] Endemic (?) resident currently known only from very few locations in N Oman, NE UAE, and SW and NE Iran; type specimen apparently collected SW Pakistan, so range may still extend outside region.

Strix hadorami
Desert Owl

OTHER NAMES Desert Tawny Owl, Hume's (Tawny) Owl.
TAXONOMY Kirwan, Schweizer and Copete, 2015. Monotypic. Formerly included within *Strix butleri* (*qv*).
DISTRIBUTION [Middle East. Stable. LC] Near-endemic. Resident E Egypt, Sinai, Israel, Palestine and Jordan, and discontinuously in W, C and E Saudi Arabia, Yemen and S Oman.

Strix uralensis
Ural Owl

TAXONOMY Pallas, 1771. Ten subspecies (three in region).
DISTRIBUTION [C and N Europe, N Asia. Stable. LC] Resident Norway, Sweden, Finland, Baltic

States and NE Poland E through Belarus and NW Russia (*liturata*), E European Russia (*uralensis*) and in C and SE Europe from Hungary to Bulgaria (*macroura*). Reintroduced S Germany; stragglers also recorded N Germany. Vagrant Albania, Austria (formerly bred) and Italy.

Strix nebulosa
Great Grey Owl

OTHER NAME Lapland Owl (*S n lapponica*).
TAXONOMY Forster, J R, 1772. Two subspecies (one in region).
DISTRIBUTION [N North America and N Eurasia. Stable. LC] Mainly resident N Europe (*lapponica*) from N Norway, NE Sweden, Finland and NW Russia S to Belarus and N Ukraine; has also bred recently S Sweden and E Poland. Bred Estonia in 18th century and again since 2009. Food-related nomadism and periodic winter irruptions W into Scandinavia well known, but do not extend S of Baltic. Vagrant Latvia and Lithuania.

Surnia ulula
Northern Hawk-Owl

OTHER NAMES (European) Hawk Owl.
TAXONOMY (Linnaeus, 1758). Three subspecies (two in region).
DISTRIBUTION [N North America and N and C Eurasia. Stable. LC] Breeds Norway, Sweden, Finland and NW Russia; rare Estonia (*ulula*). Distribution linked to availability of voles, often in four-year cycle, leading to nomadic movements and periodic irruptions S of breeding range (rarely up to 1,500 km). Vagrant Austria, Belarus, Belgium, Britain, Czech Republic, Denmark, France, Germany, Hungary, Latvia (formerly bred), Lithuania, Luxembourg, Netherlands, Poland, Romania, Serbia, Slovakia, Switzerland and Ukraine; *caparoch* (North America) vagrant Britain (4) and Canary Is (Oct 1924*, caught at sea, died in captivity).

Glaucidium passerinum
Eurasian Pygmy Owl

OTHER NAMES (European or Northern) Pygmy Owl.
TAXONOMY (Linnaeus, 1758). Two subspecies (one in region).
DISTRIBUTION [Europe and N Asia. Stable. LC] Mainly resident N Europe from Scandinavia, Baltic States, E Poland and Belarus E to Ural Mtns, and C Europe discontinuously from Germany and S Poland S to SE France, N Italy, Carpathian Mtns and Balkans S to N Greece and E to SW Ukraine (*passerinum*). Some dispersal and occasional irruptions, especially in N of range. Vagrant Belgium, Denmark and Netherlands.

Athene noctua
Little Owl

OTHER NAMES Northern Little Owl; Bactrian Little Owl or Hutton's Owlet (*A n bactriana*), Byzantine (or Caspian) Little Owl (*A n indigena*), Cucumiau (*A n noctua*), Lilith Owl/Owlet or Desert Little Owl (*A n lilith*), North African Little Owl (*A n glaux*), Saharan Little Owl (*A n saharae*).
TAXONOMY (Scopoli, 1769). Thirteen subspecies (seven in region). *Vidalii* is most genetically divergent taxon and differs significantly in vocalisations, while *lilith* also differs in *eg* mtDNA, plumage and shorter tarsus; both forms sometimes treated as specifically distinct.
DISTRIBUTION [Europe, Asia and N Africa. Stable. LC] Resident from N Mauritania and interior Morocco to W Egypt and C Arabia (*saharae*), coastal N Africa to SW Israel (*glaux*), SW and W Europe to S Baltic (*vidalii*), C, S and parts of SE Europe to NW Russia (*noctua*), S Russia and Ukraine to Romania, Greece and Turkey (*indigena*), S Turkey and Cyprus S to Sinai (*lilith*), and Iraq, Iran and Azerbaijan (*bactriana*). Introduced S Britain and Mallorca. Vagrant Andorra, Finland, Ireland, Malta and Sweden.

Athene brama
Spotted Owlet

OTHER NAMES Spotted Little Owl; Northern Spotted Owlet (*A b indica*).
TAXONOMY (Temminck, 1821). Five subspecies (one in region).
DISTRIBUTION [S and SE Asia. Stable. LC] Resident SE Iran (*indica*).

Aegolius funereus
Tengmalm's Owl

OTHER NAME Boreal Owl (IOC).
TAXONOMY (Linnaeus, 1758). Seven subspecies (two in region). Genetic studies demonstrate that mtDNA differs by 3.6 per cent between Palearctic and Nearctic populations, suggesting that they may best be treated as separate species (in which case North American name Boreal Owl would apply to latter).
DISTRIBUTION [N North America and N Eurasia. Stable. LC] Mainly resident N Europe from Scandinavia E through Baltic and NW Russia, S through E Europe and discontinuously in C, W, S and SE Europe, including Alps, Pyrenees, Balkans and Carpathians (*funereus*), also N Caucasus (*caucasicus*) and Turkey (subspecies undetermined). Some dispersal and occasional irruptions noted, especially in N of range. Vagrant Britain, Hungary and Netherlands (occasionally breeds).

Asio otus
Long-eared Owl

OTHER NAMES Common (or Northern) Long-eared Owl.
TAXONOMY (Linnaeus, 1758). Four subspecies (two in region).
DISTRIBUTION [North America, Eurasia and N Africa. Decreasing. LC] Breeds Canary Is (*canariensis*), Azores, NW Africa, Europe from Mediterranean N to Scandinavia and from Ireland E to Ural Mtns, also Caucasus, NE Iran and Asia Minor S to Israel (*otus*). Migratory in N of breeding range, wintering W and S Europe, N Middle East and N Africa. Rare passage and winter visitor Faroe Is. Vagrant Bear Is, Gibraltar, Iceland (has bred), Jan Mayen, Kuwait, Libya, Oman, Qatar, Saudi Arabia, Svalbard, Syria and UAE.

Asio flammeus
Short-eared Owl

TAXONOMY (Pontoppidan, 1763). Eleven subspecies (one in region).
DISTRIBUTION [Cosmopolitan. Decreasing. LC] Breeds from Iceland, Ireland and Britain E across much of Europe except Mediterranean basin (has bred Spain) to Ural Mtns, and S to *eg* Alps, Balkans and Caucasus (*flammeus*). Northern populations migratory, wintering S to Canary Is (irregular), N Africa and Middle East (has bred Israel). Vagrant Azores, Bear Is, Cape Verde Is, Corsica, Gibraltar, Jan Mayen, Liechtenstein, Madeira, Slovenia and Svalbard.

Asio capensis
Marsh Owl

OTHER NAMES African (or Algerian) Marsh Owl.
TAXONOMY (Smith, A, 1834). Three subspecies (one in region).
DISTRIBUTION [NW and sub-Saharan Africa. Stable. LC] Resident N Morocco (*tingitanus*), where now critically endangered; formerly also Algeria. In 19th century rare winter straggler S Iberia, especially Spain but also Portugal (two, shot); only one record since, in Spain (10 Dec 1998, shot). Precise location of single record from N Mauritania (Dec 1982, dead) unclear, but possibly outside regional boundary.

ORDER CAPRIMULGIFORMES

FAMILY CAPRIMULGIDAE

Nightjars · nine species

Chordeiles minor
Common Nighthawk (V)

OTHER NAMES (Booming) Nighthawk.
TAXONOMY (Forster, J R, 1771). Nine subspecies (one in region).
DISTRIBUTION [North and Central America, West Indies > N and C South America, West Indies. Decreasing. LC] Very rare vagrant (mainly Sep-Oct, *minor* where determined): Azores (13 records of 14 birds*); Britain (22); Canary Is (Dec 1972, dead); Faroe Is (1 Oct 1955, dead); France (17-28 Sep 1998); Iceland (23 Oct 1955, collected); Ireland (2: 24 Oct 1999; 24-25 Oct 2009); Spain (4 Nov 2005).

Caprimulgus ruficollis
Red-necked Nightjar

OTHER NAME Tunisian Red-necked Nightjar (*C r desertorum*).
TAXONOMY Temminck, 1820. Two subspecies (both in region).
DISTRIBUTION [NW Africa and Iberia > W Africa. Decreasing. LC] Breeding endemic. Summer visitor NW and N Morocco, S and E Spain and S Portugal (*C r ruficollis*) and NE Morocco, Algeria and Tunisia (*C r desertorum*). On migration regular Gibraltar and occasional N Mauritania; a few winter records from Morocco. Vagrant Balearic Is, Britain, Croatia, France, Italy, Madeira (Porto Santo) and Malta.

Caprimulgus europaeus
European Nightjar

OTHER NAMES (Common or Eurasian) Nightjar.
TAXONOMY Linnaeus, 1758. Six subspecies (three in region).
DISTRIBUTION [NW Africa and W Europe to CE Asia > sub-Saharan Africa. Decreasing. LC] Summer visitor NW Africa, from Iberia E through S Europe to Asia Minor, Caucasus and NW Iran (*meridionalis*), from Ireland and Britain E through much of C and N Europe N to S Fennoscandia and E to Ural Mtns (*europaeus*), and in N Iraq and SW and S Iran (*unwini*); has bred Israel. Occurs widely to S on migration, when regular Gibraltar and throughout N Africa and Middle East; rare Channel Is. Vagrant Azores, Canary Is, Faroe Is, Liechtenstein, Madeira and Iceland.

Caprimulgus aegyptius
Egyptian Nightjar

TAXONOMY Lichtenstein, M H C, 1823. Two subspecies (both in region).
DISTRIBUTION [NW Africa to C Asia > Sahel and N Egypt. Decreasing. LC] Summer visitor NW Sahara and W Egypt (*saharae*) and E Egypt, Iraq and SW Iran, occasionally also Israel and Jordan (*aegyptius*). Occurs on migration W Sahara, including N Mauritania, Egypt, Kuwait S through Arabia, N and SE Iran, but extent of non-breeding distribution uncertain; some winter records Egypt, few also S Sahara. Vagrant Britain, Canary Is, Denmark, Germany, Italy, Libya (has bred), Malta, Sweden and Syria.

Caprimulgus mahrattensis
Sykes's Nightjar

OTHER NAME Sind Nightjar.
TAXONOMY Sykes, 1832. Monotypic.
DISTRIBUTION [SW and S Asia. Stable. LC] Resident SE Iran. Vagrant UAE.

Caprimulgus nubicus
Nubian Nightjar

TAXONOMY Lichtenstein, M H C, 1823. Four subspecies (one in region).
DISTRIBUTION [Middle East, SW Arabia, NE and E Africa. Decreasing. LC] Breeds Israel, Jordan and discontinuously in SW Saudi Arabia and Yemen; formerly also bred Palestine (*tamaricis*). Apparently resident Israel but possibly partially migratory in S of range. Vagrant Oman.

Caprimulgus eximius
Golden Nightjar (V)

TAXONOMY Temminck, 1826. Two subspecies (one in region).
DISTRIBUTION [Sahel. Stable. LC] Extreme vagrant (*simplicior*): N Mauritania (30 Apr-6 May 2003). Claim from SW Morocco (6 Jun 1955) now regarded only as possible sight record.

Caprimulgus poliocephalus
Montane Nightjar

OTHER NAMES Abyssinian (or Mountain) Nightjar.
TAXONOMY Rüppell, 1840. Monotypic.
DISTRIBUTION [NE and E Africa, SW Arabia. Stable. LC] Breeds SW Saudi Arabia, where recorded year-round, though apparently makes seasonal altitudinal movements. One record also from W Yemen, where possibly breeds in small numbers.

Caprimulgus inornatus
Plain Nightjar

OTHER NAMES Ludovic's (or Ethiopian) Nightjar.
TAXONOMY Heuglin, 1869. Monotypic.
DISTRIBUTION [Sahel to NE Africa and SW Arabia > W, C and E Africa. Stable. LC] Summer visitor SW Saudi Arabia and Yemen.

FAMILY APODIDAE

Swifts · 12 species

Hirundapus caudacutus
White-throated Needletail (V)

OTHER NAMES Needle-tailed Swift, Northern Needletail, (White-throated, Asian or Northern) Spine-tailed Swift.
TAXONOMY (Latham, 1801). Two subspecies (one in region).
DISTRIBUTION [C Siberia to E Asia, Himalayas > E Australia, Himalayas. Stable. LC] Very rare vagrant (mainly May and June, *caudacutus*): Britain (8); Denmark (21 Oct 2012); Faroe Is (19-20 Jun 2000); Finland (4); Ireland (20 Jun 1964); Malta (mid-Nov 1971); Netherlands (22 May 1996); Norway (2: 17 May 1968; 20 May 1995); Sweden (2: 22-27 May 1994; 5 Jun 1998).

Chaetura pelagica
Chimney Swift (V)

TAXONOMY (Linnaeus, 1758). Monotypic.
DISTRIBUTION [E North America > NW South America. Decreasing. NT] Vagrant (primarily Oct-Nov): Azores (26 records involving 132 individuals*); Britain (17); Canary Is (4); France (30 Oct-1 Nov 2005); Ireland (17); Portugal (26 Oct 1999); Spain (5); Sweden (6-7 and again 12-13 Nov 1999).

Cypsiurus parvus
African Palm Swift

OTHER NAMES (Old World) Palm Swift.
TAXONOMY (Lichtenstein, M H C, 1823). Eight subspecies (one in region).
DISTRIBUTION [Sub-Saharan Africa and SW Arabia. Increasing. LC] Resident SW Saudi Arabia and W Yemen (*parvus*). Records of vagrants from N Mauritania no longer considered acceptable.

Tachymarptis melba
Alpine Swift

TAXONOMY (Linnaeus, 1758). Ten subspecies (three in region).
DISTRIBUTION [Africa, S Europe to S Asia > sub-Saharan Africa and S Asia. Stable. LC] Breeds S Europe from Iberia, Italy and Balkans N to N Alps, S Carpathian Mtns and Crimea, also Asia Minor, Caucasus and NW Iran (*melba*), from Morocco discontinuously E across N Africa to Levant and W and S Iran (*tuneti*), and from Israel and Jordan to C and SW Arabia (*archeri*). Migratory, occurring widely on passage N Africa W to Canary Is and E to much of Middle East, though uncommon Oman; some occur year-round in Arabia. Scarce overshoot Britain. Vagrant Azores, Belarus, Belgium, Cape Verde Is, Czech Republic, Denmark, Estonia, Finland, Hungary, Iceland, Ireland, Madeira, Netherlands, Norway, Poland, Slovakia, Sweden and UAE.

Apus alexandri
Cape Verde Swift (E)

OTHER NAME Alexander's Swift.
TAXONOMY Hartert, 1901. Monotypic.
DISTRIBUTION [Cape Verde Is. Stable. LC] Endemic resident Cape Verde Is (all islands except Santa Luzia).

Apus apus
Common Swift

OTHER NAMES (Eurasian or European) Swift, Northern Swift.
TAXONOMY (Linnaeus, 1758). Two subspecies (both in region).
DISTRIBUTION [N Africa and W Europe to E Asia > C, E and S Africa. Decreasing. LC] Summer visitor NW Africa, across much of Europe except far N, and Turkey, Levant and Iraq (*apus*) to Iran (*pekinensis*). Occurs widely across N Africa and Middle East on migration, W to Canary Is (possibly occasional breeder)

and N to Faroe Is, with Asian-breeding *pekinensis* also passing through Iran and C Arabia to/from African winter quarters; small numbers winter N Arabia. Vagrant Azores, Cape Verde Is, Iceland, Jan Mayen, Madeira and Svalbard.

Apus unicolor
Plain Swift (E)

OTHER NAME Plain-coloured Swift.
TAXONOMY (Jardine, 1830). Monotypic.
DISTRIBUTION [Macaronesia and NW Africa. Stable. LC] Endemic. Breeds Canary Is and Madeira, where partially resident. In winter apparently occurs coastal W Morocco, where true status unclear: probably rare winter visitor and occasional breeder (or rare resident or breeding migrant). Vagrant Cape Verde Is and N Mauritania (where may be more regular than few records suggest).

Apus pallidus
Pallid Swift

OTHER NAME Mouse-coloured Swift.
TAXONOMY (Shelley, 1870). Three subspecies (all in region).
DISTRIBUTION [N Africa, S Europe and SW Asia > W and C Africa, Sahel, SW Asia. Stable. LC] Breeding near-endemic. Breeds from N Mauritania discontinuously through interior N Africa to Egypt, Israel, Levant, Iraq, N, C and E Arabia and S Iran (*pallidus*), from Canary Is, Madeira, coastal N Africa, Iberia and S France E through S Europe to Turkey (*brehmorum*), and in coastal NW Balkans (*illyricus*). Locally resident N Libya, Egypt and S Iran but otherwise migratory, occurring S across N Africa and Middle East. Vagrant Azores, Belgium, Britain, Cape Verde Is, Channel Is, Denmark (including *illyricus*), Finland, Germany, Hungary, Ireland, Netherlands, Norway, Poland, Romania, Slovenia and Sweden.

Apus berliozi
Forbes-Watson's Swift

OTHER NAMES Watson's (or Berlioz's) Swift.
TAXONOMY Ripley, 1966. Two subspecies (one in region).
DISTRIBUTION [NE Africa, Socotra and S Arabia > E and NE Africa and Socotra. Stable. LC] Summer visitor Yemen and S Oman (*berliozi*).

Apus pacificus
Pacific Swift (V)

OTHER NAMES (Pacific) Fork-tailed Swift, Asian (or Large/Northern) White-rumped Swift.
TAXONOMY (Latham, 1802). Two subspecies (one confirmed in region). Three further extralimital Asian taxa formerly treated as subspecies now split as Salim Ali's Swift *A salimalii*, Blyth's Swift *A leuconyx* and Cook's Swift *A cooki*.
DISTRIBUTION [Siberia to E Asia. Stable. LC] Very rare vagrant: Britain (7, *pacificus*); Denmark (2: 15 Jun 2010; 25 May 2013); Sweden (4). Further records from Oman (2: 25 Dec 2010 and 4 Nov 2011) and UAE (2: 24 Sep 1999 and 20 Jan 2002) refer either to this species or to extralimital Blyth's Swift.

Apus affinis
Little Swift

TAXONOMY (Gray, J E, 1830). Six subspecies (at least one in region). Formerly included extralimital House Swift *A nipalensis* (SE Asia); ranges meet with no sign of interbreeding.
DISTRIBUTION [NW and sub-Saharan Africa to S Asia. Increasing. LC] Breeds N Mauritania (*aerobates* or *galilejensis*), NW Africa, SW Spain, SC and SE Turkey, Israel, Jordan, Syria, N Iraq, SW and NE Iran, and W Arabia; formerly also Lebanon (*galilejensis*). Populations in W resident or partially migratory, in E mainly migratory, when species occurs more widely *eg* S Iran, E Arabia and N Africa (including NW Libya). Vagrant Azerbaijan, Azores, Bahrain, Balearic Is, Britain, Bulgaria, Canary Is, Cape Verde Is, Corsica, Cyprus, France, Germany, Gibraltar, Greece, Ireland, Italy, Malta, Netherlands, Portugal, Salvage Is, Sweden.

Apus caffer
White-rumped Swift

OTHER NAME African White-rumped Swift.

TAXONOMY (Lichtenstein, M H C, 1823). Monotypic.
DISTRIBUTION [Sub-Saharan Africa, NW Africa and SW Europe. Increasing. LC] Summer visitor E Portugal, SW Spain and C Morocco. Rarely observed on migration and winter range of regional population unknown (though three winter records from S Spain). Vagrant Canary Is, Finland, France, Gibraltar, Malta, N Mauritania, Norway, Sweden, Tunisia and Yemen.

FAMILY COLIIDAE

Mousebirds · one species

Urocolius macrourus
Blue-naped Mousebird (V)

OTHER NAME Blue-naped Coly.
TAXONOMY (Linnaeus, 1766). Six subspecies (one in region).
DISTRIBUTION [Sahel, NE and E Africa. Decreasing. LC] Rare visitor N Mauritania (*macrourus*), with possibly as few as four documented records N of regional boundary (occurs generally S of 18°N). Distributed just S of regional boundary in NW Niger, where may occur as vagrant.

FAMILY CORACIIDAE

Rollers · six species

Coracias naevius
Purple Roller (V)

OTHER NAME Rufous-crowned Roller.
TAXONOMY Daudin, 1800. Two subspecies (undetermined in region).
DISTRIBUTION [Sub-Saharan Africa. Decreasing. LC] Extreme vagrant: Yemen (early 1960s, exact date unknown).

Coracias benghalensis
Indian Roller

TAXONOMY (Linnaeus, 1758). Three subspecies (one in region).
DISTRIBUTION [SW, S and SE Asia. Increasing. LC] Breeds E Iraq, S Iran, Oman and NE UAE (*benghalensis*). Mainly resident but some short-distance dispersal in autumn and winter, when also occurs Kuwait and Saudi Arabia. Vagrant Bahrain, Qatar, Turkey and Yemen.

Coracias caudatus
Lilac-breasted Roller (V)

OTHER NAME Lilac-throated Roller (*C c lorti*).
TAXONOMY Linnaeus, 1766. Two subspecies (undetermined in region).
DISTRIBUTION [NE to S Africa. Stable. LC] Extreme vagrant (two records): Oman (20 Aug 1998); Yemen (early 1960s, exact date unknown).

Coracias abyssinicus
Abyssinian Roller

TAXONOMY Hermann, 1783. Monotypic.
DISTRIBUTION [Sahel, W to E and NE Africa, SW Arabia. Increasing. LC] Resident SW Saudi Arabia and W Yemen. Vagrant Egypt, Libya and N Mauritania.

Coracias garrulus
European Roller

OTHER NAMES (Common or Eurasian) Roller, Blue Roller.
TAXONOMY Linnaeus, 1758. Two subspecies (both in region).
DISTRIBUTION [NW Africa and SW Europe to C Asia > sub-Saharan Africa. Decreasing. LC] Summer visitor NW Africa, Mediterranean islands, Iberia, S France and Italy E through Balkans and E Europe N to Estonia and E to S Ural Mtns, also Turkey, Caucasus, Levant and NW Iran (*garrulus*) and elsewhere in Iran, Iraq and UAE (*semenowi*). Occurs widely on migration S of breeding range W to Canary Is; occasional N to *eg* Finland and Sweden. Vagrant Azores, Belgium, Britain, Cape Verde Is, Channel Is, Denmark, Faroe Is, Germany, Iceland, Ireland, Liechtenstein, Luxem-

bourg, Madeira, Netherlands, Norway and Switzerland.

Eurystomus glaucurus
Broad-billed Roller (V)

OTHER NAMES African Broad-billed (or Cinnamon) Roller.
TAXONOMY (Statius Müller, 1776). Four subspecies (one in region).
DISTRIBUTION [Sub-Saharan Africa. Stable. LC] Extreme vagrant (two records, both involving *afer*): Cape Verde Is (2: 22 Nov 1897, collected; 28 Apr 1924, collected). Further record from Egypt (2010, found dead) relates to Gebel Elba, just outside regional boundary.

FAMILY ALCEDINIDAE

Kingfishers · seven species

Halcyon smyrnensis
White-throated Kingfisher

OTHER NAMES Smyrna (or White-breasted) Kingfisher.
TAXONOMY (Linnaeus, 1758). Six subspecies (one in region).
DISTRIBUTION [SW, S and SE Asia. Increasing. LC] Breeds NE Egypt, Israel, Palestine, Jordan, SW Syria, W and SC Turkey, E Iraq, Kuwait and S Iran (*smyrnensis*). Mainly resident but some winter dispersal, including to Lebanon and Saudi Arabia. Vagrant Azerbaijan, Bulgaria, Cyprus, Greece, Qatar and UAE.

Halcyon leucocephala
Grey-headed Kingfisher

OTHER NAMES Grey-hooded (or Chestnut-bellied) Kingfisher.
TAXONOMY (Statius Müller, 1776). Five subspecies (two in region).
DISTRIBUTION [Sub-Saharan Africa and S Arabia. Stable. LC] Resident Cape Verde Is (*acteon*) and summer visitor SW Saudi Arabia, Yemen and S Oman (*semicaerulea*). Vagrant UAE.

Todiramphus chloris
Collared Kingfisher

OTHER NAMES Black-masked (or Mangrove) Kingfisher, White-collared Kingfisher.
TAXONOMY (Boddaert, 1783). Fifty subspecies (two known in region).
DISTRIBUTION [NE Africa, Arabia and S Asia to Australia. Decreasing. LC] Resident NE UAE/N Oman (*kalbaensis*), SE Oman (subspecies undetermined) and SW Saudi Arabia and (marginally) NW Yemen (*abyssinica*).

Corythornis cristatus
Malachite Kingfisher

OTHER NAME Diademed Kingfisher.
TAXONOMY (Pallas, 1764). Five subspecies (undetermined in region).
DISTRIBUTION [Sub-Saharan Africa and S Arabia. Stable. LC] Rare Yemen, where apparently nomadic and irregular breeder in S (probably *stuartkeithi* but confirmation required). Vagrant Oman.

Alcedo atthis
Common Kingfisher

OTHER NAMES (Eurasian or European) Kingfisher, River Kingfisher.
TAXONOMY (Linnaeus, 1758). Seven subspecies (two in region).
DISTRIBUTION [N Africa, Eurasia and New Guinea. Population trend unknown. LC] Breeds NW Africa, S and E Iberia E through S Europe and Middle East to Iran (*atthis*) and from W and N Iberia N to Britain and Ireland and through W Europe N to S Scandinavia and E to W Russia (*ipsida*). Mainly resident but N and E populations migratory, wintering to S through *eg* Middle East (except SW Arabia) and E Egypt; scarce Libya. Vagrant Azores, Canary Is, Madeira, N Mauritania and Yemen.

Megaceryle alcyon
Belted Kingfisher (V)

TAXONOMY (Linnaeus, 1758). Monotypic.
DISTRIBUTION [North America > S North

America to N South America and West Indies. Stable. LC] Very rare vagrant (majority of records Sep-Nov): Azores (8*); Britain (3); Iceland (5); Ireland (4); Netherlands (17 Dec 1899); Spain (16 Nov 1999-9 Mar 2010).

Ceryle rudis
Pied Kingfisher

OTHER NAMES Lesser (or Small) Pied Kingfisher.
TAXONOMY (Linnaeus, 1758). Five subspecies (two in region).
DISTRIBUTION [Sub-Saharan Africa and Nile Valley to SW, S and SE Asia. Population trend unknown. LC] Breeds Nile Valley (*rudis*) and from Turkey to Levant, Israel, Iraq and SW Iran (*syriacus*). Mainly resident, but winters more widely W to *eg* Cyprus (has bred) and E along Gulf coasts to SE Iran and Qatar. Vagrant Cape Verde Is, Greece, Italy, Montenegro, Oman, Poland, UAE and Ukraine.

FAMILY MEROPIDAE

Bee-eaters · four species

Merops albicollis
White-throated Bee-eater

OTHER NAME Black-crowned Bee-eater.
TAXONOMY Vieillot, 1817. Monotypic.
DISTRIBUTION [Sahel, NE Africa and SW Arabia > W and C Africa. Stable. LC] Summer visitor SW Saudi Arabia and W Yemen. Vagrant Morocco, Oman and UAE.

Merops orientalis
Green Bee-eater

OTHER NAMES Little (or Least/Small) Green Bee-eater; Egyptian Green Bee-eater (*M o cleopatra*), Arabian Green Bee-eater (*M o cyanophrys*).
TAXONOMY Latham, 1801. Eight subspecies (five in region). Taxa divided into well-defined African, Arabian and Asian geographical groups, all of which occur in region; Arabian group (subspecies *cyanophrys* and *muscatensis*) is most distinctive and may merit species status, in which case superspecies comprised of *M viridissimus*, *M cyanophrys* and *M orientalis* could be recognised.
DISTRIBUTION [Sahel and Nile Valley to S and SE Asia. Increasing. LC] Breeds Nile Valley (*cleopatra*), S Israel, Palestine and Jordan to W Saudi Arabia and W Yemen (*cyanophrys*), C Saudi Arabia, E Yemen, Oman and UAE (*muscatensis*) and S Iran (*beludschicus*). Mainly resident but some dispersal and seasonal movements. Vagrant Bahrain, Greece (*orientalis*), Iraq and Qatar.

Merops persicus
Blue-cheeked Bee-eater

OTHER NAME Saharan Blue-cheeked Bee-eater (*M p chrysocercus*).
TAXONOMY Pallas, 1773. Two subspecies (both in region).
DISTRIBUTION [NW Africa, SW, S and C Asia > sub-Saharan Africa. Stable. LC] Summer visitor Morocco and Algeria (*chrysocercus*), and Egypt, Levant to SE Turkey, E Caucasus N to NW Kazakhstan, Iraq, Kuwait, N, S and E Iran and SE Arabia (*persicus*); has bred Libya. Occurs throughout Caucasus and Middle East on passage W to Cyprus, Egypt and Libya, also Morocco and N Mauritania (has bred Banc d'Arguin). Vagrant Balearic Is, Britain, Bulgaria, Canary Is, Cape Verde Is, Croatia, Denmark, Faroe Is, Finland, France, Germany, Gibraltar, Greece, Italy, Malta, Netherlands, Norway, Spain, Sweden, Switzerland, Tunisia and Ukraine.

Merops apiaster
European Bee-eater

OTHER NAMES (Eurasian or Golden) Bee-eater.
TAXONOMY Linnaeus, 1758. Monotypic.
DISTRIBUTION [NW Africa and SW Europe to C Asia > sub-Saharan Africa. Decreasing. LC] Summer visitor NW Africa E to W Libya, Europe from Iberia and Mediterranean N to *eg* N France, Austria, Lithuania, Belarus and Ukraine, and E to S Ural Mtns, Caucasus, Turkey, Levant, N Iraq, Iran and SE Arabia. Occurs widely on migration S through N Africa and Middle East; scarce overshoot on

passage N to *eg* Ireland, Britain, Netherlands, Denmark, N Germany, Latvia and Finland, occasionally breeding well N of usual range. Vagrant Azores*, Canary Is, Cape Verde Is, Estonia, Faroe Is, Iceland, Norway and Sweden.

FAMILY UPUPIDAE

Hoopoes · one species

Upupa epops
Eurasian Hoopoe

OTHER NAMES (Common) Hoopoe.
TAXONOMY Linnaeus, 1758. Six subspecies (at least two in region).
DISTRIBUTION [N Africa and Eurasia > N and W to E Africa, SW to E Asia. Decreasing. LC] Breeds Canary Is, Madeira (Porto Santo), NW Africa, Libya and across much of continental Europe from Mediterranean N to NW France, N Germany and Baltic States, E to Ural Mtns and Caucasus, and SE through Turkey and Levant to N Iraq and Iran (*epops*), E, C and W Arabia (*epops*, possibly also *senegalensis* in SW) and Egypt (*major*). Resident S Iberia, parts of N Africa, Levant and Arabia but otherwise highly migratory, occurring widely on passage S through N Africa and Middle East; scarce migrant NW and N Europe, very occasionally breeding outside normal range (*eg* Britain). Vagrant Azores, Cape Verde Is, Faroe Is, Iceland and Svalbard.

FAMILY BUCEROTIDAE

Hornbills · one species

Lophoceros nasutus
African Grey Hornbill

OTHER NAME Grey Hornbill.
TAXONOMY (Linnaeus, 1766). Two subspecies (one in region).
DISTRIBUTION [Sub-Saharan Africa and SW Arabia. Stable. LC] Resident SW Saudi Arabia and W Yemen (*nasutus*).

FAMILY PICIDAE

Woodpeckers · 17 species

Jynx torquilla
Eurasian Wryneck

OTHER NAMES (European or Northern) Wryneck.
TAXONOMY Linnaeus, 1758. Six subspecies (three in region).
DISTRIBUTION [N Africa and Eurasia > N and W to E Africa, S, SE and E Asia. Decreasing. LC] Breeds N Algeria and N Tunisia (*mauretanica*), Corsica, Sardinia, Italy and W Balkans (*tschusii*) and across much of rest of continental Europe from Iberia N to N Scandinavia, E to Ural Mtns, and S to Caucasus and N Turkey (*torquilla*); may also breed NE Iraq/NW Iran, and formerly did so Britain. Resident N Africa but otherwise summer visitor, occurring widely on migration S across region and W to Canary Is; regular but scarce migrant Britain and Gibraltar. Most cross Sahara but a few (including birds from C Europe) winter Iberia, Morocco and Mediterranean basin. Vagrant Faroe Is, Iceland, Ireland and Madeira.

Sphyrapicus varius
Yellow-bellied Sapsucker (V)

OTHER NAME Common Sapsucker.
TAXONOMY (Linnaeus, 1766). Monotypic.
DISTRIBUTION [N North America > E and SE North America, Central America and W Indies. Decreasing. LC] Extreme vagrant (five records, mainly in autumn): Azores (11 Oct-3 Nov 2008); Britain (26 Sep-6 Oct 1975); Iceland (2: 5 Jun 1961, dead; 7-13 Oct 2007); Ireland (16-19 Oct 1988).

Picoides tridactylus
Eurasian Three-toed Woodpecker

OTHER NAMES (Northern) Three-toed Woodpecker.
TAXONOMY (Linnaeus, 1758). Eight subspecies (three in region).
DISTRIBUTION [N and C Europe to E Asia. Stable. LC] Breeds from Jura Mtns and Alps E to Balkans and W Ukraine (*alpinus*), from Norway, Sweden and Poland E to S Ural Mtns (*tri-*

dactylus) and in N Ural Mtns (*crissoleucus*). Mainly resident, but N populations dispersive and periodically irruptive. Vagrant Denmark and Moldova.

Dendrocoptes dorae
Arabian Woodpecker (E)

TAXONOMY (Bates and Kinnear, 1935). Monotypic.
DISTRIBUTION [SW Arabia. Decreasing. VU] Endemic. Resident W and SW Saudi Arabia and W Yemen. Some seasonal altitudinal movements.

Dendrocoptes medius
Middle Spotted Woodpecker (E)

TAXONOMY (Linnaeus, 1758). Four subspecies.
DISTRIBUTION [Europe and SW Asia. Increasing. LC] Endemic. Breeds continental Europe from NW Spain and France N to Netherlands (scarce) and N Germany (formerly also S Sweden), E to Baltic States and W Russia, S to S Italy and Balkans (*medius*), also W and S Turkey and NW Syria (*anatoliae*), N Turkey and Caucasus (*caucasicus*) and NE Iraq and SW Iran (*sanctijohannis*). Resident with some short-distance dispersal. Vagrant Denmark, Liechtenstein and Finland.

Dendropicos goertae
African Grey Woodpecker

OTHER NAMES (Eastern) Grey Woodpecker.
TAXONOMY (Statius Müller, 1776). Four subspecies (one in region).
DISTRIBUTION [W and C Africa. Stable. LC] Resident N Mauritania just inside regional boundary (*goertae*).

Dryobates minor
Lesser Spotted Woodpecker

OTHER NAMES Lesser (or Little) Pied Woodpecker.
TAXONOMY (Linnaeus, 1758). Thirteen subspecies (11 in region).
DISTRIBUTION [N Africa and Eurasia. Decreasing. LC] Breeds NE Algeria and NW Tunisia (*ledouci*), S Europe (*buturlini*), C Europe (*hortorum*), S Britain (*comminutus*), Scandinavia and NE Poland E (*minor*) to Ural Mtns (*kamtschatkensis*), C and E Greece and Turkey (*danfordi*), Caucasus (*colchicus*), SE Azerbaijan (*quadrifasciatus*), N Iran (*hyrcanus*) and SW Iran and N Iraq (*morgani*). Mainly resident but N populations partially migratory, occasionally irruptive. Vagrant Andorra and Scotland.

Dendrocopos assimilis
Sind Woodpecker

OTHER NAME (Sind) Pied Woodpecker.
TAXONOMY (Blyth, 1849). Monotypic.
DISTRIBUTION [S Asia. Stable. LC] Resident SE Iran.

Dendrocopos syriacus
Syrian Woodpecker (E)

TAXONOMY (Hemprich and Ehrenberg, 1833). Three subspecies.
DISTRIBUTION [C Europe to SW Asia. Increasing. LC] Endemic. Breeds from Austria N to S Poland and E through E Europe to SW Russia and S through Balkans to Turkey, Levant, Sinai and SW Iran (*syriacus*), Caucasus and N Iran (*transcaucasicus*) and SE Iran (*milleri*). Resident and dispersive.

Dendrocopos major
Great Spotted Woodpecker

OTHER NAMES (Greater) Pied Woodpecker; Caspian (Spotted) Woodpecker (*D m poelzami*).
TAXONOMY (Linnaeus, 1758). Twenty-four subspecies (15 in region). Recent genetic analysis identified four distinct clades across Eurasia – China, Japan, Azerbaijan-Iran and Eurasia-N Africa – resulting in proposed recognition of at least four different species, including *poelzami* (S Caspian region). More research needed into *numidus* (N Algeria and N Tunisia), which is allopatric to other populations and has unique set of haplotypes, but was insufficiently sampled in same study.
DISTRIBUTION [NW Africa and Eurasia. Increasing. LC] Breeds Canary Is on Gran Canaria (*thanneri*) and Tenerife (*canariensis*), in Morocco (*mauritanus*), N Algeria and Tuni-

sia (*numidus*), across Europe from Iberia (*hispanus*), Corsica (*parroti*), Sardinia (*harterti*), Italy, Sicily and W Slovenia (*italiae*) through C Europe (*pinetorum*) to Britain and Ireland (*anglicus*), Scandinavia and NE Poland to Ural Mtns (*major*) and S Ukraine to Romania and Greece (*candidus*), and in N Turkey (*paphlagoniae*), Caucasus and Transcaucasia (*tenuirostris*), and SE Azerbaijan and N Iran (*poelzami*). Mainly resident but also dispersive; N populations partially irruptive. Vagrant Balearic Is, Faroe Is, Gibraltar and Iceland.

Dendrocopos leucotos
White-backed Woodpecker

OTHER NAME Lilford's Woodpecker (*D l lilfordi*).
TAXONOMY (Bechstein, 1802). Twelve subspecies (three in region). *Lilfordi* may represent a distinct species, being separated by altitude and ecology from parapatric *leucotos*.
DISTRIBUTION [N, C and S Europe to E Asia. Decreasing. LC] Breeds Pyrenees, C Italy, Balkans, Peleponnese, Turkey and Caucasus (*lilfordi*), Carpathians, N Serbia, Austria and E Switzerland N to Poland, Finland, S Sweden and W Norway E through W Russia (*leucotos*) to Ural Mtns (*uralensis*). Mainly resident, but some local movements. Vagrant Moldova.

Colaptes auratus
Northern Flicker (V)

OTHER NAMES Common Flicker; Yellow-shafted Flicker (*auratus* subspecies group).
TAXONOMY (Linnaeus, 1758). Ten subspecies (undetermined in region). Subspecies commonly divided into four geographical groups, each previously considered separate species; Yellow-shafted '*auratus* group' (N and E North America) and Red-shafted '*cafer* group' (W North America) intergrade in extensive stable hybrid zone.
DISTRIBUTION [North and Central America > S North and Central America. Decreasing. LC] Extreme vagrant (three records of '*auratus* group' individuals): Azores (2: 15-28 Oct 2010; 30 Sep 2013*); Denmark (18 May 1972, *auratus/luteus*). Further record from Ireland (13 Oct 1962), of ship-assisted individual which was fed during transatlantic crossing and then flew ashore, placed in Cat D.

Dryocopus martius
Black Woodpecker

OTHER NAME Great Black Woodpecker.
TAXONOMY (Linnaeus, 1758). Two subspecies (one in region).
DISTRIBUTION [W Europe to E Asia. Increasing. LC] Breeds across much of continental Europe from N Spain, Alps and Balkans N to N Scandinavia, W to North Sea coasts and E to Ural Mtns and NW Kazakhstan, also S Italy, N Turkey, Caucasus and N Iran (*martius*). Mainly resident, but dispersive and partially migratory in N of range; also some altitudinal movement in winter.

Picus viridis
European Green Woodpecker

OTHER NAMES (Eurasian) Green Woodpecker; Zagros (or Iranian/Mesopotamian) Green Woodpecker (*P v innominatus*).
TAXONOMY Linnaeus, 1758. Three subspecies (all in region). SW Asian *innominatus* is genetically and morphologically distinct. Formerly also included Iberian Green and Levaillant's Woodpeckers (*qv*).
DISTRIBUTION [Europe and SW Asia. Stable. LC] Near-endemic. Breeds from Britain, S France and C Europe to C Scandinavia and E through Baltic States and E Europe to Ural Mtns (*viridis*), from Italy E to Black Sea, Asia Minor, Caucasus and N Iran (*karelini*), and W Iran and N Iraq (*innominatus*). Mainly resident; some local winter movements. Vagrant Corsica, Crete, Finland, Ireland and Sardinia.

Picus sharpei
Iberian Green Woodpecker (E)

OTHER NAMES Iberian (or Sharpe's) Woodpecker.
TAXONOMY (Saunders, H, 1872). Monotypic. Formerly considered conspecific with European Green Woodpecker, but elevated to species level based on analysis of morphological differences between *viridis*, *sharpei* and *vaillantii*, as well as narrow hybridisation zone between *viridis* and *sharpei*.
DISTRIBUTION [SW Europe. Stable. NT] Endemic. Resident Portugal, Spain, Andorra and extreme SW France. Vagrant Gibraltar.

Picus vaillantii
Levaillant's Woodpecker (E)

OTHER NAMES Levaillant's (or North African/ Algerian) Green Woodpecker.
TAXONOMY (Malherbe, 1847). Monotypic. Formerly considered conspecific with European Green Woodpecker, but separation at species level confirmed by recent analysis of morphological differences between *viridis*, *sharpei* and *vaillantii*.
DISTRIBUTION [NW Africa. Population trend unknown. LC] Endemic. Breeds NW Morocco, N Algeria and N Tunisia. Mainly resident, but some local movements occur.

Picus canus
Grey-headed Woodpecker

OTHER NAMES Grey-faced (or Ashy) Woodpecker, Grey-headed Green Woodpecker.
TAXONOMY Gmelin, J F, 1788. Eleven subspecies (one in region).
DISTRIBUTION [W Europe to SE and E Asia. Decreasing. LC] Breeds from C Europe W to W France, Luxembourg and E Belgium, S to NE Italy, Balkans and N Turkey, E through E Europe to Ural Mtns, and N to C Germany, Poland and through Baltic States to S Finland, C Sweden and S Norway (*canus*). Mainly resident but some local and dispersive movements. Vagrant Azerbaijan and Netherlands.

ORDER FALCONIFORMES

FAMILY FALCONIDAE

Falcons · 16 species

Falco naumanni
Lesser Kestrel

TAXONOMY Fleischer, J G, 1818. Monotypic.
DISTRIBUTION [NW Africa, S Europe to N China > sub-Saharan Africa. Stable. LC] Breeds from N Africa, Iberia and S France E through S and SE Europe to S Ural Mtns, Caucasus, Iran, Iraq, Turkey and Levant. Mainly migratory, occurring more widely Mediterranean basin, N Africa (W to Canary Is) and Middle East on passage; small numbers present year-round *eg* Iberia, N Morocco and N Libya. Vagrant Austria (formerly bred), Azores, Belgium*, Britain, Cape Verde Is, Czech Republic, Denmark, Estonia, Faroe Is, Finland, Germany, Hungary (formerly bred), Ireland, Liechtenstein, Madeira, Netherlands, Poland, Slovakia, Slovenia, Svalbard, Sweden and Switzerland.

Falco tinnunculus
Common Kestrel

OTHER NAMES (Eurasian or European) Kestrel; Alexander's Kestrel (*F t alexandri*), Neglected Kestrel (*F t neglectus*).
TAXONOMY Linnaeus, 1758. Eleven subspecies (six in region). Diagnosably distinct Cape Verde Is forms *alexandri* and *neglectus* sometimes given full species status on basis of distinctive morphology (and lack of sexual dimorphism in *neglectus*), as well as isolation; *canariensis* and *dacotiae* of Madeira and Canary Is may also warrant separate treatment.
DISTRIBUTION [Widespread Eurasia and Africa. Decreasing. LC] Breeds S and E Cape Verde Is (*alexandri*), NW Cape Verde Is (*neglectus*), E Canary Is (*dacotiae*), W Canary Is and Madeira (*canariensis*), from N Africa N across Europe except extreme N Scandinavia and NW Russia, also from Asia Minor and Levant to Iran and N, C and E Arabia (*tinnunculus*), and SW Arabia and Egypt (*rupicolaeformis*). Populations in Scandinavia, E Europe, Caucasus, E Turkey and NW Iran migratory, wintering C and S Europe and Africa; very scarce on passage Faroe Is (rare in summer). Vagrant Azores, Iceland, Jan Mayen and Svalbard.

Falco sparverius
American Kestrel (V)

TAXONOMY Linnaeus, 1758. Seventeen subspecies (undetermined in region).
DISTRIBUTION [N North America to South America > C North America to South America.

Stable. LC] Extreme vagrant (seven records, most likely involving widespread nominate *sparverius*): Azores (3: 17 Feb 1968, collected; 1970, collected; 15 days in Mar 1980); Britain (2: 25-27 May 1976; 13-28 Jun 1976); Denmark (autumn 1901, collected); Malta (14 Oct 1967, collected).

Falco chicquera
Red-necked Falcon

OTHER NAMES Red-headed Falcon, Red-headed Merlin.
TAXONOMY Daudin, 1800. Three subspecies (one in region). On morphological and DNA evidence, nominate Asian form may be better treated as a species separate from the two African subspecies.
DISTRIBUTION [Sub-Saharan Africa and S Asia. Stable. LC] May have bred SE Iran in late 19th century (*chicquera*), but no records since 1898.

Falco vespertinus
Red-footed Falcon

OTHER NAME Western Red-footed Falcon.
TAXONOMY Linnaeus, 1766. Monotypic. Formerly included Amur Falcon (*qv*).
DISTRIBUTION [C Europe to C Asia > S Africa. Decreasing. LC] Summer visitor to C and E Europe, breeding W to Hungary, N to Estonia, E to Ural Mtns and S to N Balkans and N Caucasus. On migration regularly ranges W to Algeria, Italy and W Mediterranean islands, N to Denmark and SE Finland and E to Turkey and Levant; scarce overshoot W to *eg* Britain, Netherlands, Spain and Sweden. Vagrant Azerbaijan, Azores, Belgium, Canary Is, Gibraltar, Iceland, Iran, Iraq, Ireland, Kuwait, Liechtenstein, Luxembourg, Madeira, N Mauritania, Morocco, Norway, Portugal and UAE.

Falco amurensis
Amur Falcon

OTHER NAMES Eastern (or Manchurian) Red-footed Falcon.
TAXONOMY Radde, 1863. Monotypic. Frequently considered a subspecies of Red-footed Falcon, but differs in plumage, morphology and behaviour.
DISTRIBUTION [NE Asia > S Africa. Stable. LC] Uncommon passage migrant S Oman (rare in winter) and Saudi Arabia. Vagrant Azores*, Britain, Greece, Hungary, Iran, Italy, Kuwait, Qatar, Sweden, UAE and Yemen.

Falco eleonorae
Eleonora's Falcon

TAXONOMY Gené, 1839. Monotypic.
DISTRIBUTION [NW Africa and Mediterranean basin > Madagascar. Decreasing. LC] Breeding endemic, nesting late summer and early autumn Canary Is, NW Morocco and E discontinuously through Mediterranean to Greek islands and Cyprus; has bred Turkey. Wanders inland *eg* S France, Bulgaria and Syria but rarely far from Mediterranean region, through which population moves E in autumn, exiting presumably via Suez (rare Israel) and then S to winter quarters; spring migration takes more westerly route. Vagrant Armenia, Azores, Britain, Cape Verde Is, Czech Republic, Denmark, Finland, Germany, Hungary, Jordan, Madeira*, Malta, N Mauritania, Netherlands, N Niger, Poland, Romania, Slovenia, Sweden, Switzerland and UAE.

Falco concolor
Sooty Falcon

TAXONOMY Temminck, 1825. Monotypic.
DISTRIBUTION [N Africa to SW Asia > SE Africa and Madagascar. Decreasing. NT] Near-endemic breeding species. Breeds late summer from E Libya, Egypt, Israel and Jordan to Saudi Arabia, Yemen, Oman, UAE, Qatar, Bahrain and Iran (rare); also known from N Niger in late summer. Migrates S in autumn *eg* along Nile Valley, Red Sea and through Gulf of Aden. Vagrant Algeria, Cyprus, France, Italy, Kuwait, Malta, Morocco, Syria, Tunisia and Turkey.

Falco columbarius
Merlin

OTHER NAMES European Merlin (*F c subae-*

salon/aesalon), North American Merlin (*F c columbarius*), Pallid Merlin (*F c pallidus*), Taiga Merlin (*F c columbarius*).
TAXONOMY Linnaeus, 1758. Nine subspecies (five in region). Molecular data suggest that Eurasian *aesalon* and North American *columbarius* may be better treated as separate species.
DISTRIBUTION [N North America and N Eurasia > S North America to N South America, S Eurasia. Stable. LC] Breeds Iceland (*subaesalon*), Ireland, Britain and E across parts of N and C Fennoscandia, N Baltic States, N Belarus and N Russia (*aesalon*). Mainly migratory, occurring widely but often locally on passage and in winter S through Europe to NW Africa and much of Middle East. NC Siberian *insignis* winters SW to Israel, C Asian *pallidus* in N Middle East. Vagrant Azores, Bahrain, Jan Mayen, Libya, Madeira, N Mauritania, Oman and Svalbard; Nearctic *columbarius* vagrant Azores, Iceland and Ireland; *pallidus* vagrant Armenia.

Falco subbuteo
Eurasian Hobby

OTHER NAMES (European or Northern) Hobby.
TAXONOMY Linnaeus, 1758. Two subspecies (one in region).
DISTRIBUTION [Widespread Eurasia > S Africa and S Asia. Decreasing. LC] Summer visitor to much of Europe except far N and NW, also NW Africa, Levant, Turkey, Caucasus and Iran (*subbuteo*). Widespread on passage S of breeding range through Middle East and N Africa W to Canary Is. Vagrant Azores, Faroe Is, Iceland, Ireland and Madeira.

Falco biarmicus
Lanner Falcon

OTHER NAME Lanner; North African Lanner Falcon (*F b erlangeri*).
TAXONOMY Temminck, 1825. Five subspecies (four in region).
DISTRIBUTION [Africa, S Europe and SW Asia. Increasing. LC] Mainly resident from Italy and parts of Balkans to Turkey, Azerbaijan and NW Iran (*feldeggii*), NW Africa and Libya (*erlangeri*), N Niger and N Chad (*abyssinicus*) to Israel, Iraq and SW and E Arabia, formerly also Syria (*tanypterus*). Vagrant Balearic Is, Bulgaria, Corsica, Cyprus, Czech Republic, France, Gibraltar, Iraq, Malta, Portugal*, Sardinia, Slovenia, Spain, UAE and Ukraine.

Falco jugger
Laggar Falcon (V)

TAXONOMY Gray, J E, 1834. Monotypic. Sometimes considered a subspecies of Lanner Falcon.
DISTRIBUTION [S and SE Asia. Decreasing. NT] Extreme vagrant: Iran (9 Apr 1972).

Falco cherrug
Saker Falcon

OTHER NAME Saker.
TAXONOMY Gray, J E, 1834. Four subspecies (one in region).
DISTRIBUTION [C Europe to W China > E Africa, Middle East and S Asia. Decreasing. EN] Breeds C Europe (has bred Germany and Poland) and N Balkans E to S Ural Mtns, also Turkey, Iraq and Iran (*cherrug*). N populations migratory and young birds especially prone to wandering; more widespread on passage and in winter SE Europe W to S Italy, in N Africa W to Tunisia and in Middle East. Vagrant Algeria, Belarus, Estonia, Finland, France, Gibraltar, Latvia, Lithuania, Malta, N Mauritania, Morocco, Portugal, Slovenia and Spain (some extralimital records evidenced only through satellite-tracking data).

Falco rusticolus
Gyr Falcon

OTHER NAMES Gyrfalcon (IOC), Gyr; Greenland Falcon (white morph); Iceland Falcon (grey morph).
TAXONOMY Linnaeus, 1758. Monotypic. Up to seven subspecies sometimes described, but polymorphic rather than polytypic.
DISTRIBUTION [Holarctic > N North America and N Eurasia. Stable. LC] Breeds Iceland, Norway, N Sweden, N Finland and NW Russia (grey morph). Mainly sedentary, but some move S in winter to *eg* SC Finland; occasional or

rare Jan Mayen, Svalbard and Novaya Zemlya. Some extralimital and winter occurrences in NW Europe involve white-morph Greenland birds. Vagrant Austria, Britain, Channel Is, Czech Republic, Denmark, Estonia, France, Germany, Ireland, Latvia, Lithuania, Netherlands, Poland, Switzerland, Spain and Ukraine.

Falco peregrinus
Peregrine Falcon

OTHER NAMES Peregrine; Cape Verde (Peregrine) Falcon (*F p madens*), Mediterranean Peregrine Falcon (*F p brookei*), Tundra or Russian Peregrine (*F p calidus*).
TAXONOMY Tunstall, 1771. Sixteen subspecies (five in region). Formerly often treated as including Barbary Falcon (*qv*). *Madens* (Cape Verde Is) given specific status by CSNA but more typically considered a subspecies.
DISTRIBUTION [Cosmopolitan. Stable. LC] Breeds widely N, W, C and S Europe (*peregrinus*), from Lapland eastwards (*calidus*), NW Africa and Mediterranean region to Caucasus and N Iran (*brookei*), S Morocco and possibly N Mauritania (*minor*), and Cape Verde Is (*madens*); precise distribution of subspecies often unclear in N Africa, where presence of (and possible interbreeding with) Barbary Falcon is further complicating factor. N continental European populations migratory, occurring widely in winter S of Baltic, also N Africa and throughout Middle East. Exact status of *calidus* away from Lapland unclear; said to winter regularly in at least Spain, Morocco, Israel and Arabia, Russian-ringed bird recovered Sicily (Dec 2011) and partly through satellite tracking is known to have occurred *eg* France, Ukraine and Iran. Vagrant Azores, Canary Is, Faroe Is, Iceland, Jan Mayen, Madeira and Svalbard.

Falco pelegrinoides
Barbary Falcon

OTHER NAMES Red-capped (or Red-naped) Falcon; Shaheen (*F p babylonicus*).
TAXONOMY Temminck, 1829. Two subspecies (both in region). Sometimes treated as conspecific with Peregrine Falcon, but differs in proportions, structure and plumage, and sympatric in several areas where ranges overlap.
DISTRIBUTION [N Africa and Middle East to C Asia > S Asia. Stable. LC] Breeds disjunctly from Canary Is and N Africa to Israel, Jordan, N Iraq, Arabia and SW Iran (*pelegrinoides*), and in E Iran (*babylonicus*); status uncertain Syria. Some autumn and winter dispersal. Vagrant Azores, Greece, Italy, Kuwait, Lebanon, Madeira, Malta, Qatar and Sardinia.

FAMILY PSITTACIDAE

African and New World parrots · two species

Myiopsitta monachus
Monk Parakeet (I)

OTHER NAMES Grey-breasted (or Quaker) Parakeet, South American Monk Parakeet.
TAXONOMY (Boddaert, 1783). Four subspecies (at least one in region).
DISTRIBUTION [S South America. Increasing. LC] Naturalised introduction (*monachus* and/or *cotorra*), with self-sustaining populations established Belgium, Balearic Is, Canary Is, Czech Republic, Israel, Italy and Spain. Other small free-flying populations present *eg* Britain, Germany, Melilla, Morocco and Portugal.

Aratinga nenday
Nanday Parakeet (I)

OTHER NAMES Black-headed (or Black-hooded/Nanday) Conure (or Parakeet).
TAXONOMY (Vieillot, 1823). Monotypic.
DISTRIBUTION [S South America. Increasing. LC] Naturalised introduction Israel. Has also bred Canary Is (Tenerife) and Spain.

FAMILY PSITTACULIDAE

Old World parrots · two species

Psittacula eupatria
Alexandrine Parakeet (I)

TAXONOMY (Linnaeus, 1766). Five subspecies (undetermined in region).
DISTRIBUTION [S and SE Asia. Decreasing.

LC] Naturalised introduction, with self-sustaining populations established Bahrain, Iran, Kuwait, Netherlands, Qatar, Saudi Arabia, Turkey and UAE, possibly also Oman; has bred Belgium.

Psittacula krameri
Rose-ringed Parakeet (I)

OTHER NAME Ring-necked Parakeet.
TAXONOMY (Scopoli, 1769). Four subspecies (two in region).
DISTRIBUTION [Sub-Saharan Africa, S and SE Asia. Increasing. LC] Circumstantial evidence of historic Asian range reaching NE Iran, but natural occurrence in recent times not established (*contra* AERC and *BWP*); similarly, natural range in W Africa (extralimital *krameri*) not confirmed to reach N Mauritania, where records near regional boundary may have involved introduced or escaped individuals. Otherwise widely introduced (Asian *borealis* and *manillensis*), with self-sustaining populations now established *eg* Austria, Bahrain, Belgium, Britain, Canary Is, Egypt, Germany, Iran, Israel, Jordan, Kuwait, Lebanon, Netherlands, Oman, Portugal, Qatar, Saudi Arabia, Spain, Syria, Turkey, UAE and Yemen. Other free-flying populations present *eg* Azores, Cape Verde Is, Iraq, Libya, Morocco and Tunisia.

ORDER PASSERIFORMES

FAMILY PITTIDAE

Pittas · one species

Pitta brachyura
Indian Pitta (V)

OTHER NAMES Bengal Pitta (or Blue-winged/Green-winged) Pitta.
TAXONOMY (Linnaeus, 1766). Monotypic.
DISTRIBUTION [N and C Subcontinent > C and S Subcontinent. Decreasing. LC] Extreme vagrant: Iran (19 Nov 1968).

FAMILY TYRANNIDAE

Tyrant flycatchers and allies · seven species

Sayornis phoebe
Eastern Phoebe (V)

TAXONOMY (Latham, 1790). Monotypic.
DISTRIBUTION [NC to E North America > SE USA to S Mexico. Increasing. LC] Extreme vagrant: Britain (24-25 Apr 1987). Another individual found dead Britain (19 Jul 2010) not accepted as genuine vagrant.

Contopus virens
Eastern Wood Pewee (V)

OTHER NAME Eastern Pewee.
TAXONOMY (Linnaeus, 1766). Monotypic.
DISTRIBUTION [SE Canada and E USA > NW South America. Decreasing. LC] Extreme vagrant: Azores (2: 18-21 and 24 Oct 2015, photos*; 20-23 Oct 2015, photos*).

Empidonax virescens
Acadian Flycatcher (V)

TAXONOMY (Vieillot, 1818). Monotypic.
DISTRIBUTION [E USA and extreme SE Canada > SC and NW South America. Stable. LC] Extreme vagrant: Iceland (4 Nov 1967, dead).

Empidonax alnorum
Alder Flycatcher (V)

TAXONOMY Brewster, 1895. Monotypic.
DISTRIBUTION [Alaska to NE USA > NW to WC South America. Stable. LC] Extreme vagrant (three records, all within three-week period late Sep-mid Oct): Britain (2: 8-9 Oct 2008; 25-27 Sep 2010); Iceland (10 Oct 2003).

Empidonax minimus
Least Flycatcher (V)

TAXONOMY (Baird, W M, and Baird, S F, 1843). Monotypic.
DISTRIBUTION [N North America > extreme

SE USA and Middle America. Decreasing. LC] Extreme vagrant: Iceland (6 Oct 2003, caught, released 7 Oct).

Tyrannus savana
Fork-tailed Flycatcher (V)

TAXONOMY Daudin, 1802. Four subspecies (undetermined in region).
DISTRIBUTION [C and South America > C and N South America. Stable. LC] Extreme vagrant: Spain (19 Oct 2002).

Tyrannus tyrannus
Eastern Kingbird (V)

TAXONOMY (Linnaeus, 1758). Monotypic.
DISTRIBUTION [E North America > WC South America. Decreasing. LC] Extreme vagrant (two autumn records): Ireland (2: 5 Oct 2012; 24 Sep 2013).

FAMILY MALACONOTIDAE

Bush-shrikes · one species

Tchagra senegalus
Black-crowned Tchagra

OTHER NAMES Black-crowned (or Black-headed) Bushshrike; Moroccan Tchagra (*T s cucullatus*).
TAXONOMY (Linnaeus, 1766). Ten subspecies (two in region).
DISTRIBUTION [NW and sub-Saharan Africa, S Arabia. Stable. LC] Breeds Morocco, N Algeria and N Tunisia (*cucullatus*) and Yemen and S Oman (*percivali*). Mainly resident, but local seasonal movements occur. Vagrant Libya and Spain.

FAMILY LANIIDAE

Shrikes · 12 species

Lanius cristatus
Brown Shrike (V)

OTHER NAMES Red-tailed Shrike; Japanese Shrike (*L c superciliosus*), Philippine Shrike (*L c lucionensis*).
TAXONOMY Linnaeus, 1758. Four subspecies (two in region). Has been considered conspecific with Red-backed, Isabelline and Turkestan Shrikes, but genetic analyses and morphological differences support treatment as separate species.
DISTRIBUTION [C Siberia to E Asia > S and SE Asia. Decreasing. LC] Rare vagrant (mainly nominate *cristatus* where determined): Britain (17); Denmark (2); France (2); Germany (15-16 Oct 2001); Ireland (22 Nov-10 Dec 1999); Italy (29 Nov 2002-Feb 2003); Norway (2-3 Oct 2005); Oman (2); Sweden (3 Oct 1984); UAE (7, including two *lucionensis*).

Lanius collurio
Red-backed Shrike

OTHER NAME Red-tailed Shrike.
TAXONOMY Linnaeus, 1758. Two subspecies (both in region). At one time treated as conspecific with Brown, Daurian and Turkestan Shrikes (*qv*).
DISTRIBUTION [W Europe to C Asia > E and S Africa. Decreasing. LC] Summer visitor across much of continental Europe from N Iberia to C Fennoscandia (*collurio*) E to Turkey, Caucasus, Levant and NW Iran (*kobylini*). Migrates E then SE across Europe, E Mediterranean and Middle East through Egypt (W to Libya) to winter quarters; regular but scarce on passage Britain (formerly bred). Vagrant Algeria, Azores, Balearic Is, Canary Is, Gibraltar, Iceland, Ireland, Madeira, Morocco, Tunisia and Svalbard.

Lanius isabellinus
Daurian Shrike

OTHER NAMES Isabelline Shrike (IOC), Pale Brown (or Red-tailed/Rufous-tailed) Shrike; Chinese (or Kashgarian) Shrike (*L i arenarius*), Dzungarian Shrike (*L i isabellinus*).
TAXONOMY Hemprich and Ehrenberg, 1833. Three subspecies (two in region). Previously treated as conspecific with Red-backed and Brown Shrikes, and until recently also included Turkestan Shrike (*qv*). Taxonomy and nomenclature complex and controversial, but subspecies now determined as *isabellinus* (formerly

speculigerus), *arenarius* (formerly *isabellinus*) and extralimital *tsaidamensis* (see Pearson 2000 for background to changes following subspecific reidentification of type specimen, and Panov 2009 and 2011 for alternative view).
DISTRIBUTION [EC Asia to NC China > NE Africa and Arabia. Stable. LC] Passage migrant and/or winter visitor to SE of region (*isabellinus*), occurring Iran, Kuwait and E and S Arabia, but rarer farther W in Middle East; *arenarius* also winters Iran. Vagrant elsewhere, but true occurrence pattern obscured by difficulty of separation from Turkestan Shrike, and many past records indeterminate (notably so in countries where species still treated as conspecific). Extralimital *isabellinus* occurrences confirmed *eg* Azerbaijan, Belgium, Britain, Bulgaria, Estonia, Finland, France, Georgia, Germany, Greece, Iceland*, Ireland, Italy, Netherlands, Norway, Poland, Romania and Turkey, but reviews of records in some countries ongoing. Vagrancy status of *arenarius* unclear: claims from *eg* Iraq and Syria require examination.

Lanius phoenicuroides
Turkestan Shrike

OTHER NAMES Red-tailed Shrike (IOC); Rufous (or Rufous-tailed) Shrike.
TAXONOMY (Schalow, 1875). Monotypic. Formerly treated as conspecific with Daurian Shrike, but differs in plumage, wing formula, moult, breeding phenology, migration and genetics. Pale grey '*karelini*' variant is probably a colour morph, or possibly a hybrid with Red-backed Shrike.
DISTRIBUTION [S and C Asia > Sahel and NE Africa. Population trend not assessed. NE] Summer visitor E Iran; recently also discovered breeding NW Kazakhstan. Migrates through Middle East (has bred Oman, where '*karelini*' also regular on passage) W to Levant, irregularly also Cyprus; rare in winter Middle East, and also occurs N Chad. Vagrant W to N and W Europe, but see comments under Daurian Shrike on indeterminate records and differing treatments by national committees. Confirmed extralimital occurrences of this taxon include Azerbaijan, Balearic Is, Britain (including two '*karelini*'), Denmark, Estonia, France, Greece, Italy, Latvia, Malta, N Mauritania, Norway, Netherlands (two '*karelini*'), Poland, Sweden and Turkey.

Lanius vittatus
Bay-backed Shrike

TAXONOMY Valenciennes, 1826. Two subspecies (one in region).
DISTRIBUTION [WC and S Asia. Stable. LC] Scarce resident and common summer visitor SE Iran (*nargianus*). Rare migrant and winter visitor Oman (has bred) and UAE (has bred). Vagrant Saudi Arabia.

Lanius schach
Long-tailed Shrike

OTHER NAMES Rufous-backed (or Red-backed) Shrike, (Southern) Rufous-rumped Shrike, Schach Shrike; Black-headed (or Black-capped) Shrike (black-headed subspecies).
TAXONOMY Linnaeus, 1758. Nine subspecies (at least one in region). Subspecies group into two main plumage types, black masked and black headed.
DISTRIBUTION [SC, S, SE and E Asia. Population trend unknown. LC] Status uncertain NE Iran, where bred late 19th century but no records subsequently until one (migrant?) 10 Aug 2013 (*erythronotus*). Rare passage migrant and winter visitor Oman (mid Sep-early Apr). Vagrant Britain, Denmark, France, Israel, Jordan, Kuwait, Netherlands, Sweden, Turkey and UAE; most records thought to involve *erythronotus*.

Lanius minor
Lesser Grey Shrike

TAXONOMY Gmelin, J F, 1788. Monotypic.
DISTRIBUTION [Europe to C Asia > S Africa. Decreasing. LC] Summer visitor, breeding from NE Spain (very rare), S France (rare) and Italy to parts of E Europe (N to Lithuania) and E to Ural Mtns, Turkey, Caucasus, NW Iran and recently N Iraq; formerly much more widespread in continental W Europe. Migrates through E Mediterranean/Middle East (W to Libya) into/out of NE Africa. Vagrant Austria (formerly bred), Belgium (formerly bred), Britain, Channel Is, Corsica, Czech Republic (formerly bred), Denmark, Estonia, Faroe Is, Finland, Germany (formerly bred), Ireland, Latvia, Liechtenstein, Luxembourg (formerly bred), Malta, Netherlands, Norway, Sweden, Switzerland (formerly bred) and Tunisia.

Lanius excubitor
Great Grey Shrike

OTHER NAMES Homeyer's Great Grey Shrike (*L e homeyeri*), Northern (or Siberian Great) Grey Shrike (*L e sibiricus*), Northern Shrike (*L e borealis*), Steppe Grey Shrike (*L e homeyeri/'stepensis'*).
TAXONOMY Linnaeus, 1758. Seven subspecies (at least three in region). Treatment of all taxa in this and next two species follows current IOC classification, but results of new research into genetics and morphological variation require thorough revision of species limits and reassessment of subspecies. Up to six or more species, not three, now thought to be involved, with some proposed treatments variously further splitting southern taxa, returning them to present species (excluding *meridionalis*, *qv*) and/or realigning E Eurasian forms (including *sibiricus*) and Nearctic *borealis* as a separate species. No consensus likely until more research using different genetic markers has confirmed precise relationships.
DISTRIBUTION [N North America and N Eurasia. Stable. LC] Breeds from C Scandinavia E through Baltic and from E France through C Europe (formerly including Switzerland) E to N Ural Mtns (*excubitor*) and from SE Europe E to C Ural Mtns (*homeyeri*). Resident in S of breeding range but northern birds winter W to Britain and S to N Mediterranean, Black Sea, Turkey, Caucasus and N Iran. Vagrant Albania, Balearic Is, Channel Is, Corsica, Faroe Is, Iceland, Ireland, Malta and Svalbard; *sibiricus* (breeding C Siberia to NE Asia) vagrant Norway (5 Nov 1881) and Ukraine (7 specimens, though possibly occurs more frequently); *borealis/sibiricus/invictus* vagrant Finland (1-22 Mar 2000).

Lanius meridionalis
Southern Grey Shrike

OTHER NAMES Arabian (or Levant) Grey Shrike (*L m aucheri/theresae*), Asian (or Indian) Grey Shrike (*L m lahtora*), Desert (or Saharan) Grey Shrike (*L m elegans*), Iberian (Grey) Shrike (*L m meridionalis*), Mauryan Grey Shrike (*L m lahtora*), Yemen Grey Shrike (*L m buryi*).
TAXONOMY Temminck, 1820. Eleven subspecies (eight in region). As with Great Grey Shrike (*qv*), current arrangement provisional and subject to revision, recent genetic research indicating that, *contra* IOC, *meridionalis* is more closely related to E Asian/Nearctic subspecies of Great Grey Shrike than to neighbouring populations of European and N African taxa. *Meridionalis* therefore may be better treated as monotypic, with other 'southern grey' taxa either realigned with Great Grey Shrike (itself split from E Asian/Nearctic Northern Grey Shrike) or further divided into Asian, Desert and Socotra species (see *eg* Poelstra 2010); 'best fit' dependent on outcome of further genetic research. Other recent treatments of present species sometimes also include Steppe Grey Shrike (*qv*).
DISTRIBUTION [SW Europe, N Africa and Middle East to S Asia. Population trend not assessed. NE] Breeds S France and Iberia (*meridionalis*), coastal NW Africa (*algeriensis*), Canary Is (*koenigi*), Sahara from N Mauritania to Libya, Egypt and SW Israel (*elegans*), N Israel and S Lebanon (*theresae*), Jordan, Iraq, SW, S and NE Iran and much of Arabia (*aucheri*) except Yemen (*buryi*), and SE Iran (*lahtora*); formerly also bred Syria. Mainly resident but some local movements and dispersal. *Meridionalis* vagrant to Andorra, Gibraltar, Italy, Malta, Melilla and Morocco, *algeriensis* to Italy, *elegans* to Cape Verde Is, Cyprus and Malta, and *aucheri* also to Cyprus; one indeterminate *elegans/aucheri* record from Greece.

Lanius pallidirostris
Steppe Grey Shrike

OTHER NAMES Asian (or Saxaul) Grey Shrike, Baluchistan (or Turkestan) Grey Shrike, Bogdanoff's (or Grimm's Grey) Shrike. Note: common name Steppe Grey Shrike, adopted by IOC, inaccurately reflects desert habitat preferences of this species (note same English name also used for *homeyeri/'stepensis'* Great Grey Shrike), but provisionally retained here pending likely further changes to species limits, and therefore names, in large grey shrike complex.
TAXONOMY Cassin, 1851. Monotypic. Sometimes treated as conspecific with Great Grey and/or Southern Grey Shrikes (*qv*). Treatment here follows IOC but, as with other large grey shrikes, best regarded as provisional pending further research to establish true relationships within group.
DISTRIBUTION [SW and C Asia > SW Asia and

NE Africa. Population trend not assessed. NE] Breeds lower Volga region N of Caspian Sea and in NE Iran, where hybridises freely with *aucheri* Southern Grey Shrike. On passage and in winter occurs Iran and E and S Arabia. Vagrant Austria, Azerbaijan, Britain, Bulgaria, Cyprus, Denmark, Egypt, Estonia, Finland, France, Greece, Israel, Italy, Jordan, Lebanon, Netherlands, Norway, Sweden, Syria and Turkey.

Lanius senator
Woodchat Shrike

OTHER NAMES Woodchat; Balearic Woodchat Shrike (*L s badius*), Eastern Woodchat Shrike (*L s niloticus*).
TAXONOMY Linnaeus, 1758. Four subspecies (all in region).
DISTRIBUTION [S Europe, NW Africa and Middle East > Sahel, W to NE Africa, SW Arabia. Decreasing. LC] Breeding endemic. Summer visitor NW Africa, Libya and most of Iberia (*rutilans*), N Spain and France E through S Europe (discontinuously N to S Germany and Poland) to S Ukraine, W Turkey and N Africa (*senator*), W Mediterranean islands (*badius*) and from S Turkey and Cyprus E through Levant and N Iraq to Caucasus and Iran (*niloticus*). Migrates S to winter quarters, when common Gibraltar (has bred) and widespread N Africa (regular Canary Is) and Middle East; scarce migrant *eg* Britain, Luxembourg (formerly bred) and Netherlands (has bred). Some winter Algerian Sahara, also SW Arabia and Iran (rare). Vagrant Azores, Belarus, Belgium, Cape Verde Is, Channel Is, Czech Republic (formerly bred), Denmark, Estonia, Finland, Iceland, Ireland, Liechtenstein, Madeira, Norway and Sweden; *badius* vagrant NW to *eg* Britain, Finland, Ireland, Netherlands and Norway and E to Malta; *niloticus* vagrant Finland, Italy, Malta, Spain and Sweden.

Lanius nubicus
Masked Shrike

OTHER NAME Nubian Shrike.
TAXONOMY Lichtenstein, M H C, 1823. Monotypic.
DISTRIBUTION [SE Europe and Middle East > Sahel and NE Africa. Decreasing. LC] Breeding endemic. Summer visitor Greece, Macedonia, Bulgaria, W and S Turkey, Cyprus, Levant S to N Israel, NE Syria, Iraq, Kuwait and W Iran; may also breed NW Saudi Arabia. Migrates through E Mediterranean, Egypt and Middle East; some winter S Arabia, and may also occur on regional boundary in S Sahara. Vagrant Albania, Algeria, Armenia, Azerbaijan, Balearic Is, Britain, Finland, France, Gibraltar, Italy, Libya, Malta, Morocco, Norway, Spain, Sweden and Ukraine.

FAMILY VIREONIDAE

Vireos · four species

Vireo griseus
White-eyed Vireo (V)

TAXONOMY (Boddaert, 1783). Six subspecies (undetermined in region).
DISTRIBUTION [E USA and NE Mexico > S USA, West Indies and N Middle America. Increasing. LC] Extreme vagrant (four autumn records): Azores (4: 22-23 Oct and 23 Nov 2005; 24 Oct 2008; 18 Oct 2009; 6 Oct 2012*).

Vireo flavifrons
Yellow-throated Vireo (V)

TAXONOMY Vieillot, 1808. Monotypic.
DISTRIBUTION [E North America > Middle America, West Indies and N South America. Increasing. LC] Extreme vagrant (five records): Azores (3: 11-18 Oct 2008; 13 Oct 2008*; 19-20 Oct 2009*); Britain (20-27 Sep 1990); Germany (18 Sep 1998).

Vireo philadelphicus
Philadelphia Vireo (V)

TAXONOMY (Cassin, 1851). Monotypic.
DISTRIBUTION [N North America > Central America. Increasing. LC] Extreme vagrant (nine records): Azores (6: 26 Oct 2005; 6-13 Oct 2009; 30 Sep-4 Oct 2011*; 1-8 Oct 2011*; 5 Oct 2011*; 14-18 Oct 2012*; 10-11 Oct 2013*; 14 Oct 2013*); Britain (10-13 Oct 1987); Ireland (2: 12-17 Oct 1985; 13-14 Oct 2008).

Vireo olivaceus
Red-eyed Vireo (V)

TAXONOMY (Linnaeus, 1766). Ten subspecies (one in region).
DISTRIBUTION [N and E North America, NW and C South America > N and C South America. Increasing. LC] Vagrant (mainly late Sep-Oct, *olivaceus*): Azores (42*); Belgium (13 Oct 1995); Britain (127); Channel Is (12 Oct 2006); Denmark (19-27 Oct 2011); France (16); Germany (4 Oct 1957); Iceland (22); Ireland (58); Malta (29 Oct 1983); Morocco (Oct 1968); Netherlands (8); Poland (17 Oct 2000); Spain (3).

FAMILY ORIOLIDAE

Old World orioles · three species

Oriolus oriolus
Eurasian Golden Oriole

OTHER NAMES (European) Golden Oriole.
TAXONOMY (Linnaeus, 1758). Monotypic. Closely related to Indian Golden Oriole *O kundoo* (*qv*) and until recently treated as conspecific.
DISTRIBUTION [W Europe and NW Africa E to NC Asia > C and S Africa. Stable. LC] Summer visitor NW Africa and continental Europe (formerly also bred E Britain) N to S Scandinavia and E to S Ural Mtns, Caspian Sea, Caucasus, Asia Minor, Cyprus, N Iraq and N Iran. Widespread on passage Mediterranean, N Africa and Middle East. Vagrant Azores, Faroe Is, Iceland, Ireland and Madeira.

Oriolus kundoo
Indian Golden Oriole (V)

TAXONOMY Sykes, 1832. Monotypic. Formerly considered conspecific with Eurasian Golden Oriole, but differs in morphology and vocalisations, and apparently no known interbreeding.
DISTRIBUTION [SC and S Asia > S Asia. Population trend not assessed. NE] Rare or vagrant Iran; hypothetical breeder in extreme NE of country, but confirmation required. Unconfirmed UAE, but some winter records of orioles suspected possibly to relate to this species rather than Eurasian Golden Oriole.

Oriolus chinensis
Black-naped Oriole (V)

OTHER NAME Asian (or Eastern) Golden Oriole (*O c diffusus*).
TAXONOMY Linnaeus, 1766. Eighteen subspecies (undetermined in region). Geographical variation marked, and may involve several species; further study required.
DISTRIBUTION [E and SE Asia > S and SE Asia. Population trend unknown. LC] Extreme vagrant (two records): Oman (7 Dec 2011); UAE (18-25 Feb 2012).

FAMILY DICRURIDAE

Drongos · two species

Dicrurus macrocercus
Black Drongo (V)

OTHER NAMES (Black) King Crow, King Drongo.
TAXONOMY Vieillot, 1817. Seven subspecies (at least one in region). Forms a superspecies with extralimital Fork-tailed Drongo *D adsimilis* (sub-Saharan Africa) and has often been treated as conspecific, but plumage, proportions and voice differ; close relationship supported by recent DNA studies.
DISTRIBUTION [S, SE and E Asia. Population trend unknown. LC] Very rare vagrant: Iran (*albirictus* formerly bred; 4 records since 1960, 2 of which in 2010); Oman (7); UAE (3).

Dicrurus leucophaeus
Ashy Drongo (V)

OTHER NAMES Grey Drongo, Pale (Ashy) Drongo.
TAXONOMY Vieillot, 1817. Fifteen subspecies (undetermined in region). Recent DNA studies indicate that species is closest to African drongos rather than Asian congeners. Taxonomy complicated; two subspecies (*longicaudatus* and *leucogenis*) formerly treated as separate species, but many others poorly differentiated.
DISTRIBUTION [E and SE Asia > S and SE Asia. Population trend unknown. LC] Extreme vagrant

(nine records Dec-Apr): Kuwait (3: 3 Apr 2010; 7-27 Dec 2010, two; 5 Feb-4 Apr 2013*); UAE (6: 16-19 Dec 2006; 29 Feb-20 Mar 2008; 21 Dec 2009; 14-15 Jan and 2-17 Mar 2010; 27-28 Oct 2011; 18 Feb-26 Mar 2012).

FAMILY MONARCHIDAE

Monarch flycatchers and allies · two species

Hypothymis azurea
Black-naped Monarch (V)

OTHER NAMES Black-naped Blue Monarch, Black-naped Blue Flycatcher, Black-naped Monarch Flycatcher.
TAXONOMY (Boddaert, 1783). Twenty-three subspecies (undetermined in region). Extralimital SE subspecies *puella*, *aeria*, *blasii* and *catamaranensis* form well-marked '*puella* group', sometimes treated as specifically distinct.
DISTRIBUTION [S and SE Asia. Stable. LC] Extreme vagrant: Iran (13 Feb 2011, photo*).

Terpsiphone viridis
African Paradise Flycatcher

OTHER NAMES Paradise Flycatcher, Southern Paradise-Flycatcher.
TAXONOMY (Statius Müller, 1776). Ten subspecies (one in region).
DISTRIBUTION [Sub-Saharan Africa and S Arabia. Stable. LC] Breeds SW Saudi Arabia, Yemen and SW Oman (*harterti*).

FAMILY CORVIDAE

Crows, jays and magpies · 19 species

Perisoreus infaustus
Siberian Jay

OTHER NAME Grey Jay.
TAXONOMY (Linnaeus, 1758). Eight subspecies (three in region).
DISTRIBUTION [N Eurasia. Decreasing. LC] Resident from S Norway and C Sweden N and E through Fennoscandia to NW Russia (*infaustus*), NE European Russia (*ostjakorum*) and W Russia (*ruthensis*). Vagrant Belarus, Estonia (may have formerly bred), Latvia, Lithuania, Poland, Slovakia and Ukraine.

Garrulus glandarius
Eurasian Jay

OTHER NAMES Acorn (or Common/European) Jay, White-eared Jay; Black-crowned Jay ('*atricapillus* group'), Black-capped Jay ('*cervicalis* group'), Iranian Jay (*G g hyrcanus*).
TAXONOMY (Linnaeus, 1758). Thirty-four subspecies (21 in region). Variation very complex, subspecies falling into eight broad and variably defined geographical groupings; majority of mainland taxa intergrade.
DISTRIBUTION [NW Africa and W Europe to E Asia. Stable. LC] Resident across much of Europe except far N ('*glandarius* group'), including Ireland (*hibernicus*), Britain and NW France (*rufitergum*), N and C Europe to Ural Mtns (*glandarius*), Iberia (*fasciatus*), Corsica (*corsicanus*), Sardinia (*ichnusae*), mainland Italy, Sicily and Dalmatian coast of Balkans to Ionian Is (*albipectus*), W and S Balkans (*graecus*), E Bulgaria and European Turkey (*ferdinandi*), Crete (*cretorum*) and Cyprus (*glaszneri*); also N Ural Mtns (*brandtii*); NW Africa ('*cervicalis* group') in N Morocco and NW Algeria (*whitakeri*), C Morocco and Atlas Mtns of N Algeria (*minor*), and N and NE Algeria and Tunisia (*cervicalis*); N Middle East and adjacent areas ('*atricapillus* group') in Samos, SE Aegean (*samios*), W Turkey to N Iraq and W Iran (*anatoliae*), Crimea (*iphigenia*), NE Turkey and Caucasus (*krynicki*), and W Syria, W Jordan and Israel (*atricapillus*); and SE Azerbaijan and N Iran (*hyrcanus*). Mainly sedentary in W and S but irruptive migrant in N and E, where may move in substantial numbers when acorn crop fails. Vagrant Malta.

Cyanopica cooki
Iberian Magpie (E)

OTHER NAME Azure-winged Magpie.
TAXONOMY Bonaparte, 1850. Monotypic. Formerly considered conspecific with extralimital Azure-winged Magpie *C cyanus* (E Asia), but sep-

arated on basis of significant genetic and minor morphological and behavioural differences.
DISTRIBUTION [SW Europe. Population trend unknown. NE] Endemic resident Portugal and Spain; also known from fossil record in Gibraltar.

Pica pica
Eurasian Magpie

OTHER NAMES (Common) Magpie; Arabian (or Asir) Magpie (*P p asirensis*), Maghreb (or North African) Magpie (*P p mauritanica*).
TAXONOMY (Linnaeus, 1758). Eleven subspecies (six in region). Now separated from extralimital Black-billed Magpie *P hudsonia* (North America), but position of several distinct subspecies unclear, in region notably Arabian *asirensis* and NW African *mauritanica*; status of Iberian *melanotos*, intermediate between latter and nominate, also needs clarification.
DISTRIBUTION [NW Africa and Europe to SW Arabia and E Asia. Population trend unknown. NE] Resident NW Africa (*mauritanica*), across Europe from Iberia (*melanotos*) in S to N Scandinavia and NW Russia (*fennorum*) in N, and from Britain and Ireland in W to Ural Mtns in E, also Asia Minor (*pica*), Iran (*bactriana*) and SW Saudi Arabia (*asirensis*); introduced Sardinia. Vagrant Faroe Is, Israel, Lebanon and Svalbard.

Podoces pleskei
Pleske's Ground Jay (E)

OTHER NAMES Plesky's (or Iranian/Persian) Ground Jay.
TAXONOMY Zarudny, 1896. Monotypic.
DISTRIBUTION [Iran. Stable. LC] Endemic resident C and E Iran.

Nucifraga caryocatactes
Spotted Nutcracker

OTHER NAMES (Eurasian or European) Nutcracker; Siberian (or Slender-billed/Thin-billed) Nutcracker (*N c macrorhynchos*), Thick-billed Nutcracker (*N c caryocatactes*).
TAXONOMY (Linnaeus, 1758). Eight subspecies (two in region). Geographical variation complex over large range; subspecies divided into northern (including European and Siberian populations) and southern groups, which at one were time treated as separate species.
DISTRIBUTION [Scandinavia and C Europe to E Asia. Decreasing. LC] Mainly resident C and E Europe, Balkans and from S Scandinavia E through Baltic States and Russia (*caryocatactes*). Siberian population (*macrorhynchos*) periodically erupts W into region in years of poor Arolla Pine seed crop. Vagrant (almost exclusively *macrorhynchos*) Azerbaijan, Britain, Corsica, Iran, Luxembourg, Portugal, Sardinia, Spain and Turkey. Unrecorded N Africa (*contra BWP*).

Pyrrhocorax pyrrhocorax
Red-billed Chough

OTHER NAMES (Common) Chough.
TAXONOMY (Linnaeus, 1758). Eight subspecies (four in region).
DISTRIBUTION [NW Africa and W Europe to E Asia, E Africa. Decreasing. LC] Resident Canary Is, Morocco and N Algeria (*barbarus*), W Britain, Ireland and NW France (*pyrrhocorax*), Iberia, S France, Switzerland, Italy, Sardinia and Sicily (*erythroramphos*) and S Balkans and Crete E to Turkey, C Syria, N Iraq, Caucasus and Iran (*docilis*); reintroduced Channel Is. Limited altitudinal movement and some dispersal; has recently recolonised SW England, and rare but increasing winter visitor Balearic Is. Vagrant Austria (formerly bred), Bulgaria, Corsica, Germany, Gibraltar, Hungary, Israel, Lebanon (has bred), Slovenia and Tunisia.

Pyrrhocorax graculus
Alpine Chough

OTHER NAME Yellow-billed Chough.
TAXONOMY (Linnaeus, 1766). Three subspecies (two in region).
DISTRIBUTION [NW Africa and S Europe to C Asia. Stable. LC] Resident N Morocco, N Spain, S France, Corsica, C Italy, Alps through WC Europe to Balkans, Crete, parts of Turkey, Caucasus and N Iran (*graculus*), and SE Turkey S to Lebanon and E across N Iraq to SW Iran (*digitatus*). In winter disperses S in Middle East to N Israel, and rarely has been observed crossing Strait of Gibraltar. Vagrant Balearic Is, Czech Republic, Hungary, Poland, Sardinia, Slovakia, Syria, Tunisia and Ukraine.

Coloeus monedula
Western Jackdaw

OTHER NAMES (Common or Eurasian) Jackdaw.
TAXONOMY (Linnaeus, 1758). Four subspecies (all in region).
DISTRIBUTION [NW Africa and Europe to C Asia. Increasing. LC] Breeds Morocco, NW Algeria and S and W Europe (*spermologus*), NE Algeria (*cirtensis*), S and C Scandinavia (*monedula*) and E and SE Europe E to Ural Mtns, Asia Minor, Caucasus, Levant, N Iraq and N Iran (*soemmerringii*); formerly also bred Malta and Tunisia. Mainly resident, but C Scandinavian and N Russian population moves S in winter, when range also extends farther S in Middle East. Very scarce passage migrant Faroe Is, irregular Corsica. Vagrant Andorra, Azores, Balearic Is, Bear Is, Canary Is, Gibraltar, Iceland (has attempted breeding), Madeira, N Mauritania and Svalbard.

Coloeus dauuricus
Daurian Jackdaw (V)

OTHER NAMES Asian (or Chinese/Eastern) Jackdaw, Pied (or Black/Collared) Jackdaw.
TAXONOMY (Pallas, 1776). Monotypic.
DISTRIBUTION [CE to NE Asia > CE to E Asia. Stable. LC] Extreme vagrant (six records possibly involving only four birds): Denmark (12 Apr 1997, possibly same as in Netherlands); Finland (early May 1883, collected); France (22 Jun 1995, possibly same as in Netherlands); Netherlands (2: 1-15 May 1995, possibly same as in France; 8-11 May 1997, possibly same as in Denmark); Sweden (26-28 Apr 1985).

Corvus splendens
House Crow (I)

OTHER NAMES Indian (House) Crow, Grey-necked (or Colombo) Crow.
TAXONOMY Vieillot, 1817. Five subspecies (two in region).
DISTRIBUTION [SW to SE Asia, Indian Ocean. Stable. LC] Breeds S Iran (*zugmayeri*), where population established in 20th century presumed to originate from escaped individuals and/or ship-assisted birds arriving from Subcontinent; same colonisers also thought to have visited Oman across Strait of Hormuz. Subsequently established elsewhere in region partly (or mostly?) through ship-assisted passage, initially in Yemen, then spreading along coastal Arabia N to Kuwait and NW to S Jordan, S Israel and E Egypt; *zugmayeri* described from Oman and *protegatus* in UAE and Yemen, but many individuals may be intergrades. Isolated W European colony of ship-assisted birds now established Netherlands; other records of vagrants, *eg* from Belgium, Cyprus, Denmark, France, Ireland, Morocco, Poland and Spain, typically attributed to similar origin (though Polish individual probably escapee), and most placed in Cat E. Note: generally considered harmful invasive species within non-native range, and population partly controlled through eradication programmes.

Corvus frugilegus
Rook

OTHER NAMES Eurasian Rook; Eastern Rook (*C f pastinator*).
TAXONOMY Linnaeus, 1758. Two subspecies (both in region). Ranges of W and E populations apparently narrowly separated in C Asia and may merit species status, though intermediates have been reported. More research needed.
DISTRIBUTION [W Europe to E Asia. Decreasing. LC] Breeds from Ireland and Britain E through much of W, C, SE and E Europe, N to Baltic States and N European Russia (with isolated locations also in Scandinavia, Finland and N Spain), and in Caucasus, Asia Minor, NE Iraq and NW Iran (*frugilegus*). Migratory in N and NE; on passage and in winter range extends W to Faroe Is and S to Iberia (rare), N Italy, S Balkans, Levant, NE Egypt, S Iraq and SW and E Iran; in last two countries wintering population includes *pastinator* from C Asia. Vagrant Algeria, Azores, Balearic Is, Bear Is, Corsica, Cyprus, Iceland, Jordan, Kuwait, Lebanon, Liechtenstein, Madeira, Malta, Portugal (has bred), Svalbard, Tunisia and Sicily.

Corvus corone
Carrion Crow

OTHER NAMES (Common or Eurasian) Crow; Oriental (or Eastern Carrion) Crow (*C c orientalis*).

TAXONOMY Linnaeus, 1758. Two subspecies (both in region). Eastern subspecies *orientalis* is genetically distinct and potentially merits species status. Formerly also included Hooded Crow (*qv*).
DISTRIBUTION [W and C Europe and S, SE and E Asia. Decreasing. LC] Breeds Britain and W, C and SW continental Europe from S Scandinavia (scarce S and W Norway) S to Iberia (except in S) and E to C Poland, Czech Republic and Hungary (*corone*), winter range extending through N Italy; rare and irregular in winter S to Morocco. Formerly or irregularly also bred NE Iran, where now vagrant (*orientalis*). Vagrant Albania, Azores, Belarus, Canary Is, Croatia, Estonia, Faroe Is, Finland, Gibraltar, Ireland, Jan Mayen, Latvia, Lebanon, Libya, Madeira, Malta, Morocco, Romania, Serbia, Slovenia and Ukraine.

Corvus cornix
Hooded Crow

OTHER NAMES Mesopotamian (or Iraq Pied) Crow (*C c capellanus*).
TAXONOMY Linnaeus, 1758. Four subspecies (all in region). Traditionally (and sometimes still) treated as conspecific with Carrion Crow, with which it is genetically similar and hybridises in C Scotland and C Europe, but narrowness of hybrid zone suggests speciation at least partially achieved. Larger, paler and stronger-billed *capellanus* differs markedly from neighbouring *pallescens* with little evidence of interbreeding, and may have evolved to species level; more research needed.
DISTRIBUTION [NW and C Europe to C Siberia, C and SW Asia. Population trend not assessed. NE] Breeds Ireland, Isle of Man, N Scotland, Faroe Is and from Scandinavia E to Ural Mtns and S to C Europe (*cornix*), Corsica, Italy and Balkans to S Ural Mtns, Caucasus and N Iran (*sharpii*), Cyprus, Turkey and Iraq S through Levant to Nile Valley (*pallescens*), and SE Iraq and SW Iran (*capellanus*). Mainly migratory in N, winter range extending S to *eg* E England (scarce), N France (rare) and NW Kazakhstan. Vagrant Algeria, Balearic Is, Belgium, Iceland, Jan Mayen, Libya, Malta, Spain, Svalbard and Tunisia; *capellanus* vagrant Kuwait and UAE.

Corvus macrorhynchos
Large-billed Crow

OTHER NAMES Jungle (or Thick-billed) Crow; Himalayan Crow (*C m intermedius/colonorum/tibetosinensis*).
TAXONOMY Wagler, 1827. Nine subspecies (one in region). Exhibits significant variation over wide geographical area, and may actually comprise up to seven species. More research needed.
DISTRIBUTION [S, SE and E Asia. Stable. LC] Formerly bred SE Iran (*intermedius*), but no records since 1900.

Corvus albus
Pied Crow (V)

OTHER NAMES African Pied (or White-bellied) Crow.
TAXONOMY Statius Müller, 1776. Monotypic.
DISTRIBUTION [Sub-Saharan Africa and Madagascar. Stable. LC] Extreme vagrant (seven records of 13 birds): Algeria (2: 1961; Dec 1964); Egypt (11 Apr 2010*); Libya (24 Apr 1931, collected; 7-17 Jun 2013, two, then two adults and two juveniles*); N Mauritania (31 Mar-5 Apr 2003); Morocco (13 Dec 2009-29 Jan 2011, three adults, of which pair bred spring 2010, fledging one juvenile). Additional records from Canary Is and Ceuta placed in Cat D.

Corvus ruficollis
Brown-necked Raven

OTHER NAME Desert Raven.
TAXONOMY Lesson, 1831. Monotypic.
DISTRIBUTION [W Africa to C Asia. Increasing. LC] Breeds Cape Verde Is, from N Mauritania and S Morocco E discontinuously through Saharan N Africa to Sinai, N to Syria and C Iraq, S through Arabia (but absent Qatar, formerly bred Bahrain) and E through S Iran. Mainly sedentary, but some passage noted S Israel. Vagrant Turkey; record from NE Spain (11 Apr 2013) placed in Cat D.

Corvus corax
Northern Raven

OTHER NAMES (Common) Raven; Punjab Raven (*C c laurencei*).

TAXONOMY Linnaeus, 1758. Eleven subspecies (six in region). Recent molecular studies indicate different evolutionary paths for New World and Old World populations, which are morphologically and genetically distinct, and for isolated Canary Is subspecies of latter. Other Old World populations said to intergrade widely.
DISTRIBUTION [North and Central America, N Africa and Eurasia. Increasing. LC] Breeds Canary Is (*canariensis*), N Africa (*tingitanus*), W Mediterranean islands and Iberia to Italy (*hispanus*), Iceland and Faroe Is (*varius*), Ireland, Britain and NW France to Fennoscandia, C Europe and E to Ural Mtns (*corax*), and from Greece, Cyprus and Turkey to Levant, N Iraq and Iran (*laurencei*). Mainly resident, but some winter movement *eg* Iberia, C Europe, Balkans and N Caspian region; occasional Luxembourg (formerly bred). Vagrant Jan Mayen, NW Kazakhstan, Malta (*corax* and *tingitanus*), Svalbard and UAE.

Corvus rhipidurus
Fan-tailed Raven

TAXONOMY Hartert, 1918. Two subspecies (both in region).
DISTRIBUTION [Sahel, NE Africa and Middle East. Decreasing. LC] Resident N Chad (*rhipidurus*) and from Sinai, E Israel, Palestine and Jordan S through W and parts of C Saudi Arabia to Yemen and S Oman (*stanleyi*). Vagrant Libya, Turkey and UAE.

FAMILY BOMBYCILLIDAE

Waxwings · two species

Bombycilla garrulus
Bohemian Waxwing

OTHER NAMES (Common or Greater) Waxwing.
TAXONOMY (Linnaeus, 1758). Two subspecies (one in region).
DISTRIBUTION [N North America and N Eurasia. Decreasing. LC] Breeds N Scandinavia, Finland, Estonia and NW Russia (*garrulus*). Northernmost population migrates S in winter, normal range reaching *eg* Britain and Ireland (scarce), Low Countries, C Europe, N Balkans and Caucasus. Occasionally irrupts S in large numbers from farther E, range then extending farther S in Europe (rarely to Mediterranean). Rare and irregular Croatia, Iran and Turkey. Vagrant Albania, Algeria, Azores, Cyprus, France, Greece, Iceland (has bred), Israel, Jan Mayen, Liechtenstein, Macedonia, Malta, Portugal, Sardinia, Serbia, Spain and Svalbard.

Bombycilla cedrorum
Cedar Waxwing (V)

TAXONOMY Vieillot, 1808. Monotypic.
DISTRIBUTION [C and S Canada to W, N and NE USA > C USA to extreme N South America. Increasing. LC] Extreme vagrant (nine records): Azores (2: 13-19 and 23* Oct 2010; 12-15 Oct 2013*); Britain (3: 25-26 Jun 1985; 20 Feb-18 Mar 1996; 21-29 Sep 2013); Iceland (2: mid Apr-late Jul 1989; 8 Oct 2003); Ireland (2: 14 Oct 2009; 10 Nov 2012*).

FAMILY HYPOCOLIIDAE

Grey Hypocolius · one species

Hypocolius ampelinus
Grey Hypocolius

OTHER NAME Hypocolius.
TAXONOMY Bonaparte, 1850. Monotypic.
DISTRIBUTION [SW Asia. Population trend unknown. LC] Breeds Iraq and S Iran. Winters S Iraq, Kuwait, C and E Saudi Arabia, Bahrain, Qatar, UAE, Oman and SE Iran. Vagrant Cyprus, Egypt, Israel, Turkey and Yemen.

FAMILY PARIDAE

Tits and chickadees · 12 species

Periparus ater
Coal Tit

OTHER NAMES British Coal Tit (*P a britannicus*), Common Coal Tit (*P a ater*), Continental Coal Tit (*P a ater/abietum*), Irish Coal Tit (*P a hibernicus*).

TAXONOMY (Linnaeus, 1758). Twenty-four subspecies (15 in region). Geographical variation well marked; widespread *abietum* genetically distinct from other regional subspecies, and marked genetic variation also found within Scandinavian *ater*.
DISTRIBUTION [NW Africa and Europe to E Asia. Stable. LC] Breeds Morocco (*atlas*), N Algeria and N Tunisia (*ledouci*), Iberia except SW (*vieirae*), Corsica and Sardinia (*sardus*), continental W, C and SE Europe and Asia Minor (*abietum*), Britain and NE Ireland (*britannicus*), elsewhere in Ireland (*hibernicus*), Scandinavia E to Ural Mtns (*ater*), S Crimean Peninsula (*moltchanovi*), Cyprus (*cypriotes*), NE Turkey and SW Caucasus (*derjugini*), elsewhere in Caucasus (*michalowskii*), SE Azerbaijan and N Iran (*gaddi*), NE Iran (*chorassanicus*) and SW Iran (*phaeonotus*). Mainly resident, but some movement (occasionally irruptive) N Europe; extreme N Russian population migrates S in winter, when range extends *eg* S through Ukraine to NE Balkans. Vagrant Israel.

Lophophanes cristatus
European Crested Tit

OTHER NAME Crested Tit.
TAXONOMY (Linnaeus, 1758). Seven subspecies (all in region).
DISTRIBUTION [Europe and extreme W Siberia. Decreasing. LC] Near-endemic. Breeds NC Scotland (*scoticus*), from NW, N and E Europe to Carpathian Mtns (*cristatus*), W France (*abadiei*), W and S Iberia (*weigoldi*), NE Spain E to Alps, C Europe and N Balkans (*mitratus*), S Balkans (*bureschi*) and SW and C Ural Mtns (*baschkirikus*). Mainly sedentary. Vagrant Armenia, Azerbaijan, Channel Is, England, Georgia, Gibraltar and Morocco.

Poecile lugubris
Sombre Tit (E)

TAXONOMY (Temminck, 1820). Five subspecies in two groups, differing in tail/wing ratio and possibly vocalisations: *lugubris*, *lugens* and *anatoliae* (W '*lugubris* group'), and *dubius* and *kirmanensis* (E '*dubius* group'). Formerly also included Caspian Tit (*qv*).
DISTRIBUTION [SE Europe and SW Asia. Population trend unknown. LC] Endemic. Breeds from N Balkans to N Greece (*lugubris*), C and S Greece and Crete (*lugens*), Asia Minor, Levant, N Iraq, Caucasus and NW Iran (*anatoliae*), W Iran (*dubius*) and SE Iran (*kirmanensis*). Largely sedentary. Vagrant Italy, Kuwait and Slovenia.

Poecile palustris
Marsh Tit

OTHER NAME Eurasian Marsh Tit (*P p palustris*).
TAXONOMY (Linnaeus, 1758). Ten subspecies (five in region). Two widely separated and morphologically distinct populations, but little molecular divergence so far apparent.
DISTRIBUTION [W Europe to W Siberia, EC to E Asia. Decreasing. LC] Breeds Britain (*dresseri*), from N Iberia to C Scandinavia and E to Poland and Balkans (*palustris*), French Alps, Italy and Sicily (*italicus*), E Europe to S Urals and NW Turkey (*stagnatilis*) and NE Turkey to Caucasus (*kabardensis*). Generally sedentary, with limited dispersal mainly in N. Vagrant Andorra, Finland, Ireland and Sardinia.

Poecile hyrcanus
Caspian Tit (E)

OTHER NAMES Caucasian (or Hyrcan/Hyrcanian) Tit, Hyrcanian Chickadee.
TAXONOMY Zarudny and Loudon, 1905. Monotypic. Has been treated as conspecific with Sombre Tit, but more recently found to be a member of the Marsh/Willow Tit complex.
DISTRIBUTION [SW Asia. Population trend not evaluated. NE] Endemic. Breeds SE Azerbaijan and N Iran. Largely sedentary, but some evidence of post-breeding dispersal and altitudinal movement.

Poecile montanus
Willow Tit

OTHER NAME Northern Willow Tit (*P m borealis*).

TAXONOMY (Conrad von Baldenstein, 1827). Fourteen subspecies (six in region).
DISTRIBUTION [W Europe to E Asia. Decreasing. LC] Breeds Britain (*kleinschmidti*), from NW France to W Germany, Switzerland and N Italy (*rhenanus*), SE France and C Italy to Balkans (*montanus*), Germany and W Poland to NE Switzerland and Austria (*salicarius*), Scandinavia S to Ukraine (*borealis*) and SE European Russia (*uralensis*). Highly sedentary W Europe and mainly resident elsewhere, but N populations partially dispersive and periodically irruptive; *borealis* vagrant Britain.

Poecile cinctus
Siberian Tit

OTHER NAMES Grey-headed Chickadee (IOC), Taiga Chickadee (or Tit).
TAXONOMY (Boddaert, 1783). Four subspecies (two in region).
DISTRIBUTION [N Eurasia and NW North America. Decreasing. LC] Breeds C and N Scandinavia E to N European Russia (*lapponicus*) and NE European Russia (*cinctus*). Largely sedentary but some limited movement S in winter; occasionally irruptive, and has reached *eg* Russian Baltic and S Ural Mtns. Vagrant Estonia.

Cyanistes teneriffae
African Blue Tit (E)

OTHER NAMES Canary Blue (or Tenerife) Tit; Ultramarine Blue Tit (*C t ultramarinus*).
TAXONOMY (Lesson, 1831). Seven subspecies (all in region). Previously treated as conspecific with Eurasian Blue Tit, and mainland N African subspecies sometimes still included within that species, but all differ from it in morphology, genetics and vocalisations, as do mainland N African taxa from those in Canary Is.
DISTRIBUTION [Canary Is, N Africa and extreme S Europe. Population trend not assessed. NE] Endemic. Resident Canary Is on La Palma (*palmensis*), El Hierro (*ombriosus*), La Gomera and Tenerife (*teneriffae*), Gran Canaria (*hedwigae*) and Lanzarote and Fuerteventura (*degener*), and from Morocco to N Tunisia and Pantalleria Is, extreme S Italy (*ultramarinus*), and NE Libya (*cyrenaicae*). Vagrant Malta; first record for Iberia, from NE Spain (19 Feb 2013, trapped), placed in Cat D.

Cyanistes caeruleus
Eurasian Blue Tit

OTHER NAMES (European) Blue Tit.
TAXONOMY (Linnaeus, 1758). Nine subspecies (all in region). Formerly included African Blue Tit (*qv*).
DISTRIBUTION [W Europe to W Asia. Increasing. LC] Near-endemic. Breeds Britain and Ireland (*obscurus*), across much of N, C and E continental Europe S to Sicily and N Balkans and E to W and N Asia Minor (*caeruleus*), Mallorca (*balearicus*), Iberia, Corsica and Sardinia (*ogliastrae*), S Greece (*calamensis*), SE European Russia (*orientalis*), Crimea, Caucasus, E Turkey, Levant, N Iraq and NW Iran (*satunini*), N Iran (*raddei*) and SW Iran (*persicus*). Mainly resident but some movement (more rarely irruptions) of N and C populations to S and W. Vagrant Malta.

Cyanistes cyanus
Azure Tit

OTHER NAMES Yellow-breasted (Azure) Tit (*C c flavipectus*); Pleske's Tit ('*pleskii*' Azure × Eurasian Blue Tit hybrids).
TAXONOMY (Pallas, 1770). Eight subspecies (three in region). *Flavipectus* (breeding extralimitally SC Asia) sometimes treated as full species, but minimal genetic differentiation from *cyanus*. Sporadically hybridises with Eurasian Blue Tit.
DISTRIBUTION [E Europe to E Asia. Stable. LC] Breeds Belarus, extreme N Ukraine and from W and C European Russia to C Ural Mtns (*cyanus*) and SW Ural Mtns (*hyperrhiphaeus*). Undertakes periodic but impermanent range expansions to W of normal range. Vagrant Austria, Croatia, Czech Republic, Estonia, Finland (has bred), France, Germany, Iran (*flavipectus*), Latvia, Lithuania, Poland, Romania, Slovakia and Sweden; '*pleskii*' hybrids recorded Austria, Belgium, Croatia, Finland, France, Germany, Hungary, Latvia, Netherlands, Poland, Slovakia, Sweden and Ukraine.

Parus major
Great Tit

OTHER NAMES Northern Great Tit; Turkestan (or Bokhara Great) Tit (*P m bokharensis*).
TAXONOMY Linnaeus, 1758. Fifteen subspecies (12 in region). Includes *bokharensis*, sometimes separated at species level, but not Cinereous Tit (*qv*), now treated as specifically distinct.
DISTRIBUTION [NW Africa and Europe to CE Asia. Increasing. LC] Breeds NW Africa (*excelsus*), Portugal, S Spain and Corsica (*corsus*), Sardinia (*ecki*), Balearic Is (*mallorcae*), Britain and Ireland (*newtoni*), much of mainland Europe E to Ural Mtns, Caucasus and Asia Minor (*major*) except S Italy, S Greece, Aegean Is and Cyprus (*aphrodite*), Crete (*niethammeri*), Syria to NE Egypt (*terraesanctae*), SE Azerbaijan and NW Iran (*karelini*), N Iraq and NC and SW Iran (*blanfordi*) and marginally NE Iran (*bokharensis*). Mainly resident but some short-distance and altitudinal movement; N and E populations periodically irruptive. Vagrant Iceland, Kuwait, Malta and Svalbard.

Parus cinereus
Cinereous Tit

OTHER NAME Grey Tit.
TAXONOMY Vieillot, 1818. Thirteen subspecies (one in region). Formerly treated as conspecific with Great Tit, but now considered distinct on basis of morphological, genetic and vocal differences.
DISTRIBUTION [SC, S and SE Asia. Population trend not assessed. NE] Resident NE Iran (*intermedius*).

FAMILY REMIZIDAE

Penduline tits · three species

Remiz pendulinus
Eurasian Penduline Tit

OTHER NAMES (European or Western) Penduline Tit.
TAXONOMY (Linnaeus, 1758). Four subspecies (three in region). Formerly included Black-headed and White-crowned Penduline Tits (*qv*).
DISTRIBUTION [W and S Europe to C Asia > S Europe to SW Asia. Increasing. LC] Patchily distributed from parts of Iberia, Italy and Balkans N through C Europe to S Scandinavia, E through Baltic States and E Europe to Ural Mtns, Caucasus and W Turkey (*pendulinus*), S and E Turkey, Levant, Syria and recently N Iraq to Armenia, Azerbaijan and NW Iran (*menzbieri*), and SE European Russia and NW Kazakhstan (*caspius*). N populations mainly migratory, occurring widely on passage continental Europe (scarce Finland) to winter S to N Mediterranean, NE Egypt, Jordan, Kuwait (scarce) and S Iran. Vagrant Bahrain, Britain, Channel Is, Malta, Morocco, Norway, Qatar, Saudi Arabia and Tunisia; *caspius* vagrant Hungary.

Remiz macronyx
Black-headed Penduline Tit

TAXONOMY (Severtzov, 1873). Five subspecies (four in region, one of which possibly extinct; validity of fifth, extralimital *ssaposhnikowi* of E Kazakhstan, questionable). Formerly treated as conspecific with Eurasian Penduline Tit and limited hybridisation reported in areas where ranges overlap in Caspian region and C Asia, but in general exhibits constant morphological and ecological differences, latter often supported by habitat-partitioning where sympatric.
DISTRIBUTION [SW and C Asia. Stable. LC] Breeds Ural River in NW Kazakhstan (*macronyx*), extreme SE Azerbaijan and NW Iran (*altaicus*), N Iran (*neglectus*) and SE Iran (*nigricans*, possibly extinct). Mainly resident but some winter immigration N Iran. Vagrant Oman and Russia.

Remiz coronatus
White-crowned Penduline Tit

OTHER NAME Crowned Penduline Tit.
TAXONOMY (Severtzov, 1873). Two subspecies (one in region). Formerly considered conspecific with Eurasian Penduline Tit and sometimes with also-split extralimital Chinese Penduline Tit *Remiz consobrinus* (N China), but differs from former in morphology, size and breeding ecology, and from latter also in vocalisations.

DISTRIBUTION [SC Asia to S Siberia and Mongolia > SC and SW Asia. Decreasing. LC] Breeds NE Iran; occurs on passage and in winter more widely E Iran (*coronatus*). Formerly accepted record of vagrant in Austria (1 Aug 1987) rejected after review.

FAMILY PANURIDAE

Bearded Reedling · one species

Panurus biarmicus
Bearded Reedling

OTHER NAMES Bearded Parrotbill, Bearded Tit (or Tit-babbler).
TAXONOMY (Linnaeus, 1758). Three subspecies (all in region; one apparently now extinct).
DISTRIBUTION [W Europe to E Asia. Population trend unknown. LC] Highly fragmented and localised distribution across Europe and Asia Minor from C Iberia, W France and England through C and S Europe to SW Russia, Balkans and W Turkey (*biarmicus*), SC Turkey (*kosswigi*, now considered extinct) and EC and E Europe through S Russia and S to Caucasus, C Turkey and Syria (*russicus*). Vagrant Algeria, Corsica, Iran, Ireland, Israel, Kuwait, Lebanon, Liechtenstein, Morocco and Sardinia.

FAMILY ALAUDIDAE

Larks · 28 species

Alaemon alaudipes
Greater Hoopoe-Lark

OTHER NAMES Bifasciated Lark, Hoopoe (or Large Desert) Lark.
TAXONOMY (Desfontaines, 1789). Four subspecies (all in region).
DISTRIBUTION [W Africa to SW Asia. Decreasing. LC] Breeds Cape Verde Is (*boavistae*), from N Mauritania and S Morocco E through N Africa to Sinai and N to Syria (*alaudipes*), W and S Arabia (*desertorum*) and C and E Arabia N to Iraq and E to Iran (*doriae*). Vagrant Canary Is, Italy, Lebanon, Malta and Turkey.

Rhamphocoris clotbey
Thick-billed Lark

OTHER NAMES Clotbey's (or Clot Bey) Lark.
TAXONOMY (Bonaparte, 1850). Monotypic.
DISTRIBUTION [NW Africa to Middle East. Stable. LC] Breeds from N Mauritania discontinuously E through NW Africa to Libya, and in N Arabia and Jordan (rare). Some breeding areas occupied throughout year, but dispersive and partly nomadic, wandering widely; in Middle East, winters S to C Arabia. Vagrant Israel (has bred), Kuwait (has bred), Oman, Syria and Yemen.

Ammomanes deserti
Desert Lark

OTHER NAMES Desert Finch-lark, Sand Lark.
TAXONOMY (Lichtenstein, M H C, 1823). Twenty-four subspecies (19 in region). Plumage varies according to underlying rock colour of breeding areas, leading to description of colour morphs as multiple subspecies which do not have continuous ranges, and whose populations are widely separated geographically; thorough review needed.
DISTRIBUTION [N and NE Africa, Middle East and SW and C Asia. Increasing. LC] Resident from N Mauritania to S Algeria (*geyri*), S Morocco and SW Algeria (*payni*), N Algeria, Tunisia, NW Libya and NW Chad (*algeriensis*), SE Algeria and SW Libya (*whitakeri*), C Algeria (*mya*), E Egypt, Israel, parts of Jordan and N Arabia (*deserti*), N Egypt to C Arabia and SW Iraq (*isabellina*), N Oman and UAE (*taimuri*), EC Arabia (*azizi*), S Arabia (*samharensis* and *saturata*), Bahrain (*insularis*), NW Arabia, parts of Jordan and S Syria (*annae*), Syria, S Turkey and N Iraq (*coxi*), E Iraq and W Iran (*cheesmani*), SW Iran (*darica*), E Iran (*iranica*) and NE Iran (*orientalis* and *parvirostris*). Vagrant Cyprus and Lebanon (has bred).

Ammomanes cinctura
Bar-tailed Lark

OTHER NAMES Bar-tailed (or Black-tailed) Desert Lark, Sand Lark.
TAXONOMY (Gould, 1839). Three subspecies (all in region).

DISTRIBUTION [N Africa, Middle East and SW Asia. Decreasing. LC] Breeds Cape Verde Is (*cinctura*), from N Mauritania discontinuously E through N Africa to Sinai, Levant, Iraq and much of Arabia (*arenicolor*), and in E Iran (*zarudnyi*). Mainly resident but also nomadic, and some winter dispersal. Vagrant Balearic Is, Canary Is, Cyprus, Italy, Malta, Spain and Turkey.

Eremopterix nigriceps
Black-crowned Sparrow-Lark

OTHER NAMES Black-crowned Finch-Lark, White-fronted (or White-crested) Sparrow-Lark (or Finch-lark); Arabian Sparrow-Lark (*E n melanauchen*).
TAXONOMY (Gould, 1839). Five subspecies (four in region).
DISTRIBUTION [W Sahara, Sahel and NE Africa to Arabia and NW India. Increasing. LC] Breeds Cape Verde Is (*nigriceps*), N Mauritania, S Morocco, N Niger and possibly N Chad (*albifrons*), SE Egypt, W, S, C and E Arabia N to SE Iraq and S Iran (*melanauchen*) and SE Iran (*affinis*). Vagrant Algeria, Israel and Jordan.

Eremopterix signatus
Chestnut-headed Sparrow-Lark (V)

OTHER NAME Chestnut-headed Finch-Lark.
TAXONOMY (Oustalet, 1886). Two subspecies (one in region).
DISTRIBUTION [NE and E Africa. Increasing. LC] Extreme vagrant: Israel (1 May 1983, *harrisoni*).

Mirafra cantillans
Singing Bush Lark

OTHER NAMES Singing Lark, Western Singing Bushlark.
TAXONOMY Blyth, 1845. Four subspecies (one in region).
DISTRIBUTION [Sahel to S Asia. Decreasing. LC] Breeds extreme SW Saudi Arabia, W Yemen and S Oman (*simplex*). Mainly resident, but in Oman perhaps largely summer visitor.

Lullula arborea
Woodlark

TAXONOMY (Linnaeus, 1758). Two subspecies (both in region).
DISTRIBUTION [NW Africa and Europe to SW Asia > N Africa, W and S Europe and SW Asia. Population trend unknown. LC] Near-endemic. Breeds NW Africa and S Europe E to Asia Minor, Levant, Caucasus and Iran (*pallida*), N to S Britain, S Scandinavia and Finland, and E to Ukraine and W Russia (*arborea*). N, C and E continental European populations migratory, wintering S Europe and N Middle East S to *eg* N Libya, NE Egypt and Kuwait (rare). Vagrant Bahrain, Faroe Is, Ireland, Liechtenstein and Saudi Arabia.

Alauda leucoptera
White-winged Lark

TAXONOMY Pallas, 1811. Monotypic.
DISTRIBUTION [Extreme SE Europe to C Asia. Decreasing. LC] Breeds N Caspian region of Russia and NW Kazakhstan. Resident and short-distance migrant, vacating N breeding range and wintering to SW and S as far as Azerbaijan (irregular). Vagrant Austria*, Britain, Bulgaria, Finland, Germany, Greece, Iran, Italy, Malta, Norway, Poland, Romania, Serbia, Spain*, Sweden, Switzerland, Turkey (formerly more frequent) and Ukraine.

Alauda razae
Raso Lark (E)

OTHER NAME Razo Lark.
TAXONOMY (Alexander, 1898). Monotypic.
DISTRIBUTION [Cape Verde Is. Stable. CE] Resident endemic Cape Verde Is, where confined to island of Raso. Sole 'extralimital' record concerns one on nearby São Nicolau in same archipelago.

Alauda gulgula
Oriental Skylark

OTHER NAMES Eastern (or Indian) Skylark, Small (or Lesser/Little) Skylark.
TAXONOMY Franklin, 1831. Thirteen subspe-

cies (at least one in region). Sometimes considered conspecific with Eurasian Skylark.
DISTRIBUTION [C, S and SE Asia > SW to SE Asia. Decreasing. LC] Breeds E Iran (*inconspicua*). On passage and in winter occurs S Iran, Bahrain, Israel, Kuwait, Oman, Saudi Arabia and UAE; individuals trapped in Israel found to differ from typical *inconspicua*, but subspecific identity requires confirmation. Vagrant Azerbaijan, Egypt, Jordan and Lebanon.

Alauda arvensis
Eurasian Skylark

OTHER NAMES (Common or Northern) Skylark (or Sky Lark).
TAXONOMY Linnaeus, 1758. Eleven subspecies (six in region). Sometimes considered to include Oriental Skylark; extralimital *pekinensis* (NE Siberia to Kamchatka and Kuril Is) separated by some authors.
DISTRIBUTION [NW Africa and Eurasia > N Africa, W and S Europe, SW to E Asia. Decreasing. LC] Breeds NW Africa (*harterti*), Portugal, S and C Spain (*sierrae*), W, N and C Europe (*arvensis*), S Europe from NE Spain to Turkey and Caucasus (*cantarella*), SE Turkey to N Iran (*armenica*) and SW European Russia (*dulcivox*). N and E populations migratory, winter range extending S to Canary Is, from S Morocco across northernmost N Africa to Sinai, C and SE Arabia and S Iran. Vagrant Azores, Iceland, Jan Mayen, Madeira, N Mauritania and Svalbard.

Galerida theklae
Thekla Lark

TAXONOMY Brehm, A E, 1857. Twelve subspecies (six in region). Current IOC treatment followed here does not reflect deep genetic divergences among three extralimital E African taxa and Mediterranean populations, suggesting that several species may be involved (in which case regional population may become endemic species). Has also been treated as conspecific with Crested Lark, but proof of close relationship lacking.
DISTRIBUTION [N Africa and SW Europe, NE and E Africa. Decreasing. LC] Breeds SW Morocco to Mauritanian border (*theresae*), E Morocco through N Sahara to NW Egypt (*carolinae*), E Morocco, N Algeria and E Tunisia (*superflua*), C and NE Morocco, N Algeria and N Tunisia (*ruficolor*), N Morocco (*erlangeri*) and Portugal, Spain, Balearic Is and extreme SW France (*theklae*). Mainly sedentary but some dispersive movement. Scarce winter visitor Gibraltar. Presumed vagrant (or rare resident?) N Mauritania.

Galerida cristata
Crested Lark

OTHER NAME Common Crested Lark.
TAXONOMY (Linnaeus, 1758). Thirty-four subspecies (at least 24 in region). Taxonomy highly complex, with variation in plumage of numerous localised subspecies linked to soil types, climate and other environmental factors; thorough review needed. Formerly included Maghreb Lark (*qv*).
DISTRIBUTION [N Africa, W Europe to SW, C and E Asia. Decreasing. LC] Breeds N Mauritania (*balsaci* and/or *senegallensis*), W Morocco (*riggenbachi*), NW Morocco (*kleinschmidti*), NE Morocco E to N Tunisia (*carthaginis*), NE Algeria, S Tunisia and NW Libya (*arenicola*), SE Algeria and SW Libya (*helenae*), coastal NE Libya (*festae*), inland NE Libya to S Iraq and N Arabia (*brachyura*), Nile Delta (*nigricans*), C Egypt, W and S Arabia (*maculata*), S Egypt (*halfae*), Iberia (*pallida*), from France N to S Baltic and Belarus and E to Hungary and Ukraine (*cristata*), WC Italy (*neumanni*), S Italy and Sicily (*apuliae*), E Croatia to C Greece and W Turkey (*meridionalis*), Rhodes, Karpathos and Cyprus (*cypriaca*), E Hungary and Romania to S Russia (*tenuirostris*), NW Kazakhstan and discontinuously in E Arabia (*magna*), E Aegean Is, N Turkey and Caucasus (*caucasica*), NW Israel and W Lebanon (*cinnamomina*), NE Israel to S Turkey (*zion*), C Turkey to N Iran (*subtaurica*) and C Iran (*iwanowi*). May also occur NW Chad (*alexandri*?) but confirmation required. Mainly resident, but summer migrant *eg* parts of Caucasus, E Turkey and NW Iran. Vagrant Azores*, Britain, Corsica, Finland, Malta, Sardinia and Switzerland (formerly bred).

Galerida macrorhyncha
Maghreb Lark (E)

OTHER NAME Long-billed Crested Lark.

TAXONOMY Tristram, 1859. Two subspecies (both in region). Previously considered conspecific with Crested Lark, but differs both in mtDNA and in preference for more arid habitats; in very limited areas of range overlap some hybridisation does occur.
DISTRIBUTION [NW Africa. Population trend not assessed. NE] Treated here provisionally as endemic, but S limit of range around regional boundary in Mauritania imprecisely known. Breeds N Mauritania and SW Morocco to W Algeria (*macrorhyncha*) and E Morocco to NW Algeria (*randonii*). Presumed resident or mainly so.

Eremophila alpestris
Horned Lark

OTHER NAMES Shore Lark; Atlas Horned Lark (*E a atlas*), Caucasian Horned Lark (*E a penicillata*), Steppe Horned Lark (*E a brandti*).
TAXONOMY (Linnaeus, 1758). Forty-two subspecies (eight in region) under current IOC arrangement followed here, but latest genetic data suggest five Palearctic clades, all representing independent evolutionary units which should be treated as distinct species: *flava*, *brandti*, *atlas* (regional endemic), *penicillata* (including *albigula* and *balcanica*) and extralimital *elwesi*. Distribution of some clades and alignment to presently defined subspecies require more work. In addition, all Nearctic taxa best treated as further extralimital species, *alpestris*, pending more detailed research to reconstruct clade ranges and elucidate evolutionary history.
DISTRIBUTION [N America, NW South America, Eurasia and NW Africa. Decreasing. LC] Breeds Morocco (*atlas*), C and N Scandinavia and NW Russia (*flava*), SW Russia (*brandti*), NE Iran (*albigula*), elsewhere in Iran, Caucasus, N Iraq and E Turkey (*penicillata*), W and C Asia Minor (*kumerloevei*), Levant (*bicornis*) and S Balkans (*balcanica*). N European populations migratory, passing through N and E Baltic to winter S to North Sea coasts of Britain and continental Europe, S Baltic and locally inland S to C Europe and N Black Sea region; some altitudinal and short-distance migration also in SE populations, wintering *eg* N Levant and C and S Iran. Vagrant Algeria, Channel Is, Cyprus, Czech Republic, Faroe Is, Iceland, Ireland, Italy, Luxembourg, Malta, Slovakia, Slovenia, Spain, Svalbard and Switzerland.

Eremophila bilopha
Temminck's Lark (E)

OTHER NAME Temminck's Horned Lark.
TAXONOMY (Temminck, 1823). Monotypic.
DISTRIBUTION [N Africa and Middle East. Decreasing. LC] Endemic. Breeds from Morocco discontinuously E across N Africa to Sinai, Israel and Levant, and E to C Iraq and Kuwait, and S to N and E Saudi Arabia. Mainly resident but some dispersive movements. Vagrant Cyprus, Malta, N Mauritania, UAE and Yemen.

Calandrella acutirostris
Hume's Short-toed Lark

OTHER NAMES Hume's (or Slender-billed) Lark.
TAXONOMY Hume, 1873. Two subspecies (one in region).
DISTRIBUTION [SW and SC to CE Asia > SC and S Asia. Stable. LC] Scarce summer visitor NE Iran (*acutirostris*). Vagrant Israel.

Calandrella blanfordi
Blanford's Lark

OTHER NAMES Blanford's Short-toed Lark, Blanford's (or Arabian) Red-capped Lark.
TAXONOMY (Shelley, 1902). Three subspecies (one in region). Previously treated as conspecific with both Greater Short-toed Lark and extralimital Red-capped Lark *C cinerea* (E, C and S Africa).
DISTRIBUTION [NE Africa and SW Arabia. Population trend not assessed. NE] Breeds SW Saudi Arabia and W Yemen (*eremica*). Mainly resident, but some altitudinal movement in winter. Vagrant Oman.

Calandrella brachydactyla
Greater Short-toed Lark

OTHER NAMES (European) Short-toed Lark.

TAXONOMY (Leisler, 1814). Nine subspecies (seven in region).
DISTRIBUTION [S Europe and N Africa to CE Asia > N Africa, Sahel to SE Asia. Decreasing. LC] Summer visitor Canary Is, N Africa, Malta (*rubiginosa*), SW and S Europe, Mediterranean islands and apparently parts of Mediterranean coast of NW Africa (*brachydactyla*), Hungary and N Serbia (*hungarica*), NE Egypt, NW Arabia and Levant to S Turkey (*hermonensis*), S Turkey and NW Syria (*woltersi*), C Turkey, Caucasus, C and E Arabia and NW Iran (*artemisiana*), and Ukraine and SW Russia (*longipennis*). Occurs widely on passage S of breeding areas, some wintering *eg* interior N Africa, Arabia and Iran; scarce migrant Austria (has bred), Britain, Finland (has bred), Gibraltar and Switzerland (has bred). Vagrant Azores, Belgium, Cape Verde Is, Denmark, Estonia, Germany, Iceland, Ireland, Madeira, Moldova, Netherlands, Norway, Poland, Slovenia and Sweden.

Melanocorypha bimaculata
Bimaculated Lark

OTHER NAME Eastern Calandra Lark.
TAXONOMY (Ménétriés, 1832). Three subspecies (all in region).
DISTRIBUTION [SW to C Asia > SW, S and SC Asia. Stable. LC] Breeds E Turkey, Caucasus and N Iran (*bimaculata*), C and S Turkey, W Syria, Lebanon and possibly Iraq (*rufescens*), and NE Iran (*torquata*); has bred Kuwait. Resident Iran but otherwise migratory, occurring widely on passage Middle East W to Cyprus and E Egypt and S to C and SE Arabia, where some overwinter; also winters Iraq. Vagrant Britain, Denmark, Finland, Greece, Italy, Norway, Sweden, Switzerland and Yemen.

Melanocorypha calandra
Calandra Lark

OTHER NAMES European (or Western) Calandra Lark.
TAXONOMY (Linnaeus, 1766). Four subspecies (all in region).
DISTRIBUTION [NW Africa and S Europe to C Asia. Decreasing. LC] Breeds from NW Africa, Libya and Iberia E through S Europe to Black Sea, N Caspian region, Caucasus, N and W Turkey and NW Iran (*calandra*), SC Turkey and NW Syria to Israel and W Jordan (*hebraica*), E Syria and SE Turkey to SW Iran (*gaza*), and N Iraq and N Iran (*psammochroa*). Mainly resident in W of range, partial migrant in E, wintering S through N Middle East to N Egypt. Vagrant Austria, Bahrain, Belgium, Britain, Canary Is, Czech Republic, Estonia, Finland, Germany, Gibraltar, Hungary, Kuwait, Lithuania, Luxembourg, Madeira, Netherlands, Norway, Poland, Saudi Arabia, Slovenia, Sweden, Switzerland and UAE.

Melanocorypha yeltoniensis
Black Lark

TAXONOMY (Forster, J R, 1767). Monotypic.
DISTRIBUTION [Extreme SE Europe to C Asia. Decreasing. LC] Breeds N Caspian region of Russia and NW Kazakhstan. Resident and short-distance migrant, vacating N breeding range and wintering to S and SW; sporadic winter visitor Azerbaijan and N Iran. Vagrant Austria, Britain, Bulgaria, Finland, Germany, Greece, Italy, Lebanon*, Malta, Moldova, Poland, Sweden, Turkey and Ukraine.

Chersophilus duponti
Dupont's Lark (E)

OTHER NAME Koenig's Lark (*C d margaritae*).
TAXONOMY (Vieillot, 1824). Two subspecies (both in region).
DISTRIBUTION [Iberia and N Africa. Decreasing. NT] Endemic. Breeds Spain, N Morocco, N Algeria and C Tunisia (*duponti*) and C Algeria, SE Tunisia, N Libya and coastal W Egypt (*margaritae*). Resident, but probably makes irregular movements in winter, when recorded *eg* N Africa outside known breeding areas. Vagrant Cyprus, France, Greece, Italy and Malta.

Eremalauda dunni
Dunn's Lark

OTHER NAMES African Dunn's Lark (*E d dunni*), Arabian Dunn's Lark (*E d eremodites*).
TAXONOMY (Shelley, 1904). Two subspecies (both in region).

DISTRIBUTION [S Sahara and Arabia. Stable. LC] Breeds N Mauritania, N Mali and possibly S Algeria (*dunni*), from Jordan irregularly W to S Israel and Sinai and N to Syria, E to Kuwait and widely S through N and C Arabia, locally also S Arabia (*eremodites*). Mainly resident, but dispersive and partially nomadic. Vagrant Cyprus, Egypt (has bred), Lebanon, Morocco and UAE.

Alaudala cheleensis
Asian Short-toed Lark

OTHER NAMES Asiatic (or Eastern/Mongolian) Short-toed Lark, Grey Short-toed Lark.
TAXONOMY Swinhoe, 1871. Eight subspecies (two in region). Taxonomy not settled. Sometimes included with Lesser Short-toed Lark because of apparent overlap of subspecies and uncertainty over which taxa belong to which species (see *eg* Kirwan *et al* 2008). IOC treatment as two species provisionally followed here; more research needed and revision likely.
DISTRIBUTION [SW to E Asia. Stable. LC] Breeds C Turkey (*niethammeri*), E and S Iraq, and across much of Iran except NW (*persica*). Partly migratory, but non-breeding distribution poorly understood; *persica* reportedly winters widely Syria and S to Kuwait, and specimens resembling this taxon have been collected E Saudi Arabia.

Alaudala rufescens
Lesser Short-toed Lark

OTHER NAMES Common (or Rufous) Short-toed Lark.
TAXONOMY (Vieillot, 1819). Six subspecies (all in region). Analysis of mtDNA from seventh subspecies listed by IOC, *polatzeki* (C and E Canary Is, formerly including S Tenerife), and nominate *rufescens* (formerly N Tenerife) found no basis for existence of two subspecies, hence former has been synonymised with latter.
DISTRIBUTION [N Africa and Iberia to C Asia > Iberia and N Africa to Middle East. Decreasing. LC] Breeds C and E Canary Is (*rufescens*), Spain and Portugal (*apetzii*), from Morocco to NW Egypt, Sinai, Levant, S Turkey, E Iraq and (probably this taxon) E Arabia (*minor*), Nile Delta (*nicolli*), E Turkey, Caucasus and N Iran (*pseudobaetica*), and Ukraine and N Caspian region (*heinei*). Mainly resident Iberia, N Africa and Middle East, but migratory SE Europe, much of Turkey, Caucasus and N Iran, occurring on passage and in winter S throughout Middle East. Irregular Italy, rare migrant N Mauritania. Vagrant Austria, Balearic Is, Britain, Belgium, Bulgaria, Finland, Germany, Gibraltar, Greece, Madeira, Malta, Norway, Serbia, Sweden and Switzerland.

Alaudala raytal
Sand Lark

OTHER NAMES Indian Sand (or Raytal) Lark, Indian (or Asian) Short-toed Lark.
TAXONOMY (Blyth, 1845). Three subspecies (one in region).
DISTRIBUTION [S Asia. Stable. LC] Resident SE Iran (*adamsi*).

FAMILY PYCNONOTIDAE

Bulbuls · four species

Pycnonotus leucotis
White-eared Bulbul

OTHER NAME White-cheeked Bulbul.
TAXONOMY (Gould, 1836). Two subspecies (both in region). Formerly treated as conspecific with extralimital Himalayan Bulbul *P leucogenys* (Himalayas) and hybridises where ranges meet ('*humii*'), but differs morphologically and hybrid zone is relatively abrupt.
DISTRIBUTION [Middle East to S Asia. Decreasing. LC] Resident Syria, Iraq, NE Arabia and SW Iran (*mesopotamia*), S Iran (*leucotis*) and C and E Arabia, where range expansion driven partly by introductions (likely involving *leucotis* from NW India) and population probably racially mixed. Vagrant Israel, Turkey* and Jordan (where latterly small population also introduced in Azraq area).

Pycnonotus cafer
Red-vented Bulbul (I)

TAXONOMY (Linnaeus, 1766). Eight subspecies (undetermined in region).

DISTRIBUTION [S and SE Asia. Increasing. LC] Non-native. Self-sustaining naturalised populations established Bahrain, Kuwait, Oman, E Saudi Arabia and UAE; recently established SE Iran (Hormozgan).

Pycnonotus xanthopygos
White-spectacled Bulbul

OTHER NAMES Spectacled Bulbul, White-eyed (or Black-capped) Bulbul, Yellow-vented Bulbul.
TAXONOMY (Hemprich and Ehrenberg, 1833). Monotypic.
DISTRIBUTION [Middle East. Increasing. LC] Resident from S Turkey S through Levant and Israel to Sinai and E to Iraq and W, C, S and E Arabia.

Pycnonotus barbatus
Common Bulbul

OTHER NAMES Black-eyed (or Black-capped) Bulbul, Common Garden Bulbul; Egyptian Bulbul (*P b arsinoe*).
TAXONOMY (Desfontaines, 1789). Five subspecies (three in region).
DISTRIBUTION [Widespread Africa. Increasing. LC] Resident Morocco, Algeria and Tunisia (*barbatus*) and E Egypt (*arsinoe*); has also bred just inside regional boundary in N Mauritania, N Niger and N Chad (*inornatus*). Possibly vagrant S Spain, where records currently placed in Cat D and breeding took place in 2013.

FAMILY HIRUNDINIDAE

Swallows and martins · 18 species

Riparia paludicola
Brown-throated Martin

OTHER NAMES Brown-throated (or African) Sand Martin, Plain Martin.
TAXONOMY (Vieillot, 1817). Seven subspecies (one in region). Formerly included Grey-throated Martin (*qv*).
DISTRIBUTION [NW and sub-Saharan Africa, Madagascar. Decreasing. LC] Resident W Morocco (*mauritanica*). May undertake erratic local movements, and possibly post-breeding dispersal to N of normal range. Vagrant Algeria, Cape Verde Is, France and N Mauritania, but see comments under Grey-throated Martin for other possible extralimital occurrences.

Riparia chinensis
Grey-throated Martin (V)

OTHER NAME Indian Sand Martin.
TAXONOMY (Gray, J E, 1830). Two subspecies (one in region). Often considered conspecific with Brown-throated Martin, but much smaller and paler, with markedly different vocalisations.
DISTRIBUTION [SC and S Asia to E and SE Asia. Population trend not assessed. NE] Extreme vagrant (at least 10 records, *chinensis*): Iran (1 Feb 2000, four); Israel (6 May 1985); Oman (8). Further undetermined records from Iran (27 Jan 2000, four, and 21 Jan 2004) most likely to refer to this Asian species rather than Brown-throated Martin, but similar occurrences from Israel (8 Apr 1989), Saudi Arabia and Yemen could relate to either species.

Riparia riparia
Sand Martin

OTHER NAMES Bank Swallow, Collared (or Gorgeted) Sand Martin, Common (or European) Sand Martin.
TAXONOMY (Linnaeus, 1758). Six subspecies (five in region).
DISTRIBUTION [North America, Europe and N Asia > South America, sub-Saharan Africa and SE Asia. Decreasing. LC] Summer visitor to much of Europe (though range fragmented in S) from Mediterranean to N Scandinavia, W to Britain and Ireland and E to Caucasus, also Asia Minor, Syria, Iraq and Iran (*riparia*) and Egypt (*shelleyi*), where small numbers winter; has bred NW Africa. Widespread on passage throughout N Africa (W to Canary Is) and Middle East, through which *eilata*, *innominata* and *taczanowskii* (Asia) migrate to African winter quarters. Vagrant Azores, Cape Verde Is, Faroe Is, Iceland, Madeira and Svalbard.

Riparia diluta
Pale Martin

OTHER NAMES Eastern (or Pale) Sand Martin.
TAXONOMY (Sharpe and Wyatt, 1893). Six subspecies (undetermined in region). Until recently considered conspecific with Sand Martin, but differs morphologically and vocally, and found in separate breeding colonies (or breeding separately in mixed colonies) in wide area of sympatry.
DISTRIBUTION [C to E Asia > SC, S and SE Asia. Population trend not assessed. NE] Uncommon migrant and winter visitor UAE. Apparently unconfirmed Israel, where has been claimed as regular on migration despite rarity elsewhere in region; formerly accepted records from Egypt now considered erroneous. Vagrant Iran and Oman.

Riparia cincta
Banded Martin (V)

OTHER NAME Banded Sand Martin.
TAXONOMY (Boddaert, 1783). Five subspecies (undetermined in region).
DISTRIBUTION [C, E and NE Africa > widespread sub-Saharan Africa. Increasing. LC] Extreme vagrant (three records): Egypt (15 Nov 1988); Saudi Arabia (3 Oct 1996); Yemen (12 Mar 1982).

Tachycineta bicolor
Tree Swallow (V)

TAXONOMY (Vieillot, 1808). Monotypic.
DISTRIBUTION [Widespread North America > S North America and West Indies to N South America. Stable. LC] Extreme vagrant (10 records): Azores (7: 26 Oct-2 Nov 2005; 2 Nov 2005*, two; 2 Nov 2005; 5-6 Nov 2005; 19-20 Oct 2007; 23-26 Oct 2007*; 8 Oct 2012); Britain (2: 6-10 June 1990; 29 May 2002); Iceland (15-16 May 2012*).

Progne subis
Purple Martin (V)

TAXONOMY (Linnaeus, 1758). Three subspecies (undetermined in region).
DISTRIBUTION [S Canada, USA and N Mexico > SE South America. Stable. LC] Extreme vagrant (four records of five birds): Azores (3: 28 Sep 1996, two*; 6 Sep 2004; 13-18 Oct 2011); Britain (5-6 Sep 2004).

Hirundo rustica
Barn Swallow

OTHER NAMES (Common) Swallow; American Barn Swallow (*H r erythrogaster*), Eurasian (or European) Swallow (*H r rustica*), Egyptian (Barn)/Nile Swallow (*H r savignii*).
TAXONOMY Linnaeus, 1758. Seven subspecies (four in region). Analysis of mtDNA indicates existence of three major clades (North America, Europe to Kazakhstan, and Asia), with separation of Nearctic and Palearctic populations equivalent to that between species, but within North American grouping a subclade of Siberian birds; more research needed.
DISTRIBUTION [North America and Eurasia > Central and South America, sub-Saharan Africa, S Asia and N Australasia. Decreasing. LC] Breeds N Africa from Morocco to Libya, across Europe W to Britain and Ireland, N to Faroe Is and N Scandinavia and E to Ural Mtns, through Turkey and N Middle East (*rustica*), Levant (*transitiva*) and Nile Valley (*savignii*). Migratory except S Iberia, Levant and Egypt, occurring widely on passage through N Africa (W to Canary Is) and Middle East. Vagrant Azores, Cape Verde Is (has bred), Iceland (has bred), Jan Mayen, Madeira and Svalbard; *erythrogaster* (North America) vagrant Azores and Iceland*.

Hirundo aethiopica
Ethiopian Swallow (V)

TAXONOMY Blanford, 1869. Two subspecies (undetermined in region).
DISTRIBUTION [W to NE and E Africa. Increasing. LC] Extreme vagrant: Israel (22-23 Mar 1991, ringed).

Hirundo smithii
Wire-tailed Swallow (V)

OTHER NAMES African Wire-tailed Swallow

(*H s smithii*), Asian Wire-tailed Swallow (*H s filifera*).
TAXONOMY Leach, 1818. Two subspecies (undetermined in region).
DISTRIBUTION [Sub-Saharan Africa, S and SE Asia. Increasing. LC] Very rare vagrant: Oman (8); UAE (14). Formerly accepted record from Egypt (30 Mar 1995) now considered not proven.

Ptyonoprogne rupestris
Eurasian Crag Martin

OTHER NAMES (European or Mountain/Northern) Crag Martin.
TAXONOMY (Scopoli, 1769). Monotypic.
DISTRIBUTION [N Africa, S Europe, SW, C, S and SE Asia. Increasing. LC] Breeds NW Africa, NE Libya and from Iberia N to C Europe, E through S Europe to Turkey, Levant, Caucasus, N Iraq and Iran. Partly resident and altitudinal migrant, but many northern populations more migratory, occurring more widely on passage and/or in winter S Europe, N Africa and Middle East (rarer in S). Vagrant Belgium, Britain, Canary Is, Czech Republic, Denmark, Estonia, Finland, Malta, Netherlands, Norway, Salvage Is, Sweden and Ukraine.

Ptyonoprogne obsoleta
Pale Crag Martin

OTHER NAME Western Pale Crag Martin (*P o presaharica*).
TAXONOMY (Cabanis, 1850). Seven subspecies (six in region). Sometimes treated as conspecific with extralimital Rock Martin *P fuligula* (sub-Saharan Africa), but differs in paler plumage, vocalisations and larger size.
DISTRIBUTION [N Africa to SW Asia. Increasing. LC] Breeds N Mauritania and from S Morocco to NC Algeria (*presaharica*), S Algeria, SW Libya and NW Chad (*spatzi*), SW Arabia (*arabica*), NE Saudi Arabia and Iraq (*perpallida*), Egypt, Sinai, Israel, Jordan, N, C and E Arabia and SW Iran (*obsoleta*), and E Iran (*pallida*). Mainly resident but some dispersal; partly migratory Iran and Iraq. Vagrant Cape Verde Is and Kuwait.

Delichon urbicum
Common House Martin

OTHER NAMES (European, Northern or Western) House Martin.
TAXONOMY (Linnaeus, 1758). Three subspecies (two in region).
DISTRIBUTION [NW Africa, W Europe to E Asia > sub-Saharan Africa, S Arabia and SE Asia. Decreasing. LC] Summer visitor NW Africa and S Europe E to Asia Minor, Levant, N Iraq and Iran (*meridionale*), and W, C and N Europe N to N Scandinavia and E to Ural Mtns (*urbicum*). Widespread S of breeding range on migration in N Africa (W to Canary Is) and Middle East, some also wintering S Arabia. Vagrant Azores, Cape Verde Is (has bred), Iceland (has bred), Jan Mayen, Madeira and Svalbard (has bred).

Delichon dasypus
Asian House Martin (V)

OTHER NAMES Asian (or Asiatic House) Martin; Kashmir House Martin (*D d cashmeriense*).
TAXONOMY (Bonaparte, 1850). Three subspecies (undetermined in region).
DISTRIBUTION [SC, E and SE Asia > S and SE Asia. Increasing. LC] Extreme vagrant (three records): UAE (3: 8-21 Oct 1999; 25 Nov 2001; 20 Feb-8 Mar 2001).

Cecropis abyssinica
Lesser Striped Swallow (V)

OTHER NAME Striped Swallow.
TAXONOMY (Guérin-Méneville, 1843). Six subspecies (undetermined in region).
DISTRIBUTION [Sub-Saharan Africa. Increasing. LC] Extreme vagrant: Oman (4 Dec 1986).

Cecropis daurica
Red-rumped Swallow

OTHER NAMES Amur Red-rumped Swallow (*C d japonica*), Asian Red-rumped Swallow (*C d daurica/japonica*), Lesser Striated

Swallow (*C d daurica/japonica/erythropygia*).
TAXONOMY (Laxmann, 1769). Eight subspecies (at least three in region).
DISTRIBUTION [S Europe, Africa and Asia. Increasing. LC] Summer visitor NW Africa and from Iberia and S France E through S Europe and N Middle East to Iran, and in E, C and SW Arabia (*rufula*). On passage more widespread S Europe, parts of N Africa (W to Canary Is) and Middle East, where has bred UAE; scarce on migration Britain and Netherlands. Vagrant Austria, Azerbaijan, Azores, Belgium, Cape Verde Is, Channel Is, Czech Republic, Denmark, Estonia, Faroe Is, Finland, Germany, Hungary, Iceland, Ireland, Latvia, Luxembourg, Madeira, Poland, Norway, Salvage Is, Slovakia, Sweden and Ukraine. Asian *japonica* collected Norway (31 May 1905); *daurica* documented Norway (22 Aug 2007); undetermined *daurica/japonica* recorded Britain (9, 17 and 29 Jun 2011), Finland (2: 18 May 2007; 19 May 2011), Netherlands (26 May 2013*) and Norway (3 May 2006).

Petrochelidon fluvicola
Streak-throated Swallow (V)

OTHER NAMES Indian Cliff Martin, Indian Cliff Swallow.
TAXONOMY (Blyth, 1855). Monotypic. Sometimes considered conspecific with extralimital Fairy Martin *P ariel* (Australia).
DISTRIBUTION [S Asia. Increasing. LC] Very rare vagrant: Egypt (19 Nov 2003); Kuwait (6 Dec 2013); Oman (5); UAE (9).

Petrochelidon pyrrhonota
American Cliff Swallow (V)

OTHER NAME Cliff Swallow.
TAXONOMY (Vieillot, 1817). Five subspecies (undetermined in region).
DISTRIBUTION [North America > C South America. Increasing. LC] Very rare vagrant: Azores (7+ records of 12+ individuals*); Britain (9); France (2); Iceland (2: 12 Oct 1992; 12 Oct 2004, two); Ireland (2*); Madeira (27 Oct 2001); Sweden (15-20 Nov 2012*). Additional record from Canary Is (30 Sep 1991) accepted as either this species or extralimital Cave Swallow *P fulva* (S USA, Caribbean and Mexico).

FAMILY CETTIIDAE

Bush warblers · one species

Cettia cetti
Cetti's Warbler

OTHER NAMES Cetti's Bush Warbler; Eastern Cetti's Warbler (*C c albiventris*).
TAXONOMY (Temminck, 1820). Three subspecies (all in region). Eastern subspecies *orientalis* may constitute separate species, but relationships with other taxa unclear at present.
DISTRIBUTION [NW Africa, W Europe to C Asia > NW Africa, W Europe to SW Asia. Increasing. LC] Breeds from NW Africa and Iberia N to S Britain and SW Netherlands, and E through S Europe and Mediterranean to S Ukraine and W Turkey (*cetti*), C Turkey (possibly farther W) S to Levant, N through Caucasus to N Caspian region, E to Iraq and NW Iran (*orientalis*), and in N Iran (*albiventris*). Mainly resident but partial migrant in E, vacating N Turkey, Caucasus and N Caspian region during winter, when occurs farther S in Levant and also reaches *eg* Kuwait and SW/SE Iran. Periodic attempts at breeding-range expansion in NW Europe partly curtailed by severe winters. Vagrant Austria, Czech Republic, Germany (has bred), Gibraltar, Hungary (has bred), Ireland, Libya, Luxembourg, Oman, Poland, Sweden, UAE and Ukraine.

FAMILY SCOTOCERCIDAE

Streaked Scrub Warbler · one species

Scotocerca inquieta
Streaked Scrub Warbler

OTHER NAMES Scrub (or Streaked) Warbler; Levant Scrub Warbler (*S i inquieta*), Saharan Scrub Warbler (*S i saharae*), West Saharan Scrub Warbler (*S i theresae*).

TAXONOMY (Cretzschmar, 1830). Eight subspecies (seven in region). Significant variation in morphology, vocalisations and genetics between populations in N Africa, Middle East and C Asia; further research needed.
DISTRIBUTION [N Africa, SW and C Asia. Decreasing. LC] Resident SW and C Morocco (*theresae*), E Morocco to Libya (*saharae*), NE Libya to the Levant and NW Arabia (*inquieta*), W Saudi Arabia, E Yemen and Oman (*grisea*), S Saudi Arabia and W Yemen (*buryi*), SC Iraq and S Iran (*striata*), and E and N Iran (*montana*). Vagrant Azerbaijan.

FAMILY AEGITHALIDAE

Bushtits · one species

Aegithalos caudatus
Long-tailed Tit

OTHER NAMES Long-tailed Bushtit; British Long-tailed Tit (*A c rosaceus*), Central European Long-tailed Tit (*A c europaeus*), Iberian Long-tailed Tit (*A c irbii*), Sicilian Long-tailed Tit (*A c siculus*), White-headed (or Fennoscandian/Northern) Long-tailed Tit (*A c caudatus*).
TAXONOMY (Linnaeus, 1758). Seventeen subspecies (14 in region).
DISTRIBUTION [W Europe to E Asia. Stable. LC] Breeds S Portugal, S Spain and Corsica (*irbii*), N Portugal, C and N Spain, SW and S France (*taiti*), W France and Channel Is (*aremoricus*), Britain and Ireland (*rosaceus*), Fennoscandia and Poland E to Ural Mtns (*caudatus*), S Crimea (*tauricus*), NE France and Germany to N Italy and Turkey (*europaeus*), SW Slovenia and C and S Italy (*italiae*), Sicily (*siculus*), Albania and Greece to Bulgaria and NW Turkey (*macedonicus*), E Greece to C Turkey, N Iraq and Syria (*tephronotus*), NE Turkey and Caucasus (*major*), SE Turkey and SW Iran (*passekii*) and SE Azerbaijan and N Iran (*alpinus*). Mainly resident but some dispersal; irregular irruptive movements mainly involve *caudatus* (N and E Europe) occurring W and S of breeding range, rarely SW to Britain, Belgium, France, Luxembourg and Netherlands, and S to Italy. Vagrant Gibraltar, Morocco and Tunisia.

FAMILY PHYLLOSCOPIDAE

Leaf warblers · 21 species

Phylloscopus umbrovirens
Brown Woodland Warbler

OTHER NAME Brown Warbler.
TAXONOMY (Rüppell, 1840). Nine subspecies (one in region).
DISTRIBUTION [E Africa and SW Arabia. Stable. LC] Resident SW Saudi Arabia and W Yemen (*yemensis*).

Phylloscopus trochilus
Willow Warbler

OTHER NAME Northern Willow Warbler (*P t acredula*).
TAXONOMY (Linnaeus, 1758). Three subspecies (all in region).
DISTRIBUTION [W Europe to NE Asia > sub-Saharan Africa. Decreasing. LC] Summer visitor from C France, Britain and Ireland N to S Sweden and E to S Poland and N Romania (*trochilus*) and from C and N Scandinavia E to Ural Mtns (*acredula*). Widespread on migration S Europe, also Faroe Is (has bred), N Africa (W to Canary Is) and Middle East, where *yakutensis* (breeding C and E Siberia) also known on passage. Vagrant Azores, Cape Verde Is, Iceland (has bred), Madeira and Svalbard.

Phylloscopus collybita
Common Chiffchaff

OTHER NAMES (Eurasian) Chiffchaff; Anatolian Chiffchaff (*P c brevirostris*), Scandinavian Chiffchaff (*P c abietinus*), Siberian (or Eastern) Chiffchaff (*P c tristis*).
TAXONOMY (Vieillot, 1817). Six subspecies (all in region). Taxonomy complex; *tristis* hybridises with *abietinus* in narrow NW Russian contact zone but differs in diagnostic call, song and plumage, and sometimes recognised at species level.
DISTRIBUTION [W Europe to CE Asia > Sahel, N and NE Africa, W and S Europe to S Asia. Increasing. LC] Breeds much of W, C and SE

Europe from C Iberia N to SW Scandinavia and E to Poland and Balkans (*collybita*), from C and N Scandinavia E to S Ural Mtns and SE European Russia (*abietinus*), N European Russia E to Ural Mtns (*tristis*), NW and N Turkey (*brevirostris*), Caucasus to N Iran (*caucasicus*) and NE Iran (*menzbieri*). Largely migratory away from SW and extreme S of breeding range, winter distribution extending through S Europe, much of N Africa and Middle East; *tristis* mainly migrates SE, some reaching Arabia in winter, but also scarce late autumn W Europe, especially Britain and Netherlands, and overwinters in small numbers. Rare but regular Azores. Vagrant Cape Verde Is, Iceland (has bred), Jan Mayen, Madeira and Svalbard; *tristis* rare/vagrant *eg* Belgium, Denmark, Estonia, Finland, France, Germany, Gibraltar, Ireland, Latvia, Malta, Netherlands, Norway, Spain, Svalbard and Sweden.

Phylloscopus ibericus
Iberian Chiffchaff

TAXONOMY Ticehurst, 1937. Two subspecies (both in region).
DISTRIBUTION [SW Europe and NW Africa > W Africa. Stable. LC] Breeding endemic. Summer visitor to extreme SW France, N Spain and N Portugal (*biscayensis*), W and S Portugal, S Spain, N Morocco, N Algeria and NW Tunisia (*ibericus*). Recorded on migration E Spain and Gibraltar, rarer Balearic Is, and to S in Canary Is, N Mauritania and Morocco, where some may winter. Vagrant Belgium, Britain, Czech Republic, Denmark, Finland, Germany, Ireland, Italy, Malta, Netherlands, Poland, Sweden and Switzerland.

Phylloscopus canariensis
Canary Islands Chiffchaff (E)

OTHER NAME Canary Chiffchaff.
TAXONOMY (Hartwig, 1886). Two subspecies (one presumed extinct). Preliminary mtDNA analysis suggests that population on Gran Canaria could represent a new subspecies.
DISTRIBUTION [Canary Is. Stable. LC] Endemic resident Canary Is on El Hierro, La Palma, La Gomera, Tenerife and Gran Canaria (*canariensis*); formerly also Lanzarote (*exsul*†).

Phylloscopus sindianus
Mountain Chiffchaff

OTHER NAMES Caucasian (Mountain) Chiffchaff (*P s lorenzii*), Lorenz's Warbler (*P s lorenzii*), Sind (Mountain) Chiffchaff (*P s sindianus*).
TAXONOMY Brooks, W E, 1880. Two subspecies (one in region). Taxonomy not yet fully resolved, *lorenzii* sometimes being treated separately as a full species as it differs in plumage and vocalisations from *sindianus*, yet species-level separation not confirmed by DNA analysis.
DISTRIBUTION [Caucasus and C Asia > SW Asia. Stable. LC] Breeds NE Turkey, Caucasus and NW Iran (*lorenzii*). Non-breeding range poorly known, but winters at lower elevations S of breeding areas; occurs on passage Iraq, and in winter elsewhere in Iran and in Kuwait (rare). Vagrant Cyprus, Gibraltar and Israel.

Phylloscopus neglectus
Plain Leaf Warbler

OTHER NAMES Plain (Willow) Warbler.
TAXONOMY Hume, 1870. Monotypic.
DISTRIBUTION [SW Asia. Stable. LC] Summer visitor S, W and N Iran; has bred Iraq. Short-distance migrant, wintering S of breeding range to E Saudi Arabia, Oman and UAE. Vagrant Bahrain, Israel, Jordan, Lebanon, Qatar, Sweden and Turkey.

Phylloscopus bonelli
Western Bonelli's Warbler

OTHER NAME Bonelli's Warbler.
TAXONOMY (Vieillot, 1819). Monotypic. Formerly included Eastern Bonelli's Warbler (*qv*).
DISTRIBUTION [NW Africa and SW Europe > W and C Sahel. Population trend not assessed. NE] Breeding endemic. Summer visitor from Morocco E to Tunisia, and from Iberia N to N France, S Germany and E to Austria, Italy and Slovenia. Migrates to winter quarters S through N Africa, W to Canary Is and E to Egypt, also Malta. Vagrant Belgium, Britain, Cape Verde Is, Czech Republic, Denmark, Finland, Hungary, Ireland, Luxembourg,

Macedonia, Madeira, Netherlands, Norway, Poland, Serbia, Slovakia and Sweden. Not all vagrant Bonelli's warblers *sensu lato* have been identified to species level (*eg* in Britain, of all *bonelli/orientalis* recorded to 2009, 45 per cent were undetermined, though most probably relate to Western), and records in some countries (*eg* Slovakia and Ukraine) not distinguished from *P orientalis*.

Phylloscopus orientalis
Eastern Bonelli's Warbler

OTHER NAME Balkan Warbler.
TAXONOMY (Brehm, C L, 1855). Monotypic. Previously treated as conspecific with Western Bonelli's Warbler, but molecular divergence between them is as great as that between either and Wood Warbler; clear vocal differences also supported by slight, but constant, morphological distinctions.
DISTRIBUTION [SE Europe and Middle East > E Sahel. Population trend not assessed. NE] Breeding endemic. Summer visitor from Croatia S and E through Balkans to W, C and SE Turkey, Lebanon, possibly Syria, N Israel, Jordan (irregular) and W Iran. Occurs on migration Levant, NW Saudi Arabia and N Africa W to E Libya, also Malta; irregular Italy. Vagrant Britain, Finland, Iraq, Kuwait, Netherlands, Norway, Oman, E Saudi Arabia, Sweden, Tunisia and UAE; see also comments under Western Bonelli's Warbler on indeterminate vagrants.

Phylloscopus sibilatrix
Wood Warbler

TAXONOMY (Bechstein, 1793). Monotypic.
DISTRIBUTION [W Europe to C Siberia > equatorial Africa. Decreasing. LC] Widespread summer visitor to Europe, breeding W to E Ireland (rare) and Britain, N to N Fennoscandia and NW Russia, E to Ural Mtns and in NW Caucasus, and S to Balkans, Italy and Pyrénées; has bred Turkey. Migrates S through N Africa (W to Canary Is) and parts of Middle East (scarce Arabia) to winter quarters. Vagrant Azerbaijan, Faroe Is, Iceland, Iran, Madeira, Portugal, Qatar and Svalbard.

Phylloscopus fuscatus
Dusky Warbler

TAXONOMY (Blyth, 1842). Two subspecies (one in region).
DISTRIBUTION [C Siberia to E Asia > SE Asia. Stable. LC] Scarce migrant Britain (annual mean 2000-09 of 14) and Netherlands (recorded annually 1999-2011 with max of 12 in latter year, since when rarity status removed); most occur late Oct-Nov but some into Dec, and occasionally overwinters (*fuscatus*). Vagrant Austria, Azerbaijan, Balearic Is, Belgium, Channel Is, Cyprus, Czech Republic, Denmark, Egypt, Estonia, Finland, France, Germany, Gibraltar, Greece, Hungary, Iceland*, Ireland, Israel, Italy, Kuwait, Latvia, Lithuania, Luxembourg, Malta, Morocco, Norway, Oman, Poland, Portugal, Romania, Saudi Arabia, Slovenia, Spain, Sweden, Switzerland, UAE, Ukraine and Yemen.

Phylloscopus schwarzi
Radde's Warbler

OTHER NAMES Radde's Bush Warbler, Thick-billed Warbler.
TAXONOMY (Radde, 1863). Monotypic.
DISTRIBUTION [S Siberia to E Asia > SE Asia. Stable. LC] Scarce autumn migrant Britain (annual mean 2000-09 of 14), mainly late Sep-Oct. Vagrant Balearic Is, Belgium, Channel Is, Czech Republic, Denmark, Estonia, Faroe Is, Finland, France, Germany, Hungary, Ireland, Israel, Italy, Latvia, Lithuania, Malta, Netherlands, Norway, Poland, Spain, Sweden, Switzerland, Ukraine and UAE.

Phylloscopus proregulus
Pallas's Leaf Warbler

OTHER NAMES Lemon-rumped (or Pale-rumped) Warbler, Pallas's Warbler.
TAXONOMY (Pallas, 1811). Monotypic.
DISTRIBUTION [S Siberia to E Asia > SE Asia. Stable. LC] Scarce migrant N and NW Europe, from Fennoscandia SW to Britain and Netherlands; most records Oct and Nov. Vagrant Austria, Balearic Is, Belgium, Bulgaria, Channel Is, Croatia, Czech Republic, Faroe Is, France, Germany, Gibraltar, Hungary, Iran,

Ireland, Israel, Italy, Latvia, Lithuania, Luxembourg, Malta, Morocco, Norway, Poland, Portugal, Romania, Slovakia, Slovenia, Spain, Tunisia, Turkey and Ukraine.

Phylloscopus inornatus
Yellow-browed Warbler

OTHER NAME Inornate Warbler.
TAXONOMY (Blyth, 1842). Monotypic. Formerly included Hume's Leaf Warbler (*qv*).
DISTRIBUTION [W Siberia to E and NE Asia > SE Asia. Stable. LC] Summer visitor Ural Mtns. Regular but uncommon or rare migrant N and W Europe, from Fennoscandia and Faroe Is SW to Britain, Ireland, Germany, Netherlands, Belgium, France, Spain, Italy and Malta; most occur from mid-Sep to late Oct, but occasionally also winter/spring, when also regular but rare Israel and Kuwait. Vagrant Algeria, Austria, Azores, Bahrain, Balearic Is, Bulgaria, Canary Is, Cape Verde Is, Corsica, Croatia, Cyprus, Czech Republic, Egypt, Gibraltar, Greece, Hungary, Iceland, Iran, Jan Mayen, Latvia, Libya, Lithuania, Luxembourg, Madeira, Morocco, Oman, Poland, Portugal, Romania, Saudi Arabia, Salvage Is, Slovakia, Slovenia, Svalbard, Switzerland, Turkey, UAE and Ukraine.

Phylloscopus humei
Hume's Leaf Warbler

OTHER NAMES Buff-browed Warbler, Hume's (Yellow-browed) Warbler.
TAXONOMY (Brooks, W E, 1878). Two subspecies (one in region). Previously considered conspecific with Yellow-browed Warbler, but differs in mtDNA, plumage, song and call. Extralimital subspecies *mandellii* (China) possibly a separate species, in which case Hume's becomes monotypic.
DISTRIBUTION [SC Asia > S and SE Asia. Stable. LC] Winters SE Iran; rare migrant and winter visitor Oman, UAE and Iraq, very rare Israel (*humei*). Vagrant (mainly late autumn and winter) Bahrain, Belgium, Britain, Croatia, Czech Republic, Denmark, Estonia, Finland, France, Germany, Greece, Hungary, Ireland, Italy, Kuwait, Latvia, Netherlands, Norway, Poland, Qatar, Romania, Spain, Sweden, Turkey and Ukraine.

Phylloscopus borealis
Arctic Warbler

TAXONOMY (Blasius, J H, 1858). Two subspecies (one in region). Extralimital *kennicotti* (Alaska) sometimes split, as are two further extralimital taxa previously considered conspecific and now treated at species level, Kamchatka Leaf Warbler *P examinandus* (Kamchatka) and Japanese Leaf Warbler *P xanthodryas* (Japan).
DISTRIBUTION [NE Europe to NE Asia and NW North America > SE Asia. Stable. LC] Summer visitor N Norway, N Sweden, E Finland and NW Russia E to Ural Mtns (*borealis*). Migrates SE to winter quarters, but reaches W Europe annually (Sep-Oct) as vagrant, especially Britain, but also recorded Belgium, Bulgaria, Channel Is, Denmark, Estonia, Faroe Is, France, Germany, Gibraltar, Greece, Iceland, Ireland, Italy, Latvia, Malta, Netherlands, Oman, Poland, Portugal, Saudi Arabia, Spain, Turkey and Ukraine. Vagrants to region generally assumed to be *borealis*, but possibility of now-split *examinandus* and *xanthodryas* should also be considered; similarly, extralimital *kennicotti* has apparently reached Bermuda.

Phylloscopus nitidus
Green Warbler

OTHER NAMES Bright Green (or Yellowish-breasted) Warbler, Green Leaf Warbler.
TAXONOMY Blyth, 1843. Monotypic. Has been considered conspecific with Greenish Warbler (and thus Two-barred Warbler), but mtDNA differs from that of *viridanus* Greenish, and Green is reproductively isolated from all populations of that species, as well as differing in plumage characters and song.
DISTRIBUTION [Asia Minor to C Asia > S Asia. Population trend not assessed. NE] Summer visitor to N Turkey, Caucasus and N Iran. Migrates SE to winter quarters, occurring E Iran and (rare) Kuwait and Oman on passage. Vagrant Bahrain, Britain, Faroe Is, Finland, Germany, Greece, Israel, Kuwait, Lebanon, Saudi Arabia, Sweden, UAE and Ukraine.

Phylloscopus trochiloides
Greenish Warbler

OTHER NAMES Dull Green (Leaf) Warbler, Greenish Leaf Warbler.
TAXONOMY (Sundevall, 1837). Four subspecies (one in region). Formerly also included Green and Two-barred Warblers (*qv*).
DISTRIBUTION [NE Europe to E Asia > SE Asia. Increasing. LC] Summer visitor, breeding from E Europe, Baltic States and S Finland E through NW Russia, Belarus and N Ukraine to Ural Mtns (*viridanus*). Migrates SE to winter quarters, occurring S of breeding range to N Caspian region. Regularly overshoots to Denmark and Sweden (has bred), also in autumn to Britain, where very scarce migrant Aug-Sep. Vagrant Austria, Balearic Is, Bahrain, Belgium, Faroe Is, France, Iran, Ireland, Italy, Lebanon, Malta, Netherlands, Norway, Oman, Romania, Saudi Arabia, Spain and Switzerland; also five records of undetermined *trochiloides/plumbeitarsus* from Channel Is.

Phylloscopus plumbeitarsus
Two-barred Warbler (V)

OTHER NAMES Grey-legged Warbler, Two-barred Greenish Warbler.
TAXONOMY Swinhoe, 1861. Monotypic. Often treated as conspecific with Greenish Warbler, but mtDNA indicates genetic split with *viridanus* subspecies of Greenish; additionally, *plumbeitarsus* and *viridanus* do not recognise each other's songs, differ in plumage and structure and behave as separate species where ranges meet in C Siberia.
DISTRIBUTION [C Siberia to E Asia > SE Asia. Population trend not assessed. NE] Very rare vagrant: Britain (4); Finland (2-3 Oct 2002); France (26 Oct 2004); Germany (29 Sep 2011); Netherlands (3); Norway (24 Sep 2013); Sweden (2).

Phylloscopus magnirostris
Large-billed Leaf Warbler (V)

OTHER NAMES Large-billed Tree Warbler, Large-billed Willow Warbler.
TAXONOMY Blyth, 1843. Monotypic.
DISTRIBUTION [Himalayas to C China > S and SE Asia. Stable. LC] Extreme vagrant: UAE (11-12 Oct 2014).

Phylloscopus coronatus
Eastern Crowned Warbler (V)

OTHER NAMES Eastern Crowned Leaf Warbler, Temminck's Crowned (Willow) Warbler.
TAXONOMY (Temminck and Schlegel, 1847). Monotypic.
DISTRIBUTION [NE and E Asia > SE Asia. Stable. LC] Extreme vagrant (seven records): Britain (2: 22-24 Oct 2009; 30 Oct 2011); Finland (23 Oct 2004); Germany (2: 4 Oct 1843; 16 Oct 2012); Netherlands (5 Oct 2007); Norway (30 Sep 2002).

FAMILY ACROCEPHALIDAE

Reed warblers and allies · 21 species

Acrocephalus griseldis
Basra Reed Warbler

TAXONOMY (Hartlaub, 1891). Monotypic. Relationship with congeners unclear; has been treated as conspecific with Great Reed Warbler, but differs in size, structure, plumage, song and mtDNA sequences.
DISTRIBUTION [SW Asia > E Africa. Decreasing. EN] Breeding endemic. Poorly known but largely confined to C and S Iraq in summer; may breed SW Iran and Kuwait, where regular on passage. Migrants also recorded spring and autumn in Saudi Arabia, where possibly breeds. Vagrant Bahrain*, Cyprus, Israel (has bred), Oman and Syria.

Acrocephalus brevipennis
Cape Verde Warbler (E)

OTHER NAMES Cape Verde Cane Warbler, Cape Verde (or Dohrn's) Swamp Warbler.
TAXONOMY (Keulemans, 1866). Monotypic.
DISTRIBUTION [Cape Verde Is. Decreasing. EN] Endemic resident Cape Verde Is on Fogo, Santiago and São Nicolau, formerly also Brava.

Acrocephalus arundinaceus
Great Reed Warbler

OTHER NAME European Great Reed Warbler.
TAXONOMY (Linnaeus, 1758). Two subspecies (both in region).
DISTRIBUTION [NW Africa, W Europe to CE Asia > sub-Saharan Africa. Decreasing. LC] Summer visitor NW Africa, from Iberia through parts of W, C, SE and E continental Europe N to S Fennoscandia and E to S Ural Mtns, W Caspian, Caucasus, and fragmented areas of N Middle East to NW Iran (*arundinaceus*), also NW Kazakhstan, adjacent SE European Russia and N Iran (*zarudnyi*). Occurs widely on migration N Africa and Middle East, where some may winter. Vagrant Britain, Canary Is, Channel Is, Faroe Is, Gibraltar, Ireland, Liechtenstein, Madeira and Norway.

Acrocephalus orientalis
Oriental Reed Warbler (V)

OTHER NAMES Oriental (or Eastern) Great Reed Warbler.
TAXONOMY (Temminck and Schlegel, 1847). Monotypic. Sometimes treated as conspecific with Great Reed Warbler, but slightly smaller with less pointed wing, and investigation of mtDNA indicates closer relationship to Clamorous Reed Warbler.
DISTRIBUTION [E Asia > SE Asia. Population trend not assessed. NE] Extreme vagrant (two records): Israel (2: 28 Feb-13 Apr 1988; 2 May 1990, trapped).

Acrocephalus stentoreus
Clamorous Reed Warbler

OTHER NAMES Southern Great Reed Warbler; Indian (Great) Reed Warbler (*A s brunnescens*).
TAXONOMY (Hemprich and Ehrenberg, 1833). Ten subspecies (three in region). *Brunnescens* may merit specific status, differing strikingly in morphology and song from *stentoreus*; mtDNA separation limited but other molecular aspects unknown. Further study needed.
DISTRIBUTION [Egypt, NE Africa and SW Asia to SE Asia and New Guinea. Stable. LC] Breeds Egypt (*stentoreus*), Israel, Syria and perhaps NW Saudi Arabia, Jordan and Lebanon (*levantinus*), and W, S and E coasts of Arabia and SW Iran (*brunnescens*). Mainly resident, but some passage/winter movement of *brunnescens* into *eg* S Iraq (uncommon), Bahrain, Oman and SE Iran; this taxon also vagrant Israel.

Acrocephalus melanopogon
Moustached Warbler

OTHER NAMES Eurasian Moustached Warbler, Moustached Sedge Warbler; Eastern Moustached Warbler (*A m mimicus*).
TAXONOMY (Temminck, 1823). Three subspecies (all in region).
DISTRIBUTION [NW Africa and SW Europe to C Asia. Stable. LC] Fragmented breeding distribution from NW Africa, C and E Spain and Balearic Is E through parts of S and C Europe (very rare Switzerland) to Balkans, S Ukraine and W Turkey (*melanopogon*), E Turkey, Levant, C and E Saudi Arabia, Iran (perhaps also Iraq), parts of Caucasus, S European Russia and NW Kazakhstan (*mimicus*), and SE Ukraine and extreme SW Russia (*albiventris*). Some populations sedentary but others partly or wholly migratory, vacating *eg* Austria, N Black Sea and N Caspian regions to winter parts of S Europe, Malta, Cyprus and N Middle East, very locally also N Africa. Uncommon migrant Armenia. Vagrant Bahrain, Germany (has bred), Libya, Lithuania, Luxembourg, Oman, Poland, Portugal and UAE.

Acrocephalus paludicola
Aquatic Warbler

TAXONOMY (Vieillot, 1817). Monotypic.
DISTRIBUTION [E Europe > W Africa. Decreasing. VU] Breeding endemic. Summer visitor mainly to Belarus, also NE Germany, Lithuania, Poland and Ukraine; breeds irregularly Russia and probably Latvia, until recently also Hungary and formerly Austria. Migrates W and SW to reach winter quarters, occurring W to Switzerland, Netherlands, Belgium, Britain, Channel Is, France and Iberia en route, less frequently Italy; spring route poorly known but noted Morocco and E Spain, also *eg* N Italy and Switzerland. Vagrant Albania, Alge-

ria, Balearic Is, Bulgaria, Corsica, Croatia, Czech Republic, Denmark, Estonia, Finland, Greece, Ireland, Jordan, Libya, Luxembourg, Macedonia, Malta, N Mauritania, Morocco, Norway, Sardinia, Serbia, Slovakia, Slovenia, Sweden, Tunisia and Turkey.

Acrocephalus schoenobaenus Sedge Warbler

OTHER NAME European Sedge Warbler.
TAXONOMY (Linnaeus, 1758). Monotypic.
DISTRIBUTION [W Europe to C Siberia. Decreasing. LC] Summer visitor across much of Europe (except SW and most of S), from W France, Britain and Ireland N to N Norway and far N European Russia, and E through C and E Europe, N Balkans and Ukraine to Ural Mtns and Caucasus, also W and C Turkey and NW Iran; has recently bred Iraq. Widespread on passage through S Europe, N Africa and Middle East; occasional in winter S Arabia. Vagrant Azores*, Faroe Is, Gibraltar, Iceland, Madeira, Salvage Is and Svalbard.

Acrocephalus agricola Paddyfield Warbler

OTHER NAME Jerdon's Reed Warbler.
TAXONOMY (Jerdon, 1845). Two subspecies (both in region). Western *septimus* has been advocated as potentially cryptic species in one study on basis of 4.5% genetic divergence from *agricola*, but more recent analysis found no distinction in mtDNA sequences.
DISTRIBUTION [SE Europe to CE Asia > S Asia. Decreasing. LC] Summer visitor W and N Black Sea from NE Bulgaria to S Ukraine and SW European Russia, also N Caspian region of SE European Russia, NW Kazakhstan, Georgia, E and SE Turkey (rare) and Armenia (*septimus*), and NE Iran (*agricola*). Migrates SE to winter quarters in SE Iran and beyond. Vagrant Austria, Azerbaijan, Azores, Bahrain, Belgium, Britain, Channel Is, Croatia, Cyprus, Denmark, Estonia, Faroe Is, Finland (has bred), France, Germany, Greece, Hungary, Iceland, Ireland, Israel, Italy, Jordan, Kuwait, Latvia (has bred), Lithuania, Luxembourg, Malta, Netherlands (has bred), Norway, Oman, Poland, Portugal, Qatar, Spain, Syria, Sardinia, Slovenia, Sweden, Switzerland and UAE.

Acrocephalus dumetorum Blyth's Reed Warbler

TAXONOMY Blyth, 1849. Monotypic.
DISTRIBUTION [NE Europe to C Siberia > S Asia. Increasing. LC] Summer visitor, breeding from S Finland and N Baltic States E through Belarus, N Ukraine and European Russia to Ural Mtns, possibly also NE Iran; range spreading W to S Sweden and SE Norway (has also bred Denmark) and S to Poland. Migrates SE to winter quarters, when regular E Iran and UAE (rare). Vagrant Azerbaijan, Balearic Is, Bahrain, Belgium, Britain, Bulgaria, Channel Is, Croatia, Cyprus, Czech Republic, Denmark, Faroe Is, France, Germany, Gibraltar, Iceland, Ireland, Israel, Italy, Jordan, Kuwait, Luxembourg, Malta, Netherlands, Oman, Sardinia, Saudi Arabia, Slovakia, Slovenia, Spain, Switzerland, Syria and Turkey.

Acrocephalus scirpaceus Eurasian Reed Warbler

OTHER NAMES (Common or European) Reed Warbler; Caspian Reed Warbler (*A s fuscus*), Mangrove (or Red Sea) Reed Warbler (*A s avicenniae*), Siwa Reed Warbler (*A s 'ammon'*).
TAXONOMY (Hermann, 1804). Three subspecies (all in region). Treatment here follows current IOC classification, but delineation of and boundaries between taxa in this species and extralimital African Reed Warbler *A baeticatus* (sub-Saharan Africa) in need of revision. Recent study found geographically structured divergence for eight distinct lineages in *scirpaceus–baeticatus* complex (Olsson *et al* 2016), grouped into northern (SW France to Black Sea), Iberian/NW African, Sahel and tropical/southern African clades, but all taxa considered as subspecies of *scirpaceus* pending more detailed research into *eg* interactions and level of gene flow in contact zones. *Cinnamomeus* African Reed Warbler (*sensu* IOC) recently claimed breeding Libya and Morocco, but this population accords with Iberian/NW African clade of Eurasian Reed Warbler (named '*ambiguus*' by Olsson *et al* 2016).

DISTRIBUTION [NW Africa and W Europe to C Asia > sub-Saharan Africa. Decreasing. LC] Summer visitor NW Africa, NC Libya, Mediterranean Europe N to C Fennoscandia and adjacent European Russia, W to Britain and Ireland, and E to parts of SE European Russia and W Turkey (*scirpaceus*), N Middle East, Caucasus, Caspian region, N and E Iran and C Arabia (*fuscus*), and W and SW Arabia (*avicenniae*). *Scirpaceus* widespread on passage N Africa (W to Canary Is) and W Middle East, rare in winter Morocco; vagrant Cape Verde Is, Faroe Is, Iceland, Madeira and Salvage Is. *Fuscus* common on migration Middle East; vagrants occasionally reported Europe and once from Libya, but just one record west of known range, from Germany, confirmed by genetic analysis.

Acrocephalus palustris
Marsh Warbler

OTHER NAMES European Marsh Warbler, Marsh Reed Warbler.
TAXONOMY (Bechstein, 1798). Monotypic.
DISTRIBUTION [W Europe to C Siberia > E and S Africa. Increasing. LC] Summer visitor to continental Europe, breeding W to N France (a few pairs also Britain, and has bred Channel Is), Benelux and Germany, N to C Sweden, Finland and NW Russia, S to SE France, N Italy, Balkans and Ukraine, and E to Ural Mtns, Caucasus, E Turkey and NW Iran. In autumn migrates to winter quarters SE through Europe and SW via Caucasus through Middle East; uncommon on passage E Egypt and Israel, rare Cyprus and Iraq. Vagrant Algeria, Balearic Is, Corsica, Faroe Is, Iceland, Ireland, Jordan, Libya, Madeira, Malta, Morocco and Spain.

Iduna aedon
Thick-billed Warbler (V)

OTHER NAME Thick-billed Reed Warbler.
TAXONOMY (Pallas, 1776). Two subspecies (undetermined in region).
DISTRIBUTION [C Siberia to E Asia > S and SE Asia. Decreasing. LC] Very rare vagrant: Britain (5); Egypt (20 Nov 1991); Finland (11 Oct 1994); Oman (2); Norway (2).

Iduna caligata
Booted Warbler

TAXONOMY (Lichtenstein, M H C, 1823). Monotypic. Formerly included Sykes's Warbler (*qv*).
DISTRIBUTION [NE Europe to CE Asia > S Asia. Increasing. LC] Summer visitor NE Europe, breeding from SE Finland and Russian Karelia E to Ural Mtns and S to SE European Russia (but absent N Caspian region); has bred Belarus, Estonia and Latvia. Migrates SE to winter quarters beyond region, occurring Georgia, Azerbaijan, E Iran and Oman (possibly also UAE?) on passage. Vagrant Austria, Balearic Is, Belgium, Britain, Croatia, Czech Republic, Denmark, France, Germany, Greece, Hungary, Ireland, Israel, Italy, Jordan, Kuwait, Lebanon, Lithuania, Luxembourg, Malta, Netherlands, Norway, Poland, Saudi Arabia, Slovenia, Spain, Syria, Sweden, Switzerland, Turkey and Ukraine.

Iduna rama
Sykes's Warbler

OTHER NAME Sykes's Booted Warbler.
TAXONOMY (Sykes, 1832). Monotypic. Previously treated as conspecific with Booted Warbler, but breeds sympatrically S Kazakhstan and differs in mtDNA, song, habitat and morphology. Relict population with atypical wing formula in coastal mangroves of SE Arabia requires further study.
DISTRIBUTION [SW and C Asia > S Asia. Stable. LC] Summer visitor N Caspian region of SE European Russia and NW Kazakhstan, and C, N and E Iran; resident N Oman and NE UAE. Migratory population moves SE to main winter quarters, when scarce on passage Kuwait and NE Saudi Arabia. Vagrant Belgium, Britain, Faroe Is*, Finland, France, Germany, Iceland, Ireland, Israel, Netherlands, Norway, Qatar and Sweden.

Iduna pallida
Eastern Olivaceous Warbler

OTHER NAMES Olivaceous Warbler; Chad Olivaceous Warbler (*H p laeneni*), Egyptian Olivaceous Warbler (*H p pallida*), Mangrove

Olivaceous Warbler (*H p alulensis*), Saharan Olivaceous Warbler (*H p reiseri*).
TAXONOMY (Hemprich and Ehrenberg, 1833). Five subspecies (at least four in region). Formerly included Western Olivaceous Warbler (*qv*).
DISTRIBUTION [N Africa, SE Europe, SW to C Asia > W, N, NE and E Africa. Stable. LC] Patchily distributed from Morocco to W Libya (*reiseri*), possibly N Chad (*laeneni*), from E Libya to Nile Valley (*pallida*), Red Sea coast of Egypt and perhaps W Arabia (*alulensis*), and from Levant, C and E Arabia, Iran and Iraq to Caucasus, Turkey and Balkans, N to Hungary and Ukraine (*elaeica*). Resident C Sahara but otherwise migratory, occurring on passage N Africa W to N Mauritania and E to E Egypt, and widely in Middle East; irregular Italy. Vagrant Austria, Belgium, Britain, Czech Republic, Denmark, Estonia, Finland, France, Germany, Iceland, Ireland, Malta, Norway, Slovakia, Slovenia and Sweden.

Iduna opaca
Western Olivaceous Warbler

OTHER NAMES Cabanis's (or Isabelline) Warbler, Olivaceous Warbler.
TAXONOMY (Cabanis, 1851). Monotypic. Formerly treated as conspecific with Eastern Olivaceous Warbler, but differs in mtDNA, song, behaviour and morphology.
DISTRIBUTION [NW Africa and Iberia > W Africa. Stable. LC] Breeding endemic. Summer visitor from S Morocco E through NW Africa, and in S, E and NE Spain. On migration recorded W to Canary Is, S to N Mauritania and E apparently to W Egypt, but exceptional elsewhere. Status uncertain Lybia. Vagrant Balearic Is, Cape Verde Is, Egypt, France, Greece, Italy, Malta, Portugal (has bred), Salvage Is and Sweden.

Hippolais languida
Upcher's Warbler

TAXONOMY (Hemprich and Ehrenberg, 1833). Monotypic.
DISTRIBUTION [SW and C Asia > E Africa. Stable. LC] Summer visitor, breeding from SC Turkey E to Armenia, Azerbaijan, N Iraq and Iran, and S through Syria, Lebanon, N Israel, W Jordan and Palestine. Widespread Iran and Arabia on passage, when scarce Egypt. Vagrant Cyprus and European Turkey.

Hippolais olivetorum
Olive-tree Warbler

TAXONOMY (Strickland, 1837). Monotypic.
DISTRIBUTION [SE Europe, Asia Minor and Levant > E and S Africa. Stable. LC] Breeding endemic, occurring in summer W, S, C and E Balkans, NW, W, S and SE Turkey, and NW Syria, Lebanon and N Israel (rare). Migrates to winter quarters initially on SE heading, then through Levant, Israel, Jordan and NE Egypt. Vagrant Algeria, Britain, Cyprus, Iraq, Italy, Kuwait, Oman, Saudi Arabia, Ukraine and Yemen.

Hippolais polyglotta
Melodious Warbler

TAXONOMY (Vieillot, 1817). Monotypic.
DISTRIBUTION [NW Africa and SW Europe > W and WC Africa. Increasing. LC] Breeding endemic. Summer visitor NW Africa and from Iberia N to Belgium and S Netherlands (rare but recent colonist), SW Germany and SW Switzerland, and E to Corsica, Italy, Slovenia, Croatia and Montenegro (rare). Migrates S to winter quarters through NW Africa (regular W to Canary Is); scarce on passage Britain. Vagrant Albania, Azores, Czech Republic, Denmark, Faroe Is, Greece, Iceland, Ireland, Libya, Madeira, Malta, Norway, Poland, Salvage Is, Sweden and Turkey.

Hippolais icterina
Icterine Warbler

TAXONOMY (Vieillot, 1817). Monotypic.
DISTRIBUTION [N and C Europe to C Asia > EC and S Africa. Decreasing. LC] Summer visitor continental Europe, breeding from NE France and Belgium N to S Fennoscandia, E to Ural Mtns and S to C Europe, C Balkans, Crimea and SW Russia, with outlying popu-

lation S Caspian region. Migrates S to winter quarters; scarce on passage *eg* Britain, Ireland, E Spain, Balearic Is and Algeria. Vagrant Azerbaijan, Bahrain, Faroe Is, Gibraltar, Iceland, Iraq, Morocco, Portugal, Oman, Qatar, Salvage Is, Syria, UAE and Yemen.

FAMILY LOCUSTELLIDAE

Grass and bush warblers · six species

Locustella naevia
Common Grasshopper Warbler

OTHER NAMES (European or Pale) Grasshopper Warbler; Eastern Grasshopper Warbler (*L n straminea/mongolica*).
TAXONOMY (Boddaert, 1783). Four subspecies (at least three in region).
DISTRIBUTION [W Europe to CE Asia > W and NE Africa, India. Decreasing. LC] Summer visitor W, C and E Europe, breeding from N Spain, W France, Britain and Ireland N to S Scandinavia and E to Ukraine and W European Russia (*naevia*), E European Russia (*straminea*) and in Caucasus and NE Turkey (*obscurior*). Moves S/SW from European range to main wintering grounds beyond region in W Africa, but relatively infrequently recorded *eg* N Africa (W to Canary Is) and W Middle East. Vagrant Cape Verde Is, Cyprus, Faroe Is, Iceland, Iraq, Israel, Jordan, Lebanon, Madeira, Malta, Qatar, Salvage Is, Syria and Yemen; *straminea/mongolica* vagrant Britain.

Locustella fluviatilis
River Warbler

OTHER NAME Eurasian River Warbler.
TAXONOMY (Wolf, 1810). Monotypic.
DISTRIBUTION [C Europe to W Siberia > E and SE Africa. Stable. LC] Summer visitor C and E Europe, breeding W to Germany and Liechtenstein, N to S Fennoscandia, E to Ural Mtns and S to N Balkans and Ukraine. Migration concentrated through SE Europe and E Mediterranean to African winter quarters. Vagrant Algeria, Armenia, Bahrain, Belgium, Britain, Corsica, Cyprus, France, Iceland, Iraq, Iran, Italy, Lebanon, Malta, Morocco, Netherlands, Norway, Oman, Salvage Is, Saudi Arabia, Spain, Svalbard, Switzerland, UAE and Yemen.

Locustella luscinioides
Savi's Warbler

TAXONOMY (Savi, 1824). Three subspecies (all in region).
DISTRIBUTION [NW Africa and W Europe to C Asia > Sahel and NE Africa. Decreasing. LC] Summer visitor to NW Africa and continental Europe from parts of Iberia discontinuously N to S Baltic, S Finland (rare) and E to Balkans, Crimea and Ural Mtns (*luscinioides*) and E Ukraine to N Caspian region (*sarmatica*), also Turkey, Syria, Lebanon and Jordan (either *sarmatica* or *fusca*). Occurs more widely on passage parts of N Africa, N Middle East and Gulf. Vagrant Armenia, Azerbaijan, Britain (has bred), Cape Verde Is, Channel Is, Gibraltar, Iraq, Ireland, Liechtenstein, Luxembourg, Norway and Oman.

Locustella lanceolata
Lanceolated Warbler

OTHER NAMES Streaked (Grasshopper) Warbler.
TAXONOMY (Temminck, 1840). Two subspecies (one in region).
DISTRIBUTION [NE Europe, W Siberia to E Asia > SE Asia. Stable. LC]. Summer visitor E European Russia, mainly to W Urals but possibly more widespread; also rare but regular late spring S Finland and adjacent Russian Karelia, sometimes holding territory. Migrates SE to Asian winter quarters. Vagrant Bear Is (at sea), Belgium, Britain, Denmark, France, Germany, Georgia*, Iceland, Italy, Latvia, Montenegro, Netherlands, Norway and Sweden.

Locustella certhiola
Pallas's Grasshopper Warbler (V)

OTHER NAMES Pallas's Warbler, Rusty-rumped (or Grey-naped) Warbler.
TAXONOMY (Pallas, 1811). Four subspecies (at least two in region).
DISTRIBUTION [W Siberia to E Asia > S and SE Asia. Decreasing. LC] Vagrant (mainly

late Sep-early Oct, *certhiola*): Bear Is (24 Jul 2013*); Belgium (3); Britain (52, *rubescens*); Faroe Is (2); Finland (6 Sep 2013, trapped); France (6); Germany (13 Aug 1856); Ireland (2); Israel (25 Feb 1983); Latvia (15 Sep 1971); Morocco (12 Oct 2009); Netherlands (10, including one putative *centralasiae*); Norway (22); Poland (12 Sep 1989); Svalbard; Sweden (12-13 Oct 2010).

Locustella fasciolata
Gray's Grasshopper Warbler (V)

OTHER NAMES Gray's Warbler, Large (Grasshopper) Warbler.
TAXONOMY (Gray, G R, 1861). Monotypic.
DISTRIBUTION [SC Siberia to E Asia > Philippines, Indonesia and W New Guinea. Stable. LC] Extreme vagrant (two records): Denmark (25 Sep 1955, found dead); France (26 Sep 1913, found dead).

FAMILY CISTICOLIDAE

Cisticolas and allies · three species

Cisticola juncidis
Zitting Cisticola

OTHER NAMES Fan-tailed Cisticola, Fan-tailed Warbler.
TAXONOMY (Rafinesque, 1810). Seventeen subspecies (four in region).
DISTRIBUTION [Africa, S Europe, SW to E Asia and N Australasia. Increasing. LC] Resident NW Africa, Libya, Iberia, Balearic Is and W and NW France (*cisticola*), from S France E through S and SE Europe to Asia Minor and Syria, also Egypt and most large Mediterranean islands (*juncidis*), Cyprus, Lebanon and Israel to W Iran (*neuroticus*), and SW Saudi Arabia and W Yemen (*uropygialis*). Periodic range-expansion attempts in NW Europe mostly halted by cold winters; remains very scarce/rare but occasionally breeding Channel Is, Belgium and Netherlands. Vagrant Austria, Belgium, Britain, Bulgaria, Denmark, Germany, Ireland, Kuwait, Liechtenstein, Oman, Sweden and Switzerland (has bred).

Prinia gracilis
Graceful Prinia

OTHER NAME Graceful Warbler.
TAXONOMY (Lichtenstein, M H C, 1823). Twelve subspecies (10 in region). Three endemic Arabian subspecies (*yemenensis*, *hufufae* and *carpenteri*) may be colour variations rather than taxonomically distinct; further population in C Arabia does not conform to any described subspecies.
DISTRIBUTION [NE Africa, Middle East and S Asia. Stable. LC] Resident Egypt in Natron Valley (*natronensis*), Nile Valley (*gracilis*) and Nile Delta to W Israel (*deltae*), W Saudi Arabia, Yemen and S Oman (*yemenensis*), N Oman and UAE (*carpenteri*), NE Saudi Arabia and Bahrain (*hufufae*), NW Saudi Arabia to S Syria (*palaestinae*), N Syria and S Turkey (*akyildizi*), E Syria, Iraq and SW Iran (*irakensis*) and SE Iran (*lepida*). Vagrant Cyprus.

Spiloptila clamans
Cricket Warbler

OTHER NAMES Cricket (or Scaly) Longtail, Cricket Prinia, Scaly-fronted Warbler.
TAXONOMY (Cretzschmar, 1826). Monotypic.
DISTRIBUTION [Sahel. Stable. LC] Local resident N Mauritania and SW Morocco.

FAMILY LEIOTHRICHIDAE

Laughingthrushes and allies · five species

Turdoides altirostris
Iraq Babbler (E)

TAXONOMY (Hartert, 1909). Monotypic.
DISTRIBUTION [N Middle East. Decreasing. LC] Endemic resident SE Turkey, N Syria, Iraq and SW Iran.

Turdoides huttoni
Afghan Babbler

TAXONOMY (Blyth, 1847). Two subspecies

(both in region). Sometimes considered conspecific with extralimital Common Babbler *T caudata* (S Asia), but morphology and noticeably different vocalisations suggest better considered a separate species.
DISTRIBUTION [SW Asia. Population trend not assessed. NE] Near-endemic resident Kuwait, Iraq and W Iran (*salvadorii*) and E through S Iran (*huttoni*).

Turdoides squamiceps
Arabian Babbler (E)

OTHER NAME Brown Babbler.
TAXONOMY (Cretzschmar, 1827). Three subspecies (all in region).
DISTRIBUTION [Middle East. Increasing. LC] Endemic resident from NE Egypt, Israel, Palestine and Jordan to C Saudia Arabia, E Yemen and W Oman (*squamiceps*), SW Saudi Arabia and W Yemen (*yemensis*) and E UAE and N Oman (*muscatensis*).

Turdoides fulva
Fulvous Babbler

OTHER NAME Fulvous Chatterer.
TAXONOMY (Desfontaines, 1789). Four subspecies (all in region).
DISTRIBUTION [N Africa. Stable. LC] Resident S Morocco and NW Algeria (*maroccana*), perhaps same taxon also parts of S Algeria and SW Libya, then N Algeria, Tunisia and NW Libya (*fulva*), the Hoggar, S Algeria (*buchanani*), N Niger, N Chad and (at least formerly) extreme S and SE Egypt (*acaciae*).

Leiothrix lutea
Red-billed Leiothrix (I)

OTHER NAME Pekin Robin.
TAXONOMY (Scopoli, 1786). Five subspecies (undetermined in region).
DISTRIBUTION [S and E Asia. Stable. Decreasing. LC] Introduced NE Spain, France and Italy (Cat C). Also known from *eg* Germany and Portugal, where populations not (yet) considered self-sustaining.

FAMILY SYLVIIDAE

Sylviid babblers · 26 species

Sylvia atricapilla
Eurasian Blackcap

OTHER NAMES Black-capped Warbler, Blackcap.
TAXONOMY (Linnaeus, 1758). Five subspecies (all in region).
DISTRIBUTION [NW Africa and Europe to NC Asia > N, W and E Africa and W Europe. Increasing. LC] Breeds Azores and Cape Verde Is (*gularis*), Canary Is, Madeira, Morocco and W and S Iberia (*heineken*), E Spain to C and S Italy, and NE Morocco to N Tunisia (*pauluccii*), Ireland and Britain E through much of continental Europe (except far N) to Ural Mtns and S to NW Turkey (*atricapilla*), and NE Turkey, Caucasus and N Iran (*dammholzi*); has bred Faroe Is and Malta. Migratory in N, C and E of range and widespread on passage, including throughout Middle East, wintering mainly S Europe, N Africa and S Turkey; small numbers also move NW/N to winter Britain, Ireland and S Scandinavia. Vagrant Iceland, Jan Mayen and Svalbard.

Sylvia borin
Garden Warbler

TAXONOMY (Boddaert, 1783). Two subspecies (both in region).
DISTRIBUTION [Europe to NC Asia > sub-Saharan Africa. Decreasing. LC] Summer visitor, breeding from Ireland, Britain, France and N Iberia E to Scandinavia, Poland, Hungary and NW Balkans (*borin*), and E through rest of E Europe to Ural Mtns, Caucasus and N Turkey (*woodwardi*). Widespread on passage S Europe, N Africa (W to Canary Is) and Middle East (except SW Arabia) E to W Iran; scarce Faroe Is (has bred). Vagrant Azores, Cape Verde Is, Iceland, Madeira, Jan Mayen, Svalbard and Yemen.

Sylvia nisoria
Barred Warbler

OTHER NAME European Barred Warbler.

TAXONOMY (Bechstein, 1792). Two subspecies (both in region).
DISTRIBUTION [C and N Europe to C Asia > E Africa. Stable. LC] Breeds from N Italy, Switzerland (scarce), NE Germany and S Fennoscandia E through E and SE Europe to Ural Mtns, Caucasus, parts of Turkey and NW Iran (*nisoria*), and in NE Iran (*merzbacheri*). Migrates SE in autumn, when widespread Middle East and Iran (except E), and rare to uncommon Egypt; also scarce but regular in autumn W to Britain (where very rare spring) and Netherlands, scarcer still NW France and Faroe Is. Vagrant Balearic Is, Corsica, Iceland, Ireland, Jan Mayen, Luxembourg and Malta; unrecorded Iberian Peninsula and NW Africa.

Sylvia curruca
Lesser Whitethroat

OTHER NAMES Grey Whitethroat; Blyth's (or Siberian) Lesser Whitethroat (*S c 'blythi'*), Caucasian Lesser Whitethroat (*S c caucasica*), Central Asian (or Kazakh) Lesser Whitethroat (*S c halimodendri*).
TAXONOMY (Linnaeus, 1758). Four subspecies (three in region). Formerly included Desert Whitethroat and Hume's Whitethroat (*qv*), but currently considered a superspecies comprised of three allospecies. However, IOC treatment followed here does not currently recognise *'blythi'* (N Siberia to E edge of region), validity of which confirmed by new molecular research; this taxon also placed in separate clade from *curruca*. On basis of this and other recent findings, as well as unresolved issues, limits of (and subspecies within) this and next two species best regarded as provisional.
DISTRIBUTION [W Europe to CE Asia > Sahel, NE Africa, SW and S Asia. Increasing. LC] Summer visitor to much of Europe (except SW, Ireland and far N) E to Ural Mtns (*curruca*) and C Turkey, Levant, Caucasus and N Iran (*caucasica*). Migrates SE to winter quarters, occurring widely Middle East, Egypt and E Libya; some winter S Arabia. Scarce or rare migrant Faroe Is (has bred), Ireland and Malta. Vagrant Balearic Is, Canary Is, Corsica, Gibraltar, Iceland, Madeira, N Mauritania, Morocco, Portugal, Salvage Is, Sardinia, Spain, Svalbard and Tunisia. *Halimodendri* (breeding extralimitally C and CE Asia) recorded on passage Iran and Oman; several autumn/winter vagrants to W Europe (*eg* Belgium, Britain, France, Ireland, Netherlands and Sweden) involve this taxon. *'Blythi'* also recorded Qatar on passage and confirmed by mtDNA analysis W Europe (*eg* Britain and Netherlands); either this taxon or *halimodendri* documented Malta and Poland.

Sylvia minula
Desert Whitethroat

OTHER NAMES (Desert) Lesser Whitethroat, Small Whitethroat; Margelanic Lesser Whitethroat (*S c margelanica*).
TAXONOMY Hume, 1873. Three subspecies (at least one in region). Formerly included within Lesser Whitethroat, but acoustically and genetically appears sufficiently divergent to warrant species status; subspecies *margelanica* treated as an allospecies by Shirihai *et al* 2001.
DISTRIBUTION [SC Asia to N China > S Arabia to S Asia. Stable. LC] Occurs on passage and in winter E and S Iran and SE and S Arabia, and on migration N to Kuwait (scarce). Vagrant Israel (*minula*) and Qatar; either this species or *halimodendri* Lesser Whitethroat recorded Denmark and Germany.

Sylvia althaea
Hume's Whitethroat

OTHER NAMES (Hume's) Lesser Whitethroat, Mountain (Lesser) Whitethroat.
TAXONOMY Hume, 1878. Two subspecies (one in region). Formerly considered conspecific with Lesser Whitethroat, but differs significantly in genetics, morphology and vocalisations.
DISTRIBUTION [C Asia > E Arabia to S Asia. Stable. LC] Summer visitor NE and SE Iran (*althaea*). Rare in winter E and S Iran and EC Arabia; regular on passage Bahrain and Kuwait (rare). Status requires confirmation Armenia, where 12 *althaea* reportedly mistnetted May 1995 but no other records, and UAE, where formerly considered rare/very rare migrant and winter visitor but all records currently under review. Vagrant Azerbaijan*, Israel, Oman, Qatar and Turkey.

Sylvia hortensis
Western Orphean Warbler

OTHER NAME Orphean Warbler.
TAXONOMY (Gmelin, J F, 1789). Two subspecies (both in region). Formerly treated as conspecific with Eastern Orphean Warbler (*qv*).
DISTRIBUTION [SW Europe and NW Africa > Sahel. Population trend not assessed. NE] Breeding endemic. Summer visitor Iberia, S and E France and Italy (formerly also Switzerland, but now irregular), and from Morocco E to Tunisia and W Libya (*hortensis*) and NE Libya (*cyrenaicae*). Occurs on passage W Mediterranean islands and S of breeding range in NW Africa; a few winter in S Saharan oases. Vagrant Austria, Britain, Canary Is, Germany, Luxembourg, Malta and Netherlands; some vagrant orphean warblers *sensu lato* have not been identified to species level.

Sylvia crassirostris
Eastern Orphean Warbler

OTHER NAME Orphean Warbler.
TAXONOMY Cretzschmar, 1830. Three subspecies (all in region). Previously considered conspecific with Western Orphean Warbler, but differs markedly in mtDNA, plumage, moult pattern, bill morphology and song.
DISTRIBUTION [SE Europe to C Asia > NE Africa to S Asia. Population trend not assessed. NE] Breeds from extreme NE Italy and Slovenia S through Balkans, Turkey and Levant E to Caucasus and NW Iran (*crassirostris*) and in N Iran and NE Iraq (*balchanica*); has bred Oman. Occurs throughout Middle East on passage; winters extreme S Egypt (rare), S Iran and E Arabia, where *jerdoni* (C and S Asia) has also occurred. Vagrant Italy, Malta, Norway, Slovenia, Ukraine and Yemen, but see also under Western Orphean Warbler.

Sylvia leucomelaena
Arabian Warbler

OTHER NAMES Red Sea Warbler; Blanford's Warbler (*S l blanfordi*).
TAXONOMY (Hemprich and Ehrenberg, 1833). Four subspecies (three in region).
DISTRIBUTION [Middle East and NE Africa. Decreasing. LC] Breeds extreme SE Egypt (*blanfordi*), SE Israel and SW Jordan (*negevensis*), and W Saudi Arabia, Yemen and W Oman (*leucomelaena*). Largely resident, with limited short-range dispersal.

Sylvia nana
Asian Desert Warbler

OTHER NAMES Desert Warbler, Desert Whitethroat.
TAXONOMY (Hemprich and Ehrenberg, 1833). Monotypic. Formerly included African Desert Warbler (*qv*).
DISTRIBUTION [SW to CE Asia > NE Africa and SW Asia. Population trend not assessed. NE] Breeds N and E Iran. Migrates SW to winter S Iran and Middle East from Iraq and Syria S through Arabia and W to Nile Valley. Vagrant Azerbaijan, Britain, Bulgaria, Cyprus, Czech Republic, Denmark, Estonia, Finland, Germany, Latvia, Libya, Netherlands, Romania, Spain, Sweden, Turkey and Ukraine; undetermined *nana/deserti* also recorded Czech Republic and France.

Sylvia deserti
African Desert Warbler

OTHER NAMES Desert (or Sahara) Warbler.
TAXONOMY (Loche, 1858). Monotypic. Traditionally treated as conspecific with Asian Desert Warbler, but differs in vocalisations, plumage and mtDNA.
DISTRIBUTION [NW Africa. Population trend not assessed. NE] Breeds from N Mauritania and SE Morocco E through Saharan NW Africa to W Libya. Mainly sedentary but some dispersal; winters more widely NW Africa. Vagrant Canary Is, Cape Verde Is, Italy, Madeira, Malta and Spain; see Asian Desert Warbler for undetermined *nana/deserti* records.

Sylvia communis
Common Whitethroat

OTHER NAMES (Greater) Whitethroat; Eastern Whitethroat (*S c icterops*).
TAXONOMY Latham, 1787. Four subspecies (all in region).

DISTRIBUTION [NW Africa, W Europe to C Asia > sub-Saharan Africa. Decreasing. LC] Summer visitor NW Africa, Ireland, Britain and across much of continental Europe (except far N) to N Turkey (*communis*), C Turkey, Levant, S Caucasus, Iraq and Iran (*icterops*), and SE Russia and NW Kazakhstan (*volgensis*). Widespread on passage N Africa and Middle East, when *rubicola* (breeding extralimitally C and CE Asia) also presumed to occur in extreme SE of region. Vagrant Azores*, Iceland and Madeira.

Sylvia undata
Dartford Warbler (E)

TAXONOMY (Boddaert, 1783). Three subspecies.
DISTRIBUTION [SW Europe and NW Africa. Decreasing. NT] Endemic. Breeds S Britain, Channel Is, W France and NW Spain (*dartfordiensis*), elsewhere in Iberia, S France, Italy and some W Mediterranean islands (*undata*), and N Morocco, N Algeria and Tunisia (*toni*). Mainly resident but more widespread in winter, especially NW Africa, also *eg* Balearic Is, France and Italy; occurs on passage Gibraltar, scarce Malta. Vagrant Albania, Belgium, Croatia, Czech Republic, Finland, Germany, Greece, Ireland, Libya, Netherlands, Sweden and Switzerland.

Sylvia sarda
Marmora's Warbler (E)

TAXONOMY Temminck, 1820. Monotypic. Formerly included Balearic Warbler (*qv*).
DISTRIBUTION [W Mediterranean Is > N Africa. Population trend not assessed. NE] Endemic. Mainly resident Corsica, Sardinia and islands off W Italy (formerly also Pantelleria); reports of breeding on islands off Tunisia unconfirmed. Some dispersal/migration to Sicily and NE Algeria, Tunisia and NW Libya; irregular migrant mainland Italy. Vagrant Balearic Is, Belgium, Britain, Denmark, mainland France, Gibraltar, Malta and Spain; records from Greece and Morocco of vagrant Marmora's warblers *sensu lato* probably also relate to this species rather than to sedentary Balearic Warbler.

Sylvia balearica
Balearic Warbler (E)

OTHER NAME (Balearic) Marmora's Warbler.
TAXONOMY von Jordans, 1913. Monotypic. Previously treated as conspecific with Marmora's Warbler, but differs markedly in vocalisations, structure, size, plumage (especially juvenile), migratory habits and mtDNA.
DISTRIBUTION [W Mediterranean islands. Population trend not assessed. NE] Endemic. Resident Balearic Is except Menorca, where apparently displaced by Dartford Warbler in 1980s. Unrecorded with certainty elsewhere, but see comments under Marmora's Warbler.

Sylvia deserticola
Tristram's Warbler

OTHER NAMES Moroccan Warbler; Meinertzhagen's Warbler (*S d 'ticehursti'*).
TAXONOMY Tristram, 1859. Two subspecies, *contra* IOC (both in region); claimed third (*'ticehursti'*) known only from type specimen (S Morocco), and latterly considered an aberrant plumage or possibly either hybrid with African Desert Warbler or example of now-extinct subspecies.
DISTRIBUTION [NW Africa > W and N Africa. Stable. LC] Breeding endemic, ranging from W Morocco to NW Algeria (*maroccana*) and from N Algeria to N Tunisia (*deserticola*). Altitudinal and partial migrant, moving W, S and E into Sahara, reaching N Mauritania, W Libya and Egypt in winter. Vagrant Canary Is, Gibraltar, Italy and Malta.

Sylvia conspicillata
Spectacled Warbler

TAXONOMY Temminck, 1820. Two subspecies (both in region).
DISTRIBUTION [N Africa, S Europe and N Middle East. Decreasing. LC] Near-endemic. Resident Cape Verde Is, Canary Is and Madeira (*orbitalis*), from N Mauritania E through NW Africa to Libya, Malta and Sicily, also Cyprus, SE Anatolia and Levant, and in summer Iberia, S France, Italy and W Mediterranean islands (*conspicillata*). European range largely vacated in winter, when more widespread N

Africa; regular migrant Gibraltar (has bred). Some Middle Eastern birds winter S to NE Egypt and lower Nile Valley. Vagrant Albania, Belgium, Britain, Germany, Greece (has bred), Iraq, Macedonia, Netherlands, Saudi Arabia and Switzerland (has bred).

Sylvia inornata
Western Subalpine Warbler

OTHER NAME Subalpine Warbler (IOC).
TAXONOMY Tschusi, 1906. Two subspecies (both in region). Formerly included Eastern Subalpine and Moltoni's Warblers (*qv*). Split under consideration by IOC.
DISTRIBUTION [NW Africa and SW Europe > W and C Sahel. Population trend not assessed. NE] Breeding endemic. Summer visitor Morocco, Algeria, Tunisia and coastal NW Libya (*inornata*), and Portugal, Spain, S France and extreme NW Italy (*iberiae*). Some winter S Algeria, and possibly also resident in NW. Widely recorded on passage S of breeding range and W to Canary Is; regularly overshoots to NW Europe (especially Britain) in spring, less so autumn. Rare migrant Malta. Vagrant Belgium, Cape Verde Is, Denmark, Faroe Is, Finland, Germany, Ireland, Madeira, Netherlands, Norway and Switzerland (has bred); many other regional records of subalpine warblers *sensu lato*, including from Austria, Channel Is, Finland, Iceland, Lithuania, Poland and Sweden, may involve this and/or recently split *cantillans/subalpina*, and reviews under way in some countries.

Sylvia cantillans
Eastern Subalpine Warbler

OTHER NAME Subalpine Warbler (IOC).
TAXONOMY (Pallas, 1764). Two subspecies (both in region). Recently separated from Western Subalpine Warbler, from which it differs genetically, morphologically and vocally; also breeds sympatrically with Moltoni's Warbler (*qv*) in N Italy. Split under consideration by IOC.
DISTRIBUTION [SE Europe and W Turkey > C and E Sahel. Population trend not assessed. NE] Breeding endemic. Summer visitor C and S Italy and Sicily (*cantillans*), and from extreme NE Italy E through W and S Balkans to Crete and W Turkey (*albistriata*). On passage recorded from NE Algeria and Malta E to Israel, Jordan and NW Saudi Arabia; some winter extreme SE Libya and SW Egypt. Vagrant Austria, Balearic Is, Britain, Denmark, France, Germany, Hungary, Netherlands, Norway, Spain, Sweden, Switzerland and Yemen; presence of subalpine warbler *sensu lato* on Ukrainian list very likely relates to this taxon, but see also comments under Western Subalpine Warbler.

Sylvia subalpina
Moltoni's Warbler

OTHER NAME (Moltoni's) Subalpine Warbler.
TAXONOMY Temminck, 1920. Monotypic. Separated at species level from both Western and Eastern Subalpine Warblers because of significant differences in mtDNA, plumage, vocalisations and moult strategy; known to breed sympatrically with latter in N Italy.
DISTRIBUTION [S Europe > C Sahel? Population trend not assessed. NE] Breeding endemic, restricted in summer to Mallorca, Cabrera, Corsica, Sardinia and N mainland Italy; recently also reported breeding Menorca and Ibiza. Under-recorded on migration but has occurred Tunisia (two specimens); irregular in spring S France. Vagrant Belgium, Britain, Denmark, N France, Germany, Malta, Morocco and Netherlands (under review), but extralimital subalpine warblers *sensu lato* (thus including Moltoni's) not always identified to species level.

Sylvia melanocephala
Sardinian Warbler

TAXONOMY (Gmelin, J F, 1789). Five subspecies (all in region; sixth subspecies extinct).
DISTRIBUTION [N Africa, S Europe and N Middle East. Increasing. LC] Breeding endemic Canary Is (*leucogastra*), S Morocco (*valverdei*), from N Morocco E to N Libya and from Iberia E through much of Mediterranean Europe to E Balkans and N and W Turkey (*melanocephala*), Crete, Karpathos and Rhodes (*pasiphae*), and from SC Turkey through Levant to NE Egypt (*momus*); formerly also NC Egypt (*norrisae*†). Mainly resident but summer visitor E Balkans and NW Turkey; outside breeding season more widespread Mediterranean Europe, Levant, Egypt

and NW Africa, winter range also extending to parts of Sahara and NW Saudi Arabia (uncommon). Vagrant Andorra, Austria, Belgium, Britain, Cape Verde Is, Channel Is, Czech Republic, Denmark, Finland, Germany, Hungary, Ireland, Netherlands, Norway, Lithuania, Madeira, Netherlands, Oman, Poland, Romania, Sweden, Switzerland and Ukraine.

Sylvia mystacea
Ménétriés's Warbler

OTHER NAMES Menetries's Warbler (IOC), Ménétries's Warbler.
TAXONOMY Ménétriés, 1832. Three subspecies (all in region).
DISTRIBUTION [SW and C Asia > S Asia, Arabia and NE Africa. Stable. LC] Breeds NE Turkey to E Caucasus, Caspian region of SE European Russia and NW Iran (*mystacea*), SE Turkey and N Syria to Iraq and SW Iran (*rubescens*), and NE Iran (*turcmenica*). More widespread but generally local Middle East on passage; scarce Lebanon (has bred) and E Egypt. Winters S Iran and Arabia (except interior N and S). Vagrant Cyprus, Israel, Italy, Malta and Ukraine.

Sylvia ruppeli
Rüppell's Warbler

OTHER NAME Rueppell's Warbler.
TAXONOMY Temminck, 1823. Monotypic.
DISTRIBUTION [SE Europe and Asia Minor > C Sahel to NE Africa. Stable. LC] Breeding endemic, confined in summer to S Greece, Crete, W and S Turkey and extreme NW Syria (rare). On passage occurs from Libya, Egypt and Cyprus to Israel, Jordan, Lebanon and NW Saudi Arabia; some winter N Chad. Vagrant Albania, Algeria, Balearic Is, Britain, Bulgaria, Denmark, Faroe Is, Finland, France, Iran, Italy, Malta, Romania and Ukraine.

Sylvia melanothorax
Cyprus Warbler

TAXONOMY Tristram, 1872. Monotypic.
DISTRIBUTION [Cyprus > Middle East and NE Africa. Decreasing. LC] Breeding endemic restricted to Cyprus. Some remain year round but majority migrate S, occurring on passage and/or in winter Lebanon, S Israel, Jordan, Sinai and E Egypt. Vagrant Italy, Saudi Arabia and Turkey.

Sylvia buryi
Yemen Warbler (E)

OTHER NAMES Arabian Tit-Babbler, Arabian (Tit-)Warbler, Yemen Parisoma, Yemen Tit-warbler.
TAXONOMY (Ogilvie-Grant, 1913). Monotypic. Formerly placed in genus *Parisoma*, but recent studies of morphology, behaviour, voice and mtDNA confirm correct position in *Sylvia*.
DISTRIBUTION [SW Arabia. Decreasing. VU] Endemic resident SW Saudi Arabia and W Yemen.

Sinosuthora webbiana
Vinous-throated Parrotbill (I)

OTHER NAME Webb's Parrotbill.
TAXONOMY (Gould, 1852). Six subspecies (undetermined in region). Another taxon in same genus is currently treated by IOC as separate species, Ashy-throated Parrotbill *S alphonsiana*, but recent DNA-based research examining introduced populations of these taxa in region (Crottini *et al* 2010) concluded that *alphonsiana* should be treated as a subspecies of *webbiana*.
DISTRIBUTION [E Asia. Stable. LC] Naturalised resident NW Italy (Cat C); self-sustaining population also now established S Netherlands.

FAMILY ZOSTEROPIDAE

White-eyes · two species

Zosterops palpebrosus
Oriental White-eye

OTHER NAMES (Small) White-eye.
TAXONOMY (Temminck, 1824). Eleven subspecies (one in region).
DISTRIBUTION [S and SE Asia. Decreasing. LC] Very local resident SE Iran and S Oman (*palpebrosus*).

Zosterops abyssinicus
Abyssinian White-eye

OTHER NAME White-breasted White-eye.
TAXONOMY Guérin-Méneville, 1843. Six subspecies (one in region).
DISTRIBUTION [E Africa to S Arabia. Population trend unknown. LC] Breeds SW Saudi Arabia, W and E Yemen and SW Oman (*arabs*). Note: population of white-eyes in mangroves on S Red Sea coast of Saudi Arabia and Yemen, isolated from mountain range of present species, remains unidentified.

FAMILY REGULIDAE

Kinglets · four species

Regulus ignicapilla
Common Firecrest (E)

OTHER NAMES (Eurasian) Firecrest.
TAXONOMY (Temminck, 1820). Four subspecies (all in region). Formerly included Madeira Firecrest (*qv*).
DISTRIBUTION [N Africa, Europe and extreme SW Asia. Stable. LC] Breeds NW Africa and Balearic Is (*balearicus*), W, C, S and E Europe from Iberia and Mediterranean N to *eg* S Britain, Denmark, Poland and Lithuania, and E to Asia Minor (*ignicapilla*), Crimea (*tauricus*), and SW European Russia and Georgia (*caucasicus*). NE populations migratory, when scarce/rare Ireland, Scotland and Sweden. Winter range more extensive: NW Africa (E to Libya), S Europe (including Malta) and Turkey. Vagrant Belarus, Cyprus, Egypt, Estonia, Finland, Israel, Lebanon, Norway and Syria.

Regulus regulus
Goldcrest

OTHER NAMES Azores Kinglet (*R r azoricus/sanctaemariae/inermis*), Canary Islands (or Tenerife) Kinglet (*R r teneriffae/ellenthalerae*).
TAXONOMY (Linnaeus, 1758). Fourteen subspecies (eight in region). *Teneriffae* (and by extension newly described *ellenthalerae*) sometimes separated at species level as Canary Islands Kinglet, but more study of all populations needed before species limits can be redrawn.
DISTRIBUTION [Widespread Europe and Asia. Decreasing. LC] Breeds Azores on W and C islands (*inermis*), São Miguel (*azoricus*) and Santa Maria (*sanctaemariae*), Canary Is on La Gomera and Tenerife (*teneriffae*) and La Palma and El Hierro (*ellenthalerae*), and much of Europe except far N and parts of S (*regulus*) to Crimea, Caucasus and Asia Minor (*buturlini*), and N Iran (*hyrcanus*). Recent colonist Iceland; may also now breed Faroe Is. N Fennoscandian and NW Russian populations move S in winter, when more widespread W and S Europe (including Malta), also in small numbers N Algeria, N Tunisia and Levant S to Israel (irregular). Vagrant Iraq, Jan Mayen, Jordan, Libya, Morocco and Syria.

Regulus madeirensis
Madeira Firecrest (E)

OTHER NAME Madeira Kinglet.
TAXONOMY Harcourt, 1851. Monotypic. Previously treated as a subspecies of Common Firecrest, but differs genetically and in plumage and vocalisations.
DISTRIBUTION [Madeira. Decreasing. LC] Resident endemic Madeira.

Regulus calendula
Ruby-crowned Kinglet (V)

TAXONOMY (Linnaeus, 1766). Three subspecies (undetermined in region).
DISTRIBUTION [N North America > S North America. Increasing. LC] Extreme vagrant (four records, all Oct-Nov): Azores (21 Oct 2013*); Iceland (2: 23 Nov 1987; 10-11 Oct 1998); Ireland (27 Oct 2013, trapped).

FAMILY TROGLODYTIDAE

Wrens · one species

Troglodytes troglodytes
Eurasian Wren

OTHER NAMES (Northern or Winter) Wren.

TAXONOMY (Linnaeus, 1758). Twenty-nine subspecies (15 in region). Formerly included extralimital Winter *T hiemalis* and Pacific *T pacificus* Wrens (North America), but these are now considered specifically distinct on basis of differences in vocalisations and mtDNA.
DISTRIBUTION [Europe and N Africa to E Asia. Population trend not assessed. NE] Breeds NW Africa, S Spain and Balearic Is (*kabylorum*), NE Libya (*juniperi*), Corsica and Sardinia (*koenigi*), much of mainland Europe (*troglodytes*), Ireland and Britain (*indigenus*) except most of Outer Hebrides (*hebridensis*), St Kilda (*hirtensis*), Fair Isle (*fridariensis*) and Shetland (*zetlandicus*), also Faroe Is (*borealis*), Iceland (*islandicus*), Crimea, Caucasus, N Turkey and N Iran (*hyrcanus*), NE Iran (*subpallidus*), W Iran (*zagrossiensis*), and S and W Turkey to Cyprus and N Israel (*cypriotes*). Resident in most of range but mainly migratory from Fennoscandia and Baltic States E to Ural Mtns, wintering S to Turkey, Levant and C Iraq (has bred). Vagrant Azores*, Egypt, Kuwait, Madeira and Malta.

FAMILY SITTIDAE

Nuthatches · seven species

Sitta europaea
Eurasian Nuthatch

OTHER NAMES (Common or Wood) Nuthatch; Asian Nuthatch (*S e asiatica*), European Nuthatch (*S e europaea*), Southern Nuthatch (*S e caesia*).
TAXONOMY Linnaeus, 1758. Fifteen subspecies (nine in region). Eastern group of subspecies, including *asiatica*, differs markedly in mitochondrial and nuclear DNA from western taxa; formerly also included extralimital *arctica* (breeding NE Siberia), now separated as Siberian Nuthatch.
DISTRIBUTION [NW Africa and W Europe to E Asia. Stable. LC] Breeds N Morocco and Iberia (*hispaniensis*), Switzerland, Italy and Croatia (*cisalpina*), from Britain E through W, C and SE Europe to NW Turkey (*caesia*), S Scandinavia and NE Europe to E Ukraine and SE European Russia (*europaea*), extreme E European Russia (*asiatica*), NE Turkey to Caucasus and SW Russia (*caucasica*), S Turkey, N Lebanon and N Syria (*levantina*), SE Turkey, N Iraq and W Iran (*persica*) and N Iran and SE Azerbaijan (*rubiginosa*). Mainly sedentary, but populations in N and E prone to periodic irruptive movements away from breeding areas, sometimes involving *asiatica* moving W to Finland and Norway, more rarely Baltic States and Sweden, from breeding range E of region. Vagrant Channel Is, also Heligoland and Isle of Man.

Sitta whiteheadi
Corsican Nuthatch (E)

TAXONOMY Sharpe, 1884. Monotypic. Has been treated as conspecific with extralimital Red-breasted Nuthatch *S canadensis* (North America) and Chinese Nuthatch *S villosa* (E Asia), and recent mtDNA studies indicate closer relationship to those species than to Algerian and Krüper's Nuthatches.
DISTRIBUTION [Corsica. Decreasing. VU] Resident endemic Corsica.

Sitta ledanti
Algerian Nuthatch (E)

OTHER NAMES Kabylie (or Kaybylian) Nuthatch.
TAXONOMY Vielliard, 1976. Monotypic.
DISTRIBUTION [Algeria. Decreasing. EN] Resident endemic NE Algeria.

Sitta krueperi
Krüper's Nuthatch (E)

OTHER NAME Krueper's Nuthatch.
TAXONOMY Pelzeln, 1863. Monotypic.
DISTRIBUTION [SE Europe and SW Asia. Decreasing. LC] Resident endemic Turkey to W Caucasus region of Georgia and extreme S European Russia, with outlying population Greece on Lesvos. Some dispersal and altitudinal migration undertaken. Vagrant mainland Greece.

Sitta canadensis
Red-breasted Nuthatch (V)

TAXONOMY Linnaeus, 1766. Monotypic.

DISTRIBUTION [W to E North America > W, S and E North America. Increasing. LC] Extreme vagrant (two records): Britain (13 Oct 1989-6 May 1990); Iceland (21 May 1970, dead).

Sitta neumayer
Western Rock Nuthatch (E)

OTHER NAMES Lesser Rock (or Neumayer's) Nuthatch; Syrian Rock Nuthatch (*S n syriaca*).
TAXONOMY Michahelles, 1830. Five subspecies.
DISTRIBUTION [SE Europe and SW Asia. Stable. LC] Endemic. Resident Balkans (*neumayer*), N Turkey to Caucasus, N Iraq and N Iran (*rupicola*), SE Turkey to N Israel (*syriaca*), W Iran (*tschitscherini*) and SC Iran (*plumbea*). Mainly resident, but some dispersal and altitudinal movements.

Sitta tephronota
Eastern Rock Nuthatch

OTHER NAMES Great (or Large) Rock Nuthatch, Persian Nuthatch; Armenian Rock Nuthatch (*S t obscura*).
TAXONOMY Sharpe, 1872. Four subspecies (all in region).
DISTRIBUTION [SW and C Asia. Stable. LC] Resident SE Turkey to N Iraq and W Iran (*dresseri*), NE Turkey to Caucasus and N Iran (*obscura*) and in NE Iran (*tephronota* and *iranica*).

FAMILY TICHODROMADIDAE

Wallcreeper · one species

Tichodroma muraria
Wallcreeper

TAXONOMY (Linnaeus, 1766). Two subspecies (both in region).
DISTRIBUTION [S Europe to E Asia. Stable. LC] Breeds Cantabrian Mtns, Pyrenees, Alps, Corsica, Apennines, Balkans, Carpathians, Turkey, Caucasus and N Iran (*muraria*), and E Iran (*nepalensis*). Partly resident but some winter at lower altitudes and to S of breeding range, including S Spain, Cyprus, Israel and N Iraq. Vagrant Algeria, Balearic Is, Belgium, Britain, Channel Is, Czech Republic, Gibraltar, Jordan, Luxembourg, Malta, Morocco, Netherlands, Portugal, Syria and Ukraine.

FAMILY CERTHIIDAE

Treecreepers · two species

Certhia familiaris
Eurasian Treecreeper

OTHER NAMES (Common or Northern) Treecreeper.
TAXONOMY Linnaeus, 1758. Ten subspecies (six in region). Recent phylogenetic study found that species is comprised of two distinct lineages which diverged *c* 1 million years ago, one of which is now restricted to Corsica and Caucasus; more research needed.
DISTRIBUTION [W Europe to E Asia. Stable. LC] Resident Ireland and Britain (*britannica*), from N Spain and parts of France through W and C Europe (*macrodactyla*) to Fennoscandia (except far N), E and SE Europe (*familiaris*), Corsica (*corsa*), N Turkey, Crimea and Caucasus (*caucasica*), SE Azerbaijan and N Iran (*persica*). Migratory in far NE Europe, wintering to S, and some dispersal and movement elsewhere (*eg* continental *familiaris* recorded N Scotland). Vagrant Channel Is and Faroe Is.

Certhia brachydactyla
Short-toed Treecreeper (E)

OTHER NAMES Dorothy's (or Cyprus Short-toed) Treecreeper (*C b dorotheae*).
TAXONOMY Brehm, C L, 1820. Six subspecies.
DISTRIBUTION [NW Africa, Europe and SW Asia. Decreasing. LC] Endemic. Resident NW Africa (*mauritanica*), NW Iberia N to W Germany (*megarhynchos*), rest of Iberia and E France N to Denmark and E to Poland, W Ukraine and Balkans (*brachydactyla*), S Greece, Crete and Cyprus (*dorotheae*), W Turkey (*stresemanni*) and NW Caucasus (*rossocaucasica*). Largely sedentary but some evidence of local movements, *eg* occasional migrant and winter visitor Gibraltar. Vagrant Balearic Is, Belarus, Britain, Corsica (records

perhaps including *mauritanica*), Latvia, Lithuania, Sardinia and Sweden.

FAMILY MIMIDAE

Mockingbirds and thrashers · three species

Dumetella carolinensis
Grey Catbird (V)

OTHER NAMES (Common or Northern) Catbird.
TAXONOMY (Linnaeus, 1766). Monotypic.
DISTRIBUTION [S Canada to E and SE USA > E and SE USA to West Indies and Central America. Stable. LC] Extreme vagrant (10 records): Azores (3: 7 Oct 2010*; 12 Oct 2010*; 18 Oct 2011*); Belgium (15-16 Dec 2006); Britain (4-6 Oct 2001); Canary Is (4 Nov 1999); Channel Is (mid-Oct 1975, trapped, later died in captivity); Germany (2: 28 Oct 1840, collected; 2 May 1908); Ireland (4 Nov 1986). Another individual Oct 1998 arrived Mediterranean on ship from New York, USA, but was fed during journey and not accepted onto any national lists.

Mimus polyglottos
Northern Mockingbird (V)

OTHER NAMES (Common) Mockingbird.
TAXONOMY (Linnaeus, 1758). Two subspecies (undetermined in region).
DISTRIBUTION [S Canada to Mexico and West Indies. Stable. LC] Extreme vagrant (three records): Britain (2: 30 Aug 1982; 17-23 May 1988); Netherlands (16-23 Oct 1988). Other occurrences, considered unlikely to have involved genuine vagrants, placed in Cat D (one each Britain and Canary Is) and Cat E (another Britain), but treatment inconsistent and ship assistance probably a factor in most records, majority of which come from locations close to ports and/or shipping routes.

Toxostoma rufum
Brown Thrasher (V)

TAXONOMY (Linnaeus, 1758). Two subspecies (one in region).
DISTRIBUTION [SC and SE Canada to SE USA > SE USA. Decreasing. LC] Extreme vagrant: Britain (18 Nov 1966-5 Feb 1967, trapped, *rufum*). Another collected Germany in autumn 1836 not accepted because specimen cannot be traced.

FAMILY STURNIDAE

Starlings and allies · 15 species

Acridotheres cristatellus
Crested Myna (I)

OTHER NAMES Chinese Jungle (or Crested) Myna, Chinese Starling, Tufted Myna.
TAXONOMY (Linnaeus, 1758). Three subspecies (undetermined in region).
DISTRIBUTION [SE and E Asia. Stable. LC] Introduced Portugal (Cat C). Additional population Austria not (yet) considered self-sustaining.

Acridotheres ginginianus
Bank Myna (I)

TAXONOMY (Latham, 1790). Monotypic.
DISTRIBUTION [S Asia. Increasing. LC] Introduced Kuwait, Oman and UAE (Cat C). Reported also Bahrain, Qatar and Saudi Arabia, but viable populations not known to exist in those countries.

Acridotheres tristis
Common Myna

OTHER NAMES Indian (or House) Myna.
TAXONOMY (Linnaeus, 1766). Two subspecies (one in region).
DISTRIBUTION [S and SE Asia. Increasing. LC] Resident SE Iran (*tristis*). Introduced elsewhere, with naturalised populations (Cat C) established mainly in parts of Middle East, including Bahrain, Egypt, W Iran, Iraq, Israel, Kuwait, Oman, Qatar, Saudi Arabia, SE European Russia, Turkey and UAE; occurrences elsewhere, including Canary Is (Tenerife), France, Germany, Jordan and Lebanon, relate to escapes and/or

wandering individuals from feral populations.

Acridotheres burmannicus
Vinous-breasted Starling (I)

OTHER NAMES Jerdon's Starling, Vinous-breasted Myna.
TAXONOMY (Jerdon, 1862). Two subspecies (undetermined in region).
DISTRIBUTION [SE Asia. Increasing. LC] Introduced Israel (Cat C).

Gracupica contra
Pied Myna (I)

OTHER NAMES Asian Pied Myna, Asian Pied Starling.
TAXONOMY (Linnaeus, 1758). Five subspecies (undetermined in region).
DISTRIBUTION [S and SE Asia. Increasing. LC] Introduced UAE (Cat C).

Agropsar sturninus
Daurian Starling (V)

OTHER NAME Purple-backed Starling.
TAXONOMY (Pallas, 1776). Monotypic.
DISTRIBUTION [E Asia > SE Asia. Population trend unknown. LC] Extreme vagrant: Netherlands (11-12 Oct 2005). Further records currently treated as of uncertain origin and placed in Cat D (Britain: 7-28 May 1985 and 24-27 Sep 1998, both under review; Norway: 29 Sep 1985, collected) or as relating to escapes (Britain: 26 Aug-5 Sep 1997; Netherlands: 15 May 1999 and 5 Nov 1999).

Sturnia malabarica
Chestnut-tailed Starling (V)

OTHER NAMES Grey-headed (or Ashy-headed) Myna, Grey-headed (or Ashy-headed) Starling, White-winged (or Blyth's) Myna.
TAXONOMY (Gmelin, J F, 1789). Two subspecies (undetermined in region).
DISTRIBUTION [S and SE Asia. Population trend unknown. LC.] Extreme vagrant: Oman (11-14 Nov 2010).

Sturnia pagodarum
Brahminy Starling (I)

OTHER NAMES Black-headed Starling, Brahminy Myna.
TAXONOMY (Gmelin, J F, 1789). Monotypic.
DISTRIBUTION [S Asia. Population trend unknown. LC.] Naturalised population established UAE (Cat C); rare passage migrant and winter visitor Oman, presumably involving birds originating from UAE.

Pastor roseus
Rosy Starling

OTHER NAMES Rose-coloured Starling, Rosy Pastor.
TAXONOMY (Linnaeus, 1758). Monotypic.
DISTRIBUTION [SE Europe, W and C Asia > S Asia. Population trend unknown. LC] Summer visitor to Black Sea coast of Ukraine, Caucasus and parts of Turkey, E through Caspian region of SE European Russia, NW Kazakhstan and N Iran; semi-nomadic and occasionally irruptive, when may breed in numbers W of usual range in Turkey, Balkans and exceptionally Hungary. On passage occurs S to Egypt (irregular/rare), Israel, Jordan and from N and E Arabia to Oman; very scarce but regular wanderer (mainly summer and autumn) W through Europe, notably Austria, Britain, Faroe Is, Finland, France, Italy (has bred), Netherlands, Norway and Slovakia, where now classed as scarce migrant. Some winter Arabia. Vagrant Algeria, Balearic Is, Belarus, Belgium, Canary Is, Channel Is, Czech Republic, Denmark, Estonia, Germany, Iceland, Ireland, Latvia, Libya, Lithuania, Madeira, Malta, Morocco, Poland, Portugal, Sardinia, Slovenia, Spain, Sweden, Switzerland, Tunisia and Yemen.

Sturnus vulgaris
Common Starling

OTHER NAMES (European or Northern) Starling, Purple-winged Starling; Shetland Starling (*S v zetlandicus*).
TAXONOMY Linnaeus, 1758. Thirteen subspecies (10 in region). Analysis of *granti* (Azores) has shown it to be genetically isolated from other regional populations.

DISTRIBUTION [Europe to C Asia > N Africa, Middle East and S Asia. Population trend unknown. LC] Breeds Azores (*granti*), Faroe Is (*faroensis*), Shetland Is (*zetlandicus*), Canary Is, Ireland, Britain, Iceland and much of continental Europe from NE Spain N to Fennoscandia and E to Ural Mtns (*vulgaris*), S and E Ukraine, SW European Russia and most of Turkey (*tauricus*), SE European Russia and NW Kazakhstan (*poltaratskyi*), E Turkey, Georgia and Armenia (*purpurascens*), SE Turkey and N Iraq (*oppenheimi*), Caucasus to S Iran (*caucasicus*) and NE Iran (*nobilior*). N and E continental European populations mainly migratory, wintering SW and S through Europe and Mediterranean region to N Africa and Middle East (except SW Arabia). Scarce visitor Jan Mayen and Svalbard; has bred Bear Is. Vagrant Cape Verde Is, N Mauritania and Yemen.

Sturnus unicolor
Spotless Starling (E)

TAXONOMY Temminck, 1820. Monotypic. Genetic differentiation from Common Starling slight; hybridisation frequent where ranges overlap in N/NE Spain.
DISTRIBUTION [NW Africa and SW Europe. Increasing. LC] Endemic resident Morocco, N Algeria and Tunisia, and Portugal, Gibraltar, Spain, S France, Corsica, Sardinia and Sicily. Vagrant Balearic Is (has bred), Denmark, Greece, Libya, Madeira, Malta and Switzerland.

Creatophora cinerea
Wattled Starling

TAXONOMY (Meuschen, 1787). Monotypic.
DISTRIBUTION [S and E Africa. Stable. LC] Irregular on passage and in winter Yemen and Oman. Vagrant UAE.

Lamprotornis pulcher
Chestnut-bellied Starling (V)

OTHER NAME Red-bellied Starling.
TAXONOMY (Statius Müller, 1776). Monotypic.
DISTRIBUTION [Sahel. Stable. LC] Recorded N Mauritania, where documented on regional boundary at 20°N and presumed rare/vagrant, but precise data lacking; may be more frequent than records suggest, and known to occur more regularly just S of area. Similar situation may exist N Niger, where range extends to just outside regional boundary.

Cinnyricinclus leucogaster
Violet-backed Starling

OTHER NAMES Amethyst (or Plum-coloured) Starling.
TAXONOMY (Boddaert, 1783). Three subspecies (one in region).
DISTRIBUTION [Sub-Saharan Africa and SW Arabia. Decreasing. LC] Summer visitor SW Saudi Arabia and W Yemen (*arabicus*). Vagrant Israel, Oman and UAE.

Onychognathus tristramii
Tristram's Starling (E)

OTHER NAMES Tristram's Grackle, Tristram's Red-winged Starling.
TAXONOMY (Sclater, P L, 1858). Monotypic.
DISTRIBUTION [Middle East. Stable. LC] Endemic to Sinai, Israel, Jordan, W Saudi Arabia, Yemen and W Oman.

FAMILY BUPHAGIDAE

Oxpeckers · one species

Buphagus erythrorynchus
Red-billed Oxpecker (V)

TAXONOMY (Stanley, 1814). Monotypic.
DISTRIBUTION [E and S Africa. Decreasing. LC] Extreme vagrant: Yemen (5 Apr 1998).

FAMILY TURDIDAE

Thrushes · 22 species

Geokichla sibirica
Siberian Thrush (V)

OTHER NAME Siberian Ground Thrush.

TAXONOMY (Pallas, 1776). Two subspecies (one in region).
DISTRIBUTION [E Asia > SE Asia. Decreasing. LC] Rare vagrant (mainly Sep-Oct, *sibirica*): Belgium (3); Britain (8); France (6); Germany (*c* 10); Ireland (2); Italy (3); Malta (Oct 1912, trapped); Netherlands (2); Norway (8); Poland (*c* 7); Sweden (11 Sep 1990); Switzerland (8 Dec 1978). Four additional records from Poland, three of them involving flocks of 17-25 birds, now rejected after review.

Zoothera aurea
White's Thrush

OTHER NAMES (Eurasian) Scaly Thrush.
TAXONOMY (Holandre, 1825). Two subspecies (one in region). Formerly treated as conspecific with three extralimital Asian taxa which are now separated as polytypic *Z dauma*, to which name Scaly Thrush latterly applies.
DISTRIBUTION [E Europe to E Asia > SE Asia. Population trend unknown. NE] Breeds E European Russia (*aurea*). Vagrant Austria, Balearic Is, Belgium, Britain, Denmark, Faroe Is, Finland, France, Germany, Greece, Iceland, Ireland, Italy, Latvia, Netherlands, Norway, Oman, Poland, Slovenia and Sweden.

Ixoreus naevius
Varied Thrush (V)

TAXONOMY (Gmelin, J F, 1789). Four subspecies (undetermined in region). Geographical variation slight and mostly clinal.
DISTRIBUTION [NW and W North America > W North America. Decreasing. LC] Extreme vagrant (two records): Britain (14-23 Nov 1982); Iceland (3-8 May 2004).

Catharus fuscescens
Veery (V)

TAXONOMY (Stephens, 1817). Four subspecies (undetermined in region). Geographical variation rather slight and clinal.
DISTRIBUTION [S Canada and N USA > C South America. Decreasing. LC] Very rare vagrant (mainly Sep-Oct): Britain (10); Sweden (26 Sep 1978).

Catharus minimus
Grey-cheeked Thrush (V)

TAXONOMY (Lafresnaye, 1848). Two subspecies (one in region). Formerly included extralimital Bicknell's Thrush *C bicknelli* (breeding NE North America), claimed but unproven in region.
DISTRIBUTION [NE Siberia and N North America > N South America. Population trend unknown. LC] Vagrant (mainly Oct-Nov): Azores (8*); Britain (56); France (6*); Germany (18 Oct 1937); Iceland (3); Ireland (7); Italy (Nov 1901, *minimus*); Norway (2).

Catharus ustulatus
Swainson's Thrush (V)

OTHER NAMES Olive-backed Thrush (eastern '*swainsoni* group'), Russet-backed Thrush (western '*ustulatus* group').
TAXONOMY (Nuttall, 1840). Six subspecies (one in region). Two groups of subspecies differ not only in morphology but also in breeding habitat, certain vocalisations and wintering areas; more study needed.
DISTRIBUTION [N and W North America > Central America, N and W South America. Decreasing. LC] Very rare vagrant (usually Oct; *swainsoni* confirmed *eg* Belgium and Ireland): Azores (3); Belgium (4); Britain (31); Denmark (21 and 24-30 Oct 2012); Finland (2); France (8); Germany (2); Iceland (5); Ireland (5); Italy (3); Norway (4); Sweden (12-13 Nov 1995); Ukraine (Oct or Nov 1893).

Catharus guttatus
Hermit Thrush (V)

TAXONOMY (Pallas, 1811). Nine subspecies (undetermined in region). Subspecies divided into three groups (Pacific coast, W interior mountain and E groups), but geographical variation to some extent clinal.
DISTRIBUTION [N and W North America > S and W North America. Increasing. LC] Very rare vagrant (mainly Oct): Azores (2*); Britain (9); Germany (4 in 19th century); Iceland (11); Ireland (2); Italy (*c* 1854); Sweden (27 Apr 1988).

Hylocichla mustelina
Wood Thrush (V)

TAXONOMY (Gmelin, J F, 1789). Monotypic.
DISTRIBUTION [E North America > Middle America. Decreasing. LC] Extreme vagrant (four records): Azores (2: prior to 1903, collected; 9-17 Oct 2012*); Britain (7 Oct 1987); Iceland (23 Oct 1967, dead).

Turdus menachensis
Yemen Thrush (E)

TAXONOMY Ogilvie-Grant, 1913. Monotypic.
DISTRIBUTION [SW Arabia. Decreasing. VU] Endemic resident (or partially altitudinal migrant) SW Saudi Arabia and W Yemen.

Turdus unicolor
Tickell's Thrush (V)

OTHER NAME Indian Grey Thrush.
TAXONOMY Tickell, 1833. Monotypic.
DISTRIBUTION [W and C Himalayas > S Asia. Population trend unknown. LC] Extreme vagrant: Germany (15 Oct 1932, collected). Further published claim from Malta not submitted for assessment.

Turdus torquatus
Ring Ouzel

OTHER NAMES Alpine Ring Ouzel (*T t alpestris*), Caucasian Ring Ouzel (*T t amicorum*).
TAXONOMY Linnaeus, 1758. Three subspecies (all in region).
DISTRIBUTION [W Europe to WC Asia > S Europe, N Africa and N Middle East. Decreasing. LC] Near-endemic. Summer visitor Ireland, Britain, Norway, N Sweden, N Finland (rare) and Kola Peninsula (*torquatus*), NW Spain, Pyrenees, from French Alps E through C Europe and Carpathians to Greece and W Turkey (*alpestris*), and from C Turkey to Caucasus and N Iran (*amicorum*). Migratory or partly so, occurring across much of Europe S and W of breeding range but often locally scarce. Some winter S Europe, notably in Spain and Balkans, but majority in Morocco and Algeria (where also several breeding-season records), and in Tunisia, Libya (rare) and parts of N Middle East; irregular/rare in winter Canary Is, Egypt and Oman. Very scarce Faroe Is (has bred). Vagrant Azores, Bahrain, Cyprus, Iceland, Iraq, Israel, Jan Mayen (has bred), Jordan, Kuwait, Madeira, Moldova, Qatar, Svalbard, Syria and UAE.

Turdus merula
Common Blackbird

OTHER NAMES (European or Eurasian) Blackbird.
TAXONOMY Linnaeus, 1758. Nine subspecies (six in region).
DISTRIBUTION [NW Africa and Europe to S and E Asia. Stable. LC] Breeds widely Europe (*merula*) except far NE Fennoscandia and NW/extreme SE Russia, also Azores (*azorensis*), Madeira and Canary Is (*cabrerae*), NW Africa (*mauritanicus*), Balkans, much of Turkey, Caucasus and N Iran (*aterrimus*) and S Greek islands, S Turkey to S Iran, N Iraq and Egypt (*syriacus*). N Fennoscandian and E populations migratory, wintering S through W and S Europe, N Africa and N Middle East. Occasional Jan Mayen and Svalbard. Vagrant Bahrain, Qatar and UAE.

Turdus obscurus
Eyebrowed Thrush (V)

OTHER NAME Dark Thrush.
TAXONOMY Gmelin, J F, 1789. Monotypic.
DISTRIBUTION [NC to NE Siberia > SE Asia. Population trend unknown. LC] Vagrant (mainly Oct but rarely also spring): Belgium (9); Britain (20); Czech Republic (2); Finland (6); France (15); Germany (16); Israel (2); Italy (28); Malta (5); Morocco (17 Dec 2008); Netherlands (7); Norway (4); Oman (3); Poland (6); Portugal (28 Oct 1991); Sweden (3); UAE (3-4 Oct 1988).

Turdus atrogularis
Black-throated Thrush

OTHER NAME Dark-throated Thrush.
TAXONOMY Jarocki, 1819. Monotypic. Genetic divergence from Red-throated Thrush prob-

ably neglible, but recently separated at species level on basis of distinct plumage, vocalisations and habitat preferences; birds showing characters of both forms in range overlap zone indicate some gene flow between taxa.
DISTRIBUTION [E Europe to W Siberia > N India to China. Population trend unknown. NE] Breeds E European Russia. Occurs on passage and/or in winter Iran, Georgia and S Caucasus (especially Azerbaijan); scarcer or rare E Iraq, Kuwait and parts of E Arabia S to W Oman. Vagrant Armenia, Austria, Belgium, Britain, Bulgaria, Czech Republic, Denmark, Egypt, Estonia, Finland, France, Germany, Greece, Hungary, Iceland, Israel, Italy, Latvia, Lebanon, Netherlands, Norway, Poland, Qatar, Sardinia, Sweden, Turkey, Ukraine and Yemen; apparent/possible hybrids with Red-throated Thrush recorded *eg* France and UAE.

Turdus ruficollis
Red-throated Thrush (V)

OTHER NAMES Dark-throated (or Rufous-throated) Thrush.
TAXONOMY Pallas, 1776. Monotypic. Formerly included Black-throated Thrush (*qv*).
DISTRIBUTION [SC Siberia to NE China > NE India to W China. Population trend unknown. NE] Very rare vagrant (autumn through to spring): Britain (29 Sep-7 Oct 1994); France (3); Germany (*c* 4); Iran (2); Italy (21 Sep 1999); Poland (4 Mar 1979); Qatar (10 Apr 1991). Formerly accepted records from Poland (5 or 6) and Sweden (30 Apr 2001) now deleted, and published record from Madeira (1903) also rejected.

Turdus naumanni
Naumann's Thrush (V)

OTHER NAME Rufous-tailed Thrush.
TAXONOMY Temminck, 1820. Monotypic. Formerly considered conspecific with Dusky Thrush (*qv*).
DISTRIBUTION [C Siberia to NE Asia > E Asia. Population trend unknown. NE] Very rare vagrant: Austria (2); Azores (16 Jan 2008*); Belgium (26 Oct 1951); Britain (2); Czech Republic (4); Estonia (15-21 Nov 2012); Finland (4); France (4); Germany (4); Hungary (winter 1820; record not certainly within present-day national boundaries); Italy (5); Norway (5); Poland (5); Spain (23-24 Nov 2005); Ukraine (reported in winter pre-1840).

Turdus eunomus
Dusky Thrush (V)

TAXONOMY Temminck, 1831. Monotypic. Previously treated with Naumann's Thrush as a single species, but separated on basis of morphology, behaviour and song; genetic differentiation unquantified. Hybridisation occurs in limited contact zone, but extent of mixed pairings unknown.
DISTRIBUTION [NC Siberia to NE Siberia > Japan to S China and NE India. Population trend unknown. NE] Rare vagrant (mainly autumn and winter): Austria (3); Belgium (5); Britain (11); Croatia (Dec 1906, collected); Denmark (2); Finland (7); France (4); Germany (*c* 10); Italy (records under review); Kuwait (9 Jan-13 Feb 1987); Netherlands (2); Norway (7); Oman (4); Poland (21 Oct 1966); Sweden (3); UAE (3). For some other 'pre-split' records, *eg* Belarus (17 Jun 1951, shot but specimen lost), Bulgaria (May 2003), Cyprus (25 Apr 1993 and 28 Dec 1994), Faroe Is (8 Dec 1947), France (4) and Israel (12-13 Nov 1982), attribution to Dusky/Naumann's not determined or unknown; birds showing intermediate characters recorded Cyprus (10-11 Nov 1958) and France (3).

Turdus pilaris
Fieldfare

TAXONOMY Linnaeus, 1758. Monotypic.
DISTRIBUTION [N Eurasia > W Europe to SW Asia. Stable. LC] Breeds Scotland, from Scandinavia S to E France, and E across C and E Europe to Ural Mtns. Mainly migratory Fennoscandia, Belarus and European Russia, wintering across Europe W to Iceland (has bred), Faroe Is (has bred) and Ireland, and S to Mediterranean and Middle East as far S as N Libya (rare), N Egypt and N Arabia. Occasional N to Svalbard; rare Gibraltar and Tunisia. Vagrant Algeria, Azores, Bahrain, Canary Is, Madeira, Morocco and UAE.

Turdus iliacus
Redwing

TAXONOMY Linnaeus, 1758. Two subspecies (both in region).
DISTRIBUTION [N Eurasia > S Europe, N Africa and SW Asia. Decreasing. NT] Breeds Iceland and Faroe Is (*coburni*) and from Scotland, Fennoscandia, Poland, Czech Republic (rare) and Baltic States E to Ural Mtns (*iliacus*); has bred Svalbard. Small numbers remain all year in a few breeding areas, but bulk of population winters S through Europe to N Africa, Mediterranean, N Egypt, Turkey, Levant, N Iraq, N Iran, Kuwait (rare) and NE Saudi Arabia. Vagrant Azores, Bahrain, Jordan, Libya, Madeira, Salvage Is and UAE.

Turdus philomelos
Song Thrush

OTHER NAMES European (or Common) Song Thrush.
TAXONOMY Brehm, C L, 1831. Four subspecies (three in region).
DISTRIBUTION [Europe to N Siberia > S Europe, N Africa to SW Asia. Decreasing. LC] Breeds NW Scotland and W Ireland (*hebridensis*), through rest of Britain and Ireland and across much of W Europe (*clarkei*) and rest of Europe (except S Iberia, parts of Italy and Balkans and SE European Russia) to N Turkey, Caucasus and N Iran (*philomelos*). Scandinavian and E populations migratory, occurring on passage Faroe Is (scarce) and wintering S through Europe to parts of N Africa and Middle East (scarce S Arabia). Rare visitor Jan Mayen. Vagrant Azores, Cape Verde Is, Iceland, Madeira, N Mauritania and Svalbard.

Turdus viscivorus
Mistle Thrush

TAXONOMY Linnaeus, 1758. Three subspecies (two in region).
DISTRIBUTION [Europe, N Africa to W Siberia > W and S Europe to SW Asia. Decreasing. LC] Breeds across much of Europe (except far N, S Ukraine and SE European Russia) E to Turkey, Caucasus and N Iran (*viscivorus*) and in NW Africa, Corsica and Sardinia (*deichleri*). Migratory from Scandinavia and N Germany E to Ural Mtns, wintering S in Europe to N Africa (rare Libya and Egypt), parts of N Middle East (rare in S), and NW and NE Arabia. Vagrant Azores, Bahrain, Canary Is, Faroe Is, Iceland, Madeira, Oman, Saudi Arabia and UAE.

Turdus migratorius
American Robin (V)

TAXONOMY Linnaeus, 1766. Seven subspecies (one in region).
DISTRIBUTION [Widespread North America > W and S North America. Increasing. LC] Rare vagrant, mainly late autumn and winter (*migratorius* where identified): Austria (3); Azores (2*); Belgium (2); Britain (27); Czech Republic (one between 1857 and 1874, shot); Denmark (16 Nov 1994); Germany (6); Iceland (5); Ireland (10); Spain (10-15 Dec 1999); Sweden (2).

FAMILY MUSCICAPIDAE

Chats and Old World flycatchers · 64 species

Cercotrichas podobe
Black Scrub Robin

OTHER NAMES Black Bush Robin.
TAXONOMY (Statius Müller, 1776). Two subspecies (both in region).
DISTRIBUTION [Sahel, NE Africa and Arabia. Stable. LC] Mainly resident Saudi Arabia and Yemen (*melanoptera*, but at least 25% of W Arabian population also show characters of *podobe*, breeding extralimitally Africa, and are either of that subspecies or intergrades). Rare visitor Israel (mainly spring, has bred; *podobe* or intergrades) and N Mauritania (*podobe*), where breeds just outside regional boundary. Vagrant Algeria (*podobe*), Bahrain, Egypt, Jordan (*melanoptera* and *podobe*/intergrade), Kuwait, Oman, Qatar and UAE.

Cercotrichas galactotes
Rufous-tailed Scrub Robin

OTHER NAMES Rufous Bush (or Scrub) Robin, Rufous Bush Chat, Rufous Warbler.
TAXONOMY (Temminck, 1820). Five subspecies

(four in region). Extralimital *minor* (Sahel) and *hamertoni* (E Africa) sometimes separated as African Scrub Robin.
DISTRIBUTION [SW Europe, N and E Africa and Sahel to C Asia > Sahel to NE Africa. Stable. LC] Breeds N Mauritania (*minor*), S Iberia and from Morocco E to SW Syria (*galactotes*), SW and S Balkans to S Turkey, Lebanon and W Syria (*syriaca*), and SE Turkey, Iraq, N and E Arabia and Iran (*familiaris*). Widespread N Africa and Middle East on passage, when possibly also regular Italy. Vagrant Bosnia-Herzegovina, Britain, Canary Is, Channel Is, Croatia, Finland, France, Germany, Hungary, Ireland, Madeira, Malta, Netherlands, Norway, Romania, Switzerland and Ukraine.

Muscicapa striata
Spotted Flycatcher

TAXONOMY (Pallas, 1764). Seven subspecies (at least six in region). Recent research indicates that mitochondrial genetic distance between mainland and insular W Mediterranean lineages is about 3.5% and that latter taxa, which further differ in morphology and migration strategy, could be treated as separate polytypic species *M tyrrhenica*, with subspecies *tyrrhenica* and *balearica* (in which case English name of Mediterranean Flycatcher is appropriate).
DISTRIBUTION [NW Africa and W Europe to CE Asia > sub-Saharan Africa. Decreasing. LC] Summer visitor NW Africa, Britain, Ireland and across continental Europe except extreme N Scandinavia and NW Russia (*striata*), Crimea (*inexpectata*), Balearic Is (*balearica*), Corsica, Sardinia and possibly Tyrrhenian coast of mainland Italy (*tyrrhenica*), from Aegean islands through parts of Turkey, Caucasus, Levant, N Iraq, W and N Iran (*neumanni*) to E Iran (*sarudnyi*). Widespread S of breeding range in N Africa and Middle East on passage, when *mongola* (breeding extralimitally SC Siberia and Mongolia) may occur in E of region; scarce migrant Faroe Is. Vagrant Azores, Cape Verde Is, Iceland, Madeira and Svalbard.

Muscicapa gambagae
Gambaga Flycatcher

OTHER NAMES (Gambaga) Dusky Flycatcher, Gambaga Spotted Flycatcher.
TAXONOMY (Alexander, 1901). Monotypic. Sometimes treated as conspecific with Spotted Flycatcher.
DISTRIBUTION [W, C and E Africa, SW Arabia. Stable. LC] Summer visitor SW Saudi Arabia and W Yemen.

Muscicapa sibirica
Dark-sided Flycatcher (V)

OTHER NAMES Siberian Flycatcher, (Asian/Siberian) Sooty Flycatcher.
TAXONOMY Gmelin, J F, 1789. Four subspecies (undetermined in region).
DISTRIBUTION [S Siberia to E Asia > S and SE Asia. Stable. LC] Extreme vagrant: Iceland (1-5 Oct 2012, photos*).

Muscicapa dauurica
Asian Brown Flycatcher (V)

OTHER NAME Brown Flycatcher.
TAXONOMY Raffles, 1822. Three subspecies (undetermined in region).
DISTRIBUTION [S Siberia to E Asia > S and SE Asia. Stable. LC] Extreme vagrant (10 records): Britain (4: 1-2 July 1992; 3-4 Oct 2007; 24-25 Sep 2008; 5 Sep 2010); Denmark (24-25 Sep 1959); Greece (4 Sep 1993); Oman (3: 25-26 Oct 2006; 3 Nov 2007; 14 Oct 2012); Sweden (27 and 30 Sep 1986). Further tentative but convincing report from Iran (25 Sep 2001*) appears to have involved this species.

Cyanoptila cyanomelana
Blue-and-white Flycatcher (V)

TAXONOMY (Temminck, 1829). Two subspecies (undetermined in region).
DISTRIBUTION [SE Siberia to E Asia > SE Asia. Stable. LC] Extreme vagrant (four records): Oman (2: 3-8 Jan 1982; 14 Nov 2006); UAE (2: 17-30 Nov 1980; 11 Feb 1999).

Eumyias thalassinus
Verditer Flycatcher (V)

OTHER NAMES Asian (or Indian) Verditer Flycatcher.

TAXONOMY (Swainson, 1838). Two subspecies (undetermined in region).
DISTRIBUTION [Himalayas, CE and SE Asia > S and SE Asia. Stable. LC] Extreme vagrant: Iran (28 Mar 2001).

Erithacus rubecula
European Robin

OTHER NAMES Robin; Tenerife Robin (*E r superbus*).
TAXONOMY (Linnaeus, 1758). Eight subspecies (all in region). *Superbus* of C Canary Is has been proposed as separate species on basis of subtle differences in song and plumage, and latterly genetically isolated Gran Canaria population also proposed as new subspecies '*marionae*'; species-level distinctions currently considered insufficient for adoption.
DISTRIBUTION [N Africa, Europe to C Asia > N Africa, W and S Europe to SW Asia. Increasing. LC] Breeds Britain and Ireland (*melophilus*), C Canary Is (*superbus*), W Canary Is, Azores, Madeira, NW Morocco and much of continental Europe from Mediterranean to N Fennoscandia, E European Russia and W Turkey (*rubecula*), S Crimea (*valens*), Ural Mtns (*tataricus*), E Turkey and Caucasus (*caucasicus*), SE Azerbaijan and N Iran (*hyrcanus*), and N Algeria and N Tunisia (*witherbyi*). N and E continental populations mainly migratory, winter range extending W through Europe and S to N Africa and N and E Arabia. Scarce visitor (has bred) Faroe Is and Iceland. Vagrant Cape Verde Is, Jan Mayen, N Mauritania, Oman, Qatar and Svalbard.

Larvivora cyane
Siberian Blue Robin (V)

OTHER NAME Siberian Blue Chat.
TAXONOMY (Pallas, 1776). Three subspecies (undetermined in region).
DISTRIBUTION [C Siberia to E Asia > SE Asia. Decreasing. LC] Extreme vagrant (five records): Britain (3: 23 Oct 2000; 2 Oct 2001; 1 Oct 2011, dead); Channel Is (27 Oct 1975); Spain (18 Oct 2000).

Larvivora sibilans
Rufous-tailed Robin (V)

OTHER NAMES Swinhoe's (Rufous-tailed) Robin, Whistling Nightingale.
TAXONOMY Swinhoe, 1863. Monotypic.
DISTRIBUTION [C Siberia to E Asia > SE Asia. Stable. LC] Extreme vagrant (five records): Britain (3: 23 Oct 2004; 2 Oct 2010, dead; 14 Oct 2011); Denmark (14-21 Oct 2012); Poland (30-31 Dec 2005).

Luscinia svecica
Bluethroat

OTHER NAMES Red-spotted Bluethroat (*L s svecica*, *pallidogularis* and extralimital red-spotted forms), Unspotted (or Caucasian/Turkish) Bluethroat (*L s magna*), White-spotted Bluethroat (*L s cyanecula*, *namnetum* and *volgae*).
TAXONOMY (Linnaeus, 1758). Ten subspecies (six in region). Variation marked but, while some populations geographically and ecologically isolated, others intergrade where ranges overlap.
DISTRIBUTION [Eurasia and NW North America > S Eurasia and Africa. Stable. LC] Summer visitor SW and C France (*namnetum*), C Spain, E France and Netherlands to Belarus and NW Ukraine (*cyanecula*), N Fennoscandia E to N Russia (*svecica*), W Russia and NE Ukraine (*volgae*), and E Turkey, Caucasus and N Iran (*magna*, possibly also *pallidogularis* in N Iran). Widespread on migration elsewhere in continental Europe, much of N Africa and Middle East, but scarce Britain and Malta, rare Faroe Is. Winters locally *eg* S Iberia, N Africa and Middle East, but some populations from N and E of region now known to winter S Asia. Vagrant Canary Is, Iceland, Ireland, Liechtenstein, Montenegro and Svalbard.

Luscinia luscinia
Thrush Nightingale

OTHER NAMES Eastern Nightingale, Sprosser.
TAXONOMY (Linnaeus, 1758). Monotypic.
DISTRIBUTION [N and E Europe to C Siberia > S and E Africa. Increasing. LC] Summer visitor S Fennoscandia, NE Germany and E

Europe E to Ural Mtns and S to Black Sea and Caucasus; formerly also bred C Europe. Migrates to winter quarters mainly through SE Europe, Egypt and Middle East (except SE). Uncommon/rare Italy, Netherlands (has bred) and UAE. Vagrant Azerbaijan, Belgium, Britain, Channel Is, France, Iceland, Ireland, Libya, Malta, Montenegro, Oman and Switzerland.

Luscinia megarhynchos
Common Nightingale

OTHER NAMES (European, Rufous or Western) Nightingale; Caucasian Nightingale (*L m africana*), Eastern Nightingale (*L m golzii*).
TAXONOMY Brehm, C L, 1831. Three subspecies (all in region). For nomenclatural reasons subspecific name *hafizi* now replaced by *golzii*.
DISTRIBUTION [NW Africa, Europe to EC Asia > W, C and E Africa. Increasing. LC] Summer visitor NW Africa, from Iberia N to S Britain and Germany, E through C and SE Europe to Balkans, C Turkey and Jordan (*megarhynchos*), E Turkey (taxon uncertain), Caucasus, Syria, Iraq and NW and S Iran (*africana*), and E Iran (*golzii*). Widespread on passage through N Africa (W to Canary Is) and Middle East; occasional in winter S Arabia. Vagrant Cape Verde Is, Denmark, Finland, Iceland, Ireland, Lithuania, Madeira, Norway and Sweden; *golzii* very rare vagrant W of breeding range, *eg* Britain, France, Iceland, Kuwait and Sweden, probably this form also Faroe Is.

Irania gutturalis
White-throated Robin

OTHER NAMES (White-throated) Irania, Persian Robin.
TAXONOMY (Guérin-Méneville, 1843). Monotypic.
DISTRIBUTION [SW to C Asia > E Africa. Stable. LC] Summer visitor Turkey and Levant E to Caucasus and Iran. Widespread Middle East on passage. Vagrant Belgium, Britain, Cyprus, Egypt, Greece (has bred), Italy, Netherlands, Norway, Sweden and Switzerland.

Calliope calliope
Siberian Rubythroat

OTHER NAME (Common) Rubythroat.
TAXONOMY (Pallas, 1776). Three subspecies (one in region).
DISTRIBUTION [W Siberia to E Asia > S and SE Asia. Stable. LC] Summer visitor N Ural Mtns (*calliope*). Vagrant Britain, Denmark, Estonia, Finland, Germany, Iceland, Italy, Malta, Norway, Slovenia and Sweden*.

Tarsiger cyanurus
Red-flanked Bluetail

OTHER NAMES Orange-flanked (or Chestnut-flanked/Red-flanked) Bush Robin.
TAXONOMY (Pallas, 1773). Monotypic. Formerly included extralimital Himalayan Bluetail *T rufilatus* (S and E Asia).
DISTRIBUTION [N Eurasia > SE Asia. Stable. LC] Summer visitor Finland and NW Russia. Vagrant Belgium, Britain, Bulgaria, Channel Is, Cyprus, Czech Republic, Denmark, Estonia (has bred), Faroe Is, France, Germany, Hungary, Ireland, Israel, Italy, Lebanon, Lithuania, Malta, Netherlands, Norway (has bred), Poland, Slovenia, Spain, Sweden (has bred) and UAE.

Myophonus caeruleus
Blue Whistling Thrush (V)

OTHER NAMES Large (or Violet) Whistling Thrush; Himalayan Whistling Thrush (*M c temminckii*).
TAXONOMY (Scopoli, 1786). Six subspecies (one in region).
DISTRIBUTION [C, SE and E Asia > E and SE Asia. Population trend unknown. LC] Extreme vagrant: Iran (28 Nov 1900, collected, *temminckii*).

Ficedula hypoleuca
European Pied Flycatcher

OTHER NAMES (Eurasian or Western) Pied Flycatcher; Iberian Pied Flycatcher (*F h iberiae*).
TAXONOMY (Pallas, 1764). Three subspecies (all in region). Formerly included Atlas Pied Flycatcher (*qv*); despite morphological and

geographical closeness of *iberiae* European Pied to that species, genetic research indicates close relationship to nominate.
DISTRIBUTION [W Europe to W Siberia > W to C Africa. Decreasing. LC] Breeds from Spain (*iberiae*), Ireland (rare), Britain and parts of France across C and N Europe to N Fennoscandia and E to Ural Mtns and Ukraine (*hypoleuca*). Widespread on passage S Europe, N Africa (W to Canary Is) and E to Turkey, Caucasus and Levant; *sibirica* (breeding Siberia) regular migrant in E, also confirmed *eg* Britain. Vagrant Azores, Cape Verde Is, Iceland, Madeira, Oman, Svalbard and UAE.

Ficedula speculigera
Atlas Pied Flycatcher

OTHER NAMES Atlas (or North African Pied) Flycatcher.
TAXONOMY (Bonaparte, 1850). Monotypic. Formerly treated as conspecific with European Pied Flycatcher, but found to be as distinct genetically from that species and Collared Flycatcher as they are from each other.
DISTRIBUTION [NW Africa > W Africa. Population trend not assessed. NE] Breeding endemic N Morocco, N Algeria and N Tunisia. Rarely observed on migration; postulated to occur N Mauritania but no confirmed records. Vagrant Italy, Malta and Spain*.

Ficedula albicollis
Collared Flycatcher

TAXONOMY (Temminck, 1815). Monotypic. Hybridises with European Pied Flycatcher in some areas of sympatry, notably Swedish Baltic.
DISTRIBUTION [C Europe to W Russia > SC and S Africa. Increasing. LC] Breeding endemic, ranging in summer from E France, Italy and W Balkans through C and E Europe N to Swedish Baltic, and E to C European Russia and Ukraine. Migrates through C, S and SE Europe (E to Georgia), N Africa and NW Middle East on passage, when irregular *eg* Corsica and Finland. Vagrant Algeria, Armenia, Bahrain, Balearic Is, Belgium, Britain, Canary Is, Estonia, Iraq, Ireland, NW Kazakhstan, Kuwait, Latvia, Lithuania, Luxembourg, Morocco, Netherlands, Norway and Spain.

Ficedula semitorquata
Semicollared Flycatcher

OTHER NAMES Eastern Collared (or Half-collared) Flycatcher.
TAXONOMY (Homeyer, 1885). Monotypic.
DISTRIBUTION [SE Europe and SW Asia > EC Africa. Decreasing. LC] Endemic or near-endemic breeding visitor, summer range extending from S and E Balkans and Turkey to Caucasus and N Iran. Occurs E Egypt and Middle East on passage, when scarce Cyprus and rare Malta. Vagrant Balearic Is, Corsica, Croatia, France, Italy, Libya, Romania, Switzerland and Tunisia.

Ficedula mugimaki
Mugimaki Flycatcher (V)

OTHER NAME Robin Flycatcher.
TAXONOMY (Temminck, 1836). Monotypic.
DISTRIBUTION [S Siberia to E Asia > SE Asia. Stable. LC] Extreme vagrant (at least two records): Italy (6 Oct 2011, trapped); Russia (2 Aug 2007). Further records from Britain (16-17 Nov 1991) and Italy (29 Oct 1957) placed in Cat D.

Ficedula parva
Red-breasted Flycatcher

TAXONOMY (Bechstein, 1792). Monotypic. Formerly included Taiga Flycatcher (*qv*).
DISTRIBUTION [C and E Europe > S Asia. Stable. LC] Summer visitor to C and E Europe, N to S Scandinavia and S to N Balkans, Caucasus and Turkey. Scarce/rare W Europe and N Africa on migration, when moves SE to winter quarters; some winter SE Arabia. Vagrant Algeria, Azores, Belgium, Canary Is, Corsica, Faroe Is, Gibraltar, Iceland, Iraq, Madeira, N Mauritania, Morocco, Portugal, Sardinia, Spain and Svalbard.

Ficedula albicilla
Taiga Flycatcher

OTHER NAME Red-throated Flycatcher.
TAXONOMY (Pallas, 1811). Monotypic. Formerly treated as conspecific with Red-breasted

Flycatcher, but differs in song, plumage pattern, moult sequence and mtDNA.
DISTRIBUTION [W Siberia to E Asia > S and SE Asia. Stable. LC] Summer visitor E European Russia. Vagrant Britain, France, Israel, Kuwait, Oman, Sweden and UAE.

Ficedula superciliaris
Ultramarine Flycatcher (V)

OTHER NAMES White-browed (Blue) Flycatcher.
TAXONOMY (Jerdon, 1840). Two subspecies (undetermined in region).
DISTRIBUTION [SC and EC Asia > S and SE Asia. Stable. LC] Extreme vagrant: Iran (25 Apr 2009, three).

Phoenicurus erythronotus
Eversmann's Redstart

OTHER NAMES Rufous-backed (or Red-backed) Redstart.
TAXONOMY (Eversmann, 1841). Monotypic.
DISTRIBUTION [C Asia > SW and S Asia. Stable. LC] Migrant and winter visitor Iran, N Oman (rare) and UAE (rare). Vagrant Azerbaijan, Bahrain, Iraq, Israel, Kuwait, Qatar and Saudi Arabia.

Phoenicurus ochruros
Black Redstart

OTHER NAMES Caucasian Black Redstart (*P o ochruros*), Eastern (or Central Asian) Black Redstart (*P o phoenicuroides*, sometimes including *ochruros*, *semirufus* and extralimital Asian *rufiventris*), Levant Black Redstart (*P o semirufus*), Western Black Redstart (*P o gibraltariensis*, sometimes including *ochruros* and *semirufus*).
TAXONOMY (Gmelin, S G, 1774). Five subspecies (at least four in region). Several well-differentiated geographical forms, but genetic analysis gives little molecular justification for revising species limits. Birds from S Iberia sometimes described as *aterrimus*.
DISTRIBUTION [W Europe to E Asia > S Europe, N and E Africa to S Asia. Increasing. LC] Breeds NW Africa, Europe from Mediterranean N to S Britain, S Scandinavia and E to SW Russia (*gibraltariensis*), Turkey, Caucasus, N Iraq and NW Iran (*ochruros*), NE Iran (*phoenicuroides*), and W Syria, Lebanon and NE Israel (*semirufus*). Present year-round S Europe, parts of NW Africa, Asia Minor and Levant but N and E populations mainly migratory, occurring widely on passage and wintering N Africa (*gibraltariensis*), N Middle East (*gibraltariensis* and *ochruros*), and Arabia and S Iran, few also Israel (*phoenicuroides*). Vagrant Azores, Cape Verde Is, Faroe Is, Iceland and Svalbard; *ochruros* vagrant UAE; *phoenicuroides* vagrant Belgium, Channel Is, Cyprus, Egypt, Finland, France, Germany, Georgia, Netherlands, Norway and Spain. Note that extralimital 'Eastern Black Redstart' occurrences not always identified to taxon (and this subspecies group in any case variably defined); indeterminate vagrant eastern taxa also recorded Britain (*phoenicuroides/rufiventris*), Norway (*ochruros/semirufus/phoenicuroides*) and Sweden (*ochruros/semirufus/phoenicuroides/rufiventris*).

Phoenicurus phoenicurus
Common Redstart

OTHER NAMES White-fronted (or Eurasian/European) Redstart; Eastern (or Ehrenberg's) Redstart (*P p samamisicus*).
TAXONOMY (Linnaeus, 1758). Two subspecies (both in region).
DISTRIBUTION [Europe and N Africa to C Asia > W to E Africa. Increasing. LC] Summer visitor NW Africa and much of Europe N to Scandinavia and NW Russia and E to Balkans (*phoenicurus*), and from NE Greece, Crimea, Caucasus, Turkey and Levant E to Iran (*samamisicus*). Widespread to S on passage; a few remain NW Africa in winter. Vagrant Azores, Cape Verde Is, Faroe Is, Iceland, Madeira and Svalbard.

Phoenicurus auroreus
Daurian Redstart (V)

OTHER NAME Siberian Redstart.
TAXONOMY (Pallas, 1776). Two subspecies (undetermined in region).
DISTRIBUTION [Siberia to E and NE Asia > N

SE Asia to Japan. Stable. LC] Extreme vagrant: Russia (18-30 Sep 2006).

Phoenicurus moussieri
Moussier's Redstart (E)

OTHER NAME Coroneted Redstart.
TAXONOMY (Olphe-Galliard, 1852). Monotypic.
DISTRIBUTION [NW Africa. Stable. LC] Endemic. Breeds N Morocco, N Algeria and N Tunisia. Some altitudinal and short-distance migration, wintering S of breeding range and E to NW Libya. Vagrant Balearic Is, Britain, France, Greece, Italy, Malta, Portugal and Spain.

Phoenicurus erythrogastrus
Güldenstädt's Redstart

OTHER NAMES (Large) White-winged Redstart.
TAXONOMY (Güldenstädt, 1775). Two subspecies (one in region).
DISTRIBUTION [SW, C and E Asia. Stable. LC] Breeds Greater Caucasus Mtns in Azerbaijan, Georgia and S Russia (*erythrogastrus*). Altitudinal migrant, dispersing to lower areas in winter. Vagrant Armenia. Formerly also considered to have occurred Iran, but deleted from national list after review.

Monticola saxatilis
Common Rock Thrush

OTHER NAMES European (or Mountain) Rock Thrush, Rufous-tailed (or Chestnut-tailed) Rock Thrush.
TAXONOMY (Linnaeus, 1766). Monotypic.
DISTRIBUTION [NW Africa and S Europe to E Asia > Sahel, NE and E Africa. Stable. LC] Summer visitor NW Africa, Mediterranean Europe N to Alps and Carpathian Mtns (rare), also Ukraine, Caucasus, Asia Minor, Levant, N Iraq and Iran. Widespread on passage N Africa and Middle East; very rare in winter Iran and S Arabia. Vagrant Belgium, Britain, Canary Is, Czech Republic, Denmark, Estonia, Finland, Germany, Ireland, Liechtenstein, Luxembourg, Madeira, Netherlands, Norway and Sweden.

Monticola rufocinereus
Little Rock Thrush

TAXONOMY (Rüppell, 1837). Two subspecies (one in region).
DISTRIBUTION [E and NE Africa, SW Arabia. Stable. LC] Resident SW Saudi Arabia and W Yemen (*sclateri*). All previously accepted occurrences in Oman now deleted after review.

Monticola solitarius
Blue Rock Thrush

TAXONOMY (Linnaeus, 1758). Five subspecies (two in region).
DISTRIBUTION [NW Africa, SW Europe to E Asia > W Africa, SW Europe to SE and E Asia. Stable. LC] Mainly resident NW Africa and in Mediterranean Europe, W and S Turkey and Levant (*solitarius*). In summer range extends N to Alps and from interior Asia Minor to Caucasus (also *solitarius*), NE Iraq and Iran (*longirostris*). On passage or in winter occurs locally in parts of N Africa S to N Chad, and throughout Middle East. Vagrant Austria (formerly bred), Belgium, Britain, Canary Is, Finland, Hungary, Madeira*, Netherlands, Romania, Slovakia, Sweden and Ukraine.

Saxicola rubetra
Whinchat

TAXONOMY (Linnaeus, 1758). Monotypic.
DISTRIBUTION [W Europe to C Siberia > W to E Africa. Decreasing. LC] Summer visitor across much of Europe (patchily distributed or absent in large areas of S) and E Turkey, Caucasus and NW Iran. Widespread on migration S through N Africa and Middle East; scarce migrant Faroe Is. Vagrant Azores, Cape Verde Is, Iceland, Jan Mayen, Madeira and Svalbard.

Saxicola dacotiae
Canary Islands Stonechat (E)

OTHER NAMES Canarian Stonechat, Canary Islands (Bush) Chat, Fuerteventura Chat, Fuerteventura Stonechat, Meade-Waldo's Chat; Muriel's Chat (*S d murielae*).

TAXONOMY (Meade-Waldo, 1889). Two subspecies (one extinct).
DISTRIBUTION [Canary Is. Decreasing. NT] Endemic resident Canary Is on Fuerteventura (*dacotiae*); formerly also on Montaña Clara and Alegranza, off N Lanzarote (*murielae*†).

Saxicola rubicola
European Stonechat

OTHER NAMES (Common or Eurasian) Stonechat, Collared Stonechat; British Stonechat (*S r hibernans*), Continental Stonechat (*S r rubicola*).
TAXONOMY (Linnaeus, 1766). Two subspecies (both in region). Formerly included Siberian, Stejneger's and African Stonechats (*qv*).
DISTRIBUTION [Europe, NW Africa and Asia Minor > W and S Europe, N and NE Africa, Middle East. Population trend not assessed. NE] Breeding endemic W Iberia, W France, Channel Is, Britain, Ireland and SW Norway (*hibernans*), and NW Africa, rest of Iberia and Mediterranean islands N to S Denmark and E to Belarus, Ukraine, SE European Russia, Balkans, W and N Turkey (*rubicola*). Continental N and E populations migratory (partly so Britain and Ireland), occurring widely on passage N to Sweden and Finland and wintering S to Canary Is, N Africa and Middle East (mainly N, rare SE Arabia). Vagrant Azores*, Estonia, Faroe Is, Finland, Iceland, Latvia (has bred), Madeira, N Mauritania and Svalbard.

Saxicola maurus
Siberian Stonechat

OTHER NAMES Asian (or Eastern) Stonechat; Armenian (or Byzantine) Stonechat (*S m variegatus*), Caspian (or Caucasian) Stonechat (*S m hemprichii*, sometimes also including *variegatus*), West Siberian Stonechat (*S m maurus*).
TAXONOMY (Pallas, 1773). Five subspecies (three, possibly four, in region). Separated at species level from European and African Stonechats on basis of significant genetic differences, allopatric distribution, ecology, physiology and morphology; see Svensson *et al* (2012) for background to revision of subspecific scientific names. Formerly also included Stejneger's Stonechat (*qv*).
DISTRIBUTION [NE Europe, W and SW Asia to C Siberia > SW Asia to SE Asia. Population trend not assessed. NE] Breeds NW Russia and NE Iran (*maurus*), NW Kazakhstan, NW and W Caspian region and N Caucasus (*hemprichii*), and S Caucasus, E Turkey and N Iraq to N and W Iran (*variegatus*). Occurs annually E Finland, and has bred. Winters from E Egypt through Middle East (except much of Saudi Arabia) to Iran, but non-breeding subspecific distribution poorly understood; *maurus* is most widespread, *hemprichii* winters S to *eg* Egypt, Israel, Jordan, Kuwait and parts of Arabia, and *variegatus* uncommon on passage S to wintering grounds in Egypt, SW Arabia and possibly elsewhere. Extralimital occurrences mainly involve *maurus*, but now-split Stejneger's Stonechat not always excluded: vagrant Armenia, Austria, Belgium, Bulgaria, Britain, Channel Is, Cyprus, Denmark, Estonia, France, Germany, Greece, Ireland, Italy, Lithuania, Netherlands, Norway, Poland*, Portugal, Romania, Spain and Sweden; *hemprichii* vagrant Britain, Cyprus, Egypt, Estonia, Finland, France, Greece, Italy, Jordan, Netherlands, Norway, Oman, Qatar, Spain, Sweden, Turkey and UAE; *variegatus* vagrant *eg* Egypt and Qatar, but likely under-recorded. Extralimital *indicus* (S Asia) hypothetical Iran.

Saxicola stejnegeri
Stejneger's Stonechat (V)

OTHER NAMES (East) Siberian Stonechat.
TAXONOMY (Parrot, 1908). Monotypic. Formerly included within Siberian Stonechat, but despite close morphological similarities *maurus* and *stejnegeri* are highly divergent genetically and not each other's closest relative.
DISTRIBUTION [E Asia > SE Asia. Population trend not evaluated. NE] Provisionally classified as extreme vagrant (difficulty of separation from Siberian Stonechat may obscure greater frequency of records): four occurrences of three individuals confirmed by analysis of DNA samples, in Britain (24-26 Oct 2012, same as Netherlands), Finland (31 Oct-26 Nov 2013*), Netherlands (8-23 Oct 2012, same as Britain) and Sweden (2 Oct 2008*). Further DNA testing and specimen examination, as well as refinement of field ID criteria, may confirm suspected status as more

frequent late-autumn vagrant to W Europe, possibly also Middle East, where hypothetical migrant and winter visitor E Saudi Arabia but no records yet confirmed by genetic sampling. See also comments under Siberian Stonechat.

Saxicola torquata
African Stonechat

OTHER NAMES (Common) Stonechat.
TAXONOMY (Linnaeus, 1766) Thirteen subspecies (one in region). Specifically distinct from European, Siberian and Stejneger's Stonechats, with which formerly considered conspecific, on basis of significant differences in genetics, distribution, ecology, physiology and morphology.
DISTRIBUTION [Sub-Saharan Africa and SW Arabia. Population trend not evaluated. NE] Resident SW Saudi Arabia and W Yemen (*felix*).

Saxicola caprata
Pied Stonechat

OTHER NAMES Pied Bush Chat (IOC), Black Bush Chat, Pied Chat.
TAXONOMY (Linnaeus, 1766). Sixteen subspecies (two in region).
DISTRIBUTION [SW and C Asia to SE Asia and New Guinea. Stable. LC] Resident and partial migrant NE Iran (*rossorum*) and SE Iran (*bicolor*). Vagrant Cyprus, Egypt, Finland, Iraq, Israel, Kuwait, Oman, Qatar, Saudi Arabia and UAE.

Myrmecocichla aethiops
Anteater Chat (V)

OTHER NAMES (Northern) Ant Chat, Northern Ant-eater Chat.
TAXONOMY Cabanis, 1850. Three subspecies (undetermined in region).
DISTRIBUTION [Sahel and E Africa. Stable. LC] Extreme vagrant: N Chad (1954, collected).

Oenanthe oenanthe
Northern Wheatear

OTHER NAMES (Common) Wheatear, Eurasian (or European) Wheatear; Greenland Wheatear (*O o leucorhoa*).
TAXONOMY (Linnaeus, 1758). Three subspecies (all in region). Formerly included Seebohm's Wheatear (*qv*), currently treated as conspecific by IOC.
DISTRIBUTION [Eurasia and N North America > sub-Saharan Africa. Decreasing. LC] Widespread summer visitor, breeding from parts of Iberia E through S Europe to N Middle East (*libanotica*), and across much of W, C and N Europe E to Ural Mtns and N irregularly to Svalbard (*oenanthe*) and Iceland (*leucorhoa*). Occurs widely on passage S in non-breeding areas of Europe, N Africa and throughout Middle East, with *leucorhoa* regular in W of region, including Atlantic islands (scarce).

Oenanthe seebohmi
Seebohm's Wheatear

OTHER NAME Black-throated Wheatear.
TAXONOMY (Dixon, 1882). Monotypic. Previously treated as a subspecies of Northern Wheatear, but separated on basis of distinct male plumage and allopatric range.
DISTRIBUTION [NW Africa > W Africa. Population trend not assessed. NE] Breeding endemic Morocco and Algeria; also occasional in winter Morocco. Vagrant Canary Is, Gibraltar and Tunisia.

Oenanthe bottae
Red-breasted Wheatear

OTHER NAMES Botta's Chat, Botta's (or Buff-breasted) Wheatear.
TAXONOMY (Bonaparte, 1854). Two subspecies (one in region). Previously included extralimital Heuglin's Wheatear *O heuglini* (Sahel), and occasionally also treated as conspecific with Isabelline Wheatear.
DISTRIBUTION [NE Africa and S Arabia. Stable. LC] Resident SW Saudi Arabia and W Yemen (*bottae*).

Oenanthe isabellina
Isabelline Wheatear

OTHER NAME Isabelline Chat.

TAXONOMY (Temminck, 1829). Monotypic.
DISTRIBUTION [SE Europe to E Asia > Sahel, NE Africa, Arabia and India. Stable. LC] Summer visitor E Balkans, S Ukraine, SW Russia and Caucasus S through N Middle East to Turkey in W and Iran in E. Widespread on passage SE Europe, N Africa (W to Algeria), Cyprus and Arabia; rare but regular Malta (mainly spring). Common Arabia in winter and occasional N Mauritania. Vagrant Azores, Belgium, Britain, Canary Is, Croatia, Denmark, Estonia, Finland, France, Germany, Hungary, Ireland, Italy, Lithuania*, Madeira, Montenegro, Morocco, Netherlands, Norway, Poland, Romania, Spain and Sweden.

Oenanthe monacha
Hooded Wheatear

OTHER NAME Hooded Chat.
TAXONOMY (Temminck, 1825). Monotypic.
DISTRIBUTION [NE Africa to SW Asia. Stable. LC] Near-endemic resident, breeding E Egypt, S Israel, S Jordan and widely (but locally) Arabia, and in E and S Iran. Some movement into Gulf in winter, when recorded Qatar. Vagrant Bahrain, Cyprus, Greece, Kuwait and Turkey.

Oenanthe deserti
Desert Wheatear

TAXONOMY (Temminck, 1825). Three subspecies (all in region).
DISTRIBUTION [N Africa to CE Asia > Sahel, N and NE Africa, SW Asia. Stable. LC] Breeds from Morocco and perhaps N Mauritania through parts of Saharan N Africa E to NW Egypt (*homochroa*), NE Egypt to Syria and N Saudi Arabia, and E Iran (*deserti*). Occurs more widely Middle East and N Africa on passage and in winter, when *oreophila* (breeding extralimitally CE Asia) also present in S and E Arabia. Vagrant Armenia, Azerbaijan, Balearic Is, Belgium, Britain, Bulgaria, Canary Is, Channel Is, Corsica, Denmark, Estonia, Finland, France, Georgia, Germany, Gibraltar, Greece, Hungary, Ireland, Italy, Latvia, Madeira, Malta, Netherlands, Norway, Poland, Portugal, Slovakia, Spain, Sweden, Switzerland, Turkey (has bred) and Ukraine.

Oenanthe hispanica
Black-eared Wheatear

OTHER NAMES Black-throated Wheatear; Eastern Black-eared Wheatear (*O h melanoleuca*), Western Black-eared (or Spanish) Wheatear (*O h hispanica*).
TAXONOMY (Linnaeus, 1758). Two subspecies (both in region) occasionally treated as separate species; alternatively, considered conspecific with Pied Wheatear, with extensive hybridisation where ranges meet in E of region.
DISTRIBUTION [NW Africa and S Europe to C Asia > Sahel. Decreasing. LC] Summer visitor NW Africa (E to N Libya), Iberia, S France and N Italy (*hispanica*), and from S Italy and Balkans E to Turkey, Levant, N Iraq, Caucasus and W Iran (*melanoleuca*). Widespread on passage N Africa (W to Canary Is) and through Middle East except SE Arabia. Vagrant Austria, Azores, Belgium, Britain, Channel Is, Denmark, Finland, Germany, Hungary, Ireland, Netherlands, Norway, Oman, Poland, Slovakia, Slovenia, Sweden and Switzerland; record from Czech Republic under review.

Oenanthe cypriaca
Cyprus Wheatear

OTHER NAME Cyprus Pied Wheatear.
TAXONOMY (Homeyer, 1884). Monotypic. Formerly considered conspecific with Pied Wheatear, but differs in morphology (including only slight sexual dimorphism), vocalisations, behaviour and habitat.
DISTRIBUTION [Cyprus > NE Africa. Stable. LC] Breeding endemic restricted to Cyprus. On migration recorded E Egypt, Israel, Jordan, Saudi Arabia and Syria; regular spring overshoot S Turkey. Vagrant Germany, Lebanon and Oman.

Oenanthe pleschanka
Pied Wheatear

OTHER NAMES Common Pied (or Siberian) Wheatear, Pleschanka's Chat, Pleschanka's Wheatear.
TAXONOMY (Lepechin, 1770). Monotypic. Formerly included Cyprus Wheatear (*qv*). Closely related to Black-eared Wheatear and freely interbreeds where ranges overlap (*eg*

Azerbaijan and E Balkans); sometimes considered conspecific.
DISTRIBUTION [SE Europe to CE Asia > SW Arabia, NE and E Africa. Stable. LC] Summer visitor, breeding from NE Balkans E through S Ukraine and S European Russia to Ural Mtns, and in SE Caucasus, E Turkey and Iran. White-throated *'vittata'* form of male rare in Middle East, and reportedly unknown in European breeding population. Migrates SE or S to wintering grounds, occurring from Turkey (uncommon) to Iran and S through N Middle East and Arabia, where occasional in S in winter. Vagrant Austria, Belgium, Britain, Croatia, Cyprus, Denmark, Egypt, Estonia, Finland, France, Germany, Hungary, Ireland, Italy, Latvia, Libya, Lithuania, Malta, Netherlands, Norway, Poland, Sweden and Tunisia; *'vittata'* vagrant *eg* Israel and Oman.

Oenanthe moesta
Red-rumped Wheatear (E)

OTHER NAMES Buff-rumped (or Grey-headed) Wheatear, Tristram's Wheatear.
TAXONOMY (Lichtenstein, M H C, 1823). Two subspecies (both in region).
DISTRIBUTION [N Africa and Middle East. Stable. LC] Endemic resident from Morocco discontinuously E through N Africa (*moesta*) to NE Egypt, Jordan and S Syria (*brooksbanki*). Formerly known also from Iraq, but no records since 1920s. Very rare Israel (has bred). Vagrant N Mauritania and Saudi Arabia (breeding postulated in NW but not confirmed).

Oenanthe melanura
Blackstart

OTHER NAMES Black-tailed Cercomela, Black-tailed (Rock) Chat.
TAXONOMY (Temminck, 1824). Six subspecies (three in region). In S Arabia, subspecies *melanura* and *neumanni* breed in close proximity but are ecologically separated, former favouring drier arid regions, latter more precipitous areas of higher rainfall and humidity.
DISTRIBUTION [Sahel, NE Africa and Middle East. Stable. LC] Resident N Niger and N Chad (*airensis*), NE Egypt, Israel, Palestine, Syria and Jordan to C Saudi Arabia, E Yemen and S Oman (*melanura*) and SW Saudi Arabia, W Yemen and S Oman (*neumanni*). Vagrant Libya and UAE.

Oenanthe picata
Variable Wheatear

OTHER NAME Eastern Pied Wheatear; Gould's Chat (*O p 'capistrata'*), Pied Chat (*O p 'picata'*), Strickland's Chat (*O p 'opistholeuca'*).
TAXONOMY (Blyth, 1847). Monotypic. Intraspecific variation has led to description of three main plumage types – *'capistrata'*, *'picata'* and *'opistholeuca'* – at various times classified as species, subspecies and colour morphs, but in parts of range populations intergrade and variability in plumage and genetics complex. Variable Wheatear may be good example of stable genetic polymorphism originating between originally independent taxa (see Panov 2005 for background); currently considered monotypic by IOC.
DISTRIBUTION [SW to C Asia > E Arabia to S Asia. Stable. LC] Breeds NE and SE Iran. Northern population migratory; winters Oman and UAE. Vagrant Qatar and Saudi Arabia; also reported Lebanon but record due for review. Previously accepted occurrence in Israel reassessed and confirmed as *warriae* Mourning Wheatear.

Oenanthe leucura
Black Wheatear (E)

TAXONOMY (Gmelin, J F, 1789). Two subspecies.
DISTRIBUTION [NW Africa and SW Europe. Stable. LC] Endemic resident from N Mauritania and Morocco E to NW Libya (*syenitica*), and Portugal, Spain and formerly Gibraltar and S France (*leucura*). Vagrant Balearic Is, Bulgaria, Egypt, Greece, Italy (bred in historic times), Malta, Sardinia and Ukraine; either this species or White-crowned Wheatear also recorded Ireland.

Oenanthe leucopyga
White-crowned Wheatear

OTHER NAMES White-crowned Black Chat, White-crowned Black Wheatear, White-tailed (or White-rumped) Wheatear.

TAXONOMY (Brehm, C L, 1855). Three subspecies (all in region).
DISTRIBUTION [N Africa and Middle East. Stable. LC] Breeds from N Mauritania and Morocco eastwards discontinuously through Saharan N Africa to W Egypt (*aegra*), Nile Valley (*leucopyga*) and E through Sinai to Israel, Palestine, Jordan and Saudi Arabia (*ernesti*). Largely sedentary, but some short-distance movements undertaken. Vagrant Bahrain, Britain, Canary Is, Cape Verde Is, Cyprus, Denmark, Germany, Greece, Iran, Iraq, Italy, Kuwait, Malta, Oman, Qatar, Portugal, Spain, Syria, Turkey and UAE.

Oenanthe albonigra
Hume's Wheatear

TAXONOMY (Hume, 1872). Monotypic.
DISTRIBUTION [SW Asia. Stable. LC] Breeds E Iraq, widely in Iran (except N), and in UAE and Oman. Disperses N and W, also with some altitudinal migration, but mainly winters close to breeding areas. Vagrant Bahrain, Kuwait, Saudi Arabia and Qatar.

Oenanthe finschii
Finsch's Wheatear

OTHER NAMES Barnes's Chat, White-backed Wheatear.
TAXONOMY (Heuglin, 1869). Two subspecies (both in region). Has been treated as conspecific with Mourning Wheatear.
DISTRIBUTION [SW and C Asia > SW Asia. Stable. LC] Breeds from SC and SE Turkey S through Levant to N Israel and E to N Iraq and W and SW Iran (*finschii*), and from extreme E Turkey N to S and E Caucasus and E to N and C Iran (*barnesi*). Winters from S of breeding range through N Middle East W to Cyprus and S to NE Egypt and N Saudi Arabia; rare elsewhere in Arabia. Vagrant Bahrain, Greece, Oman, Qatar and UAE (where records currently under review).

Oenanthe lugens
Mourning Wheatear (E)

OTHER NAMES Eastern Mourning Wheatear, Mourning Chat; Basalt (or Black-bellied) Wheatear (*O l warriae*), Iranian Wheatear (*O l persica*).
TAXONOMY (Lichtenstein, M H C, 1823). Three subspecies. Formerly included Arabian and Maghreb Wheatears (*qv*) and extralimital Abyssinian Wheatear *O lugubris* (NE and E Africa). Subspecies *persica* is genetically distinct and may deserve specific status, but more complete sampling of taxa required. Recently described *warriae* (formerly treated as '*basalti*' black morph) considered closely related to nominate *lugens* but mixed pairings rarely documented.
DISTRIBUTION [SW Asia. Stable. LC] Endemic. Breeds from NE Egypt to NW Saudi Arabia (*lugens*), NE Jordan and S Syria, possibly also adjacent NW Saudi Arabia (*warriae*), and NE Iraq and Iran (*persica*). Disperses *eg* W across Nile (*lugens*), but also winters S to N, E and S Arabia (*persica*). Vagrant Cyprus (taxon undetermined but probably *lugens*) and Turkey (*lugens*), Yemen (probably *persica*), and Egypt, Israel and Turkey (*warriae*, including apparent nesting with female Finsch's Wheatear in Turkey, spring 2011*).

Oenanthe halophila
Maghreb Wheatear (E)

OTHER NAMES Western Mourning (or North African) Wheatear.
TAXONOMY (Tristram, 1859). Monotypic. Traditionally considered conspecific with Mourning Wheatear, but now treated as specifically distinct on basis of clear sexual dimorphism and allopatric range.
DISTRIBUTION [N Africa. Population trend not assessed. NE] Endemic, breeding N Morocco, Algeria, S Tunisia, N and possibly SW Libya, and W Egypt. Partial migrant, moving into Sahara in winter. Vagrant N Mauritania and Turkey.

Oenanthe lugentoides
Arabian Wheatear (E)

OTHER NAMES South Arabian (Mourning) Wheatear.
TAXONOMY (Seebohm, 1881). Two subspecies. Formerly considered conspecific with Mourn-

ing Wheatear, but significantly different in morphology, including marked sexual dimorphism, and habitat.
DISTRIBUTION [S Arabia. Stable. LC] Endemic resident SW Saudi Arabia and W Yemen (*lugentoides*) and E Yemen and S Oman (*boscaweni*).

Oenanthe xanthoprymna
Kurdish Wheatear

OTHER NAMES Kurdistan (or Chestnut-rumped) Wheatear, Red-rumped (or Red-tailed/Rufous-tailed) Wheatear.
TAXONOMY (Hemprich and Ehrenberg, 1833). Monotypic. Formerly included Persian Wheatear (*qv*).
DISTRIBUTION [SW Asia > NE Africa. Stable. LC] Summer visitor E Turkey, SW Iran and N Iraq. Scarce migrant Kuwait and Saudi Arabia, very rare migrant and winter visitor Israel; winters S of breeding areas as far as Sinai and E Egypt, but very local and range poorly known. Vagrant Cyprus, Georgia, Jordan, Libya, Oman, Qatar, Syria and UAE.

Oenanthe chrysopygia
Persian Wheatear

OTHER NAMES Red-tailed Wheatear (IOC), Afghan Wheatear, Red-tailed Chat, Rufous-tailed (or Rusty-tailed) Wheatear.
TAXONOMY (de Filippi, 1863). Monotypic. Traditionally treated as conspecific with Kurdish Wheatear, but differs in *eg* lack of sexual dimorphism and almost wholly allopatric range (found breeding sympatrically N Iraq 2009); claims of hybridisation as yet unproven, apparent hybrids ('*cummingi*') showing characters shared with first-year Kurdish Wheatear.
DISTRIBUTION [SW to SC Asia > E Africa. Stable. LC] Summer visitor S Armenia, Azerbaijan and disjunct areas of N, S and E Iran; may also breed regularly NE Iraq. Status unconfirmed E Turkey, where multiple reports but no officially accepted records. On passage and/or in winter occurs Bahrain, S Iran, S Iraq, Kuwait, Qatar, Saudi Arabia, Oman, UAE and Yemen (rare). Vagrant Israel.

FAMILY CINCLIDAE

Dippers · one species

Cinclus cinclus
White-throated Dipper

OTHER NAMES (Eurasian or European) Dipper, White-breasted (or White-fronted) Dipper; Black-bellied Dipper (*C c cinclus*), Irish Dipper (*C c hibernicus*).
TAXONOMY (Linnaeus, 1758). Fourteen subspecies (10 in region, one of which extinct).
DISTRIBUTION [NW Africa and W Europe to C Asia. Stable. LC] Breeds Morocco and Algeria (*minor*), from SE and E Iberia and E France E through Italy to Greece and N to Belgium, Germany and S Poland (*aquaticus*), W and N Iberia, C France, Corsica, Sardinia, and Fennoscandia, Baltic States and NW Russia (*cinclus*), Ireland and W Scotland (*hibernicus*), Scotland (except W), Wales and N and W England (*gularis*), Ural Mtns (*uralensis*), W Syria and Lebanon (*rufiventris*), Turkey, Caucasus, N Iraq and N Iran (*caucasicus*), and SW Iran (*persicus*); formerly also Cyprus (*olympicus*†, last recorded 1945). Mainly resident but locally and altitudinally dispersive; some populations, especially N Europe, undertake medium- or long-distance movements, wintering near to or well S of breeding range. Regular in winter Netherlands, where has bred irregularly. Vagrant Faroe Is, Malta, Tunisia and Svalbard.

FAMILY NECTARINIIDAE

Sunbirds · six species

Hedydipna platura
Pygmy Sunbird

OTHER NAMES Pygmy Long-tailed Sunbird, Western (or Southern) Pygmy Sunbird.
TAXONOMY (Vieillot, 1819). Monotypic.
DISTRIBUTION [Sahel to NC and E Africa. Stable. LC] Breeds (or has bred) N Chad. Apparently scarce or rare visitor N Mauritania in rainy season, possibly also N Niger.

Hedydipna metallica
Nile Valley Sunbird

OTHER NAMES Bronzed (or Eastern/Northern) Pygmy Sunbird, Metallic Sunbird.
TAXONOMY (Lichtenstein, M H C, 1823). Monotypic.
DISTRIBUTION [NE Africa and S Arabia. Stable. LC] Breeds Egypt in Nile Valley and Delta, SW Saudi Arabia, Yemen and S Oman. Some mainly local movements reported. Vagrant Red Sea coast of Egypt and N Oman.

Cinnyris pulchellus
Beautiful Sunbird

OTHER NAME Beautiful Long-tailed Sunbird.
TAXONOMY (Linnaeus, 1766). Two subspecies (one in region).
DISTRIBUTION [W, C and E Africa. Stable. LC] Occurs on SW fringe of region in N Mauritania, reaching upland areas in rainy season at 20°N (*pulchellus*).

Cinnyris osea
Palestine Sunbird

OTHER NAMES (Northern) Orange-tufted Sunbird.
TAXONOMY Bonaparte, 1856. Two subspecies (one in region).
DISTRIBUTION [NC Africa, Levant and Arabia. Stable. LC] Breeds from S Lebanon, Jordan, Palestine, Israel and Sinai SE through W Saudi Arabia to Yemen and S Oman (*osea*). Mainly resident but dispersive, nomadic and short-distance migrant, wintering N to extreme SW Syria.

Cinnyris habessinicus
Shining Sunbird

OTHER NAME Abyssinian Sunbird.
TAXONOMY (Hemprich and Ehrenberg, 1828). Five subspecies (two in region).
DISTRIBUTION [NE Africa and S Arabia. Stable. LC] Resident W Saudi Arabia (*kinneari*) and SW Saudi Arabia, Yemen and S Oman (*hellmayri*).

Cinnyris asiaticus
Purple Sunbird

TAXONOMY (Latham, 1790). Three subspecies (one in region).
DISTRIBUTION [E Arabia to SE Asia. Stable. LC] Breeds N and E Oman, UAE and S Iran (*brevirostris*). Mainly resident but some local movement, *eg* S in Oman in winter. Vagrant Kuwait.

FAMILY PASSERIDAE

Old World sparrows and snowfinches · 17 species

Passer ammodendri
Saxaul Sparrow (V)

TAXONOMY Gould, 1872. Three subspecies (one in region).
DISTRIBUTION [C Asia. Stable. LC] Extreme vagrant (two records, but may wander more regularly from nearest breeding areas north of regional boundary): Iran (2, *ammodendri*: 3 Nov 1900; 27 Apr 2008, two males and possibly one female, showing behaviour suggestive of breeding).

Passer domesticus
House Sparrow

OTHER NAMES English Sparrow; Indian (or Oriental House) Sparrow ('*indicus* group').
TAXONOMY (Linnaeus, 1758). Twelve subspecies (10 in region) divided into *domesticus* (Palearctic) and *indicus* (Oriental) groups. Latter sometimes given full species rank, but subspecies from both intergrade in some areas where ranges meet.
DISTRIBUTION [N and NE Africa, Eurasia; introduced worldwide. Decreasing. LC] Resident Azores, Canary Is and Cape Verde Is (introduced), NW Africa and NW Libya (*tingitanus*), NE Libya and Egypt (*niloticus*), Iberia, Balearic Is, S France, Balkans and W and C Turkey (*balearoibericus*), W, C, N and E Europe from Ireland, Britain and C France N to SE Iceland and N Fennoscandia, E to Ural Mtns, Ukraine, N Balkans, Caucasus and NE Turkey (*domesticus*), Cyprus, N Israel, NW Jordan and W Syria

to SE Turkey, Iraq and NW Iran (*biblicus*), S Israel and S Palestine E through most of Arabia (*indicus*), NE Arabia (*hufufae*), SE Azerbaijan and N Iran (*hyrcanus*), C and E Iran (*persicus*) and NE Iran (*bactrianus*). Largely sedentary, but some short-distance movements recorded (mostly involving juveniles) and sporadic larger-scale migration known mainly from N of range. Vagrant Corsica (has bred), Madeira, N Niger and Svalbard.

Passer italiae
Italian Sparrow (E)

TAXONOMY Vieillot, 1817. Monotypic. Has been treated as a subspecies of both House and Spanish Sparrows, but latterly recognised as a species derived from hybrid House × Spanish origin. Now largely reproductively isolated from both parent species, breeding sympatrically with Spanish Sparrow in SE Italy, and with only limited interbreeding with House Sparrow in contact zone in Italian Alps.
DISTRIBUTION [S Europe. Population trend not assessed. NE] Endemic resident Italy (including Sardinia and parts of Sicily), adjacent fringes of France, Switzerland, Austria and Slovenia, and Corsica, where population in S shows some Spanish Sparrow characteristics. Local movements may occur in winter, as demonstrated by Corsican recovery of individual ringed N Italy. Populations resembling 'pure' Italian Sparrow also found *eg* Crete, Algeria and Tunisia, but at least in N Africa these probably derive from more recent hybridisation events and may not be stable *italiae*. Existence of House × Spanish Sparrow hybrids in some areas (generally uncommon but occasionally abundant) also makes genuine vagrancy difficult to establish; single *italiae* records from Balearic Is and Hungary placed in Cat D, and apparently also recorded Bulgaria (has bred).

Passer hispaniolensis
Spanish Sparrow

OTHER NAME Willow Sparrow.
TAXONOMY (Temminck, 1820). Two subspecies (both in region). Geographical variation slight, and some populations (*eg* parts of Turkey and Middle East) not readily assignable to subspecies.
DISTRIBUTION [N Africa and S Europe to C Asia. Stable. LC] Breeds E Atlantic islands, N Africa, Iberia, Malta and some other Mediterranean islands (where hybridises with House Sparrow), SE Italy, Balkans and W Turkey to C Arabia and UAE (*hispaniolensis*), and from E Turkey and Levant to Iran (*transcaspius*). Migratory in SE Europe, Asia Minor and Iran, wintering S through N Middle East and W to Egypt; some migration/dispersal also Iberia, when recorded Gibraltar, and NW Africa. Vagrant Bahrain, Balearic Is, Britain, Corsica, Finland, France, Hungary (has bred), Netherlands, Norway, Slovenia and Sweden.

Passer pyrrhonotus
Sind Sparrow

OTHER NAMES (Sind) Jungle Sparrow, Rufous-backed Sparrow.
TAXONOMY Blyth, 1845. Monotypic.
DISTRIBUTION [NW Subcontinent. Stable. LC] Resident extreme SE Iran.

Passer moabiticus
Dead Sea Sparrow

OTHER NAMES Scrub Sparrow; Afghan Scrub Sparrow (*P m yatii*).
TAXONOMY Tristram, 1864. Three subspecies (all in region). Geographically isolated and larger *yatii* differs constantly in plumage and sometimes treated as distinct species; separation of *mesopotamicus* from nominate questionable.
DISTRIBUTION [SW Asia. Stable. LC] Breeds Israel and Jordan (*moabiticus*), from S Turkey and N Syria to Iraq and SW Iran (*mesopotamicus*), and in SE Iran (*yatii*). Some winter dispersal. Vagrant Bahrain, Cyprus (formerly bred), Greece, Kuwait, Lebanon* and UAE.

Passer iagoensis
Iago Sparrow (E)

OTHER NAMES Cape Verde (or Rufous-backed) Sparrow.
TAXONOMY (Gould, 1838). Monotypic.

DISTRIBUTION [Cape Verde Is. Stable. LC] Resident endemic Cape Verde Is. One record away from Cape Verde of ship-assistance of a small flock, first to Desertas Is, then Madeira and finally Netherlands, in May 2013.

Passer simplex
Desert Sparrow

OTHER NAME African Desert Sparrow.
TAXONOMY (Lichtenstein, M H C, 1823). Two subspecies (both in region). Formerly included Zarudny's Sparrow (*qv*).
DISTRIBUTION [N Africa. Decreasing. LC] Resident Sahara discontinuously from N Mauritania and SE Morocco E to Libya (*saharae*), and on S Saharan fringe of region (*simplex*). Vagrant Egypt.

Passer zarudnyi
Zarudny's Sparrow

OTHER NAMES (Asian) Desert Sparrow.
TAXONOMY Pleske, 1896. Monotypic. Formerly treated as conspecific with Desert Sparrow, but separated on basis of much-reduced sexual dimorphism, larger size and other differences in plumage and structure.
DISTRIBUTION [C Asia. Decreasing. LC] Formerly known from E Iran, but no confirmed records since 1901.

Passer montanus
Eurasian Tree Sparrow

OTHER NAMES (European) Tree Sparrow.
TAXONOMY (Linnaeus, 1758). Nine subspecies (two in region).
DISTRIBUTION [Widespread Eurasia. Stable. LC] Breeds Canary Is (Gran Canaria), N Morocco (scarce), N Tunisia and across much of Europe from Iberia and Mediterranean islands N to Ireland, Britain and S Fennoscandia, E to Ural Mtns, Ukraine and Balkans (*montanus*), and in Asia Minor, Caucasus and N Iran (*transcaucasicus*). Migratory in NE of range, wintering S to N Middle East, locally also Morocco; occasional on passage Gibraltar. Vagrant Algeria, Cyprus, Egypt, Faroe Is, Iceland, Israel, Libya, Svalbard and UAE.

Passer luteus
Sudan Golden Sparrow

OTHER NAME Golden Sparrow.
TAXONOMY (Lichtenstein, M H C, 1823). Monotypic. Has been regarded as conspecific with Arabian Golden Sparrow.
DISTRIBUTION [Sahel. Stable. LC] Nomadic. Presence along S fringe of region erratic, but breeds or has bred N Mauritania, N Chad and possibly N Niger. Vagrant S Morocco.

Passer euchlorus
Arabian Golden Sparrow

OTHER NAME Golden Sparrow.
TAXONOMY (Bonaparte, 1850). Monotypic. Has been treated as conspecific with Sudan Golden Sparrow.
DISTRIBUTION [NE Africa and SW Arabia. Stable. LC] Near-endemic and nomadic resident SW Saudi Arabia and W Yemen. Records from Kuwait (including reported breeding) considered to relate to escapes from captivity.

Carpospiza brachydactyla
Pale Rockfinch

OTHER NAMES Hill Sparrow, Pale Petronia, Pale Rock Sparrow.
TAXONOMY (Bonaparte, 1850). Monotypic.
DISTRIBUTION [Middle East to C Asia > S Arabia and NE Africa. Stable. LC] Nomadic breeder, variably distributed from SE Turkey and S Caucasus E to Iran and S through Middle East and locally across Arabia (except SW). Migratory, occurring widely Middle East on passage; some winter S Arabia. Vagrant Cyprus, Egypt, SW Russia (Dagestan) and Yemen.

Petronia petronia
Rock Sparrow

OTHER NAMES Common (or Eurasian/Streaked) Rock Sparrow, Rock Petronia.
TAXONOMY (Linnaeus, 1766). Seven subspecies (five in region).
DISTRIBUTION [S Europe and N Africa to CE Asia. Increasing. LC] Breeds NW Africa (*bar-*

bara), Canary Is, Madeira and from Iberia N to C France and E through parts of S Europe and Mediterranean to W Turkey (*petronia*), S Turkey to Jordan (*puteicola*), C Turkey to Caucasus, N Iraq and N Iran (*exigua*) and W and S Iran (*intermedia*). Migratory in E Turkey, Caucasus and W Iran, leaving mountains to winter to S. Vagrant Austria (formerly bred), Azores, Belgium, Britain, Cyprus, Germany (formerly bred), Malta, Montenegro, Poland (formerly bred), Slovenia, Switzerland and Ukraine.

Gymnoris dentata
Bush Petronia

OTHER NAMES Lesser Petronia, Lesser Rock (or Bush) Sparrow, Russet-browed Bush-sparrow.
TAXONOMY (Sundevall, 1850). Monotypic.
DISTRIBUTION [Sahel and SW Arabia. Stable. LC] Resident W Yemen.

Gymnoris xanthocollis
Yellow-throated Sparrow

OTHER NAMES Chestnut-shouldered Petronia, Chestnut-shouldered (Bush-)Sparrow.
TAXONOMY (Burton, 1838). Two subspecies (one in region).
DISTRIBUTION [SW Asia > S Asia. Stable. LC] Summer visitor SE Turkey, N Syria, Iraq, Iran, Kuwait, N Oman and UAE (*transfuga*). Migrates SE to winter quarters, when regular on passage Qatar and E Saudi Arabia (rare). Vagrant Bahrain, Egypt, Israel, Jordan and Lebanon.

Montifringilla nivalis
White-winged Snowfinch

OTHER NAMES (Common or Eurasian) Snow Finch.
TAXONOMY (Linnaeus, 1766). Seven subspecies (four in region).
DISTRIBUTION [SW Europe to CE Asia. Stable. LC] Resident N Spain, Alps, Corsica, C Italy and Balkans (*nivalis*), S and E Turkey (*leucura*), Caucasus, N Iraq and N Iran (*alpicola*), and SW (possibly also SE) Iran (*gaddi*). Some movement in winter, mainly altitudinal. Vagrant Balearic Is, Croatia, Cyprus, Czech Republic, N Germany, Hungary, Malta, Poland, Portugal, Sardinia, Slovakia and Ukraine.

FAMILY PLOCEIDAE

Weavers and allies · nine species

Ploceus galbula
Rüppell's Weaver

TAXONOMY Rüppell, 1840. Monotypic.
DISTRIBUTION [NE Africa and S Arabia. Stable. LC] Resident SW Saudia Arabia, Yemen and S Oman.

Ploceus cucullatus
Village Weaver (V)

OTHER NAMES Black-headed Weaver, Layard's (Black-headed) Weaver, Spotted-backed Weaver.
TAXONOMY (Statius Müller, 1776). Ten subspecies (undetermined in region).
DISTRIBUTION [Sub-Saharan Africa. Stable. LC] Extreme vagrant: Egypt (1 May 2006).

Ploceus melanocephalus
Black-headed Weaver (I)

OTHER NAMES Yellow-backed (or Yellow-collared) Weaver.
TAXONOMY (Linnaeus, 1758). Five subspecies (undetermined in region).
DISTRIBUTION [W, C and NE Africa. Stable. LC] Naturalised population established Portugal (Cat C).

Ploceus jacksoni
Golden-backed Weaver (I)

OTHER NAMES Eastern Golden (or Golden-backed) Weaver, Jackson's (Golden-backed) Weaver.
TAXONOMY Shelley, 1888. Monotypic.
DISTRIBUTION [EC and E Africa. Stable. LC] Naturalised population established UAE (Cat C).

Ploceus benghalensis
Black-breasted Weaver (I)

OTHER NAMES Bengal (or Black-throated) Weaver.
TAXONOMY (Linnaeus, 1758). Monotypic.
DISTRIBUTION [S Asia. Stable. LC] Naturalised population established Qatar (Cat C).

Ploceus manyar
Streaked Weaver (I)

OTHER NAMES Striated (or Manyar) Weaver.
TAXONOMY (Horsfield, 1821). Four subspecies (undetermined in region).
DISTRIBUTION [S and SE Asia. Stable. LC] Naturalised populations established N Egypt, Qatar, Saudi Arabia and UAE (Cat C), recently also Oman.

Quelea quelea
Red-billed Quelea (V)

OTHER NAMES Black-faced (or Black-fronted/ Sudan) Dioch, Black-faced (or Pink-billed/ Sudan) Quelea.
TAXONOMY (Linnaeus, 1758). Three subspecies (one in region).
DISTRIBUTION [Sub-Saharan Africa. Stable. LC] Extreme vagrant: N Mauritania (7 May 2004, *quelea*).

Foudia madagascariensis
Red Fody (I)

OTHER NAMES Madagascar Red (or Cardinal) Fody.
TAXONOMY (Linnaeus, 1766). Monotypic.
DISTRIBUTION [Madagascar. Stable. LC] Naturalised population established Bahrain.

Euplectes afer
Yellow-crowned Bishop (V, I)

OTHER NAMES Black-winged (or Napoleon/ Golden) Bishop.
TAXONOMY (Gmelin, J F, 1789). Four subspecies (at least one in region).
DISTRIBUTION [Sub-Saharan Africa. Stable. LC] Extreme vagrant: N Mauritania (7 May 2004, *afer*). Naturalised population also established Portugal (Cat C) and has spread to Spain, where may now be sufficiently well established for admission to Cat C.

FAMILY ESTRILDIDAE

Waxbills, munias and allies · nine species

Lagonosticta senegala
Red-billed Firefinch (I)

OTHER NAMES Senegal (or Common) Firefinch.
TAXONOMY (Linnaeus, 1766). Six subspecies (one in region).
DISTRIBUTION [Sub-Saharan Africa. Stable. LC] Naturalised population established S Algeria (*rhodopsis*). Record from Morocco (3 May 1964, origin unknown) placed in Cat D.

Estrilda rufibarba
Arabian Waxbill (E)

TAXONOMY (Cabanis, 1851). Monotypic.
DISTRIBUTION [S Arabia. Decreasing. LC] Endemic resident SW Saudi Arabia and W and C Yemen.

Estrilda astrild
Common Waxbill (I)

OTHER NAMES Brown (or Barred) Waxbill, Black-lored (or Black-faced) Waxbill.
TAXONOMY (Linnaeus, 1758). Seventeen subspecies (undetermined in region).
DISTRIBUTION [Sub-Saharan Africa. Stable. LC] Naturalised populations established Azores, Canary Is, Cape Verde Is, Madeira, Portugal and Spain. Vagrant Gibraltar (from nearby Spanish population).

Amandava amandava
Red Avadavat (I)

OTHER NAMES Avadavat, Red Munia, Strawberry Finch.

TAXONOMY (Linnaeus, 1758). Three subspecies (one in region).
DISTRIBUTION [S and SE Asia. Stable. LC] Naturalised populations established Bahrain, Egypt, Iran, Israel, Italy, Portugal, Saudi Arabia and Spain (Cat C), and formerly also bred UAE (*amandava*). One record NE Libya.

Amandava subflava
Orange-breasted Waxbill (I?)

OTHER NAMES Zebra (or Golden-breasted) Waxbill.
TAXONOMY (Vieillot, 1819). Three subspecies (undetermined in region).
DISTRIBUTION [Sub-Saharan Africa. Stable. LC] Resident SW Yemen, where first recorded 1962 and now very numerous; considered unlikely to have reached SW Arabia naturally from Africa, and therefore probably introduced.

Euodice cantans
African Silverbill

OTHER NAMES (Warbling or Black-rumped) Silverbill.
TAXONOMY (Gmelin, J F, 1789). Two subspecies (both in region). Has been considered conspecific with Indian Silverbill.
DISTRIBUTION [Sahel, NE and E Africa to S Arabia. Stable. LC] Breeds very locally N Mauritania and S Algeria (*cantans*), where colonised as recently as 1996, and in SW Saudi Arabia, Yemen and S Oman (*orientalis*). Some dispersal outside breeding season.

Euodice malabarica
Indian Silverbill

OTHER NAME (Warbling or White-throated) Silverbill, White-throated Munia.
TAXONOMY (Linnaeus, 1758). Monotypic. Has been considered conspecific with African Silverbill.
DISTRIBUTION [SW and S Asia. Stable. LC] Resident S Iran; also breeds (by natural colonisation and/or introduction) Oman and UAE, with naturalised populations established Bahrain, Egypt, SE France, Israel, Jordan, Kuwait, Qatar and Saudi Arabia.

Lonchura punctulata
Scaly-breasted Munia (I)

OTHER NAMES Spice Finch, Nutmeg Mannikin, White-throated (or Spotted) Munia.
TAXONOMY (Linnaeus, 1758). Twelve subspecies (undetermined in region).
DISTRIBUTION [S and SE Asia. Stable. LC] Naturalised population established UAE (Cat C).

Lonchura malacca
Tricolored Munia (I)

OTHER NAMES Black-headed (or Chestnut) Munia, Chestnut (or Tricolored) Mannikin.
TAXONOMY (Linnaeus, 1766). Monotypic.
DISTRIBUTION [S and SE Asia. Stable. LC] Naturalised population established Portugal (Cat C); has also bred Spain.

FAMILY PRUNELLIDAE

Accentors · six species

Prunella collaris
Alpine Accentor

OTHER NAMES Collared (or Rock) Accentor.
TAXONOMY (Scopoli, 1769). Nine subspecies (three in region).
DISTRIBUTION [NW Africa and SW Europe to E Asia. Stable. LC] Breeds N Morocco, N Algeria (rare) and from Iberia discontinuously through montane S Europe and Alps to Slovenia and Carpathians (*collaris*), Croatia to Bulgaria, Greece, Crete and SW Turkey (*subalpina*) and from N and E Turkey to Caucasus and Iran (*montana*). Resident and partial migrant, some dispersing to winter more widely at lower altitudes *eg* C and S Europe, NW Africa, Israel and N Iraq (may also breed). Vagrant Belgium, Britain, Channel Is, Czech Republic, Denmark, Estonia, Finland, Jordan, Lebanon, Luxembourg,

Malta, Netherlands, Norway, Sweden and Syria.

Prunella montanella
Siberian Accentor

OTHER NAME Mountain Accentor.
TAXONOMY (Pallas, 1776). Two subspecies (one in region).
DISTRIBUTION [W Siberia to NE Russia > SE Asia. Stable. LC] Summer visitor N Ural Mtns (*montanella*). Vagrant Austria, Belarus, Czech Republic, Denmark, Finland, Italy, Lebanon, Lithuania, Luxembourg, Norway, Poland, Slovakia, Sweden, Turkey and Ukraine.

Prunella ocularis
Radde's Accentor (E)

OTHER NAMES Spot-throated (or Spotted) Accentor.
TAXONOMY (Radde, 1884). Monotypic.
DISTRIBUTION [SW Asia. Stable. LC] Endemic. Breeds S and E Turkey, Caucasus and N, SW and E Iran. Some altitudinal and short-distance migration; winters N Iraq (also recorded May) and N Israel (rare). Vagrant Jordan, Kuwait, Lebanon, Oman, Syria and UAE.

Prunella fagani
Arabian Accentor (E)

OTHER NAME Yemen Accentor.
TAXONOMY (Ogilvie-Grant, 1913). Monotypic.
DISTRIBUTION [SW Arabia. Stable. NT] Endemic resident W Yemen.

Prunella atrogularis
Black-throated Accentor

TAXONOMY (von Brandt, J F, 1843). Two subspecies (one in region).
DISTRIBUTION [NW Russia and C Asia > C and SW Asia. Stable. LC] Summer visitor N Ural Mtns (*atrogularis*); scarce winter visitor Iran. Vagrant Bulgaria, Croatia, Finland, France, Germany, Israel, Kuwait, Norway, Oman, Slovakia and Sweden.

Prunella modularis
Dunnock

OTHER NAMES Hedge Accentor, Hedge Sparrow.
TAXONOMY (Linnaeus, 1758). Eight subspecies (all in region).
DISTRIBUTION [Europe and N Middle East > W and S Europe, N Middle East. Stable. LC] Breeding endemic, ranging from SE, C and N Iberia to SC France, Italy and possibly Greece (*mabottii*), Ireland and Hebrides, NW Scotland (*hebridium*), elsewhere in Scotland, England, Wales and W France (*occidentalis*), C and N Europe E to N Balkans and Ural Mtns (*modularis*), S Balkans (*meinertzhageni*), S Crimea (*fuscata*), NW and N Turkey (*euxina*) and NE Turkey, Caucasus and N Iran (*obscura*). Populations from C Europe to Scandinavia and eastwards are migratory, wintering S through S Europe and N Middle East to coastal N Africa (local). Very scarce migrant Faroe Is (mainly autumn). Vagrant Algeria, Canary Is, Iceland, Jan Mayen, Kuwait, Morocco (occasionally winters) and Svalbard.

FAMILY MOTACILLIDAE

Wagtails and pipits · 23 species

Dendronanthus indicus
Forest Wagtail

OTHER NAME Tree Wagtail.
TAXONOMY (Gmelin, J F, 1789). Monotypic.
DISTRIBUTION [E Asia > S and SE Asia. Stable. LC] Rare migrant and winter visitor UAE. Vagrant Kuwait and Oman.

Motacilla flava
Western Yellow Wagtail

OTHER NAMES Yellow Wagtail; Ashy-headed (or Italian) Wagtail (*M f cinereocapilla*), Black-headed Wagtail (*M f feldegg*), Blue-headed Wagtail (*M f flava*), East Siberian Yellow Wagtail (*M f 'plexa'*), Egyptian Wagtail (*M f pygmaea*), Grey-headed Wagtail (*M f thunbergi*), Romanian Black-headed Wagtail (*M f 'dombrowskii'*), Spanish (or Iberian) Wagtail

(*M f iberiae*), Sykes's Wagtail (*M f beema*), White-headed Wagtail (*M f leucocephala*), White-throated Wagtail (*M f cinereocapilla/iberiae*), (British) Yellow Wagtail (*M f flavissima*), Yellow-headed (or Kirghiz Steppe/Yellow-backed) Wagtail (*M f lutea*).
TAXONOMY Linnaeus, 1758. Ten subspecies (all in region). Formerly included Eastern Yellow Wagtail (*qv*). Relationships among *flava-tschutschensis-citreola* taxa are complex and not yet settled; treatment here follows IOC but inevitably provisional. Several subspecies have previously been treated as full species, and in some regional countries still are (*eg* Germany and Netherlands), but intergradation frequent between certain populations, leading to variety of distinctive hybrid types, *eg* 'Channel Wagtail' (*flavissima* × *flava*, resembling *beema*), '*superciliaris*' (*feldegg* × *flava* or *beema*), '*dombrowskii*' (*feldegg* × *flava* or *beema*) and '*xanthophrys*' (*feldegg* × *lutea*). Eastern-breeding '*plexa*' is synonym of *thunbergi*.
DISTRIBUTION [N Africa and Europe to W and N Asia > sub-Saharan Africa to S Asia. Decreasing. LC] Widespread. Mainly summer visitor and passage migrant, breeding ranges including (very broadly) extreme S France, Balearic Is, Iberia, Morocco and N Mauritania (*iberiae*, also *cinereocapilla* SW Morocco; probably partly resident NW Africa), France, Benelux countries and Germany (*flava*), Britain and Ireland (*flavissima*), Scandinavia and N Russia (*thunbergi*), Italy and NW Balkans (*cinereocapilla*), Balkans, Turkey and Middle East (*feldegg*), and Egypt (*pygmaea*; probably mainly resident). Occurs widely outside breeding areas on migration throughout region, when *lutea* also regular on passage Levant and Arabia (vagrant Cyprus and Ukraine), *beema* on migration throughout Middle East and in small numbers S Arabia in winter, and *leucocephala* vagrant Azerbaijan, Iran, Israel, Kuwait, Turkey and UAE. Vagrant Western Yellow Wagtails also recorded Azores, Bear Is, Cape Verde Is, Iceland, Madeira and Svalbard.

Motacilla tschutschensis
Eastern Yellow Wagtail (V)

OTHER NAMES Yellow Wagtail; Alaskan (or American) Yellow Wagtail (*M t tschutschensis*), Kuril (or Green-headed/Green-backed) Wagtail (*M t taivana*), Manchurian (or Ussuri) Yellow Wagtail (*M t macronyx*).
TAXONOMY Gmelin, J F, 1789. Four subspecies (at least one in region). Formerly considered conspecific with Western Yellow Wagtail, but does not share common ancestry. Relationship of both to respective W and E forms of Citrine Wagtail, and corresponding shared DNA history, need full elucidation, as does suggestion of specific status for Eastern Yellow subspecies *taivana* (breeding SE Siberia to Japan).
DISTRIBUTION [E and NE Asia, Alaska > SE Asia to Australia. Population trend not assessed. NE] Extreme vagrant (nine records): Britain (3, of which two confirmed by mtDNA analysis, but see Collinson *et al* 2013 for caveats about (sub)specific identification: 9 Oct 1909; 4-19 Dec 2010; 10-14 Oct 2011); Ireland (12-28 Oct 2013*); Oman (5, *taivana*: 27 Jan 2002; 20 Feb 2003; 25 Feb 2010; 17-18 Mar 2011; 20 Sep 2013-22 Feb 2014, at least two).

Motacilla citreola
Citrine Wagtail

OTHER NAMES Black-backed Wagtail (*M c calcarata*), Yellow-headed (or Yellow-hooded) Wagtail (*M c citreola*).
TAXONOMY Pallas, 1776. Two subspecies (both in region). Genetically divergent W and E populations separated from each other at same time as did respective populations of 'Yellow Wagtail', latter group sharing some DNA history with Citrine, but full position unclear and *calcarata* not yet studied; more research needed.
DISTRIBUTION [E Europe to N and CE Asia > SW to SE Asia. Stable. LC] Summer visitor E Europe from Baltic States and Belarus E through European Russia and N Ukraine to Ural Mtns, also parts of Turkey, Caucasus, N Syria (*citreola*) and E Iran (*calcarata*); rare/irregular but increasing Finland and Poland. More widespread on passage E Balkans (fewer W to Italy), S Ukraine and Caucasus to Middle East; some winter Iran, Iraq and E Arabia. Vagrant Austria (has bred), Azores, Balearic Is, Belgium, Britain, Canary Is, Cape Verde Is, Channel Is, Croatia, Czech Republic, Denmark, Egypt, Faroe Is, France, Germany, Hungary, Iceland, Ireland, Italy, Madeira, Malta,

Morocco, Netherlands (has bred), Norway, Portugal, Slovakia (has bred), Slovenia, Spain, Switzerland (has bred) and Tunisia; *calcarata* vagrant Oman and Turkey.

Motacilla cinerea
Grey Wagtail

TAXONOMY Tunstall, 1771. Three subspecies (all in region).
DISTRIBUTION [W Europe to E Asia > W Europe, N and E Africa to S Asia. Stable. LC] Breeds Azores (*patriciae*), Madeira (*schmitzi*), Canary Is, NW Africa, S and W Europe N to W Scandinavia and E to Balkans, also Turkey, Caucasus and Iran (*cinerea*). Migratory in NE of range, wintering S to N Africa, Levant and Arabia. Vagrant Belarus, Iceland, Lithuania and Svalbard.

Motacilla alba
White Wagtail

OTHER NAMES Amur (or White-faced) Wagtail (*M a leucopsis*), Masked (or Turkestan) Wagtail (*M a personata*), Moroccan (Pied) Wagtail (*M a subpersonata*), (British) Pied Wagtail (*M a yarrellii*).
TAXONOMY Linnaeus, 1758. Nine subspecies (five in region). With very few exceptions contiguously distributed taxa interbreed, hence all subspecies currently usually treated as conspecific.
DISTRIBUTION [NW Africa and Eurasia > W, N and NE Africa and S Eurasia. Stable LC] Breeds Morocco (*subpersonata*), Britain, Ireland and locally adjacent coasts of continental Europe (*yarrellii*), Iceland, Faroe Is and across continental Europe from Mediterranean N to N Scandinavia and E to Ural Mtns, also Turkey, Caucasus, N Iraq and W Iran (*alba*), and E Iran (*personata*). Migratory in N and much of E of range, with *alba* wintering widely S and W Europe, N Africa and Middle East, *yarrellii* S to Iberia and NW Africa, and *personata* Oman and less commonly Qatar and UAE. Vagrant Azores, Bear Is, Cape Verde Is, Jan Mayen, Madeira and Svalbard (has bred); *leucopsis* (breeding extralimitally E Asia) vagrant Britain, Norway and Oman; *personata* vagrant Cyprus, Israel, Kuwait, Norway and Sweden; *subpersonata* vagrant Corsica, Gibraltar, Portugal and Spain; *yarrellii* vagrant N to Iceland, S to Canary Is and E to Cyprus.

Motacilla aguimp
African Pied Wagtail

OTHER NAME African Wagtail.
TAXONOMY Temminck, 1820. Two subspecies (one in region).
DISTRIBUTION [Sub-Saharan Africa. Stable. LC] Breeds extreme S Egypt (*vidua*).

Tmetothylacus tenellus
Golden Pipit (V)

TAXONOMY (Cabanis, 1878). Monotypic.
DISTRIBUTION [NE and E Africa. Stable. LC] Extreme vagrant: Oman (17-24 Jun 1983). Published occurrence in Yemen (Porter and Aspinall 2010) is erroneous.

Anthus richardi
Richard's Pipit

TAXONOMY Vieillot, 1818. Monotypic. Several subspecies have been described, but geographical variation slight and mainly clinal. May form superspecies with African Pipit, Paddyfield Pipit and extralimital New Zealand Pipit *A novaeseelandiae* (New Zealand, with *australis* latterly separated as Australian Pipit), all of which previously considered conspecific, but DNA studies do not support close relationship.
DISTRIBUTION [W Siberia to E Asia > SW to SE Asia. Stable. LC] Scarce migrant NW and W Europe (chiefly Sep-Nov, few spring), mainly Britain (annual mean 1990-2012 of 122), Fennoscandia and North Sea coasts of continental Europe from Germany S to France (annual mean 2003-08 of 54). Also regular passage and/or winter parts of SW Europe (notably W Iberia, S France and Italy), Middle East (including Turkey, Cyprus, Israel, Jordan, Kuwait and especially SE Arabia), and perhaps N Egypt. Vagrant Albania, Algeria, Austria, Azerbaijan, Balearic Is, Bulgaria, Canary Is, Corsica, Croatia, Czech Republic, Estonia, Faroe Is, Gibraltar, Greece, Hungary, Iran, Ire-

land, Latvia, Lebanon, Libya, Lithuania, Luxembourg, Malta, N Mauritania, Montenegro, Morocco (wintering in recent years), Poland, Qatar, Sardinia, Slovenia, Svalbard, Switzerland, Syria, Tunisia and Ukraine.

Anthus rufulus
Paddyfield Pipit (V)

OTHER NAMES Indian (or Oriental) Pipit.
TAXONOMY Vieillot, 1818. Five subspecies (undetermined in region). Has been treated as conspecific with Richard's Pipit (*qv*).
DISTRIBUTION [S and SE Asia. Stable. LC] Extreme vagrant (two records of single birds in different locations on same date, possibly indicating more regular presence in area): Iran (2: 12 Dec 2010; 12 Dec 2010).

Anthus cinnamomeus
African Pipit

OTHER NAMES Grassland (or Grassveld) Pipit.
TAXONOMY Rüppell, 1840. Fifteen subspecies (one in region). See comments under Richard's Pipit.
DISTRIBUTION [Sahel, NE, E and S Africa, SW Arabia. Population trend not assessed. NE] Resident SW Saudi Arabia (rare) and W Yemen (*eximius*).

Anthus godlewskii
Blyth's Pipit

OTHER NAME Godlewski's Pipit.
TAXONOMY (Taczanowski, 1876). Monotypic.
DISTRIBUTION [S Siberia to NW China > S Asia. Stable. LC] Migrant and winter visitor UAE (late Sep-mid Apr). Vagrant Bahrain, Belgium, Britain, Cyprus, Denmark, Estonia, Finland, France, Germany, Israel, Jordan, Kuwait, Netherlands, Norway, Oman, Portugal, Spain, Sweden and Turkey.

Anthus campestris
Tawny Pipit

TAXONOMY (Linnaeus, 1758). Monotypic.
DISTRIBUTION [NW Africa and SW Europe to CE Asia > Sahel to S Asia. Stable. LC] Summer visitor to NW Africa, continental Europe from Mediterranean discontinuously N to S Sweden and Baltic States (absent large areas of W Europe), E to Ukraine and N Caspian region, also Caucasus, Turkey, Levant and N Iran. Widespread on passage continental Europe S of Scandinavia, also Canary Is, N Africa and Middle East; regular on migration Sweden, very scarce migrant Britain. Winters Egypt (Nile Valley), most of Arabia, S Iraq and S Iran, a few also N Mauritania. Vagrant Azores, Cape Verde Is, Finland (has bred), Iceland, Ireland, Liechtenstein, Madeira and Norway.

Anthus similis
Long-billed Pipit

OTHER NAMES Brown Rock Pipit; Persian Rock Pipit (*A s decaptus*).
TAXONOMY (Jerdon, 1840). Twenty subspecies (four in region).
DISTRIBUTION [Sub-Saharan Africa and Middle East to SE Asia. Stable. LC] Breeds SE Egypt (*nivescens*), Israel, Jordan, Palestine, Syria and Lebanon (*captus*), SW, S and SE Arabia (*arabicus*) and S Iran (*decaptus*). Mainly resident but some altitudinal movement Arabia, where scarce winter visitor to lowland N Oman and coastal UAE. Vagrant Cyprus and Iraq.

Anthus pratensis
Meadow Pipit

TAXONOMY Linnaeus, 1758. Monotypic. Two subspecies have been described, but variation now regarded as mainly clinal.
DISTRIBUTION [W and N Europe to W Siberia > N Africa, W and S Europe and Middle East. Decreasing. NT] Breeds across W and N Europe from W and C France N to Iceland, Jan Mayen and N Scandinavia, and E to Ural Mtns and Ukraine. Except in Britain, Ireland, continental W Europe and Caucasus, breeding population wholly migratory, occurring widely on passage and wintering through S Europe to N Africa and Middle East S to C and E Arabia (rare Oman). Vagrant Azores, Bear Is, Madeira, N Mauritania and Svalbard (has bred).

Anthus trivialis
Tree Pipit

OTHER NAMES Eurasian (or Brown) Tree Pipit.
TAXONOMY (Linnaeus, 1758). Two subspecies (one in region).
DISTRIBUTION [W Europe to E Asia > sub-Saharan Africa and S Asia. Decreasing. LC] Summer visitor, breeding across much of Europe (though absent Iceland, Ireland, S Iberia and most Mediterranean islands), also N Turkey, Caucasus and N Iran (*trivialis*). Widespread S of breeding range on passage, when also occurs *eg* Faroe Is, Gibraltar and Ireland. Rare in winter S Arabia. Vagrant Azores*, Cape Verde Is, Iceland, Jan Mayen, Madeira and Svalbard.

Anthus hodgsoni
Olive-backed Pipit

OTHER NAMES Indian (or Hodgson's/Oriental) Tree Pipit; Siberian Tree Pipit (*A h yunnanensis*).
TAXONOMY Richmond, 1907. Two subspecies (one in region).
DISTRIBUTION [NE Europe to E Asia > S and SE Asia. Stable. LC] Summer visitor NE European Russia (*yunnanensis*). Reaches W Europe annually in autumn and now considered scarce migrant Britain; rare migrant and winter visitor Israel, Oman and UAE. Vagrant Austria, Bahrain, Balearic Is, Belgium, Canary Is, Channel Is, Cyprus, Denmark, Estonia, Faroe Is, Finland, France, Germany, Iceland, Iran, Ireland, Israel, Italy, Jordan, Kuwait, Malta, Morocco, Netherlands, Norway, Poland, Portugal, Saudi Arabia, Spain, Sweden, Switzerland and Turkey.

Anthus gustavi
Pechora Pipit

OTHER NAME Siberian Pipit.
TAXONOMY Swinhoe, 1863. Two subspecies (one in region).
DISTRIBUTION [NW Russia to NE Russia > Philippines, Borneo and Wallacea. Stable. LC] Summer visitor NE European Russia (*gustavi*). Vagrant Britain, Faroe Is, Finland, France, Germany, Iceland, Ireland, Lithuania, Norway and Sweden.

Anthus cervinus
Red-throated Pipit

TAXONOMY (Pallas, 1811). Monotypic.
DISTRIBUTION [N Eurasia and W Alaska > Mediterranean basin, W, NC, NE and E Africa, SE Asia. Stable. LC] Summer visitor N Norway, Sweden, N Finland (rare) and far N Russia. Occurs widely on passage S through region W to Canary Is, but scarce or rare W Europe spring and autumn. Winters locally in C and E Mediterranean (rarer in W) and parts of N Africa (irregular Canary Is) and Middle East. Vagrant Azerbaijan, Azores, Cape Verde Is, Faroe Is, Gibraltar, Ireland, Liechtenstein, Luxembourg, Madeira, Moldova, Portugal and Svalbard.

Anthus rubescens
Buff-bellied Pipit

OTHER NAMES American (Buff-bellied) Pipit (*A r rubescens*), Asian (or Siberian) Buff-bellied Pipit (*A r japonicus*).
TAXONOMY (Tunstall, 1771). Three subspecies (two in region). Has been considered conspecific with Water and Rock Pipits, but differs in genetics, morphology and ecology. Populations in America (*rubescens* and extralimital *alticola*) are distinct in mtDNA, plumage and perhaps voice from those in Asia (*japonicus*), and may represent separate species.
DISTRIBUTION [N and W North America, NE Siberia > SW and S North America, S and E Asia. Decreasing. LC] Rare migrant and winter visitor Israel (*japonicus*); may be more regular in Middle East, where separation from Water Pipit is problematic. *Japonicus* vagrant Azerbaijan, Cyprus, Egypt, Georgia, Iran, Italy, Jordan, Kuwait (possibly now regular in winter), Norway, Oman, Svalbard, Syria, Tunisia, Turkey and UAE (perhaps rare winter visitor). *Rubescens* vagrant (mainly autumn, rare winter) Azores, Britain, Channel Is, Faroe Is, France, Germany, Iceland, Ireland, Norway, Spain and Sweden.

Anthus spinoletta
Water Pipit

OTHER NAMES Mountain Pipit; Asian Water Pipit (*A s blakistoni*), Caucasian Water Pipit (*A s coutellii*), European Water Pipit (*A s spinoletta*).

TAXONOMY (Linnaeus, 1758). Three subspecies (all in region). Formerly included European Rock and Buff-bellied Pipits (*qv*). Based on distinctive call, plumage differences and preliminary genetic analysis, *coutellii* may represent separate species; *blakistoni* (breeding extralimitally C and CE Asia) also genetically distinct. Further sampling required to confirm whether species status merited for all three taxa.
DISTRIBUTION [W Europe to C Asia > W Europe and N Africa to E Asia. Stable. LC] Breeds montane SW, S, C, SE and parts of E Europe (*spinoletta*), and Turkey, Caucasus and N Iran (*coutellii*). Winters lowland W Europe from Iberia N to Britain and Denmark, rare S Sweden; also EC, S and SE Europe, Mediterranean islands, N Africa and much of Middle East (except SW Arabia) to Iran, where Asian *blakistoni* also winters. Vagrant Belarus, Canary Is, Estonia, Gibraltar, Ireland, Jan Mayen, Latvia, Lithuania, Norway and Svalbard; this species or Rock Pipit also recorded N Mauritania.

Anthus petrosus
European Rock Pipit (E)

OTHER NAMES Eurasian Rock Pipit (IOC), Rock Pipit; British Rock Pipit (*A p petrosus*), Scandinavian Rock Pipit (*A p littoralis*).
TAXONOMY (Montagu, 1798). Two subspecies.
DISTRIBUTION [N and W Europe > W Europe. Stable. LC] Breeds coastally NW France, Channel Is, Britain, Ireland and Faroe Is (*petrosus*) and Fennoscandia, Estonia (scarce) and Kola Peninsula (*littoralis*). Former subspecies mainly resident or short-distance migrant, latter largely migratory, occurring Baltic coasts on passage and wintering coastal W and SW Europe from Germany (has bred) to W Iberia. Vagrant Algeria, Balearic Is, Canary Is, Corsica, Czech Republic, Gibraltar, Greece, Iceland (has bred), Italy, Jan Mayen, Malta, Morocco, Romania, Svalbard and Tunisia; this species or Water Pipit also recorded N Mauritania.

Anthus berthelotii
Berthelot's Pipit (E)

OTHER NAMES Canarian (or Canary Islands) Pipit.
TAXONOMY Bolle, 1862. Monotypic.
DISTRIBUTION [Atlantic islands. Stable. LC] Endemic resident Canary Is, Salvage Is and Madeira.

FAMILY FRINGILLIDAE

Finches · 40 species

Fringilla coelebs
Common Chaffinch

OTHER NAMES (Eurasian or European) Chaffinch; African Chaffinch (*F c africana/spodiogenys*), Atlas Chaffinch (*F c africana*), Azores Chaffinch (*F c moreletti*), Canaries (or Canarian/Canary Islands) Chaffinch (*F c canariensis/ombriosa/palmae*), Madeiran Chaffinch (*F c maderensis*), Tunisian Chaffinch (*F c spodiogenys*).
TAXONOMY Linnaeus, 1758. Nineteen subspecies (all in region). Populations in Atlantic islands and N Africa differ from European congeners in plumage, voice and ecology, and some (*eg* Canary Is subspecies and *spodiogenys* of N Africa) are significantly distinct genetically, but further study needed to establish full extent of divergence in all populations.
DISTRIBUTION [N Africa and Europe to C Asia > W and S Europe, N Africa and Middle East. Decreasing. LC] Breeds Canary Is on El Hierro (*ombriosa*), La Palma (*palmae*) and La Gomera, Tenerife and Gran Canaria (*canariensis*), Madeira (*maderensis*), Azores (*moreletti*), Morocco to NW Tunisia (*africana*), E Tunisia and NW Libya (*spodiogenys*), NE Libya (*harterti*), Iberia and Balearic Is (*balearica*), Sardinia (*sarda*), Corsica (*tyrrhenica*), Britain and Ireland (*gengleri*), continental Europe from France N to N Scandinavia and E to Ural Mtns (*coelebs*) except most of Balkans, N Turkey, C and E Caucasus and NW Iran (*caucasica*), S Greece, Crete and W Turkey (*schiebeli*), Cyprus and SE Turkey to Jordan and N Iraq (*syriaca*), Crimea and SW Caucasus (*solomkoi*), N Iran (*alexandrovi*) and NE Iran (*transcaspia*). Scandinavian, E European and SW Asian populations mainly migratory, wintering S through Europe into N Africa and in N Middle East. Regular migrant Faroe Is

(has bred), rare Iceland (has bred). Vagrant Bear Is, Svalbard and UAE; *africana* vagrant Gibraltar, N Mauritania and Netherlands; *spodiogenys* vagrant Italy (Pantelleria).

Fringilla teydea
Blue Chaffinch (E)

OTHER NAMES Canary Islands (or Mount Teyde/Teydean) Chaffinch; Gran Canaria Blue Chaffinch (*F t polatzeki*), Tenerife Blue Chaffinch (*F t teydea*).
TAXONOMY Webb, Berthelot and Moquin-Tandon, 1836. Two subspecies. Newly identified diagnostic distinctions in plumage, songs, calls and morphometrics between geographically isolated forms on Gran Canaria and Tenerife, as well as previously reported differences in mtDNA, strongly suggest that *polatzeki* and *teydea* represent separate species; IOC yet to assess evidence, so long-standing treatment as single polytypic species maintained *pro tem*.
DISTRIBUTION [C Canary Is. Increasing. NT] Endemic resident Canary Is on Tenerife (*teydea*) and Gran Canaria (*polatzeki*).

Fringilla montifringilla
Brambling

TAXONOMY Linnaeus, 1758. Monotypic.
DISTRIBUTION [N Europe to NE Asia > N and W Europe to E Asia. Decreasing. LC] Summer visitor from C and N Scandinavia E through NE Europe. Some remain W Scandinavia, otherwise widespread on migration S of breeding range W to Faroe Is (has bred), more rarely Iceland (has bred); winters W, C and E Europe, N Morocco, N Algeria and N Middle East locally S to Levant and Gulf. Vagrant Azores, Canary Is, Jan Mayen, Madeira, N Mauritania, Svalbard, Tunisia and UAE.

Mycerobas carnipes
White-winged Grosbeak

TAXONOMY (Hodgson, 1836). Three subspecies (one in region).
DISTRIBUTION [WC Asia to W China. Stable. LC] Resident N Iran (*speculigerus*). Some altitudinal movement in winter, when may descend to lower elevations.

Hesperiphona vespertina
Evening Grosbeak (V)

TAXONOMY (Cooper, W, 1825). Three subspecies (undetermined in region).
DISTRIBUTION [N and W North America > S North America. Decreasing. LC] Extreme vagrant (four records, all in spring): Britain (2: 26 Mar 1969; 10-25 Mar 1980); Norway (2: 2-9 May 1973; 17-26 May 1975).

Coccothraustes coccothraustes
Hawfinch

TAXONOMY (Linnaeus, 1758). Six subspecies (three in region).
DISTRIBUTION [NW Africa and W Europe to E Asia. Stable. LC] Breeds NW Africa (*buryi*), Britain and continental Europe N to S Scandinavia, E to Ural Mtns, S to much of Iberia, France, Italy and Balkans (*coccothraustes*) and S Ukraine, N Turkey, Caucasus and N Iran (*nigricans*). C Swedish and E populations migratory, wintering S through Europe and N Middle East to Levant, NE Egypt and N Libya. Vagrant Canary Is, Gibraltar, Faroe Is, Iceland, Jan Mayen, Kuwait, Madeira, Qatar, Saudi Arabia, Svalbard and Syria.

Pinicola enucleator
Pine Grosbeak

OTHER NAME Pine Rosefinch.
TAXONOMY (Linnaeus, 1758). Ten subspecies (one in region).
DISTRIBUTION [W and N North America, N Eurasia. Decreasing. LC] Breeds N Fennoscandia and far N European Russia (*enucleator*). Resident and short-distance migrant, usually wintering SW as far as N and E Baltic, but occasionally irruptive when numbers augmented from farther E, and range may penetrate farther S and W into Europe. Vagrant Austria, Belarus, Britain, Czech Republic, Denmark, France, Germany, Hungary, Italy, Netherlands, Poland, Slovakia, Slovenia, Switzerland and Ukraine.

Pyrrhula pyrrhula
Eurasian Bullfinch

OTHER NAMES (Common) Bullfinch; Northern Bullfinch (*P p pyrrhula*).
TAXONOMY (Linnaeus, 1758). Eleven subspecies (seven in region). Formerly included Azores Bullfinch (*qv*).
DISTRIBUTION [Widespread Eurasia. Decreasing. LC] Breeds N Iberia and SW France (*iberiae*), Britain and Ireland (*pileata*), rest of W Europe (*europaea*), SC and E Europe N to N Fennoscandia and E to Ural Mtns (*pyrrhula*), NW Turkey (*paphlagoniae*), NE Turkey and Caucasus (*rossikowi*), and Azerbaijan and N Iran (*caspica*). Population in far NE Europe migratory and occasionally irruptive, reaching NW Europe and even Spain. Overall winter range extends through much of S Europe and more widely Turkey and Iran. Scarce on passage Faroe Is. Vagrant Algeria, Balearic Is, Ceuta, Corsica, Cyprus, Gibraltar, Iceland, Malta, Morocco, Sardinia, Svalbard and Tunisia.

Pyrrhula murina
Azores Bullfinch (E)

TAXONOMY Godman, 1866. Monotypic. Formerly treated as conspecific with Eurasian Bullfinch, but differs in genetics, plumage (including lack of sexual dimorphism), structure, voice and ecology.
DISTRIBUTION [Azores. Stable. EN] Resident endemic Azores, where confined to E São Miguel.

Rhodopechys sanguineus
Eurasian Crimson-winged Finch

OTHER NAMES Asian Crimson-winged Finch, Crimson-winged (Desert) Finch.
TAXONOMY (Gould, 1838). Monotypic. Formerly included African Crimson-winged Finch (*qv*).
DISTRIBUTION [SW and C Asia. Stable. LC] Resident from C Turkey S to Lebanon, Syria and N Israel, and E to Caucasus and Iran. Some dispersal in winter, when recorded more widely *eg* Syria and W Turkey. Vagrant Cyprus and Iraq.

Rhodopechys alienus
African Crimson-winged Finch (E)

OTHER NAME Crimson-winged (Desert) Finch.
TAXONOMY Whitaker, 1897. Monotypic. Previously considered conspecific with allopatric Eurasian Crimson-winged Finch, but shows constant differences in plumage and morphometric data.
DISTRIBUTION [NW Africa. Stable. LC] Endemic resident NW Morocco and NE Algeria.

Bucanetes githagineus
Trumpeter Finch

OTHER NAME Common Trumpeter Finch.
TAXONOMY (Lichtenstein, M H C, 1823). Four subspecies (all in region).
DISTRIBUTION [NW Africa to SW Asia. Stable. LC] Breeds C and E Canary Is (*amantum*), S Spain, NW and NC Africa (*zedlitzi*), Egypt (*githagineus*) and discontinuously through Middle East to SE Turkey, S Arabia and E Iran (*crassirostris*). Resident but nomadic, dispersive and also partial migrant in E; scarce or infrequent visitor S Italy and Malta, uncommon in winter Iraq (has bred). Vagrant Austria, Balearic Is, Britain, Bulgaria, Cape Verde Is, Channel Is, Corsica, Cyprus, Czech Republic, Denmark, Estonia, Finland, France, Gibraltar, Greece, Netherlands, Norway, Portugal, Qatar, Romania, Sweden and Switzerland.

Bucanetes mongolicus
Mongolian Finch

OTHER NAME Mongolian Trumpeter Finch.
TAXONOMY (Swinhoe, 1870). Monotypic.
DISTRIBUTION [SW, C and CE Asia. Stable. LC] Breeds Armenia, Azerbaijan, E Turkey and Iran. Vagrant Bahrain, Kuwait and Saudi Arabia.

Carpodacus erythrinus
Common Rosefinch

OTHER NAMES Scarlet Grosbeak, Scarlet Rosefinch.
TAXONOMY (Pallas, 1770). Five subspecies (two in region).

DISTRIBUTION [C Europe to E Asia > S and SE Asia. Stable. LC] Summer visitor, breeding mainly from C and E Europe (rare N Balkans) and S Scandinavia E to Ural Mtns (*erythrinus*) and Turkey, Caucasus and N Iran (*kubanensis*). Occasional spring influxes occur W of range, with irregular breeding *eg* Britain, France and Netherlands. Migrates E and SE to winter quarters, partly through N Middle East; scarce migrant W Europe and Italy (has bred). Rare in winter Arabia. Vagrant Balearic Is, Belgium, Bosnia-Herzegovina, Canary Is, Ceuta, Croatia, Cyprus, Egypt, Faroe Is, Gibraltar, Greece, Hungary, Iceland, Iraq, Ireland, Jan Mayen, Jordan, Luxembourg, Malta, Moldova, Morocco, Portugal, Spain, Svalbard and Syria.

Carpodacus rubicilla
Great Rosefinch

OTHER NAMES Caucasian Great (or Spot-crowned) Rosefinch.
TAXONOMY (Güldenstädt, 1775). Four subspecies (one in region). Nominate *rubicilla* endemic to Caucasus and sometimes considered specifically distinct from extralimital Asian subspecies, which collectively are then treated as Spotted Great Rosefinch *C severtzovi*.
DISTRIBUTION [SE Europe, SW, C and EC Asia. Stable. LC] Endemic resident and altitudinal migrant Greater Caucasus Mtns of Azerbaijan, Georgia and extreme SW Russia (*rubicilla*).

Carpodacus synoicus
Sinai Rosefinch (E)

TAXONOMY (Temminck, 1825). Monotypic. Formerly included extralimital Pale Rosefinch *C stoliczkae* (Asia).
DISTRIBUTION [Middle East. Stable. LC] Breeds NE Egypt, Palestine, S Israel, W Jordan and NW Saudi Arabia. Mainly resident but nomadic and partial altitudinal migrant, dispersing to lower levels in winter.

Carpodacus sibiricus
Long-tailed Rosefinch

OTHER NAME Siberian Rosefinch.
TAXONOMY (Pallas, 1773). Five subspecies (one in region).
DISTRIBUTION [W Siberia to E Asia. Stable. LC] Breeds extreme E European Russia in S Ural Mtns; also common migrant and scarce winter visitor C and S Urals (*sibiricus*). Other records in Europe, *eg* Britain, Denmark, Finland, Netherlands and Sweden, all now attributed to escapes from captivity.

Carpodacus roseus
Pallas's Rosefinch (V)

OTHER NAME Siberian Rosefinch.
TAXONOMY (Pallas, 1776). Two subspecies (one in region).
DISTRIBUTION [SC Siberia to NE Asia > S Siberia and EC Asia to E Asia.] Extreme vagrant (nine records of 12 individuals, *roseus*): Hungary (1 Dec 1850, caught, died in captivity); Russia (5: Nov 1900, collected; 9 Dec 1995, trapped; 24 Nov 1998, collected; 16 Nov 2005, three, one of which caught; 25-26 Jan 2011); Ukraine (3: 15 Nov 1902, collected; 13 Dec 1902, two collected; late Nov 1927, collected). Other claims from 12 countries in region considered doubtful (11 records), erroneous (8) or of dubious origin (21).

Chloris chloris
European Greenfinch

OTHER NAMES (Common or Western) Greenfinch.
TAXONOMY (Linnaeus, 1758). Ten subspecies (nine in region), but variation minor and clinal and some authors recognise as few as four.
DISTRIBUTION [N Africa and Europe to WC Asia. Decreasing. LC] Near-endemic. Breeds C Morocco and N Algeria (*voousi*), NW Morocco, Portugal and NW Spain (*vanmarli*), elsewhere in Spain and S France E through S Europe to W Greece (*aurantiiventris*), Corsica and Sardinia (*madaraszi*), Britain (except N Scotland) and Ireland (*harrisoni*), N Scotland and continental Europe from N France to N Scandinavia and E through C, N and E Europe to Ural Mtns and N Ukraine (*chloris*), S Ukraine and Caucasus to N Iran, N Iraq and probably NE Turkey (*bilkevitchi*), Balkans and W Turkey (*muehlei*), and SC Turkey

S through Levant to NE Egypt (*chlorotica*). Migratory from C and N Scandinavia to Ural Mtns and in parts of E Turkey and Caucasus; winter range extends along N African coast and through much of N Middle East. Vagrant Faroe Is, Iceland, Kuwait, N Mauritania and Svalbard.

Rhodospiza obsoleta
Desert Finch

OTHER NAMES Black-billed Finch, Lichtenstein's (Desert) Finch.
TAXONOMY (Lichtenstein, M H C, 1823). Monotypic.
DISTRIBUTION [SW to EC Asia. Stable. LC] Breeds SE Turkey, Iran, Syria, Israel, Palestine, Jordan and N and C Saudi Arabia. Resident, with some dispersal; also partial migrant, wintering Lebanon, Iraq (may breed in N) and irregularly NE Egypt* (has bred). Vagrant Armenia, Azerbaijan, Kuwait (has bred) and Qatar.

Rhynchostruthus percivali
Arabian Grosbeak (E)

OTHER NAMES Arabian Golden-winged Grosbeak (IOC), Golden-winged Grosbeak.
TAXONOMY Ogilvie-Grant, 1900. Monotypic. Previously treated as conspecific with extralimital Socotra Grosbeak *R socotranus* (Socotra) and Somali Grosbeak *R louisae* (NE Africa), but differs in plumage and size.
DISTRIBUTION [S Arabia. Decreasing. NT] Endemic resident SW Saudi Arabia, W Yemen and SW Oman.

Crithagra rothschildi
Arabian Serin (E)

OTHER NAMES Arabian Canary, Olive-rumped Serin.
TAXONOMY (Ogilvie-Grant, 1902). Monotypic. Previously included within extralimital Yellow-rumped Seedeater *C xanthopygia* (NE Africa). N population sometimes described as *C r philbyi*.
DISTRIBUTION [W Arabia. Stable. LC] Endemic resident W Saudi Arabia and W Yemen.

Crithagra menachensis
Yemen Serin (E)

OTHER NAMES Menacha Seedeater, Menacha Serin.
TAXONOMY (Ogilvie-Grant, 1913). Monotypic.
DISTRIBUTION [W and S Arabia. Stable. LC] Endemic resident SW Saudi Arabia, W Yemen and SW Oman.

Linaria flavirostris
Twite

OTHER NAMES Caucasian (or Turkish) Twite (*L f brevirostris*).
TAXONOMY (Linnaeus, 1758). Eleven subspecies (five in region).
DISTRIBUTION [NW Europe, SW to CE Asia. Stable. LC] Breeds Ireland and Britain (*pipilans*) except Outer Hebrides in W Scotland (*bensorum*), W and N Scandinavia and extreme NW Russia (*flavirostris*), N Caspian (*kirghizorum*) and Turkey, Caucasus and N Iran (*brevirostris*). Mainly migratory, especially in NW Europe, leaving uplands to winter *eg* S Baltic and North Sea coasts, E and C Europe and S locally to Balkans, but irregular Italy; also more widespread in winter E Turkey, Iran and Iraq. Vagrant Bosnia-Herzegovina, Croatia, Faroe Is (has bred), Portugal, Romania, Serbia, Slovenia, Spain, Svalbard, Switzerland and Ukraine; *brevirostris* vagrant Bulgaria.

Linaria cannabina
Common Linnet

OTHER NAMES (Eurasian) Linnet.
TAXONOMY (Linnaeus, 1758). Seven subspecies (all in region).
DISTRIBUTION [NW Africa and W Europe to C Asia. Decreasing. LC] Breeds W and C Canary Is (*meadewaldoi*), E Canary Is (*harterti*), Madeira (*nana*), N Africa, Iberia, S Italy, Balkans and Mediterranean islands (*mediterranea*), widely elsewhere in Europe from Scotland (*autochthona*), England, Wales, Ireland, France and N Italy N to C Scandinavia and E to Ukraine, Ural Mtns and NW Kazakhstan (*cannabina*), and Crimea, Caucasus, Turkey,

Levant, N Iraq and Iran (*bella*). Scandinavian, E European and some Caucasian/Middle Eastern populations migratory, wintering S through W and S Europe, N Africa and N Middle East. Vagrant Azores*, Bahrain, Faroe Is, Iceland, N Mauritania and UAE.

Linaria yemenensis
Yemen Linnet (E)

OTHER NAME Arabian Linnet.
TAXONOMY (Ogilvie-Grant, 1913). Monotypic.
DISTRIBUTION [SW Arabia. Stable. LC] Endemic resident SW Saudi Arabia and W Yemen.

Acanthis flammea
Common Redpoll

OTHER NAMES Redpoll; Greenland (or Greater) Redpoll (*C f rostrata*), Iceland (or Icelandic) Redpoll (*C f 'islandica'*), Lesser Redpoll (*C f cabaret*), Mealy Redpoll (*C f flammea*), Northwestern Redpoll (*C f rostrata/'islandica'*), Pale Iceland (or Iceland Arctic) Redpoll (unnamed pale form breeding in Iceland).
TAXONOMY (Linnaeus, 1758). Three subspecies (all in region). Taxonomy unsettled, with *C f cabaret* sometimes treated as a full species, *C f 'islandica'* often recognised as a valid subspecies (here subsumed within *rostrata*, following IOC) and further unnamed pale taxon occurring Iceland; Arctic Redpoll has also been treated as conspecific.
DISTRIBUTION [N North America and N Eurasia. Decreasing. LC] Breeds Ireland, Britain and adjacent NW continental Europe discontinuously to Alps and S Scandinavia (*cabaret*), through N Scandinavia and NW Russia (*flammea*), and in Iceland (*rostrata*, but see Taxonomy above). Partly migratory, winter range extending S across Europe to *eg* C France, N Italy, N Balkans, Crimea, Caucasus and NW Kazakhstan; occasionally irruptive, with large influxes S of normal range. *Rostrata* and *'islandica'* occur on passage and in winter Faroe Is and to lesser extent N Scotland, with pale Iceland types also provisionally documented Britain in winter. Vagrant Albania, Azerbaijan, Azores, Balearic Is, Bosnia-Herzegovina, Ceuta, Cyprus, Jan Mayen, Macedonia, Malta, Montenegro, Morocco, Portugal, Spain and Turkey; many extralimital occurrences not assigned to subspecies, but *cabaret* documented Balearic Is, Faroe Is, Macedonia, Malta, Serbia and Spain, *flammea* Albania, Azores, Bosnia-Herzegovina, Cyprus, Malta, Sardinia and Turkey, and *rostrata* Norway.

Acanthis hornemanni
Arctic Redpoll

OTHER NAMES Hoary (or Tundra) Redpoll; Coues's Arctic Redpoll (*C h exilipes*), Hornemann's (Arctic) Redpoll (*C h hornemanni*).
TAXONOMY (Holböll, 1843). Two subspecies (both in region). Sometimes treated as conspecific with Common Redpoll.
DISTRIBUTION [Circumpolar in Arctic > N North America and N Eurasia. Population trend not assessed. NE] Breeds far N Scandinavia and extreme NW Russia, possibly regularly also Svalbard (*exilipes*). Winters S through Scandinavia and E Baltic to Estonia, but occurs more widely NW Europe in irruption years, when very scarce visitor Britain. Vagrant Azores, Belgium, Bosnia-Herzegovina, Czech Republic, Faroe Is, France, Germany, Hungary, Iceland, Ireland, Jan Mayen, Latvia, Lithuania, Netherlands, Poland, Slovakia, Svalbard (has bred) and Ukraine; *hornemanni* (breeding Arctic Canada and Greenland) vagrant Belgium, Britain, France, Germany, Iceland, Netherlands, Norway and Sweden.

Loxia pytyopsittacus
Parrot Crossbill

TAXONOMY von Borkhausen, 1793. Monotypic.
DISTRIBUTION [N Europe to NW Siberia. Stable. LC] Near-endemic. Breeds Norway, Sweden, Finland, Estonia, Latvia and NW Russia E to Ural Mtns; recent colonist N Scotland (rare). Mainly resident but dispersive within Baltic region, occurring *eg* Belarus (has bred), Denmark and Lithuania. Periodically also irruptive in response to food shortages, making less extensive movements S and W of breeding range than Common Crossbill, and later in autumn. Vagrant Austria, Belgium, S Britain (has bred), Croatia, Czech Republic, France,

Germany, Iceland, Italy, Netherlands, Poland (has bred), Serbia, Slovenia and Ukraine.

Loxia scotica
Scottish Crossbill (E)

TAXONOMY Hartert, 1904. Monotypic. Has been considered conspecific with Common Crossbill (*qv*).
DISTRIBUTION [Scotland. Stable. LC] Endemic resident N Scotland; locally nomadic.

Loxia curvirostra
Common Crossbill

OTHER NAMES Red Crossbill (IOC); Atlas Crossbill (*L c poliogyna*), Balearic Crossbill (*L c balearica*), Balkan (or Cyprus) Crossbill (*L c guillemardi*), Corsican Crossbill (*L c corsicana*); additional voice-related informal names, denoted by letters A-F, applied to distinct vocal types in Europe (similar naming system also used for extralimital North American vocal types 1-8).
TAXONOMY Linnaeus, 1758. Twenty subspecies (five in region). Current arrangement of one species with 11 Eurasian and nine North American subspecies in need of review, following discovery of distinct vocal types across range, some of which may represent cryptic taxa. In Europe, six such types identified, with evidence of assortative pairing and sympatric breeding, as well as subtle but distinct morphological traits (*eg* differences in bill structure and plumage coloration). Some European vocal types may align to existing subspecies, as in North America, but others may represent new (sub)species; further research required. Basic range areas of currently recognised subspecies outlined below; see Robb (2000) for discussion and potential implications for crossbill taxonomy and distribution.
DISTRIBUTION [North America, NW Africa and Eurasia. Stable. LC] Breeds NW Africa and S Italy (*poliogyna*), Balearic Is, E, C and N Spain and irregularly Portugal (*balearica*), Corsica (*corsicana*), Ireland, Britain and parts of continental Europe from SE France, C Italy and N Balkans N to N Scandinavia and E to Ural Mtns (*curvirostra*), and E Balkans, Turkey, Cyprus, Caucasus and S Ukraine (*guillemardi*). Resident and dispersive, with periodic widespread irruptions linked to state of conifer seed crop in N of range and/or from farther E in Siberia, hence also fluid breeding distribution; in largest movements may occur across much of Europe. Vagrant Azores, Canary Is, Iceland (has bred), Iran, Israel, Jan Mayen, Jordan, Madeira, Malta and Svalbard.

Loxia leucoptera
Two-barred Crossbill

OTHER NAME White-winged Crossbill.
TAXONOMY Gmelin, J F, 1789. Two subspecies (one in region).
DISTRIBUTION [N North America and N Eurasia. Stable. LC] Breeds N Norway, N Sweden, Finland and NW Russia (*bifasciata*). Mainly resident but dispersive, when recorded *eg* Baltic States and Denmark. Periodically undertakes larger-scale food-related irruptive movements, occurring in numbers W and S of usual range. Vagrant Austria, Belarus, Belgium, Britain, Bulgaria, Czech Republic, France, Germany (has bred), Hungary, Iceland, Ireland, Italy, Netherlands, Poland, Slovakia, Slovenia, Svalbard and Ukraine.

Carduelis carduelis
European Goldfinch

OTHER NAMES Goldfinch; Eastern (or Grey-headed) Goldfinch ('*caniceps* group').
TAXONOMY (Linnaeus, 1758). Fourteen subspecies (11 in region). Grey-headed subspecies from Asia (*paropanisi* and *ultima*, breeding Iran, and extralimital *caniceps* and *subalata*) probably split from *carduelis* subspecies group during Pleistocene glaciations, and are sometimes separated at species level as Eastern Goldfinch *C caniceps*; more research needed.
DISTRIBUTION [Europe and N Africa to S Siberia > Europe, N Africa and SW Asia. Stable. LC] Breeds Azores, Madeira, Canary Is, NW Africa and N Libya, SW Europe and Balearic Is (*parva*), Ireland, Britain and coastally from W France to W Netherlands (*britannica*), continental Europe from C France and Italy N to S Scandinavia and E to N Balkans and C European Russia (*carduelis*), Corsica, Sardinia and Sicily (*tschusii*), elsewhere in Balkans, Crete and NW

Turkey (*balcanica*), S Ukraine, S Russia and NW Kazakhstan (*volgensis*), Crimea, N and SW Caucasus (*colchica*), S Caucasus, E Turkey and N Iran (*brevirostris*), NE Iran (*paropanisi*), S Iran (*ultima*) and elsewhere in Turkey, SW Iran and N Iraq S through Cyprus and Levant to NW Saudi Arabia, N Egypt and Nile Valley (*niediecki*). Partly migratory, some populations in N and especially E of range wintering to W and S in Europe and N Middle East.

Carduelis citrinella
Citril Finch (E)

OTHER NAMES Alpine Citril Finch, European Citril.
TAXONOMY (Pallas, 1764). Monotypic. Formerly included Corsican Finch (*qv*).
DISTRIBUTION [SW and C Europe. Stable. LC] Endemic. Breeds montane Spain, SW and E France and Alps E to Slovenia. Mainly resident but partially migratory and dispersive, with movement between Alps and Spain, also N Italy, confirmed by ringing. Vagrant Balearic Is, Belgium, Britain, Ceuta, Croatia, Czech Republic, Poland, Portugal, Sardinia and Slovenia; further record from Finland accepted only into Cat D. See also Corsican Finch.

Carduelis corsicana
Corsican Finch (E)

OTHER NAMES Corsican (or Mediterranean) Citril Finch, Corsican Citril.
TAXONOMY (Koenig, A F, 1899). Monotypic. Previously treated as conspecific with Citril Finch, but differs in mtDNA, morphology, voice and habitat selection.
DISTRIBUTION [W Mediterranean islands. Stable. LC] Endemic resident Corsica, Sardinia, Gorgona, Capri and Elba. Vagrant mainland Italy; undetermined *corsicana/citrinella* also recorded Algeria in 19th century.

Serinus pusillus
Red-fronted Serin

OTHER NAMES Fire-fronted (or Gold-fronted) Serin.
TAXONOMY (Pallas, 1811). Monotypic.
DISTRIBUTION [SW Asia to C Asia and Himalayas. Stable. LC] Breeds Turkey, N Syria, Caucasus, N Iraq and Iran. Disperses in winter to lower altitudes, occurring S in Levant (rare Israel). Vagrant Bulgaria, Cyprus, Egypt, Greece and Jordan.

Serinus serinus
European Serin (E)

OTHER NAMES (Common) Serin.
TAXONOMY (Linnaeus, 1766). Monotypic.
DISTRIBUTION [Europe, N Africa and W Middle East. Decreasing. LC] Breeds Canary Is, NW Africa, N Libya and N Egypt, across continental Europe from Iberia N to S Scandinavia and Baltic States, E to Ukraine and Balkans, and Asia Minor, SW Caucasus and Levant. Migratory in N and E of range; more widespread winter in W Middle East and N Africa, also Malta. Scarce migrant Britain and Finland (has bred both countries). Vagrant Azores, Iran, Ireland, Madeira and Norway.

Serinus syriacus
Syrian Serin (E)

OTHER NAME Tristram's Serin.
TAXONOMY Bonaparte, 1850. Monotypic.
DISTRIBUTION [SW Asia. Decreasing. VU] Endemic. Breeds Lebanon, S Syria, N Israel and Jordan. Mainly resident but dispersive, wintering to Palestine, S Israel and NE Egypt (Sinai).

Serinus canaria
Atlantic Canary (E)

OTHER NAMES (Common) Canary, Island Canary.
TAXONOMY (Linnaeus, 1758). Monotypic.
DISTRIBUTION [Atlantic islands. Stable. LC] Endemic resident Azores, Madeira and Canary Is.

Spinus spinus
Eurasian Siskin

OTHER NAMES (Common) Siskin.

TAXONOMY (Linnaeus, 1758). Monotypic.
DISTRIBUTION [Eurasia. Decreasing. LC] Breeds C and E Europe N through Scandinavia, E to Ural Mtns, S into Balkans, N Turkey, Caucasus and N Iran, and in W in Britain, Ireland and very locally France and Iberia; has bred Faroe Is. N and NE populations migratory; winter range extends through S Europe to NW Africa, Libya (scarce) and N Middle East, locally to Gulf. Vagrant Bahrain, Iceland (has bred), Oman and Qatar.

FAMILY PARULIDAE

New World warblers · 27 species

Seiurus aurocapilla
Ovenbird (V)

TAXONOMY (Linnaeus, 1766). Three subspecies (undetermined in region). Genetic studies indicate that species is sister to all other true parulids, with waterthrushes now separated into *Parkesia*.
DISTRIBUTION [E North America > Middle America and West Indies. Stable. LC] Very rare vagrant, typically Oct but rarely also early winter: Azores (6 records of 7 individuals*); Britain (6); Ireland (2); Norway (25-27 Oct 2003).

Parkesia noveboracensis
Northern Waterthrush (V)

TAXONOMY (Gmelin, J F, 1789). Monotypic. Formerly considered polytypic with three subspecies, but differences now judged to be clinal.
DISTRIBUTION [N North America > West Indies, Middle America and N South America. Stable. LC] Very rare vagrant (mainly late Sep and Oct): Azores (5+*); Britain (7); Canary Islands (10-26 Nov 1991); Channel Is (17 Apr 1977); France (3); Ireland (2); Netherlands (18-21 Sep 2010).

Vermivora chrysoptera
Golden-winged Warbler (V)

OTHER NAMES Brewster's/Lawrence's Warbler (distinct hybrid/back-cross types with Blue-winged Warbler).
TAXONOMY (Linnaeus, 1766). Monotypic.
DISTRIBUTION [E North America > Central America. Decreasing. NT] Extreme vagrant (two records): Azores (12-13 Oct 2012*); Britain (24 Jan-10 Apr 1989).

Vermivora cyanoptera
Blue-winged Warbler (V)

OTHER NAMES Brewster's Warbler, Lawrence's Warbler (distinct hybrid/back-cross types with Golden-winged Warbler).
TAXONOMY Olson and Reveal, 2009. Monotypic.
DISTRIBUTION [E North America > Middle America. Decreasing. LC] Extreme vagrant (two records): Azores (2-11 Oct 2011); Ireland (4-10 Oct 2000).

Mniotilta varia
Black-and-white Warbler (V)

TAXONOMY (Linnaeus, 1766). Monotypic.
DISTRIBUTION [N and E North America > S USA to NW South America. Decreasing. LC] Very rare vagrant (mainly Sep-Oct): Azores (8*); Britain (14); Canary Is (18 Dec 2009); Faroe Is (18-20 Jul 1984); Iceland (2); Ireland (2).

Leiothlypis peregrina
Tennessee Warbler (V)

TAXONOMY (Wilson, A, 1811). Monotypic.
DISTRIBUTION [N North America > Middle America and N South America. Stable. LC] Extreme vagrant (10 records, mainly Sep-Oct): Azores (4: 21 Nov 2005; 7 Oct 2010*; 7-11 Oct 2011*; 13-18 Oct 2012*); Britain (4: 6-20 Sep 1975; 24 Sep 1975; 5-7 Sep 1982; 20 Sep 1995); Faroe Is (21-28 Sep 1984); Iceland (14 Oct 1956).

Geothlypis trichas
Common Yellowthroat (V)

OTHER NAME Yellowthroat.

TAXONOMY (Linnaeus, 1766). Fourteen subspecies (undetermined in region). Geographical variation complex within huge range, and recent mtDNA research suggests that E and W populations may be closer to other congeners than to each other; further study required.
DISTRIBUTION [North America > S North America, Middle America, West Indies and N South America. Stable. LC] Very rare vagrant (mainly Oct but several winter and spring records): Azores (17*); Britain (10); Iceland (3); Ireland (3-4 Oct 2003); Portugal (18-19 Oct 2010).

Setophaga citrina
Hooded Warbler (V)

TAXONOMY (Boddaert, 1783). Monotypic.
DISTRIBUTION [E North America > West Indies and Middle America. Increasing. LC] Extreme vagrant (four records, all Sep-Oct): Azores (2: 26-27 Oct 2005; 11 Oct 2008); Britain (2: 20-23 Sep 1970; 10 Sep 1992).

Setophaga ruticilla
American Redstart (V)

TAXONOMY (Linnaeus, 1758). Monotypic.
DISTRIBUTION [N and E North America > West Indies, Central America and N and NW South America. Decreasing. LC] Very rare vagrant (mainly Oct): Azores (10*); Britain (5); France (10 Oct 1961); Iceland (10-12 Sep 1975); Ireland (2); Salvage Is (2 Oct 1981*).

Setophaga tigrina
Cape May Warbler (V)

TAXONOMY (Gmelin, J F, 1789). Monotypic.
DISTRIBUTION [N North America > West Indies. Stable. LC] Extreme vagrant (two records): Britain (2: 17 June 1977; 23 Oct-2 Nov 2013).

Setophaga cerulea
Cerulean Warbler (V)

TAXONOMY (Wilson, A, 1810). Monotypic.
DISTRIBUTION [E North America > NW South America. Decreasing. VU] Extreme vagrant: Iceland (1-7 Oct 1997).

Setophaga americana
Northern Parula (V)

OTHER NAMES (American) Parula, (Northern) Parula Warbler; Sutton's Warbler (hybrid with Yellow-throated Warbler).
TAXONOMY (Linnaeus, 1758). Monotypic.
DISTRIBUTION [E North America > West Indies and Middle America. Increasing. LC] Rare vagrant (mainly Oct): Azores (9*); Britain (15); France (3); Germany (25 Oct 1985); Iceland (8); Ireland (3).

Setophaga magnolia
Magnolia Warbler (V)

TAXONOMY (Wilson, A, 1811). Monotypic.
DISTRIBUTION [N North America > West Indies and Middle America. Increasing. LC] Extreme vagrant (seven records, most from late Sep): Azores (3: 21-22 Sep 1999*; 6 Oct 2009; 10 Oct 2012*); Britain (2: 27-28 Sep 1981; 23 Sep 2012); Iceland (2: 29 Sep-7 Dec 1995; 21-23 Oct 1995).

Setophaga castanea
Bay-breasted Warbler (V)

TAXONOMY (Wilson, A, 1810). Monotypic.
DISTRIBUTION [N North America > S Central America and N South America. Decreasing. LC] Extreme vagrant: Britain (1 Oct 1995).

Setophaga fusca
Blackburnian Warbler (V)

TAXONOMY (Statius Müller, 1776). Monotypic.
DISTRIBUTION [N and E North America > S Central America and NW South America. Increasing. LC] Extreme vagrant (four autumn records): Britain (3: 5 Oct 1961; 7 Oct 1988; 12-14 Sep 2009); Iceland (1: autumn 1987, at sea).

Setophaga aestiva
American Yellow Warbler (V)

OTHER NAMES (Northern) Yellow Warbler.
TAXONOMY (Gmelin, J F, 1789). Nine subspecies (one in region). Formerly also included extralimital Mangrove Warbler *S petechia* (West Indies, Middle America and South America; 34 subspecies in two groups, 'Golden' *petechia* and 'Mangrove' *erythachorides*).
DISTRIBUTION [North America > Middle America and N South America. Stable. LC] Very rare vagrant (mainly late Aug-early Oct): Azores (7*); Britain (5, first of which *aestiva*); France (30 Aug 2011, *aestiva*); Iceland (4); Ireland (4); Salvage Is (10-12 Sep 1993, two).

Setophaga pensylvanica
Chestnut-sided Warbler (V)

TAXONOMY (Linnaeus, 1766). Monotypic.
DISTRIBUTION [E North America > Central America. Decreasing. LC] Extreme vagrant (four autumn records): Azores (13 Oct 2009*); Britain (2: 20 Sep 1985; 18 Oct 1995); France (10-12 Oct 2010).

Setophaga striata
Blackpoll Warbler (V)

OTHER NAME Blackpoll.
TAXONOMY (Forster, J R, 1772). Monotypic.
DISTRIBUTION [N North America > N South America. Decreasing. LC] Rare vagrant (mainly late Sep-early Nov): Azores (6*); Britain (45); Channel Is (26 Oct-4 Nov 1980); Denmark (15-16 Oct 2006); Finland (23 Oct 2008); France (4); Iceland (14); Ireland (10); Norway (2).

Setophaga caerulescens
Black-throated Blue Warbler (V)

TAXONOMY (Gmelin, J F, 1789). Two subspecies (undetermined in region).
DISTRIBUTION [E North America > West Indies. Increasing. LC] Extreme vagrant (six records): Azores (4: 24 Oct-23 Nov 2005*; 24-29 Oct 2006; 28 Oct 2006; 4-5, 14-15 and 17 Oct 2013*); Iceland (2: 14-19 Sep 1988; 17-18 Oct 2003).

Setophaga palmarum
Palm Warbler (V)

OTHER NAMES Western Palm Warbler (*S p palmarum*), Yellow Palm Warbler (*S p hypochrysea*).
TAXONOMY (Gmelin, J F, 1789). Two subspecies (one in region).
DISTRIBUTION [N North America > SE USA, West Indies and NE Mexico. Increasing. LC] Extreme vagrant: Iceland (5-10 Oct 1997, *hypochrysea*). One found dead on tideline in Britain (18 May 1976) may not have arrived alive, so record not accepted onto national list.

Setophaga coronata
Myrtle Warbler (V)

OTHER NAME Yellow-rumped Warbler.
TAXONOMY (Linnaeus, 1766). Monotypic. Considered specifically distinct until 1973, when lumped as Yellow-rumped Warbler with two extralimital forms, Audubon's Warbler *S auduboni* of W North America (including NC Mexican subspecies *nigrifrons*) and Goldman's Warbler *S goldmani* of Central America, but lack of free interbreeding with Audubon's in narrow and stable hybrid zone confirms validity of original treatment.
DISTRIBUTION [N North America > SE USA, West Indies and E Central America. Increasing. LC] Rare vagrant (mainly Oct-mid Nov): Azores (16*); Britain (18); Canary Islands (25 Feb-2 Mar 1984); Iceland (17); Ireland (16); Netherlands (13-15 Oct 1996); Norway (8 Oct 1996); Spain (18 Sep 1982).

Setophaga dominica
Yellow-throated Warbler (V)

TAXONOMY (Linnaeus, 1766). Three subspecies (undetermined in region).
DISTRIBUTION [E USA > SE USA, West Indies and E Middle America. Increasing. LC] Extreme vagrant: Azores (16-17 Oct 2013*).

Setophaga discolor
Prairie Warbler (V)

TAXONOMY (Vieillot, 1809). Two subspecies (undetermined in region).
DISTRIBUTION [E North America > SE USA, West Indies and E Central America. Decreasing. LC] Extreme vagrant: Azores (20-23 Oct 2012*).

Setophaga virens
Black-throated Green Warbler (V)

TAXONOMY (Gmelin, J F, 1789). Monotypic.
DISTRIBUTION [N and E North America > West Indies, Central America and N South America. Decreasing. LC] Extreme vagrant (six records, all Oct): Azores (4: 22-23 Oct 2008*; 5-7 Oct 2009; 9-12 Oct 2009; 24-28 Oct 2013*); Germany (19 Oct 1858, collected); Iceland (27-28 Oct 2003). Another record from Iceland involving bird found dead on ship in Sep 1984 placed in Cat D.

Cardellina canadensis
Canada Warbler (V)

TAXONOMY (Linnaeus, 1766). Monotypic.
DISTRIBUTION [N North America > NW South America. Decreasing. LC] Extreme vagrant (three records): Azores (12 Oct 2009*); Iceland (29 Sep 1973, collected); Ireland (8-13 Oct 2006).

Cardellina pusilla
Wilson's Warbler (V)

OTHER NAMES Pileolated (or Black-capped) Warbler.
TAXONOMY (Wilson, A, 1811). Three subspecies (undetermined in region). Genetic distinctiveness between western and eastern groups strongly suggests two cryptic species with no known intermediates, despite morphological similarities.
DISTRIBUTION [W and N North America > Central America. Decreasing. LC] Extreme vagrant (two records): Britain (13 Oct 1985); Ireland (18-21 Sep 2013).

FAMILY ICTERIDAE

New World orioles, blackbirds and allies · five species

Xanthocephalus xanthocephalus
Yellow-headed Blackbird (V)

TAXONOMY (Bonaparte, 1826). Monotypic.
DISTRIBUTION [C and W North America > SW USA to C Mexico. Increasing. LC] Extreme vagrant (two records): Iceland (23-24 July 1983); Netherlands (18 and 20 May, 14 June, and 2-3 Jul 1982). Other regional occurrences *eg* in Britain, Channel Is, Denmark, France, Norway and Sweden placed in Cat D or E of respective national lists, though some may well relate to wild individuals.

Dolichonyx oryzivorus
Bobolink (V)

TAXONOMY (Linnaeus, 1758). Monotypic.
DISTRIBUTION [N and C North America > C South America. Decreasing. LC] Rare vagrant (mainly late Sep-early Nov): Azores (24*); Britain (32); Canary Islands (22-23 Oct 2010); France (3); Gibraltar (11-16 May 1984); Ireland (3); Italy (18 Sep 1989); Norway (6-8 Nov 1977).

Icterus galbula
Baltimore Oriole (V)

OTHER NAME Northern Oriole.
TAXONOMY (Linnaeus, 1758). Monotypic. Hybridises regularly with extralimital Bullock's Oriole *I bullockii* (W North America) in range overlap zone and thus previously treated as conspecific under name of Northern Oriole, but narrow and stable zone and possibly selection against hybrids (or their reduced fitness) support species status; molecular analysis also suggests that species are not each other's closest relatives.
DISTRIBUTION [C and E North America > Middle America and N South America. Stable. LC] Very rare vagrant (mainly late Sep-Oct, several also winter and spring): Azores (10*); Britain

(25); Iceland (4); Ireland (2); Netherlands (2); Norway (28 Sep 1986).

Molothrus ater
Brown-headed Cowbird (V)

OTHER NAME Common Cowbird.
TAXONOMY (Boddaert, 1783). Three subspecies (at least one in region).
DISTRIBUTION [Widespread North America > S North America. Decreasing. LC] Extreme vagrant (nine records): Britain (5, *ater* or *artemisiae*: 24 Apr 1988; 25 Apr-2 May 2009; 8-10 May 2009; 14-15 July 2009; 10 May 2010); France (4 May 2010); Germany (15 May 2012); Norway (2: 1 Jun 1987; 15 May 2010).

Quiscalus quiscula
Common Grackle (V)

OTHER NAMES Bronzed Grackle (*Q q versicolor*), Florida Grackle (*Q q quiscula*), Purple Grackle (*Q q stonei*).
TAXONOMY (Linnaeus, 1758). Three subspecies (one in region).
DISTRIBUTION [NC, C and E North America > E USA. Decreasing. LC] Extreme vagrant: Netherlands (8 April 2013, *versicolor*). Further record from Denmark (late Mar-*c* 20 Apr 1970) not admitted to national list.

FAMILY EMBERIZIDAE

Buntings, New World sparrows and allies · 34 species

Emberiza calandra
Corn Bunting

TAXONOMY Linnaeus, 1758. Two subspecies (both in region).
DISTRIBUTION [NW Africa and W Europe to C Asia. Decreasing. LC] Widespread from Canary Is and NW Africa across Europe N to Britain, extreme S Sweden and Latvia, E to Ukraine and SW Russia, also Turkey (except SE) and Caucasus (*calandra*), and from SE Turkey to Levant, E Saudi Arabia, UAE and parts of Iran (*buturlini*). Populations in C and E Europe, Caucasus and E Turkey migratory; winter range extends along coastal N Africa, through Middle East S to Oman and widely in Iran. Vagrant Faroe Is, Finland, Madeira, N Mauritania and Norway.

Emberiza citrinella
Yellowhammer

OTHER NAMES Yellow Bunting; Eastern Yellowhammer (*E c erythrogenys*).
TAXONOMY Linnaeus, 1758. Three subspecies (all in region). Closely related to Pine Bunting, with which it freely interbreeds in broad zone from Ural Mtns to Siberia; has been postulated that subspecies *erythrogenys* has resulted only from this hybridisation.
DISTRIBUTION [Europe to SC Siberia. Decreasing. LC] Breeds across much of Europe from W and N British Isles (*caliginosa*), SE England and W continental Europe to Scandinavia, NW Russia, Poland and Balkans (*citrinella*), and E Europe to Ural Mtns and locally in Caucasus (*erythrogenys*); absent much of Iberia, Mediterranean coasts and islands, and Arctic fringe. Northernmost population migratory; winter range extends to C Iberia and E across N Mediterranean to Turkey, Caspian region and N Middle East. Rare migrant Gibraltar. Vagrant Algeria, Bahrain, Cyprus, Faroe Is, Iceland, Kuwait, Malta, Morocco, Sardinia, Svalbard, Tunisia and UAE.

Emberiza leucocephalos
Pine Bunting

TAXONOMY Gmelin, S G, 1771. Two subspecies (one in region). Hybridises with Yellowhammer (*qv*) where breeding ranges overlap.
DISTRIBUTION [W Siberia to E Asia > SW and E Asia. Stable. LC] Summer visitor to Ural Mtns (*leucocephalos*). Occurs Iran and Iraq (uncommon) in winter, when also very rare/irregular S France, Italy, Israel, Lebanon and possibly Syria. Vagrant Albania, Armenia, Azerbaijan, Belgium, Britain, Bulgaria, Channel Is, Croatia, Cyprus, Czech Republic, Denmark, Finland, France, Germany, Gibraltar, Greece, Hungary, Iceland, Ireland, Latvia, Jordan, Malta, Netherlands, Norway, Oman,

Poland, Romania, Saudi Arabia, Slovenia, Spain, Sweden, Switzerland, Turkey, UAE and Ukraine.

Emberiza cia
Rock Bunting

OTHER NAMES Eurasian (or European/Western) Rock Bunting, Meadow Bunting.
TAXONOMY Linnaeus, 1766. Six subspecies (four in region).
DISTRIBUTION [NW Africa and SW Europe to C Asia. Increasing. LC] Breeds NW Africa, S and C Europe E to Balkans and Turkey (*cia*), Greece through S Turkey S to Israel and Jordan (*hordei*), S Ukraine, Caucasus, E Turkey and SW and N Iran (*prageri*) and NE Iran (*par*). Resident or partial and altitudinal migrant, wintering elsewhere within S Europe and in NW Middle East, *eg* Cyprus and Iraq. Vagrant Belgium, Britain, Corsica, Czech Republic, Cyprus, Denmark, Germany, Kuwait, Libya, Luxembourg, Malta, Netherlands, Poland, Sardinia and Sweden.

Emberiza stewarti
White-capped Bunting (V)

OTHER NAMES Chestnut-breasted (or Stewart's) Bunting.
TAXONOMY (Blyth, 1854). Monotypic.
DISTRIBUTION [SC and S Asia > S Asia. Stable. LC] Extreme vagrant (four dated records), but possibly more regular: Iran (3 dated records: 30 Oct 1898, collected; 20 Aug 1940, three collected; 17 Jun 1973; possibly rare summer visitor in NE, where in early 20th century also said to be winter vagrant and rare migrant); UAE (5-6 Nov 1992, two).

Emberiza buchanani
Grey-necked Bunting

OTHER NAME Grey-hooded Bunting.
TAXONOMY Blyth, 1845. Three subspecies (two in region).
DISTRIBUTION [SW to C Asia > S Asia. Stable. LC] Breeds E Turkey, S Armenia, Azerbaijan, N Iraq (rare) and discontinuously in Iran (*cerrutii*). Migrates E to winter quarters, occurring more widely Iran on passage. Vagrant Finland, Germany, Kuwait, Netherlands (*neobscura*), Norway, Oman, Russia, Sweden, Syria and Ukraine.

Emberiza cineracea
Cinereous Bunting

OTHER NAMES Ashy-headed (or Grey-headed) Bunting; Eastern Cinereous (or Smyrna) Bunting (*E c semenowi*), Western Cinereous Bunting (*E c cineracea*).
TAXONOMY Brehm, C L, 1855. Two subspecies (both in region). Differences in plumage, ecology and habitat suggest that eastern and western forms may be allospecies, but more research needed.
DISTRIBUTION [SW Asia > SW Arabia and NE Africa. Decreasing. NT] Breeding endemic, in summer ranging from E Greek islands and SW Turkey (*cineracea*) E through SE Turkey, N Iraq and SW Iran (*semenowi*). On passage occurs S through Middle East and E Egypt, *cineracea* predominating in Levant (rare Cyprus) and *semenowi* in Arabia (where also winters SW Saudi Arabia and W Yemen). Vagrant Bahrain, Denmark (*semenowi*) and Germany.

Emberiza hortulana
Ortolan Bunting

OTHER NAME Ortolan.
TAXONOMY Linnaeus, 1758. Monotypic.
DISTRIBUTION [SW Europe to C Asia > W and NE Africa. Decreasing. LC] Summer visitor mainly to S, E and NE Europe from Iberia to Balkans, N to Germany and S Scandinavia and E to Russia and Ukraine, also Caucasus, Turkey and through N Middle East to E Iran. Occurs more widely on passage C and W Europe (scarce Britain), N Africa and Middle East. Some winter Bahrain and Yemen. Vagrant Canary Is, Faroe Is, Iceland, Ireland, Liechtenstein, Luxembourg and Madeira.

Emberiza caesia
Cretzschmar's Bunting

TAXONOMY Cretzschmar, 1827. Monotypic.
DISTRIBUTION [SE Europe and SW Asia > NE Africa. Stable. LC] Breeding endemic. Summer

visitor S Albania, Greece, Crete, W and S Turkey, Cyprus and Levant, possibly also N Iraq. More widespread on passage Turkey, Levant to NW Saudi Arabia and Egypt. Vagrant Algeria, Austria, Balearic Is, Britain, Bulgaria, Finland, France, Germany, Iran, Italy, Libya, Malta, Netherlands, Oman, Qatar, Romania, Sweden, Ukraine and Yemen.

Emberiza cirlus
Cirl Bunting (E)

TAXONOMY Linnaeus, 1766. Monotypic.
DISTRIBUTION [NW Africa, W and S Europe, Asia Minor. Decreasing. LC] Endemic. Breeds NW Africa, Mediterranean islands (not Cyprus) and S Europe N to SW England, France (formerly also Luxembourg), C Europe and Balkans, also W and N Turkey. Resident and partial migrant, vacating WC Europe in winter, when more widespread *eg* S Iberia (occasional migrant Gibraltar) and S Turkey. Vagrant Belgium (has bred), Denmark, Egypt, Georgia, Ireland, Liechtenstein, Malta, Netherlands, Poland, Sweden and Ukraine; record from Czech Republic under review.

Emberiza striolata
Striolated Bunting

OTHER NAMES Striated (or Striped) Bunting.
TAXONOMY (Lichtenstein, M H C, 1823). Three subspecies (one in region). Formerly included House Bunting (*qv*).
DISTRIBUTION [NE Africa and SW Asia. Increasing. LC] Resident NE Egypt, Israel, Saudi Arabia, Yemen, Oman, UAE and S Iran (*striolata*). Vagrant Bahrain, Cyprus, Kuwait and Qatar.

Emberiza sahari
House Bunting

OTHER NAME Mountain Bunting.
TAXONOMY Levaillant, J, 1850. Two subspecies (both in region). Previously treated as conspecific with Striolated Bunting, but differs in morphology, vocalisations, ecology and behaviour.
DISTRIBUTION [W and NW Africa to C Sahel and C Sahara. Population trend not assessed. NE] Resident N Mauritania (*sanghae*) and from Morocco and Algeria to Tunisia and W Libya, and discontinuously S to C and S Sahara (*sahari*). Vagrant Spain.

Emberiza tahapisi
Cinnamon-breasted Bunting

OTHER NAMES African (or Cinnamon-breasted) Rock Bunting.
TAXONOMY Smith, A, 1836. Five subspecies (one in region). Formerly included Gosling's Bunting (*qv*).
DISTRIBUTION [S, C, E and NE Africa and S Arabia. Stable. LC] Resident SW Saudi Arabia, W and extreme SE Yemen, and SW Oman (*arabica*).

Emberiza goslingi
Gosling's Bunting

TAXONOMY (Alexander, 1906). Monotypic. Previously included within Cinnamon-breasted Bunting, but genetic evidence and differences in morphology and vocalisations suggest that evolutionary lineages are long separated.
DISTRIBUTION [W and C Africa. Population trend not assessed. NE] Non-breeding visitor to S border of region in N Mauritania, N Mali and possibly N Niger.

Emberiza fucata
Chestnut-eared Bunting (V)

OTHER NAMES Grey-headed (or Grey-hooded) Bunting.
TAXONOMY Pallas, 1776. Three subspecies (one in region).
DISTRIBUTION [E Asia > SE Asia. Stable. LC] Extreme vagrant (three records): Britain (2: 15-20 Oct 2004, *fucata*; 23-25 Oct 2012); Sweden (25 Oct 2011).

Emberiza pusilla
Little Bunting

TAXONOMY Pallas, 1776. Monotypic.

DISTRIBUTION [Scandinavia to NE Asia > E and SE Asia. Stable. LC] Summer visitor N and NE Fennoscandia and NW Russia. Migrates SE to winter quarters, but recorded on passage W to Britain and Netherlands (where classed as scarce migrant, rare in winter) and widely across Europe and Middle East, mainly in autumn (some winter records; rare spring). Vagrant Austria, Balearic Is, Belgium, Bulgaria, Canary Is, Channel Is, Corsica, Cyprus, Czech Republic, Denmark, Egypt, Estonia, Faroe Is, Germany, Gibraltar, Greece, Hungary, Iceland, Iran, Ireland, Israel, Italy, Jordan, Kuwait, Latvia, Lebanon, Lithuania, Macedonia, Malta, Oman, Poland, Portugal, Sardinia, Saudi Arabia, Serbia, Slovenia, Spain, Switzerland, Turkey, UAE and Ukraine.

Emberiza chrysophrys
Yellow-browed Bunting (V)

TAXONOMY Pallas, 1776. Monotypic.
DISTRIBUTION [C and E Siberia > S and E China. Stable. LC] Extreme vagrant (nine records): Belgium (20 Oct 1966, trapped, kept in captivity until died in winter 1969-70); Britain (5: 19 Oct 1975; 12-23 Oct 1980; 22-23 Sep 1992; 19-22 Oct 1994; 4-5 May 1998); France (autumn 1827, collected); Netherlands (19 Oct 1982, trapped); Sweden (3 Jan-27 Feb 2009). Additional record from Ukraine (Jan 1983) considered unacceptable because of insufficient documentation.

Emberiza rustica
Rustic Bunting

TAXONOMY Pallas, 1776. Monotypic.
DISTRIBUTION [N Eurasia > C and E Asia. Decreasing. LC] Summer visitor C, N and E Scandinavia, Finland and N Russia. Migrates SE to winter quarters but occurs on migration W of breeding range, and now classed as very rare migrant Netherlands. Vagrant Austria, Belgium, Britain, Bulgaria, Channel Is, Croatia, Czech Republic, Denmark, Egypt, Estonia, Faroe Is, Germany, Greece, Hungary, Iceland, Iran, Iraq, Ireland, Israel, Italy, Jordan, Kuwait, Latvia, Lebanon, Lithuania, Malta, Oman, Poland, Portugal, Qatar, Romania, Sardinia, Saudi Arabia, Slovenia, Spain, Svalbard, Syria, Switzerland, Turkey, UAE and Ukraine.

Emberiza aureola
Yellow-breasted Bunting

OTHER NAMES White-shouldered (or Golden) Bunting.
TAXONOMY Pallas, 1773. Two subspecies (one in region).
DISTRIBUTION [NE Europe to E and NE Asia > S and SE Asia. Decreasing. EN] Rapidly declining summer visitor N European Russia, formerly also Finland, and irregular/local N Ukraine (*aureola*); remaining extent of breeding areas unquantified but now largely absent from much of former range. Migrates SE in autumn, when most extralimital records have occurred in region (especially NW Europe), but now extremely rare. Vagrant Bahrain, Balearic Is, Belgium, Britain, Channel Is, Cyprus, Czech Republic, Denmark, Egypt, Estonia, Finland (formerly bred), France, Germany, Greece, Iceland, Iran, Ireland, Israel, Italy, Latvia, Malta, Netherlands, Norway, Oman, Poland, Portugal, Saudi Arabia, Spain, Sweden and UAE.

Emberiza rutila
Chestnut Bunting (V)

OTHER NAME Ruddy Bunting.
TAXONOMY Pallas, 1776. Monotypic.
DISTRIBUTION [S Siberia to E Asia > SE Asia. Stable. LC] Extreme vagrant (nine records): Finland (30 Sep-1 Oct 2002); France (2: 5-11 Oct 1995; 13 Oct 2009); Hungary (24-25 Sep 2011; Cat D_A); Malta (12 Nov 1983); Netherlands (5 Nov 1937); Norway (2: 13-15 Oct 1974; 5 Oct 2010); Slovenia (10 Oct 1987). Other regional records treated as escapes or of unknown origin, including *eg* Britain (8 in Cat E), Belgium (2 in Cat D) and France (3 in Cat D).

Emberiza melanocephala
Black-headed Bunting

TAXONOMY Scopoli, 1769. Monotypic.
DISTRIBUTION [SE Europe and SW Asia > S

Asia. Decreasing. LC] Breeding endemic. Summer visitor to S Italy, Balkans, Crete, Cyprus, Levant, Turkey, Caucasus N to SE Ukraine and SE European Russia, N Iraq, and Iran, where hybridises with Red-headed Bunting in Alborz Mtns. Also breeds occasionally (almost annually) S France. Migrates SE to winter quarters, when widespread on passage through much of Middle East S to C Saudi Arabia and Oman. Vagrant Algeria, Austria, Bahrain, Balearic Is, Britain, Channel Is, Corsica, Czech Republic, Denmark, Egypt, Estonia, Faroe Is, Finland, Germany, Hungary, Iceland, Ireland, Italy, Malta, Morocco, Netherlands, Norway, Poland, Qatar, Sardinia, Slovakia, Spain, Sweden, Switzerland and Tunisia.

Emberiza bruniceps
Red-headed Bunting

TAXONOMY von Brandt, J F, 1841. Monotypic.
DISTRIBUTION [W and C Asia > S Asia. Stable. LC] Summer visitor extreme SE European Russia, NW Kazakhstan and in N Iran, where significant hybrid zone with Black-headed Bunting established Alborz Mtns. Migrates SE to winter quarters; rare passage migrant and winter visitor Kuwait. Widely recorded elsewhere in region but true occurrence pattern obscured by birds of unknown origin (*eg* Cat D Belgium, Britain, Faroe Is, Finland, Germany, Iceland and Ireland) and/or escapes (*eg* Cat E Belgium, Britain, Norway and Switzerland) when species was popular in cagebird trade (imports from India banned 1982). Considered genuine vagrant Azerbaijan, Czech Republic, Estonia, France, Israel, Italy, Netherlands, Oman, Saudi Arabia, Spain, Turkey* and UAE; some other recent records (including in *eg* Britain, where Cat D status retained) may also involve birds of natural origin.

Emberiza spodocephala
Black-faced Bunting (V)

OTHER NAMES Grey-headed Black-faced (or Masked) Bunting.
TAXONOMY Pallas, 1776. Three subspecies (one in region).
DISTRIBUTION [C Siberia to E Asia > SE and E Asia. Stable. LC] Very rare vagrant (*spodocephala*): Britain (5); Finland (2 Nov 1981); France (8-9 Apr 2006); Germany (2); Netherlands (4); Norway (2*); Sweden (2).

Emberiza pallasi
Pallas's Reed Bunting

OTHER NAME Pallas's Bunting.
TAXONOMY (Cabanis, 1851). Four subspecies (one in region).
DISTRIBUTION [NE Europe to NE Asia > E Asia. Stable. LC] Summer visitor to extreme NE European Russia (*polaris*). Vagrant Britain and Portugal.

Emberiza schoeniclus
Common Reed Bunting

OTHER NAMES Reed Bunting; Dark Reed Bunting (*E s intermedia*, *reiseri* and *caspia*, sometimes including *witherbyi* and *tschusii*), Northern Reed Bunting (*E s schoeniclus*), Pale Reed Bunting (*E s pyrrhuloides*, sometimes including extralimital Asian subspecies *harterti*, *centralasiae* and *zaidamensis*), Thick-billed Reed Bunting (*E s witherbyi*, *intermedia*, *tschusii*, *reiseri*, *caspia*, *korejewi* and *pyrrhuloides*), Thin-billed Reed Bunting (*E s lusitanicus*, *schoeniclus* and *passerina*).
TAXONOMY (Linnaeus, 1758). Nineteen subspecies (14 in region). Regional taxa fall broadly into northern (thin-billed), southern (thick-billed) and eastern (intermediate) groups across extensive distribution, based on partially discrete ranges, bill morphology and degree of ecological separation; more than one species may be represented (*eg* in N Italy apparent lack of interbreeding between thin-billed northern *schoeniclus* and thick-billed southern *intermedia* where ranges meet; clear but weak genetic differentiation also detected). Colour cline exists throughout, populations becoming paler from W to E. Thorough review required.
DISTRIBUTION [Widespread Eurasia. Decreasing. LC] Breeds N Morocco, Spain (except NW), Balearic Is, Sardinia and S France (*witherbyi*), NW Spain and Portugal (*lusitanicus*), Ireland, Britain and much of W, N and NC continental Europe to N European Russia (*schoeniclus*), EC Europe and N Balkans (*stre-*

semanni), Italy and Adriatic coast to Albania (*intermedia*), SE Albania, NW Greece, Macedonia and W and C Turkey (*reiseri*), E Balkans and N Black Sea region (*tschusii*), from Moldova and Ukraine to C European Russia (*ukrainae*), N Ural Mtns (*passerina*), E European Russia (*pallidior*) to E of Volga River (*incognita*), N Caspian in extreme SE European Russia and NW Kazakhstan (*pyrrhuloides*), Caucasus, E Turkey and N Iran (*caspia*), and SW and E Iran (*korejewi*). Migratory in Fennoscandia and C and E Europe, to lesser extent parts of W Europe, Caucasus and E Turkey; winters through S Europe to N Africa and N Middle East. Scarce migrant Faroe Is (has bred). Vagrant Bahrain, Canary Is, Gibraltar, Iceland, Madeira, Oman, Saudi Arabia, Svalbard and UAE.

Passerella iliaca
Red Fox Sparrow (V)

OTHER NAME Fox Sparrow.
TAXONOMY (Merrem, 1786). Two subspecies (at least one in region). Formerly also included 13 other subspecies in three well-differentiated extralimital groups (all in W North America) now separated as Sooty Fox Sparrow *P unalaschensis*, Slate-coloured Fox Sparrow *P schistacea* and Thick-billed Fox Sparrow *P megarhyncha*; some hybridisation occurs, but no evidence of free interbreeding.
DISTRIBUTION [NW and N North America > C and E North America. Stable. LC] Extreme vagrant (four records involving three birds): Estonia (2-17 Dec 2012, same as Finland, *iliaca/zaboria*); Finland (20 Dec 2012-3 Apr 2013, same as Estonia); Iceland (5 Nov 1944, *iliaca*); Ireland (3-4 Jun 1961, *iliaca*). Record from Germany considered to relate to escape from captivity and placed in Cat E; further Danish 'Fox Sparrow' record, also assigned to Cat E, may have involved one of now-split western taxa rather than *P iliaca*.

Melospiza melodia
Song Sparrow (V)

TAXONOMY (Wilson, A, 1810). Twenty-four subspecies (undetermined in region).
DISTRIBUTION [S Canada and N and W USA > SW Canada and USA. Stable. LC] Very rare vagrant: Belgium (30 Sep 2004); Britain (7); Netherlands (30 April 2006); Norway (11 May 1975); Switzerland (1 May 2010); Spain (29 Mar 2013*); Sweden (28 May-1 Jun 2010).

Melospiza lincolnii
Lincoln's Sparrow (V)

TAXONOMY (Audubon, 1834). Three subspecies (one in region).
DISTRIBUTION [N and W North America > W and S North America to C America. Stable. LC] Extreme vagrant (four records, presumed *lincolnii*): Azores (3: 12-13 Oct 2010; 17-18 Oct 2012*; 31 Oct 2012*); Iceland (7 Dec 2013-19 Apr 2014*).

Zonotrichia leucophrys
White-crowned Sparrow (V)

OTHER NAME Gambel's Sparrow (*Z l gambelii*).
TAXONOMY (Forster, J R, 1772). Five subspecies (two in region).
DISTRIBUTION [W and N North America > W and S North America. Stable. LC] Very rare vagrant (most records presumed to involve *leucophrys*, but also two *gambelii*): Azores (4, including 1 *gambelii**); Britain (5); France (25 Aug 1965); Iceland (4-6 October 1978); Ireland (20-27 May 2003); Netherlands (mid-Dec 1981-mid-Feb 1982); Norway (2-9 Oct 2009, *gambelii*); another record from Iceland (28 May 2002) involving ship-assisted bird placed in Cat D.

Zonotrichia albicollis
White-throated Sparrow (V)

TAXONOMY (Gmelin, J F, 1789). Monotypic.
DISTRIBUTION [N and NE North America > E and S USA to N Mexico. Stable. LC] Rare vagrant (most often in spring): Azores (2*); Belgium (12-16 Oct 2013); Britain (46); Denmark (2); France (20 Apr-29 Jun 2008); Finland (3); Germany (30 May 2011); Iceland (6); Ireland (3); Malta (Oct 1995); Netherlands (5); Norway (2); Sweden (5-6

Dec 1963). Additionally, a Gibraltar record considered likely to have involved ship assistance (18-25 May 1986) was placed in Cat D (see also Dark-eyed Junco); another individual in Baltic Sea (Aug 2012) entered at least Estonian and Swedish waters on ship originating from (and returning to) USA via Britain.

Junco hyemalis
Dark-eyed Junco (V)

OTHER NAME Slate-coloured Junco (*'hyemalis* group').
TAXONOMY (Linnaeus, 1758). Fifteen subspecies (one in region). Taxonomy complex, subspecies falling into six groups which exhibit strong morphological differentiation but weak genetic divergence; intergrades frequent.
DISTRIBUTION [Alaska to E Canada, W and NE USA > SW and SE Canada, USA and N Mexico. Stable. LC] Rare vagrant (*hyemalis*): Azores (5 Nov 2009); Britain (38); Denmark (13 Dec 1980, trapped, kept in captivity until died 18 Mar 1993); Iceland (6 Nov 1955); Ireland (3); Netherlands (Feb 1962, trapped, kept in captivity until died 7 Nov 1968); Norway (3); Poland (4 May 1963). As with previous species, and coinciding on same dates, presumed ship-assisted Gibraltar record (18-25 May 1986) placed in Cat D.

Passerculus sandwichensis
Savannah Sparrow (V)

OTHER NAME Ipswich Sparrow (*P s princeps*).
TAXONOMY (Gmelin, J F, 1789). Seventeen subspecies (at least two in region). Taxa highly variable and fall into five groups, at least one of which (*rostratus*, SW North America) may be distinct species. In E North America represented by nominate group and distinctive *princeps*, at one time separated as Ipswich Sparrow.
DISTRIBUTION [Canada and N USA > USA and Mexico. Stable. LC] Extreme vagrant (six records): Azores (3: 31 Oct 2002; 6 Oct 2009*; 21-29 Oct 2009); Britain (3: 11-16 Apr 1982, *princeps*; 30 Sep-1 Oct 1987, *oblitus* or *labradorius*; 14-19 Oct 2003).

Chondestes grammacus
Lark Sparrow (V)

TAXONOMY (Say, 1822). Two subspecies (one in region).
DISTRIBUTION [N Mexico, W and C USA to S Canada > W and S USA, Mexico. Decreasing. LC] Extreme vagrant (two records, both *grammacus*): Britain (2: 30 Jun-8 Jul 1981; 15-17 May 1991).

Pipilo erythrophthalmus
Eastern Towhee (V)

OTHER NAME Rufous-sided Towhee.
TAXONOMY (Linnaeus, 1758). Four subspecies (undetermined in region). Formerly considered conspecific with extralimital Spotted Towhee *P maculatus* (W North America), with which it occasionally hybridises in C USA, but molecular analysis found significant genetic divergence.
DISTRIBUTION [SE Canada, C and E USA > SE USA. Stable. LC] Extreme vagrant: Britain (7 Jun 1966).

FAMILY CALCARIIDAE

Longspurs and snow buntings · two species

Calcarius lapponicus
Lapland Longspur

OTHER NAME Lapland Bunting.
TAXONOMY (Linnaeus, 1758). Five subspecies (two in region).
DISTRIBUTION [Circumpolar in Arctic and Sub-Arctic. Increasing. LC] Summer visitor W and N Fennoscandia and far N Russia (*lapponicus*). Migrates through Baltic and N Russia to winter SW Baltic and North Sea coasts from S Scandinavia to France, Britain and Ireland, where wintering population augmented by *subcalcaratus* (Greenland), and from Ukraine E to N Caspian region. Scarce migrant Hungary (occasional also in winter) and Italy. Vagrant Austria, Azores, Belarus, Bosnia-Herzegovina, Bulgaria, Canary Is*, Croatia, Czech Republic, Faroe Is, Iceland (has bred), Jan Mayen, Malta, Moldova, Por-

tugal, Romania, Sardinia, Serbia, Slovakia, Slovenia, Spain, Svalbard, Switzerland and Turkey.

Plectrophenax nivalis
Snow Bunting

OTHER NAME Icelandic Snow Bunting (*P n insulae*).
TAXONOMY (Linnaeus, 1758). Four subspecies (three in region).
DISTRIBUTION [Circumpolar in Arctic and Sub-Arctic. Decreasing. LC] Breeds Iceland (*insulae*), N Scotland (mainly *insulae*, few *nivalis*), N Scandinavia, Svalbard, Kola Peninsula (*nivalis*) and NE European Russia (*vlasowae*); has bred Faroe Is. Migrates S to winter Britain, Ireland, coastal continental Europe from S Scandinavia to W Iberia, and from S Baltic inland across E Europe to N Balkans, N Black Sea and N Caspian region. Scarce or rare migrant and winter visitor Austria, Italy and Switzerland. Vagrant Algeria, Azores, Balearic Is, Bosnia-Herzegovina, Bulgaria, Canary Is, Corsica, Croatia, Greece, Hungary, Luxembourg, Madeira, Malta, Montenegro, Morocco, Serbia, Tunisia and Turkey.

FAMILY CARDINALIDAE

Cardinals, grosbeaks and allies · five species

Piranga rubra
Summer Tanager (V)

TAXONOMY (Linnaeus, 1758). Two subspecies (probably one in region).
DISTRIBUTION [E and S USA, N Mexico > West Indies and S Mexico to N and NW South America. Stable. LC] Extreme vagrant (four records, likely *rubra*): Azores (3: 26-28 Oct 2006; 14 Oct 2010; 23-27 Oct 2011*); Britain (11-25 Sep 1957).

seen 592 in WP
1138 sp

Piranga olivacea
Scarlet Tanager (V)

TAXONOMY (Gmelin, J F, 1789). Monotypic.
DISTRIBUTION [E North America > NW South America. Stable. LC] Very rare vagrant: Azores (10*); Britain (6); France (12-18 Oct 2000); Iceland (4); Ireland (4).

Pheucticus ludovicianus
Rose-breasted Grosbeak (V)

OTHER NAME Common Grosbeak.
TAXONOMY (Linnaeus, 1766). Monotypic.
DISTRIBUTION [NW to SE Canada and NE USA > West Indies, S Mexico to N South America. Decreasing. LC] Rare vagrant: Azores (*c* 21*); Britain (26); Channel Is (2); Denmark (1-20 Dec 2008); France (2); Iceland (20-29 Oct 2001); Ireland (8); Malta (3); Norway (3); Sweden (2).

Spiza americana
Dickcissel (V)

OTHER NAME Black-throated Bunting.
TAXONOMY (Gmelin, J F, 1789). Monotypic.
DISTRIBUTION [S Canada, C and S USA > Mexico to N South America. Stable. LC] Extreme vagrant (five records of six birds): Azores (4: 6 Nov 2009, two; 26-28 Sep 2011; 14-16 Oct 2012*; 23 Oct 2012*); Norway (29 Jul 1981).

Passerina cyanea
Indigo Bunting (V)

TAXONOMY (Linnaeus, 1766). Monotypic.
DISTRIBUTION [SW, C and E USA, SE Canada > West Indies and Middle America. Decreasing. LC] Very rare vagrant: Azores (*c* 41*); Britain (2); Gibraltar (30 Apr 2004); Iceland (2); Ireland (9-19 Oct 1985); Netherlands (2). Additionally, two records of uncertain status from Russia (Kaliningrad Oblast): 21 Jun 1975 and 26 Apr 1982 (both adult males).

ANNEX A: INDETERMINATE SPECIES RECORDED IN THE REGION

Terpsiphone paradisi/affinis/incei
Indian/Oriental/Amur Paradise Flycatcher (V)

Extreme vagrant: UAE (27 Oct 2011).

Phylloscopus tenellipes/borealoides
Pale-legged/Sakhalin Leaf Warbler (V)

Extreme vagrant: Britain (22 Oct 2012).

APPENDIX 1: ENDEMIC SPECIES

Endemic species are defined here as those birds whose typical breeding and non-breeding ranges lie entirely within Europe, North Africa and the Middle East. Only exceptionally might any of them (exclusively migratory or dispersive species) occur outside the region; not included here are near-endemics (defined as those species with typically more than 90 per cent of their world range lying within the region) or breeding endemics (species whose breeding range falls entirely within the region but whose non-breeding range regularly extends to, or even lies wholly outside, the regional boundary). There are 87 'true' endemic bird species in total, comprising some 7.5 per cent of all species in the region, and they are listed below together with their primary breeding areas. Of these, the Atlantic islands are the most important for endemic birds (21 species), followed by south-west Arabia (13 species) and north-west Africa (11 species).

Key: AR (Arabian Peninsula), AI (Atlantic islands), CA (Caucasus), CE (central Europe), MA (Maghreb region of north-west Africa), ME (Middle East), MI (Mediterranean islands), NA (North Africa), NW (north-west Europe), SE (south-east Europe), SO (south Europe), SW (south-west Europe), WD (widespread).

Lyrurus mlokosiewiczi	Caucasian Grouse	CA
Tetraogallus caucasicus	Caucasian Snowcock	CA
Alectoris graeca	Rock Partridge	CE, SE, SO
Alectoris philbyi	Philby's Partridge	AR
Alectoris barbara	Barbary Partridge	MA, NA
Alectoris rufa	Red-legged Partridge	SW
Alectoris melanocephala	Arabian Partridge	AR
Puffinus yelkouan	Yelkouan Shearwater	MI, SE
Puffinus mauretanicus	Balearic Shearwater	MI, SW
Puffinus baroli	Barolo Shearwater	AI
Puffinus boydi	Boyd's Shearwater	AI
Oceanodroma monteiroi	Monteiro's Storm-petrel	AI
Oceanodroma jabejabe	Cape Verde Storm-petrel	AI
Phalacrocorax aristotelis	European Shag	WD
Aquila adalberti	Spanish Imperial Eagle	SW
Milvus milvus	Red Kite	WD
Buteo bannermani	Cape Verde Buzzard	AI
Larus argentatus	European Herring Gull	NW
Larus armenicus	Armenian Gull	CA, ME
Columba trocaz	Trocaz Pigeon	AI
Columba bollii	Bolle's Pigeon	AI

Columba junoniae	Laurel Pigeon	AI
Otus pamelae	Arabian Scops Owl	AR
Strix butleri	Omani Owl	AR, ME
Apus alexandri	Cape Verde Swift	AI
Apus unicolor	Plain Swift	AI
Dendrocoptes dorae	Arabian Woodpecker	AR
Dendrocoptes medius	Middle Spotted Woodpecker	WD
Dendrocopos syriacus	Syrian Woodpecker	WD
Picus sharpei	Iberian Green Woodpecker	SW
Picus vaillantii	Levaillant's Woodpecker	MA
Cyanopica cooki	Iberian Magpie	SW
Podoces pleskei	Pleske's Ground Jay	ME
Poecile lugubris	Sombre Tit	SE, ME
Poecile hyrcanus	Caspian Tit	CA, ME
Cyanistes teneriffae	African Blue Tit	AI, MA, NA
Alauda razae	Raso Lark	AI
Galerida macrorhyncha	Maghreb Lark	MA
Eremophila bilopha	Temminck's Lark	MA, NA, ME
Chersophilus duponti	Dupont's Lark	SW, MA
Phylloscopus canariensis	Canary Islands Chiffchaff	AI
Acrocephalus brevipennis	Cape Verde Warbler	AI
Turdoides altirostris	Iraq Babbler	ME
Turdoides squamiceps	Arabian Babbler	ME
Sylvia undata	Dartford Warbler	NW, SW
Sylvia sarda	Marmora's Warbler	SO
Sylvia balearica	Balearic Warbler	MI
Sylvia buryi	Yemen Warbler	AR
Regulus ignicapilla	Common Firecrest	WD
Regulus madeirensis	Madeira Firecrest	AI
Sitta whiteheadi	Corsican Nuthatch	MI
Sitta ledanti	Algerian Nuthatch	MA
Sitta krueperi	Krüper's Nuthatch	SE, ME
Sitta neumayer	Western Rock Nuthatch	SE, ME, CA
Certhia brachydactyla	Short-toed Treecreeper	WD
Sturnus unicolor	Spotless Starling	SW, MI, MA
Onychognathus tristramii	Tristram's Starling	AR, ME
Turdus menachensis	Yemen Thrush	AR
Phoenicurus moussieri	Moussier's Redstart	MA
Saxicola dacotiae	Canary Islands Stonechat	AI
Oenanthe moesta	Red-rumped Wheatear	MA, NA, ME
Oenanthe leucura	Black Wheatear	SW, MA, NA
Oenanthe lugens	Mourning Wheatear	NA, ME
Oenanthe halophila	Maghreb Wheatear	MA
Oenanthe lugentoides	Arabian Wheatear	AR
Passer italiae	Italian Sparrow	SO, MI
Passer iagoensis	Iago Sparrow	AI
Estrilda rufibarba	Arabian Waxbill	AR
Prunella ocularis	Radde's Accentor	CA, ME
Prunella fagani	Arabian Accentor	AR
Anthus petrosus	European Rock Pipit	NW
Anthus berthelotii	Berthelot's Pipit	AI
Fringilla teydea	Blue Chaffinch	AI
Pyrrhula murina	Azores Bullfinch	AI

Rhodopechys alienus	African Crimson-winged Finch	MA
Carpodacus synoicus	Sinai Rosefinch	ME
Rhynchostruthus percivali	Arabian Grosbeak	AR
Crithagra rothschildi	Arabian Serin	AR
Crithagra menachensis	Yemen Serin	AR
Linaria yemenensis	Yemen Linnet	AR
Loxia scotica	Scottish Crossbill	NW
Carduelis citrinella	Citril Finch	CE, SW
Carduelis corsicana	Corsican Finch	MI
Serinus serinus	European Serin	WD
Serinus syriacus	Syrian Serin	ME
Serinus canaria	Atlantic Canary	AI
Emberiza cirlus	Cirl Bunting	WD

APPENDIX 2: EXTINCT SPECIES

Extinct species are defined as those which have not been recorded anywhere in the world for at least 50 years, and which are thought to have died out. In total, 145 bird species have been classified as extinct since 1500, including five that are extinct in the wild but still have populations remaining in captivity (BirdLife International 2014b). Just two bird species which formerly occurred in the region are known to have gone extinct in this period (BirdLife International 2014a), and these are listed here. A third species, Eskimo Curlew (see page 97), is now classified as Possibly Extinct; there have also been no confirmed records anywhere for 15 years of Slender-billed Curlew (page 97), now another potential candidate for extinction.

Haematopus meadewaldoi
Canarian Oystercatcher

[Canary Islands] Formerly endemic to Fuerteventura, Lanzarote and their offshore islets. Last collected 1913 and reported to have become extinct by 1940s.

Pinguinus impennis
Great Auk

[NE North America and NW Europe > NE and SE North America and NW and SW Europe] Formerly distributed on North Atlantic coasts from NE Canada N to Greenland and E to Iceland, Faroe Is, Britain and Ireland. Last known pair killed Iceland 1844 and last live bird seen off Newfoundland Banks 1852.

APPENDIX 3: OMITTED SPECIES

The following species are sometimes considered to have occurred in the Western Palearctic, but have been omitted here for the reasons given below. This is not intended to be an exhaustive list of all such species or of records claimed but not accepted, and neither does it include species which are routinely placed by national committees in Categories D and E (see pages 14-15) or those formerly placed in Category C but no longer considered to have self-sustaining populations.

Dendrocygna viduata
White-faced Whistling-Duck

[South America and sub-Saharan Africa] Listed by Jirle (2015) but no accepted records of wild individuals within regional boundary as defined here.

Anas cyanoptera
Cinnamon Teal

[North and South America] Listed by Jirle (2015) but no accepted records of wild individuals within regional boundary as defined here.

Anas smithii
Cape Shoveler

[S Africa] Listed by Jirle (2015) but no accepted records of wild individuals within regional boundary as defined here.

Fregetta grallaria
White-bellied Storm-petrel

[S Ocean] Record from Cape Verde Is on 17 Aug 1986 (Hazevoet 1995) now considered indeterminate, relating to either this species or Black-bellied Storm-petrel (Crochet and Haas 2008, Howell 2010). Reported occurrences from Yemen involve sightings from "north and south of Socotra" in July 1969 (Porter and Stanton 2007, *Sea Swallow* 21: 41), and thus apparently outside pelagic area of region.

Phalacrocorax fuscicollis
Indian Cormorant

[S Atlantic] Listed by Jirle (2015) and historic range mapped to include SE Iran, but no traceable confirmed records and species not accepted onto national list (Nelson 2005, Scott and Adhami 2006).

Polemaetus bellicosus
Martial Eagle

[Sub-Saharan Africa] Single vagrancy record mapped by Brown *et al* (1982) and Demey and Borrow (2000) in area of Tibesti, Chad, apparently based on original map in Snow (1978) (M Woodcock 2015, pers comm). Latter map based mainly on localities of specimens from Natural History Museum, also some records from unpublished sources, but no specimen located in NHM (H van Grouw 2015, pers comm) and no other reference traceable.

Neotis denhami
Denham's Bustard

[Sub-Saharan Africa] Two published records from 1953 and 1957 (Simon 1965) are not supported by descriptions or specimens, leaving identity of species involved uncertain (Haas and Crochet 2010).

Amaurornis flavirostris
Black Crake

[Sub-Saharan Africa] Specimen from Madeira in 1895 originally accepted as this species but subsequently reidentified as Lesser Moorhen (Matias 2009).

Grus antigone
Sarus Crane

[S Asia] Unconfirmed Azerbaijan, where singles said to have been shot 6 Mar 1876 and 5 Mar 1882 but specimens probably lost, and Dagestan, Russian Caucasus, where one reportedly observed (Radde 1884 in Patrikeev 2004).

Otus bakkamoena
Indian Scops Owl

[S Asia] Reported doubtfully in SE Iran (Porter and Aspinall 2010) but no accepted records.

Campethera punctuligera
Fine-spotted Woodpecker

[Sub-Saharan Africa] One published record from just inside regional boundary in N Mauritania (Isenmann 2010), but single-observer record with no description or photo, and possibility of confusion with *eg* African Grey Woodpecker cannot be excluded (P Isenmann 2015, pers comm).

Dendrocopos leucopterus
White-winged Woodpecker

[C Asia] Said to have occurred NE Iran (Scott 2008, Roselaar and Aliabadian 2009, Porter and Aspinall 2010), but, following reassessment of old records and absence of confirmed occurrences since 1960s, species now removed from Iranian list (Khaleghizadeh *et al* 2011).

Picus squamatus
Scaly-bellied Woodpecker

[SC and S Asia] Formerly described from E Iran (Scott 2008, Porter and Aspinall 2010), but, after review of historical records and lack of confirmed occurrences since 1960s, species now removed from Iranian list (Khaleghizadeh *et al* 2011).

Lanius excubitoroides
Grey-backed Fiscal

[Sub-Saharan Africa] Published distribution in Mauritania from Sahel N to S Sahara (Isenmann 2010), and thus inside regional boundary at 20°N, now known to be incorrect, with no confirmed records from anywhere in country (Crochet and Haas 2013).

Dicrurus adsimilis
Fork-tailed Drongo

[Sub-Saharan Africa] First record of any drongo in Arabia, from Yemen on 28 Oct 1946, considered likely to have involved this species, and record now included by Blair *et al* (2016). However, published description (Browne 2015) does not correspond with *D adsimilis* on four points, and possibility of similar drongo species not fully excluded.

Mirafra cordofanica
Kordofan Lark

[Sub-Saharan Africa] Confusion with Dunn's Lark may have led to previous claims of this species being distributed within the extreme SW of region in N Mauritania and N Mali; no records have been substantiated with photos, recordings or specimens (Crochet and Haas 2013).

Bombycilla japonica
Japanese Waxwing

[E Asia] European records considered to involve escapes and all currently placed in Cat E, but at least one, involving first-winter male with Bohemian Waxwings in Poland (4-11 Jan 2009), may involve genuine vagrant (Ławicki 2013).

Cisticola aridulus
Desert Cisticola

[Sub-Saharan Africa] Some authors show breeding distribution N of 20°N in N Niger (*eg* Urban *et al* 1997, Borrow and Demey 2000) and species may possibly occur around southernmost margin of region, but in absence of unequivocal documentary evidence, and with no mapped data in West African Bird Database, provisionally treated only as hypothetical.

Mimus gilvus
Tropical Mockingbird

[Middle America, Lesser Antilles and N South America] One seen in Gibraltar in Feb 2012 and relocated across border in Spain in Aug 2012 (Pérez and Bensusan 2012) considered to have been ship-assisted and not accepted onto either national list.

Oenanthe familiaris
Familiar Chat

[NE Africa] Claimed status of vagrant to Yemen (Porter and Aspinall 2010) based on single report from 1960s, but no description, date or other details available (R Porter 2015, pers comm), and possibility of confusion with similar Persian Wheatear *O chrysopygia* cannot be excluded.

Parkesia motacilla
Louisiana Waterthrush

[North America to Middle America and West Indies] Record from Canary Is (10-26 Nov 1991) rejected after review and accepted instead as Northern Waterthrush (Rouco *et al* 2016).

Protonotaria citrea
Prothonotary Warbler

[North America to N South America] One photographed on ship in N Atlantic *c* 250 nautical miles (c 460 km) SW of Iceland in Jun or Jul 2010 (Olofson 2012) may have been close to or possibly within regional boundary, but exact position of vessel impossible to confirm. Bird remained on board for *c* 24 hours, but precise route of vessel and point of bird's departure also unknown.

Passerina caerulea
Blue Grosbeak

[North America] Various records from N and W Europe now all considered to be of uncertain or captive origin and treated as Cat D or E, though adjudication of one recent report from Azores (14 Oct 2011) not yet published by Portuguese Rarities Committee.

APPENDIX 4: NATIONAL LISTS

This appendix details the total number of bird species recorded in those countries or discrete geographical recording areas whose boundaries fall wholly or largely within Europe, North Africa and the Middle East; most of those of which a substantial part lies outside, specifically Chad, Kazakhstan, Mali, Mauritania and Niger, are not included. The totals are intended as a guideline comparison only, because of the different ways in which national lists are structured and maintained, including the base taxonomy adopted, list category definitions and assessment of rarity occurrences (for example, the treatment of ship-assisted vagrants or species of uncertain origin). In most cases the figure given is the total for the officially adopted national or territorial list, with the addition in a few instances of well-documented species which have occurred since that list was last published, and official acceptance of which is considered straightforward. Some countries do not publish or regularly update a national list, and in a minority of cases where there is no such readily accessible or up-to-date reference the total researched and compiled by BirdLife International (www.birdlife.org/datazone/country) or the online *Handbook of the Birds of the World* (www.hbw.com) is given instead (indicated with an asterisk*). New species are periodically added to national lists and old records reviewed and sometimes discounted, so the totals given below are subject to ongoing change.

Albania	361
Algeria	406
Andorra	147*
Armenia	357
Austria	423
Azerbaijan	391
Azores	392
Bahrain	323
Balearic Islands	368
Belarus	322
Belgium	442
Bosnia-Herzegovina	336*
Britain	603
Bulgaria	420
Cape Verde Islands	228
Channel Islands	370
Croatia	389
Cyprus	397
Czech Republic	389
Denmark	467
Egypt	448
Estonia	385
Faroe Is	307
Finland	474
France	577
Georgia	360
Germany	533
Gibraltar	310
Greece	449
Hungary	413
Iceland	394
Iran	549
Iraq	408
Ireland	475
Israel	538
Italy	548

Jan Mayen	134	Portugal (mainland)	418
Jordan	436	Qatar	325
Kuwait	404	Romania	373
Latvia	364	Russia (European)	414*
Lebanon	395	San Marino	196*
Libya	347	Saudi Arabia	517
Liechtenstein	237*	Serbia	352
Lithuania	376	Slovakia	359
Luxembourg	298	Slovenia	380
Macedonia	327	Spain (including Balearic Islands, Canary Islands and autonomous North African territories of Ceuta and Melilla)	560
Malta	403	Svalbard	214
Moldova	273*	Sweden	508
Monaco	101*	Switzerland	411
Montenegro	377*	Syria	383
Morocco	502	Tunisia	404
Netherlands	510	Turkey	480
Norway (including Svalbard and Jan Mayen)	496	Ukraine	426
Oman	527	United Arab Emirates	462
Palestine	261	Yemen	401
Poland	451		

CHECKLIST

Scientific name	English name
☐ **Family Struthionidae**	**Ostriches**
☐ *Struthio camelus*	Common Ostrich
☐ **Family Anatidae**	**Ducks, geese and swans**
☐ *Dendrocygna bicolor*	Fulvous Whistling Duck
☐ *Dendrocygna javanica*	Lesser Whistling Duck
☑ *Anser fabalis*	Taiga Bean Goose
☑ *Anser serrirostris*	Tundra Bean Goose
☑ *Anser brachyrhynchus*	Pink-footed Goose
☑ *Anser anser*	Greylag Goose
☑ *Anser albifrons*	Greater White-fronted Goose
☑ *Anser erythropus*	Lesser White-fronted Goose
☑ *Anser indicus*	Bar-headed Goose
☑ *Anser caerulescens*	Snow Goose
☐ *Anser rossii*	Ross's Goose
☑ *Branta canadensis*	Canada Goose
☐ *Branta hutchinsii*	Cackling Goose
☑ *Branta bernicla*	Brent Goose
☑ *Branta leucopsis*	Barnacle Goose
☑ *Branta ruficollis*	Red-breasted Goose
☑ *Cygnus atratus*	Black Swan
☑ *Cygnus olor*	Mute Swan
☑ *Cygnus columbianus*	Tundra Swan
☑ *Cygnus cygnus*	Whooper Swan
☐ *Plectropterus gambensis*	Spur-winged Goose
☐ *Sarkidiornis melanotos*	Knob-billed Duck

☑	*Alopochen aegyptiaca*	Egyptian Goose
☑	*Tadorna tadorna*	Common Shelduck
☑	*Tadorna ferruginea*	Ruddy Shelduck
☑	*Cairina moschata*	Muscovy Duck
☐	*Aix sponsa*	Wood Duck
☑	*Aix galericulata*	Mandarin Duck
☐	*Nettapus coromandelianus*	Cotton Pygmy Goose
☐	*Anas capensis*	Cape Teal
☑	*Anas strepera*	Gadwall
☑	*Anas falcata*	Falcated Duck
☑	*Anas penelope*	Eurasian Wigeon
☑	*Anas americana*	American Wigeon
☐	*Anas rubripes*	American Black Duck
☑	*Anas platyrhynchos*	Mallard
☐	*Anas discors*	Blue-winged Teal
☑	*Anas clypeata*	Northern Shoveler
☑	*Anas erythrorhyncha*	Red-billed Teal
☑	*Anas acuta*	Northern Pintail
☑	*Anas querquedula*	Garganey
☑	*Anas formosa*	Baikal Teal
☑	*Anas crecca*	Eurasian Teal
☑	*Anas carolinensis*	Green-winged Teal
☑	*Marmaronetta angustirostris*	Marbled Duck
☑	*Netta rufina*	Red-crested Pochard
☐	*Netta erythrophthalma*	Southern Pochard
☐	*Aythya valisineria*	Canvasback
☐	*Aythya americana*	Redhead
☑	*Aythya ferina*	Common Pochard
☑	*Aythya nyroca*	Ferruginous Duck
☑	*Aythya collaris*	Ring-necked Duck
☑	*Aythya fuligula*	Tufted Duck
☑	*Aythya marila*	Greater Scaup
☑	*Aythya affinis*	Lesser Scaup
☑	*Polysticta stelleri*	Steller's Eider
☑	*Somateria fischeri*	Spectacled Eider
☑	*Somateria spectabilis*	King Eider
☑	*Somateria mollissima*	Common Eider
☑	*Histrionicus histrionicus*	Harlequin Duck
☐	*Melanitta perspicillata*	Surf Scoter

☑ *Melanitta fusca*	Velvet Scoter
☐ *Melanitta deglandi*	White-winged Scoter
☑ *Melanitta nigra*	Common Scoter
☐ *Melanitta americana*	Black Scoter
☑ *Clangula hyemalis*	Long-tailed Duck
☑ *Bucephala albeola*	Bufflehead
☑ *Bucephala clangula*	Common Goldeneye
☑ *Bucephala islandica*	Barrow's Goldeneye
☑ *Mergellus albellus*	Smew
☑ *Lophodytes cucullatus*	Hooded Merganser
☑ *Mergus merganser*	Common Merganser
☑ *Mergus serrator*	Red-breasted Merganser
☑ *Oxyura jamaicensis*	Ruddy Duck
☑ *Oxyura leucocephala*	White-headed Duck
☐ **Family Numididae**	**Guineafowl**
☑ *Numida meleagris*	Helmeted Guineafowl
☐ **Family Odontophoridae**	**New World quail**
☐ *Callipepla californica*	California Quail
☐ *Colinus virginianus*	Northern Bobwhite
☐ **Family Phasianidae**	**Grouse, partridges, francolins and pheasants**
☑ *Tetrastes bonasia*	Hazel Grouse
☑ *Tetrao urogallus*	Western Capercaillie
☑ *Lyrurus tetrix*	Black Grouse
☑ *Lyrurus mlokosiewiczi*	Caucasian Grouse
☑ *Lagopus muta*	Rock Ptarmigan
☑ *Lagopus lagopus*	Willow Ptarmigan
☑ *Tetraogallus caucasicus*	Caucasian Snowcock
☑ *Tetraogallus caspius*	Caspian Snowcock
☑ *Alectoris graeca*	Rock Partridge
☑ *Alectoris chukar*	Chukar Partridge
☑ *Alectoris philbyi*	Philby's Partridge
☑ *Alectoris barbara*	Barbary Partridge
☑ *Alectoris rufa*	Red-legged Partridge
☑ *Alectoris melanocephala*	Arabian Partridge
☑ *Ammoperdix griseogularis*	See-see Partridge
☑ *Ammoperdix heyi*	Sand Partridge
☑ *Francolinus francolinus*	Black Francolin
☑ *Francolinus pondicerianus*	Grey Francolin
☑ *Pternistis bicalcaratus*	Double-spurred Francolin

☐	*Pternistis erckelii*	Erckel's Francolin
☑	*Perdix perdix*	Grey Partridge
☑	*Coturnix coturnix*	Common Quail
☐	*Coturnix delegorguei*	Harlequin Quail
☐	*Syrmaticus reevesii*	Reeves's Pheasant
☑	*Phasianus colchicus*	Common Pheasant
☑	*Chrysolophus pictus*	Golden Pheasant
☐	*Chrysolophus amherstiae*	Lady Amherst's Pheasant
☐	**Family Gaviidae**	**Divers**
☑	*Gavia stellata*	Red-throated Diver
☑	*Gavia arctica*	Black-throated Diver
☐	*Gavia pacifica*	Pacific Diver
☑	*Gavia immer*	Great Northern Diver
☑	*Gavia adamsii*	White-billed Diver
☐	**Family Oceanitidae**	**Austral storm-petrels**
☑	*Oceanites oceanicus*	Wilson's Storm-petrel
☑	*Pelagodroma marina*	White-faced Storm-petrel
☐	*Fregetta tropica*	Black-bellied Storm-petrel
☐	**Family Diomedeidae**	**Albatrosses**
☐	*Diomedea dabbenena*	Tristan Albatross
☐	*Thalassarche melanophris*	Black-browed Albatross
☐	*Thalassarche cauta*	Shy Albatross
☐	*Thalassarche chlororhynchos*	Atlantic Yellow-nosed Albatross
☐	**Family Hydrobatidae**	**Northern storm-petrels**
☑	*Hydrobates pelagicus*	European Storm-petrel
☑	*Oceanodroma castro*	Band-rumped Storm-petrel
☑	*Oceanodroma monteiroi*	Monteiro's Storm-petrel
☐	*Oceanodroma jabejabe*	Cape Verde Storm-petrel
☑	*Oceanodroma monorhis*	Swinhoe's Storm-petrel
☑	*Oceanodroma leucorhoa*	Leach's Storm-petrel
☐	*Oceanodroma matsudairae*	Matsudaira's Storm-petrel
☐	**Family Procellariidae**	**Petrels and shearwaters**
☐	*Macronectes giganteus*	Southern Giant Petrel
☑	*Fulmarus glacialis*	Northern Fulmar
☐	*Daption capense*	Cape Petrel
☐	*Pterodroma incerta*	Atlantic Petrel
☐	*Pterodroma mollis*	Soft-plumaged Petrel
☑	*Pterodroma madeira*	Zino's Petrel
☑	*Pterodroma feae*	Fea's Petrel

☐ *Pterodroma deserta*	Desertas Petrel
☐ *Pterodroma cahow*	Bermuda Petrel
☐ *Pterodroma hasitata*	Black-capped Petrel
☐ *Pterodroma arminjoniana*	Trindade Petrel
☐ *Calonectris leucomelas*	Streaked Shearwater
☐ *Calonectris diomedea*	Scopoli's Shearwater
☐ *Calonectris borealis*	Cory's Shearwater
☐ *Calonectris edwardsii*	Cape Verde Shearwater
☐ *Ardenna pacificus*	Wedge-tailed Shearwater
☐ *Ardenna griseus*	Sooty Shearwater
☐ *Ardenna carneipes*	Flesh-footed Shearwater
☐ *Ardenna gravis*	Great Shearwater
☐ *Puffinus puffinus*	Manx Shearwater
☐ *Puffinus yelkouan*	Yelkouan Shearwater
☐ *Puffinus mauretanicus*	Balearic Shearwater
☐ *Puffinus persicus*	Persian Shearwater
☐ *Puffinus bailloni*	Tropical Shearwater
☐ *Puffinus baroli*	Barolo Shearwater
☐ *Puffinus boydi*	Boyd's Shearwater
☐ *Bulweria bulwerii*	Bulwer's Petrel
☐ *Bulweria fallax*	Jouanin's Petrel
☐ **Family Podicipedidae**	**Grebes**
☐ *Tachybaptus ruficollis*	Little Grebe
☐ *Podilymbus podiceps*	Pied-billed Grebe
☐ *Podiceps grisegena*	Red-necked Grebe
☐ *Podiceps cristatus*	Great Crested Grebe
☐ *Podiceps auritus*	Horned Grebe
☐ *Podiceps nigricollis*	Black-necked Grebe
☐ **Family Phoenicopteridae**	**Flamingos**
☐ *Phoenicopterus ruber*	Greater Flamingo
☐ *Phoeniconaias minor*	Lesser Flamingo
☐ **Family Phaethontidae**	**Tropicbirds**
☐ *Phaethon aethereus*	Red-billed Tropicbird
☐ *Phaethon lepturus*	White-tailed Tropicbird
☐ **Family Ciconiidae**	**Storks**
☐ *Mycteria ibis*	Yellow-billed Stork
☐ *Anastomus lamelligerus*	African Openbill
☐ *Ciconia nigra*	Black Stork
☐ *Ciconia abdimii*	Abdim's Stork

☐	*Ciconia episcopus*	Woolly-necked Stork
☑	*Ciconia ciconia*	White Stork
☐	*Leptoptilos crumenifer*	Marabou Stork
☐	**Family Threskiornithidae**	**Ibises and spoonbills**
☑	*Threskiornis aethiopicus*	African Sacred Ibis
☑	*Geronticus eremita*	Northern Bald Ibis
☑	*Plegadis falcinellus*	Glossy Ibis
☑	*Platalea leucorodia*	Eurasian Spoonbill
☐	*Platalea alba*	African Spoonbill
☐	**Family Ardeidae**	**Bitterns, herons and egrets**
☑	*Botaurus stellaris*	Eurasian Bittern
☐	*Botaurus lentiginosus*	American Bittern
☐	*Ixobrychus exilis*	Least Bittern
☑	*Ixobrychus minutus*	Little Bittern
☐	*Ixobrychus sinensis*	Yellow Bittern
☐	*Ixobrychus eurhythmus*	Von Schrenck's Bittern
☐	*Ixobrychus cinnamomeus*	Cinnamon Bittern
☐	*Ixobrychus sturmii*	Dwarf Bittern
☑	*Nycticorax nycticorax*	Black-crowned Night Heron
☐	*Nyctanassa violacea*	Yellow-crowned Night Heron
☑	*Butorides virescens*	Green Heron
☑	*Butorides striata*	Striated Heron
☑	*Ardeola ralloides*	Squacco Heron
☑	*Ardeola grayii*	Indian Pond Heron
☐	*Ardeola bacchus*	Chinese Pond Heron
☑	*Bubulcus ibis*	Western Cattle Egret
☐	*Bubulcus coromandus*	Eastern Cattle Egret
☑	*Ardea cinerea*	Grey Heron
☐	*Ardea herodias*	Great Blue Heron
☐	*Ardea melanocephala*	Black-headed Heron
☑	*Ardea goliath*	Goliath Heron
☑	*Ardea purpurea*	Purple Heron
☑	*Ardea alba*	Great Egret
☑	*Ardea intermedia*	Intermediate Egret
☑	*Egretta ardesiaca*	Black Heron
☐	*Egretta tricolor*	Tricolored Heron
☐	*Egretta caerulea*	Little Blue Heron
☐	*Egretta thula*	Snowy Egret
☑	*Egretta garzetta*	Little Egret

☑	*Egretta gularis*	Western Reef Egret
☐	**Family Scopidae**	**Hamerkop**
☑	*Scopus umbretta*	Hamerkop
☐	**Family Pelecanidae**	**Pelicans**
☑	*Pelecanus onocrotalus*	Great White Pelican
☑	*Pelecanus rufescens*	Pink-backed Pelican
☑	*Pelecanus crispus*	Dalmatian Pelican
☐	**Family Fregatidae**	**Frigatebirds**
☐	*Fregata aquila*	Ascension Frigatebird
☐	*Fregata magnificens*	Magnificent Frigatebird
☐	*Fregata minor*	Great Frigatebird
☐	*Fregata ariel*	Lesser Frigatebird
☐	**Family Sulidae**	**Gannets and boobies**
☑	*Morus bassanus*	Northern Gannet
☐	*Morus capensis*	Cape Gannet
☑	*Sula dactylatra*	Masked Booby
☐	*Sula sula*	Red-footed Booby
☑	*Sula leucogaster*	Brown Booby
☐	**Family Phalacrocoracidae**	**Cormorants and shags**
☐	*Microcarbo africanus*	Reed Cormorant
☑	*Microcarbo pygmeus*	Pygmy Cormorant
☐	*Phalacrocorax auritus*	Double-crested Cormorant
☑	*Phalacrocorax aristotelis*	European Shag
☑	*Phalacrocorax carbo*	Great Cormorant
☐	*Phalacrocorax lucidus*	White-breasted Cormorant
☑	*Phalacrocorax nigrogularis*	Socotra Cormorant
☐	**Family Anhingidae**	**Anhingas and darters**
☐	*Anhinga rufa*	African Darter
☐	**Family Pandionidae**	**Ospreys**
☑	*Pandion haliaetus*	Western Osprey
☐	**Family Accipitridae**	**Kites, hawks and eagles**
☑	*Elanus caeruleus*	Black-winged Kite
☐	*Chelictinia riocourii*	Scissor-tailed Kite
☑	*Gypaetus barbatus*	Bearded Vulture
☑	*Neophron percnopterus*	Egyptian Vulture
☑	*Pernis apivorus*	European Honey Buzzard
☑	*Pernis ptilorhyncus*	Crested Honey Buzzard
☐	*Elanoides forficatus*	Swallow-tailed Kite
☐	*Necrosyrtes monachus*	Hooded Vulture

☐	*Gyps africanus*	White-backed Vulture
☐	*Gyps bengalensis*	White-rumped Vulture
☐	*Gyps rueppelli*	Rüppell's Vulture
☐	*Gyps himalayensis*	Himalayan Vulture
☑	*Gyps fulvus*	Griffon Vulture
☑	*Aegypius monachus*	Cinereous Vulture
☑	*Torgos tracheliotos*	Lappet-faced Vulture
☑	*Circaetus gallicus*	Short-toed Snake Eagle
☐	*Circaetus cinereus*	Brown Snake Eagle
☐	*Terathopius ecaudatus*	Bateleur
☑	*Clanga nipalensis*	Lesser Spotted Eagle
☑	*Clanga clanga*	Greater Spotted Eagle
☐	*Hieraaetus wahlbergi*	Wahlberg's Eagle
☑	*Hieraaetus pennatus*	Booted Eagle
☐	*Aquila rapax*	Tawny Eagle
☑	*Aquila nipalensis*	Steppe Eagle
☑	*Aquila adalberti*	Spanish Imperial Eagle
☑	*Aquila heliaca*	Eastern Imperial Eagle
☑	*Aquila chrysaetos*	Golden Eagle
☑	*Aquila verreauxii*	Verreaux's Eagle
☑	*Aquila fasciata*	Bonelli's Eagle
☐	*Micronisus gabar*	Gabar Goshawk
☐	*Melierax metabates*	Dark Chanting Goshawk
☐	*Accipiter badius*	Shikra
☑	*Accipiter brevipes*	Levant Sparrowhawk
☑	*Accipiter nisus*	Eurasian Sparrowhawk
☑	*Accipiter gentilis*	Northern Goshawk
☑	*Circus aeruginosus*	Western Marsh Harrier
☐	*Circus spilonotus*	Eastern Marsh Harrier
☑	*Circus cyaneus*	Hen Harrier
☐	*Circus hudsonius*	Northern Harrier
☑	*Circus macrourus*	Pallid Harrier
☑	*Circus pygargus*	Montagu's Harrier
☑	*Milvus milvus*	Red Kite
☑	*Milvus migrans*	Black Kite
☐	*Milvus aegyptius*	Yellow-billed Kite
☐	*Haliastur indus*	Brahminy Kite
☐	*Haliaeetus vocifer*	African Fish Eagle
☐	*Haliaeetus leucoryphus*	Pallas's Fish Eagle

☑	*Haliaeetus albicilla*	White-tailed Eagle
☐	*Haliaeetus leucocephalus*	Bald Eagle
☐	*Butastur rufipennis*	Grasshopper Buzzard
☐	*Butastur teesa*	White-eyed Buzzard
☑	*Buteo lagopus*	Rough-legged Buzzard
☐	*Buteo hemilasius*	Upland Buzzard
☑	*Buteo rufinus*	Long-legged Buzzard
☑	*Buteo bannermani*	Cape Verde Buzzard
☑	*Buteo buteo*	Common Buzzard
☐	**Family Otididae**	**Bustards**
☑	*Otis tarda*	Great Bustard
☐	*Ardeotis arabs*	Arabian Bustard
☑	*Chlamydotis undulata*	Houbara Bustard
☑	*Chlamydotis macqueenii*	Macqueen's Bustard
☐	*Neotis nuba*	Nubian Bustard
☑	*Tetrax tetrax*	Little Bustard
☐	**Family Rallidae**	**Rails, crakes and coots**
☑	*Rallus aquaticus*	Water Rail
☐	*Crex egregia*	African Crake
☑	*Crex crex*	Corn Crake
☑	*Amaurornis phoenicurus*	White-breasted Waterhen
☑	*Porzana parva*	Little Crake
☐	*Porzana pusilla*	Baillon's Crake
☑	*Porzana porzana*	Spotted Crake
☐	*Porzana carolina*	Sora
☐	*Porzana fusca*	Ruddy-breasted Crake
☐	*Aenigmatolimnas marginalis*	Striped Crake
☐	*Gallicrex cinerea*	Watercock
☑	*Porphyrio porphyrio*	Western Swamphen
☐	*Porphyrio madagascariensis*	African Swamphen
☑	*Porphyrio poliocephalus*	Grey-headed Swamphen
☐	*Porphyrio alleni*	Allen's Gallinule
☐	*Porphyrio martinica*	Purple Gallinule
☑	*Gallinula chloropus*	Common Moorhen
☐	*Paragallinula angulata*	Lesser Moorhen
☑	*Fulica cristata*	Red-knobbed Coot
☑	*Fulica atra*	Eurasian Coot
☐	*Fulica americana*	American Coot

☐ **Family Gruidae**	**Cranes**
☐ *Grus leucogeranus*	Siberian Crane
☐ *Grus canadensis*	Sandhill Crane
☑ *Grus virgo*	Demoiselle Crane
☑ *Grus grus*	Common Crane
☐ **Family Turnicidae**	**Buttonquail**
☑ *Turnix sylvaticus*	Common Buttonquail
☐ **Family Burhinidae**	**Stone-curlews and thick-knees**
☑ *Burhinus oedicnemus*	Eurasian Stone-curlew
☐ *Burhinus senegalensis*	Senegal Thick-knee
☑ *Burhinus capensis*	Spotted Thick-knee
☐ *Esacus recurvirostris*	Great Stone-curlew
☐ **Family Haematopodidae**	**Oystercatchers**
☑ *Haematopus ostralegus*	Eurasian Oystercatcher
☐ **Family Dromadidae**	**Crab-plover**
☑ *Dromas ardeola*	Crab-plover
☐ **Family Recurvirostridae**	**Stilts and avocets**
☑ *Himantopus himantopus*	Black-winged Stilt
☑ *Recurvirostra avosetta*	Pied Avocet
☐ **Family Charadriidae**	**Lapwings and plovers**
☑ *Vanellus vanellus*	Northern Lapwing
☑ *Vanellus spinosus*	Spur-winged Lapwing
☐ *Vanellus tectus*	Black-headed Lapwing
☐ *Vanellus cinereus*	Grey-headed Lapwing
☑ *Vanellus indicus*	Red-wattled Lapwing
☑ *Vanellus gregarius*	Sociable Lapwing
☑ *Vanellus leucurus*	White-tailed Lapwing
☑ *Pluvialis apricaria*	European Golden Plover
☑ *Pluvialis fulva*	Pacific Golden Plover
☑ *Pluvialis dominica*	American Golden Plover
☑ *Pluvialis squatarola*	Grey Plover
☑ *Charadrius hiaticula*	Common Ringed Plover
☐ *Charadrius semipalmatus*	Semipalmated Plover
☑ *Charadrius dubius*	Little Ringed Plover
☐ *Charadrius vociferus*	Killdeer
☐ *Charadrius pecuarius*	Kittlitz's Plover
☐ *Charadrius tricollaris*	Three-banded Plover
☑ *Charadrius alexandrinus*	Kentish Plover
☑ *Charadrius mongolus*	Lesser Sand Plover

☑	*Charadrius leschenaultii*	Greater Sand Plover
☐	*Charadrius asiaticus*	Caspian Plover
☐	*Charadrius veredus*	Oriental Plover
☑	*Charadrius morinellus*	Eurasian Dotterel
☐	**Family Pluvianidae**	**Egyptian Plover**
☐	*Pluvianus aegyptius*	Egyptian Plover
☐	**Family Rostratulidae**	**Painted-snipes**
☑	*Rostratula benghalensis*	Greater Painted-snipe
☐	**Family Jacanidae**	**Jacanas**
☑	*Hydrophasianus chirurgus*	Pheasant-tailed Jacana
☐	**Family Scolopacidae**	**Sandpipers and snipes**
☑	*Scolopax rusticola*	Eurasian Woodcock
☐	*Scolopax minor*	American Woodcock
☑	*Lymnocryptes minimus*	Jack Snipe
☐	*Gallinago solitaria*	Solitary Snipe
☐	*Gallinago stenura*	Pin-tailed Snipe
☐	*Gallinago megala*	Swinhoe's Snipe
☑	*Gallinago media*	Great Snipe
☑	*Gallinago gallinago*	Common Snipe
☐	*Gallinago delicata*	Wilson's Snipe
☐	*Limnodromus griseus*	Short-billed Dowitcher
☑	*Limnodromus scolopaceus*	Long-billed Dowitcher
☐	*Limnodromus semipalmatus*	Asian Dowitcher
☑	*Limosa limosa*	Black-tailed Godwit
☐	*Limosa haemastica*	Hudsonian Godwit
☑	*Limosa lapponica*	Bar-tailed Godwit
☐	*Numenius minutus*	Little Curlew
☐	*Numenius borealis*	Eskimo Curlew
☑	*Numenius phaeopus*	Eurasian Whimbrel
☑	*Numenius hudsonicus*	Hudsonian Whimbrel
☐	*Numenius tenuirostris*	Slender-billed Curlew
☑	*Numenius arquata*	Eurasian Curlew
☐	*Numenius madagascariensis*	Far Eastern Curlew
☑	*Bartramia longicauda*	Upland Sandpiper
☑	*Tringa erythropus*	Spotted Redshank
☑	*Tringa totanus*	Common Redshank
☑	*Tringa stagnatilis*	Marsh Sandpiper
☑	*Tringa nebularia*	Common Greenshank
☑	*Tringa melanoleuca*	Greater Yellowlegs

☑	*Tringa flavipes*	Lesser Yellowlegs
☑	*Tringa ochropus*	Green Sandpiper
☐	*Tringa solitaria*	Solitary Sandpiper
☑	*Tringa glareola*	Wood Sandpiper
☐	*Tringa brevipes*	Grey-tailed Tattler
☐	*Tringa semipalmata*	Willet
☑	*Xenus cinereus*	Terek Sandpiper
☑	*Actitis hypoleucos*	Common Sandpiper
☑	*Actitis macularius*	Spotted Sandpiper
☑	*Arenaria interpres*	Ruddy Turnstone
☑	*Calidris tenuirostris*	Great Knot
☑	*Calidris canutus*	Red Knot
☑	*Calidris alba*	Sanderling
☑	*Calidris pusilla*	Semipalmated Sandpiper
☑	*Calidris mauri*	Western Sandpiper
☐	*Calidris ruficollis*	Red-necked Stint
☑	*Calidris minuta*	Little Stint
☑	*Calidris temminckii*	Temminck's Stint
☐	*Calidris subminuta*	Long-toed Stint
☑	*Calidris minutilla*	Least Sandpiper
☑	*Calidris fuscicollis*	White-rumped Sandpiper
☑	*Calidris bairdii*	Baird's Sandpiper
☑	*Calidris melanotos*	Pectoral Sandpiper
☑	*Calidris acuminata*	Sharp-tailed Sandpiper
☑	*Calidris ferruginea*	Curlew Sandpiper
☑	*Calidris maritima*	Purple Sandpiper
☑	*Calidris alpina*	Dunlin
☐	*Calidris himantopus*	Stilt Sandpiper
☐	*Eurynorhynchus pygmeus*	Spoon-billed Sandpiper
☑	*Limicola falcinellus*	Broad-billed Sandpiper
☑	*Tryngites subruficollis*	Buff-breasted Sandpiper
☑	*Philomachus pugnax*	Ruff
☑	*Phalaropus tricolor*	Wilson's Phalarope
☑	*Phalaropus lobatus*	Red-necked Phalarope
☑	*Phalaropus fulicarius*	Red Phalarope
☐	**Family Glareolidae**	**Coursers and pratincoles**
☑	*Cursorius cursor*	Cream-coloured Courser
☐	*Cursorius temminckii*	Temminck's Courser
☑	*Glareola pratincola*	Collared Pratincole

☑ *Glareola maldivarum*	Oriental Pratincole
☑ *Glareola nordmanni*	Black-winged Pratincole
☐ *Glareola lactea*	Small Pratincole
☐ **Family Laridae**	**Gulls, terns and skimmers**
☑ *Anous stolidus*	Brown Noddy
☐ *Anous tenuirostris*	Lesser Noddy
☐ *Rynchops flavirostris*	African Skimmer
☐ *Rynchops albicollis*	Indian Skimmer
☑ *Rissa tridactyla*	Black-legged Kittiwake
☐ *Pagophila eburnea*	Ivory Gull
☑ *Xema sabini*	Sabine's Gull
☑ *Chroicocephalus genei*	Slender-billed Gull
☑ *Chroicocephalus philadelphia*	Bonaparte's Gull
☐ *Chroicocephalus brunnicephalus*	Brown-headed Gull
☑ *Chroicocephalus ridibundus*	Black-headed Gull
☑ *Chroicocephalus cirrocephalus*	Grey-headed Gull
☑ *Hydrocoloeus minutus*	Little Gull
☑ *Rhodostethia rosea*	Ross's Gull
☑ *Leucophaeus atricilla*	Laughing Gull
☐ *Leucophaeus pipixcan*	Franklin's Gull
☐ *Ichthyaetus relictus*	Relict Gull
☑ *Ichthyaetus audouinii*	Audouin's Gull
☑ *Ichthyaetus melanocephalus*	Mediterranean Gull
☑ *Ichthyaetus ichthyaetus*	Pallas's Gull
☑ *Ichthyaetus leucophthalmus*	White-eyed Gull
☑ *Ichthyaetus hemprichii*	Sooty Gull
☑ *Larus canus*	Common Gull
☑ *Larus delawarensis*	Ring-billed Gull
☑ *Larus marinus*	Great Black-backed Gull
☐ *Larus dominicanus*	Kelp Gull
☐ *Larus glaucescens*	Glaucous-winged Gull
☑ *Larus hyperboreus*	Glaucous Gull
☑ *Larus glaucoides*	Iceland Gull
☐ *Larus thayeri*	Thayer's Gull
☑ *Larus argentatus*	European Herring Gull
☐ *Larus smithsonianus*	American Herring Gull
☑ *Larus cachinnans*	Caspian Gull
☑ *Larus michahellis*	Yellow-legged Gull
☑ *Larus armenicus*	Armenian Gull

☐	*Larus schistisagus*	Slaty-backed Gull
☑	*Larus fuscus*	Lesser Black-backed Gull
☑	*Gelochelidon nilotica*	Gull-billed Tern
☑	*Hydroprogne caspia*	Caspian Tern
☐	*Thalasseus maximus*	Royal Tern
☑	*Thalasseus bergii*	Greater Crested Tern
☑	*Thalasseus elegans*	Elegant Tern
☑	*Thalasseus bengalensis*	Lesser Crested Tern
☑	*Thalasseus sandvicensis*	Sandwich Tern
☐	*Thalasseus acuflavidus*	Cabot's Tern
☑	*Sternula albifrons*	Little Tern
☑	*Sternula saundersi*	Saunders's Tern
☐	*Sternula antillarum*	Least Tern
☐	*Onychoprion aleuticus*	Aleutian Tern
☑	*Onychoprion anaethetus*	Bridled Tern
☑	*Onychoprion fuscatus*	Sooty Tern
☐	*Sterna aurantia*	River Tern
☑	*Sterna dougallii*	Roseate Tern
☑	*Sterna hirundo*	Common Tern
☑	*Sterna repressa*	White-cheeked Tern
☑	*Sterna paradisaea*	Arctic Tern
☐	*Sterna forsteri*	Forster's Tern
☑	*Chlidonias hybrida*	Whiskered Tern
☑	*Chlidonias leucopterus*	White-winged Tern
☑	*Chlidonias niger*	Black Tern
☐	**Family Stercorariidae**	**Skuas**
☐	*Stercorarius maccormicki*	South Polar Skua
☐	*Stercorarius antarcticus*	Brown Skua
☑	*Stercorarius skua*	Great Skua
☑	*Stercorarius pomarinus*	Pomarine Skua
☑	*Stercorarius parasiticus*	Arctic Skua
☑	*Stercorarius longicaudus*	Long-tailed Skua
☐	**Family Alcidae**	**Auks**
☑	*Alle alle*	Little Auk
☑	*Uria lomvia*	Brünnich's Guillemot
☑	*Uria aalge*	Common Guillemot
☑	*Alca torda*	Razorbill
☑	*Cepphus grylle*	Black Guillemot
☐	*Brachyramphus perdix*	Long-billed Murrelet

☐	*Synthliboramphus antiquus*	Ancient Murrelet
☐	*Aethia psittacula*	Parakeet Auklet
☐	*Aethia cristatella*	Crested Auklet
☑	*Fratercula arctica*	Atlantic Puffin
☐	*Fratercula cirrhata*	Tufted Puffin
☐	**Family Pteroclidae**	**Sandgrouse**
☐	*Syrrhaptes paradoxus*	Pallas's Sandgrouse
☑	*Pterocles alchata*	Pin-tailed Sandgrouse
☑	*Pterocles exustus*	Chestnut-bellied Sandgrouse
☑	*Pterocles senegallus*	Spotted Sandgrouse
☑	*Pterocles orientalis*	Black-bellied Sandgrouse
☑	*Pterocles coronatus*	Crowned Sandgrouse
☑	*Pterocles lichtensteinii*	Lichtenstein's Sandgrouse
☐	**Family Columbidae**	**Pigeons and doves**
☑	*Columba livia*	Rock Dove
☐	*Columba guinea*	Speckled Pigeon
☑	*Columba oenas*	Stock Dove
☐	*Columba eversmanni*	Yellow-eyed Pigeon
☑	*Columba palumbus*	Common Wood Pigeon
☑	*Columba trocaz*	Trocaz Pigeon
☑	*Columba bollii*	Bolle's Pigeon
☑	*Columba junoniae*	Laurel Pigeon
☐	*Columba arquatrix*	African Olive Pigeon
☑	*Streptopelia turtur*	European Turtle Dove
☑	*Streptopelia lugens*	Dusky Turtle Dove
☑	*Streptopelia orientalis*	Oriental Turtle Dove
☑	*Streptopelia decaocto*	Eurasian Collared Dove
☑	*Streptopelia roseogrisea*	African Collared Dove
☐	*Streptopelia decipiens*	Mourning Collared Dove
☐	*Streptopelia semitorquata*	Red-eyed Dove
☐	*Streptopelia tranquebarica*	Red Turtle Dove
☑	*Spilopelia senegalensis*	Laughing Dove
☐	*Turtur abyssinicus*	Black-billed Wood Dove
☑	*Oena capensis*	Namaqua Dove
☐	*Geopelia striata*	Zebra Dove
☐	*Zenaida macroura*	Mourning Dove
☑	*Treron waalia*	Bruce's Green Pigeon
☐	**Family Cuculidae**	**Cuckoos**
☐	*Centropus senegalensis*	Senegal Coucal

☐	*Centropus superciliosus*	White-browed Coucal
☑	*Clamator glandarius*	Great Spotted Cuckoo
☐	*Clamator jacobinus*	Jacobin Cuckoo
☐	*Coccyzus americanus*	Yellow-billed Cuckoo
☐	*Coccyzus erythropthalmus*	Black-billed Cuckoo
☑	*Eudynamys scolopaceus*	Asian Koel
☐	*Chrysococcyx caprius*	Diederik Cuckoo
☐	*Chrysococcyx klaas*	Klaas's Cuckoo
☐	*Cacomantis passerinus*	Grey-bellied Cuckoo
☐	*Hierococcyx varius*	Common Hawk-Cuckoo
☐	*Cuculus optatus*	Oriental Cuckoo
☑	*Cuculus canorus*	Common Cuckoo
☐	**Family Tytonidae**	**Barn owls**
☑	*Tyto alba*	Western Barn Owl
☐	**Family Strigidae**	**Typical owls**
☑	*Otus brucei*	Pallid Scops Owl
☑	*Otus pamelae*	Arabian Scops Owl
☑	*Otus scops*	Eurasian Scops Owl
☑	*Bubo scandiacus*	Snowy Owl
☑	*Bubo bubo*	Eurasian Eagle-Owl
☑	*Bubo ascalaphus*	Pharaoh Eagle-Owl
☑	*Bubo africanus*	Spotted Eagle-Owl
☑	*Ketupa zeylonensis*	Brown Fish Owl
☑	*Strix aluco*	Tawny Owl
☐	*Strix butleri*	Omani Owl
☑	*Strix hadorami*	Desert Owl
☑	*Strix uralensis*	Ural Owl
☑	*Strix nebulosa*	Great Grey Owl
☑	*Surnia ulula*	Northern Hawk-Owl
☑	*Glaucidium passerinum*	Eurasian Pygmy Owl
☑	*Athene noctua*	Little Owl
☐	*Athene brama*	Spotted Owlet
☑	*Aegolius funereus*	Tengmalm's Owl
☑	*Asio otus*	Long-eared Owl
☑	*Asio flammeus*	Short-eared Owl
☑	*Asio capensis*	Marsh Owl
☐	**Family Caprimulgidae**	**Nightjars**
☐	*Chordeiles minor*	Common Nighthawk
☑	*Caprimulgus ruficollis*	Red-necked Nightjar

☑ *Caprimulgus europaeus*	European Nightjar
☑ *Caprimulgus aegyptius*	Egyptian Nightjar
☐ *Caprimulgus mahrattensis*	Sykes's Nightjar
☑ *Caprimulgus nubicus*	Nubian Nightjar
☐ *Caprimulgus eximius*	Golden Nightjar
☐ *Caprimulgus poliocephalus*	Montane Nightjar
☐ *Caprimulgus inornatus*	Plain Nightjar
☐ **Family Apodidae**	**Swifts**
☐ *Hirundapus caudacutus*	White-throated Needletail
☐ *Chaetura pelagica*	Chimney Swift
☐ *Cypsiurus parvus*	African Palm Swift
☑ *Tachymarptis melba*	Alpine Swift
☑ *Apus alexandri*	Cape Verde Swift
☑ *Apus apus*	Common Swift
☑ *Apus unicolor*	Plain Swift
☑ *Apus pallidus*	Pallid Swift
☑ *Apus berliozi*	Forbes-Watson's Swift
☐ *Apus pacificus*	Pacific Swift
☑ *Apus affinis*	Little Swift
☐ *Apus caffer*	White-rumped Swift
☐ **Family Coliidae**	**Mousebirds**
☑ *Urocolius macrourus*	Blue-naped Mousebird
☐ **Family Coraciidae**	**Rollers**
☐ *Coracias naevius*	Purple Roller
☑ *Coracias benghalensis*	Indian Roller
☐ *Coracias caudatus*	Lilac-breasted Roller
☐ *Coracias abyssinicus*	Abyssinian Roller
☑ *Coracias garrulus*	European Roller
☐ *Eurystomus glaucurus*	Broad-billed Roller
☐ **Family Alcedinidae**	**Kingfishers**
☑ *Halcyon smyrnensis*	White-throated Kingfisher
☑ *Halcyon leucocephala*	Grey-headed Kingfisher
☐ *Todiramphus chloris*	Collared Kingfisher
☐ *Corythornis cristatus*	Malachite Kingfisher
☑ *Alcedo atthis*	Common Kingfisher
☐ *Megaceryle alcyon*	Belted Kingfisher
☑ *Ceryle rudis*	Pied Kingfisher
☐ **Family Meropidae**	**Bee-eaters**
☐ *Merops albicollis*	White-throated Bee-eater

☑	*Merops orientalis*	Green Bee-eater
☑	*Merops persicus*	Blue-cheeked Bee-eater
☑	*Merops apiaster*	European Bee-eater
☐	**Family Upupidae**	**Hoopoes**
☑	*Upupa epops*	Eurasian Hoopoe
☐	**Family Bucerotidae**	**Hornbills**
☐	*Lophoceros nasutus*	African Grey Hornbill
☐	**Family Picidae**	**Woodpeckers**
☑	*Jynx torquilla*	Eurasian Wryneck
☐	*Sphyrapicus varius*	Yellow-bellied Sapsucker
☑	*Picoides tridactylus*	Eurasian Three-toed Woodpecker
☐	*Dendrocoptes dorae*	Arabian Woodpecker
☑	*Dendrocoptes medius*	Middle Spotted Woodpecker
☐	*Dendropicos goertae*	African Grey Woodpecker
☑	*Dryobates minor*	Lesser Spotted Woodpecker
☐	*Dendrocopos assimilis*	Sind Woodpecker
☑	*Dendrocopos syriacus*	Syrian Woodpecker
☑	*Dendrocopos major*	Great Spotted Woodpecker
☑	*Dendrocopos leucotos*	White-backed Woodpecker
☐	*Colaptes auratus*	Northern Flicker
☑	*Dryocopus martius*	Black Woodpecker
☑	*Picus viridis*	European Green Woodpecker
☑	*Picus sharpei*	Iberian Green Woodpecker
☑	*Picus vaillantii*	Levaillant's Woodpecker
☑	*Picus canus*	Grey-headed Woodpecker
☐	**Family Falconidae**	**Falcons**
☑	*Falco naumanni*	Lesser Kestrel
☑	*Falco tinnunculus*	Common Kestrel
☐	*Falco sparverius*	American Kestrel
☐	*Falco chicquera*	Red-necked Falcon
☑	*Falco vespertinus*	Red-footed Falcon
☐	*Falco amurensis*	Amur Falcon
☑	*Falco eleonorae*	Eleonora's Falcon
☑	*Falco concolor*	Sooty Falcon
☑	*Falco columbarius*	Merlin
☑	*Falco subbuteo*	Eurasian Hobby
☑	*Falco biarmicus*	Lanner Falcon
☐	*Falco jugger*	Laggar Falcon
☑	*Falco cherrug*	Saker Falcon

☐ *Falco rusticolus*	Gyr Falcon
☐ *Falco peregrinus*	Peregrine Falcon
☐ *Falco pelegrinoides*	Barbary Falcon
☐ **Family Psittacidae**	**African and New World parrots**
☐ *Myiopsitta monachus*	Monk Parakeet
☐ *Aratinga nenday*	Nanday Parakeet
☐ **Family Psittaculidae**	**Old World parrots**
☐ *Psittacula eupatria*	Alexandrine Parakeet
☐ *Psittacula krameri*	Rose-ringed Parakeet
☐ **Family Pittidae**	**Pittas**
☐ *Pitta brachyura*	Indian Pitta
☐ **Family Tyrannidae**	**Tyrant flycatchers and allies**
☐ *Sayornis phoebe*	Eastern Phoebe
☐ *Contopus virens*	Eastern Wood Pewee
☐ *Empidonax virescens*	Acadian Flycatcher
☐ *Empidonax alnorum*	Alder Flycatcher
☐ *Empidonax minimus*	Least Flycatcher
☐ *Tyrannus savana*	Fork-tailed Flycatcher
☐ *Tyrannus tyrannus*	Eastern Kingbird
☐ **Family Malaconotidae**	**Bush-shrikes**
☐ *Tchagra senegalus*	Black-crowned Tchagra
☐ **Family Laniidae**	**Shrikes**
☐ *Lanius cristatus*	Brown Shrike
☐ *Lanius collurio*	Red-backed Shrike
☐ *Lanius isabellinus*	Daurian Shrike
☐ *Lanius phoenicuroides*	Turkestan Shrike
☐ *Lanius vittatus*	Bay-backed Shrike
☐ *Lanius schach*	Long-tailed Shrike
☐ *Lanius minor*	Lesser Grey Shrike
☐ *Lanius excubitor*	Great Grey Shrike
☐ *Lanius meridionalis*	Southern Grey Shrike
☐ *Lanius pallidirostris*	Steppe Grey Shrike
☐ *Lanius senator*	Woodchat Shrike
☐ *Lanius nubicus*	Masked Shrike
☐ **Family Vireonidae**	**Vireos**
☐ *Vireo griseus*	White-eyed Vireo
☐ *Vireo flavifrons*	Yellow-throated Vireo
☐ *Vireo philadelphicus*	Philadelphia Vireo
☐ *Vireo olivaceus*	Red-eyed Vireo

☐	**Family Oriolidae**	**Old World orioles**
☐	*Oriolus oriolus*	Eurasian Golden Oriole
☐	*Oriolus kundoo*	Indian Golden Oriole
☐	*Oriolus chinensis*	Black-naped Oriole
☐	**Family Dicruridae**	**Drongos**
☐	*Dicrurus macrocercus*	Black Drongo
☐	*Dicrurus leucophaeus*	Ashy Drongo
☐	**Family Monarchidae**	**Monarch flycatchers and allies**
☐	*Hypothymis azurea*	Black-naped Monarch
☐	*Terpsiphone viridis*	African Paradise Flycatcher
☐	**Family Corvidae**	**Crows, jays and magpies**
☐	*Perisoreus infaustus*	Siberian Jay
☐	*Garrulus glandarius*	Eurasian Jay
☐	*Cyanopica cooki*	Iberian Magpie
☐	*Pica pica*	Eurasian Magpie
☐	*Podoces pleskei*	Pleske's Ground Jay
☐	*Nucifraga caryocatactes*	Spotted Nutcracker
☐	*Pyrrhocorax pyrrhocorax*	Red-billed Chough
☐	*Pyrrhocorax graculus*	Alpine Chough
☐	*Coloeus monedula*	Western Jackdaw
☐	*Coloeus dauuricus*	Daurian Jackdaw
☐	*Corvus splendens*	House Crow
☐	*Corvus frugilegus*	Rook
☐	*Corvus corone*	Carrion Crow
☐	*Corvus cornix*	Hooded Crow
☐	*Corvus macrorhynchos*	Large-billed Crow
☐	*Corvus albus*	Pied Crow
☐	*Corvus ruficollis*	Brown-necked Raven
☐	*Corvus corax*	Northern Raven
☐	*Corvus rhipidurus*	Fan-tailed Raven
☐	**Family Bombycillidae**	**Waxwings**
☐	*Bombycilla garrulus*	Bohemian Waxwing
☐	*Bombycilla cedrorum*	Cedar Waxwing
☐	**Family Hypocoliidae**	**Grey Hypocolius**
☐	*Hypocolius ampelinus*	Grey Hypocolius
☐	**Family Paridae**	**Tits and chickadees**
☐	*Periparus ater*	Coal Tit
☐	*Lophophanes cristatus*	European Crested Tit
☐	*Poecile lugubris*	Sombre Tit

☑	*Poecile palustris*	Marsh Tit
☐	*Poecile hyrcanus*	Caspian Tit
☑	*Poecile montanus*	Willow Tit
☑	*Poecile cinctus*	Siberian Tit
☑	*Cyanistes teneriffae*	African Blue Tit
☑	*Cyanistes caeruleus*	Eurasian Blue Tit
☑	*Cyanistes cyanus*	Azure Tit
☑	*Parus major*	Great Tit
☐	*Parus cinereus*	Cinereous Tit
☐	**Family Remizidae**	**Penduline tits**
☑	*Remiz pendulinus*	Eurasian Penduline Tit
☐	*Remiz macronyx*	Black-headed Penduline Tit
☐	*Remiz coronatus*	White-crowned Penduline Tit
☐	**Family Panuridae**	**Bearded Reedling**
☑	*Panurus biarmicus*	Bearded Reedling
☐	**Family Alaudidae**	**Larks**
☑	*Alaemon alaudipes*	Greater Hoopoe-Lark
☑	*Rhamphocoris clotbey*	Thick-billed Lark
☑	*Ammomanes deserti*	Desert Lark
☑	*Ammomanes cinctura*	Bar-tailed Lark
☑	*Eremopterix nigriceps*	Black-crowned Sparrow-Lark
☐	*Eremopterix signatus*	Chestnut-headed Sparrow-Lark
☑	*Mirafra cantillans*	Singing Bush Lark
☑	*Lullula arborea*	Woodlark
☐	*Alauda leucoptera*	White-winged Lark
☑	*Alauda razae*	Raso Lark
☐	*Alauda gulgula*	Oriental Skylark
☑	*Alauda arvensis*	Eurasian Skylark
☑	*Galerida theklae*	Thekla Lark
☑	*Galerida cristata*	Crested Lark
☑	*Galerida macrorhyncha*	Maghreb Lark
☑	*Eremophila alpestris*	Horned Lark
☑	*Eremophila bilopha*	Temminck's Lark
☐	*Calandrella acutirostris*	Hume's Short-toed Lark
☐	*Calandrella blanfordi*	Blanford's Lark
☑	*Calandrella brachydactyla*	Greater Short-toed Lark
☑	*Melanocorypha bimaculata*	Bimaculated Lark
☑	*Melanocorypha calandra*	Calandra Lark
☐	*Melanocorypha yeltoniensis*	Black Lark

☑ *Chersophilus duponti*	Dupont's Lark
☑ *Eremalauda dunni*	Dunn's Lark
☐ *Alaudala cheleensis*	Asian Short-toed Lark
☑ *Alaudala rufescens*	Lesser Short-toed Lark
☐ *Alaudala raytal*	Sand Lark
☐ **Family Pycnonotidae**	**Bulbuls**
☑ *Pycnonotus leucotis*	White-eared Bulbul
☑ *Pycnonotus cafer*	Red-vented Bulbul
☑ *Pycnonotus xanthopygos*	White-spectacled Bulbul
☑ *Pycnonotus barbatus*	Common Bulbul
☐ **Family Hirundinidae**	**Swallows and martins**
☑ *Riparia paludicola*	Brown-throated Martin
☑ *Riparia chinensis*	Grey-throated Martin
☑ *Riparia riparia*	Sand Martin
☑ *Riparia diluta*	Pale Martin
☐ *Riparia cincta*	Banded Martin
☐ *Tachycineta bicolor*	Tree Swallow
☐ *Progne subis*	Purple Martin
☑ *Hirundo rustica*	Barn Swallow
☐ *Hirundo aethiopica*	Ethiopian Swallow
☐ *Hirundo smithii*	Wire-tailed Swallow
☑ *Ptyonoprogne rupestris*	Eurasian Crag Martin
☑ *Ptyonoprogne obsoleta*	Pale Crag Martin
☑ *Delichon urbicum*	Common House Martin
☐ *Delichon dasypus*	Asian House Martin
☐ *Cecropis abyssinica*	Lesser Striped Swallow
☑ *Cecropis daurica*	Red-rumped Swallow
☐ *Petrochelidon fluvicola*	Streak-throated Swallow
☐ *Petrochelidon pyrrhonota*	American Cliff Swallow
☐ **Family Cettiidae**	**Bush warblers**
☑ *Cettia cetti*	Cetti's Warbler
☐ **Family Scotocercidae**	**Streaked Scrub Warbler**
☑ *Scotocerca inquieta*	Streaked Scrub Warbler
☐ **Family Aegithalidae**	**Bushtits**
☑ *Aegithalos caudatus*	Long-tailed Tit
☐ **Family Phylloscopidae**	**Leaf warblers**
☐ *Phylloscopus umbrovirens*	Brown Woodland Warbler
☑ *Phylloscopus trochilus*	Willow Warbler
☑ *Phylloscopus collybita*	Common Chiffchaff

☑	*Phylloscopus ibericus*	Iberian Chiffchaff
☑	*Phylloscopus canariensis*	Canary Islands Chiffchaff
☑	*Phylloscopus sindianus*	Mountain Chiffchaff
☑	*Phylloscopus neglectus*	Plain Leaf Warbler
☑	*Phylloscopus bonelli*	Western Bonelli's Warbler
☑	*Phylloscopus orientalis*	Eastern Bonelli's Warbler
☑	*Phylloscopus sibilatrix*	Wood Warbler
☑	*Phylloscopus fuscatus*	Dusky Warbler
☑	*Phylloscopus schwarzi*	Radde's Warbler
☑	*Phylloscopus proregulus*	Pallas's Leaf Warbler
☑	*Phylloscopus inornatus*	Yellow-browed Warbler
☑	*Phylloscopus humei*	Hume's Leaf Warbler
☐	*Phylloscopus borealis*	Arctic Warbler
☑	*Phylloscopus nitidus*	Green Warbler
☑	*Phylloscopus trochiloides*	Greenish Warbler
☐	*Phylloscopus plumbeitarsus*	Two-barred Warbler
☐	*Phylloscopus magnirostris*	Large-billed Leaf Warbler
☐	*Phylloscopus coronatus*	Eastern Crowned Warbler
☐	**Family Acrocephalidae**	**Reed warblers and allies**
☐	*Acrocephalus griseldis*	Basra Reed Warbler
☑	*Acrocephalus brevipennis*	Cape Verde Warbler
☑	*Acrocephalus arundinaceus*	Great Reed Warbler
☐	*Acrocephalus orientalis*	Oriental Reed Warbler
☑	*Acrocephalus stentoreus*	Clamorous Reed Warbler
☑	*Acrocephalus melanopogon*	Moustached Warbler
☑	*Acrocephalus paludicola*	Aquatic Warbler
☑	*Acrocephalus schoenobaenus*	Sedge Warbler
☑	*Acrocephalus agricola*	Paddyfield Warbler
☐	*Acrocephalus dumetorum*	Blyth's Reed Warbler
☑	*Acrocephalus scirpaceus*	Eurasian Reed Warbler
☑	*Acrocephalus palustris*	Marsh Warbler
☐	*Iduna aedon*	Thick-billed Warbler
☐	*Iduna caligata*	Booted Warbler
☐	*Iduna rama*	Sykes's Warbler
☑	*Iduna pallida*	Eastern Olivaceous Warbler
☑	*Iduna opaca*	Western Olivaceous Warbler
☐	*Hippolais languida*	Upcher's Warbler
☑	*Hippolais olivetorum*	Olive-tree Warbler

☑ *Hippolais polyglotta*	Melodious Warbler
☑ *Hippolais icterina*	Icterine Warbler
☐ **Family Locustellidae**	**Grass and bush warblers**
☑ *Locustella naevia*	Common Grasshopper Warbler
☑ *Locustella fluviatilis*	River Warbler
☑ *Locustella luscinioides*	Savi's Warbler
☐ *Locustella lanceolata*	Lanceolated Warbler
☐ *Locustella certhiola*	Pallas's Grasshopper Warbler
☐ *Locustella fasciolata*	Gray's Grasshopper Warbler
☐ **Family Cisticolidae**	**Cisticolas and allies**
☑ *Cisticola juncidis*	Zitting Cisticola
☑ *Prinia gracilis*	Graceful Prinia
☑ *Spiloptila clamans*	Cricket Warbler
☐ **Family Leiothrichidae**	**Laughingthrushes and allies**
☐ *Turdoides altirostris*	Iraq Babbler
☑ *Turdoides huttoni*	Afghan Babbler
☑ *Turdoides squamiceps*	Arabian Babbler
☑ *Turdoides fulva*	Fulvous Babbler
☐ *Leiothrix lutea*	Red-billed Leiothrix
☐ **Family Sylviidae**	**Sylviid babblers**
☑ *Sylvia atricapilla*	Eurasian Blackcap
☑ *Sylvia borin*	Garden Warbler
☑ *Sylvia nisoria*	Barred Warbler
☑ *Sylvia curruca*	Lesser Whitethroat
☐ *Sylvia minula*	Desert Whitethroat
☐ *Sylvia althaea*	Hume's Whitethroat
☑ *Sylvia hortensis*	Western Orphean Warbler
☑ *Sylvia crassirostris*	Eastern Orphean Warbler
☑ *Sylvia leucomelaena*	Arabian Warbler
☑ *Sylvia nana*	Asian Desert Warbler
☑ *Sylvia deserti*	African Desert Warbler
☑ *Sylvia communis*	Common Whitethroat
☑ *Sylvia undata*	Dartford Warbler
☐ *Sylvia sarda*	Marmora's Warbler
☑ *Sylvia balearica*	Balearic Warbler
☑ *Sylvia deserticola*	Tristram's Warbler
☑ *Sylvia conspicillata*	Spectacled Warbler
☑ *Sylvia inornata*	Western Subalpine Warbler
☑ *Sylvia cantillans*	Eastern Subalpine Warbler

☐	*Sylvia subalpina*	Moltoni's Warbler
☑	*Sylvia melanocephala*	Sardinian Warbler
☑	*Sylvia mystacea*	Ménétriés's Warbler
☑	*Sylvia ruppeli*	Rüppell's Warbler
☑	*Sylvia melanothorax*	Cyprus Warbler
☐	*Sylvia buryi*	Yemen Warbler
☐	*Sinosuthora webbiana*	Vinous-throated Parrotbill
☐	**Family Zosteropidae**	**White-eyes**
☐	*Zosterops palpebrosus*	Oriental White-eye
☑	*Zosterops abyssinicus*	Abyssinian White-eye
☐	**Family Regulidae**	**Kinglets**
☑	*Regulus ignicapilla*	Common Firecrest
☑	*Regulus regulus*	Goldcrest
☑	*Regulus madeirensis*	Madeira Firecrest
☐	*Regulus calendula*	Ruby-crowned Kinglet
☐	**Family Troglodytidae**	**Wrens**
☑	*Troglodytes troglodytes*	Eurasian Wren
☐	**Family Sittidae**	**Nuthatches**
☑	*Sitta europaea*	Eurasian Nuthatch
☑	*Sitta whiteheadi*	Corsican Nuthatch
☐	*Sitta ledanti*	Algerian Nuthatch
☑	*Sitta krueperi*	Krüper's Nuthatch
☐	*Sitta canadensis*	Red-breasted Nuthatch
☑	*Sitta neumayer*	Western Rock Nuthatch
☑	*Sitta tephronota*	Eastern Rock Nuthatch
☐	**Family Tichodromadidae**	**Wallcreeper**
☑	*Tichodroma muraria*	Wallcreeper
☐	**Family Certhiidae**	**Treecreepers**
☑	*Certhia familiaris*	Eurasian Treecreeper
☑	*Certhia brachydactyla*	Short-toed Treecreeper
☐	**Family Mimidae**	**Mockingbirds and thrashers**
☐	*Dumetella carolinensis*	Grey Catbird
☐	*Mimus polyglottos*	Northern Mockingbird
☐	*Toxostoma rufum*	Brown Thrasher
☐	**Family Sturnidae**	**Starlings and allies**
☐	*Acridotheres cristatellus*	Crested Myna
☑	*Acridotheres ginginianus*	Bank Myna
☑	*Acridotheres tristis*	Common Myna
☐	*Acridotheres burmannicus*	Vinous-breasted Starling

☐	*Gracupica contra*	Pied Myna
☐	*Agropsar sturninus*	Daurian Starling
☐	*Sturnia malabarica*	Chestnut-tailed Starling
☐	*Sturnia pagodarum*	Brahminy Starling
☐	*Pastor roseus*	Rosy Starling
☐	*Sturnus vulgaris*	Common Starling
☐	*Sturnus unicolor*	Spotless Starling
☐	*Creatophora cinerea*	Wattled Starling
☐	*Lamprotornis pulcher*	Chestnut-bellied Starling
☐	*Cinnyricinclus leucogaster*	Violet-backed Starling
☐	*Onychognathus tristramii*	Tristram's Starling
☐	**Family Buphagidae**	**Oxpeckers**
☐	*Buphagus erythrorynchus*	Red-billed Oxpecker
☐	**Family Turdidae**	**Thrushes**
☐	*Geokichla sibirica*	Siberian Thrush
☐	*Zoothera aurea*	White's Thrush
☐	*Ixoreus naevius*	Varied Thrush
☐	*Catharus fuscescens*	Veery
☐	*Catharus minimus*	Grey-cheeked Thrush
☐	*Catharus ustulatus*	Swainson's Thrush
☐	*Catharus guttatus*	Hermit Thrush
☐	*Hylocichla mustelina*	Wood Thrush
☐	*Turdus menachensis*	Yemen Thrush
☐	*Turdus unicolor*	Tickell's Thrush
☐	*Turdus torquatus*	Ring Ouzel
☐	*Turdus merula*	Common Blackbird
☐	*Turdus obscurus*	Eyebrowed Thrush
☐	*Turdus atrogularis*	Black-throated Thrush
☐	*Turdus ruficollis*	Red-throated Thrush
☐	*Turdus naumanni*	Naumann's Thrush
☐	*Turdus eunomus*	Dusky Thrush
☐	*Turdus pilaris*	Fieldfare
☐	*Turdus iliacus*	Redwing
☐	*Turdus philomelos*	Song Thrush
☐	*Turdus viscivorus*	Mistle Thrush
☐	*Turdus migratorius*	American Robin
☐	**Family Muscicapidae**	**Chats and Old World flycatchers**
☐	*Cercotrichas podobe*	Black Scrub Robin
☐	*Cercotrichas galactotes*	Rufous-tailed Scrub Robin

☑	*Muscicapa striata*	Spotted Flycatcher
☐	*Muscicapa gambagae*	Gambaga Flycatcher
☐	*Muscicapa sibirica*	Dark-sided Flycatcher
☐	*Muscicapa dauurica*	Asian Brown Flycatcher
☐	*Cyanoptila cyanomelana*	Blue-and-white Flycatcher
☐	*Eumyias thalassinus*	Verditer Flycatcher
☑	*Erithacus rubecula*	European Robin
☐	*Larvivora cyane*	Siberian Blue Robin
☐	*Larvivora sibilans*	Rufous-tailed Robin
☑	*Luscinia svecica*	Bluethroat
☑	*Luscinia luscinia*	Thrush Nightingale
☑	*Luscinia megarhynchos*	Common Nightingale
☑	*Irania gutturalis*	White-throated Robin
☐	*Calliope calliope*	Siberian Rubythroat
☑	*Tarsiger cyanurus*	Red-flanked Bluetail
☐	*Myophonus caeruleus*	Blue Whistling Thrush
☑	*Ficedula hypoleuca*	European Pied Flycatcher
☑	*Ficedula speculigera*	Atlas Pied Flycatcher
☑	*Ficedula albicollis*	Collared Flycatcher
☑	*Ficedula semitorquata*	Semicollared Flycatcher
☐	*Ficedula mugimaki*	Mugimaki Flycatcher
☑	*Ficedula parva*	Red-breasted Flycatcher
☐	*Ficedula albicilla*	Taiga Flycatcher
☐	*Ficedula superciliaris*	Ultramarine Flycatcher
☐	*Phoenicurus erythronotus*	Eversmann's Redstart
☑	*Phoenicurus ochruros*	Black Redstart
☑	*Phoenicurus phoenicurus*	Common Redstart
☐	*Phoenicurus auroreus*	Daurian Redstart
☑	*Phoenicurus moussieri*	Moussier's Redstart
☑	*Phoenicurus erythrogastrus*	Güldenstädt's Redstart
☑	*Monticola saxatilis*	Common Rock Thrush
☐	*Monticola rufocinereus*	Little Rock Thrush
☑	*Monticola solitarius*	Blue Rock Thrush
☑	*Saxicola rubetra*	Whinchat
☑	*Saxicola dacotiae*	Canary Islands Stonechat
☑	*Saxicola rubicola*	European Stonechat
☑	*Saxicola maurus*	Siberian Stonechat
☐	*Saxicola stejnegeri*	Stejneger's Stonechat
☐	*Saxicola torquata*	African Stonechat

☐	*Saxicola caprata*	Pied Stonechat
☐	*Myrmecocichla aethiops*	Anteater Chat
☑	*Oenanthe oenanthe*	Northern Wheatear
☑	*Oenanthe seebohmi*	Seebohm's Wheatear
☐	*Oenanthe bottae*	Red-breasted Wheatear
☑	*Oenanthe isabellina*	Isabelline Wheatear
☑	*Oenanthe monacha*	Hooded Wheatear
☑	*Oenanthe deserti*	Desert Wheatear
☑	*Oenanthe hispanica*	Black-eared Wheatear
☑	*Oenanthe cypriaca*	Cyprus Wheatear
☑	*Oenanthe pleschanka*	Pied Wheatear
☑	*Oenanthe moesta*	Red-rumped Wheatear
☑	*Oenanthe melanura*	Blackstart
☐	*Oenanthe picata*	Variable Wheatear
☑	*Oenanthe leucura*	Black Wheatear
☑	*Oenanthe leucopyga*	White-crowned Wheatear
☑	*Oenanthe albonigra*	Hume's Wheatear
☑	*Oenanthe finschii*	Finsch's Wheatear
☑	*Oenanthe lugens*	Mourning Wheatear
☑	*Oenanthe halophila*	Maghreb Wheatear
☑	*Oenanthe lugentoides*	Arabian Wheatear
☑	*Oenanthe xanthoprymna*	Kurdish Wheatear
☑	*Oenanthe chrysopygia*	Persian Wheatear
☐	**Family Cinclidae**	**Dippers**
☑	*Cinclus cinclus*	White-throated Dipper
☐	**Family Nectariniidae**	**Sunbirds**
☐	*Hedydipna platura*	Pygmy Sunbird
☑	*Hedydipna metallica*	Nile Valley Sunbird
☐	*Cinnyris pulchellus*	Beautiful Sunbird
☑	*Cinnyris osea*	Palestine Sunbird
☑	*Cinnyris habessinicus*	Shining Sunbird
☑	*Cinnyris asiaticus*	Purple Sunbird
☐	**Family Passeridae**	**Old World sparrows and snowfinches**
☐	*Passer ammodendri*	Saxaul Sparrow
☑	*Passer domesticus*	House Sparrow
☑	*Passer italiae*	Italian Sparrow
☑	*Passer hispaniolensis*	Spanish Sparrow
☐	*Passer pyrrhonotus*	Sind Sparrow
☑	*Passer moabiticus*	Dead Sea Sparrow

☑ *Passer iagoensis*	Iago Sparrow
☑ *Passer simplex*	Desert Sparrow
☐ *Passer zarudnyi*	Zarudny's Sparrow
☑ *Passer montanus*	Eurasian Tree Sparrow
☑ *Passer luteus*	Sudan Golden Sparrow
☐ *Passer euchlorus*	Arabian Golden Sparrow
☐ *Carpospiza brachydactyla*	Pale Rockfinch
☑ *Petronia petronia*	Rock Sparrow
☐ *Gymnoris dentata*	Bush Petronia
☐ *Gymnoris xanthocollis*	Yellow-throated Sparrow
☑ *Montifringilla nivalis*	White-winged Snowfinch
☐ **Family Ploceidae**	**Weavers and allies**
☑ *Ploceus galbula*	Rüppell's Weaver
☐ *Ploceus cucullatus*	Village Weaver
☐ *Ploceus melanocephalus*	Black-headed Weaver
☐ *Ploceus jacksoni*	Golden-backed Weaver
☐ *Ploceus benghalensis*	Black-breasted Weaver
☐ *Ploceus manyar*	Streaked Weaver
☐ *Quelea quelea*	Red-billed Quelea
☐ *Foudia madagascariensis*	Red Fody
☐ *Euplectes afer*	Yellow-crowned Bishop
☐ **Family Estrildidae**	**Waxbills, munias and allies**
☐ *Lagonosticta senegala*	Red-billed Firefinch
☐ *Estrilda rufibarba*	Arabian Waxbill
☑ *Estrilda astrild*	Common Waxbill
☑ *Amandava amandava*	Red Avadavat
☐ *Amandava subflava*	Orange-breasted Waxbill
☑ *Euodice cantans*	African Silverbill
☑ *Euodice malabarica*	Indian Silverbill
☐ *Lonchura punctulata*	Scaly-breasted Munia
☐ *Lonchura malacca*	Tricolored Munia
☐ **Family Prunellidae**	**Accentors**
☑ *Prunella collaris*	Alpine Accentor
☐ *Prunella montanella*	Siberian Accentor
☑ *Prunella ocularis*	Radde's Accentor
☐ *Prunella fagani*	Arabian Accentor
☐ *Prunella atrogularis*	Black-throated Accentor
☑ *Prunella modularis*	Dunnock

☐	**Family Motacillidae**	**Wagtails and pipits**
☐	*Dendronanthus indicus*	Forest Wagtail
☑	*Motacilla flava*	Western Yellow Wagtail
☐	*Motacilla tschutschensis*	Eastern Yellow Wagtail
☑	*Motacilla citreola*	Citrine Wagtail
☑	*Motacilla cinerea*	Grey Wagtail
☑	*Motacilla alba*	White Wagtail
☐	*Motacilla aguimp*	African Pied Wagtail
☐	*Tmetothylacus tenellus*	Golden Pipit
☐	*Anthus richardi*	Richard's Pipit
☐	*Anthus rufulus*	Paddyfield Pipit
☐	*Anthus cinnamomeus*	African Pipit
☐	*Anthus godlewskii*	Blyth's Pipit
☑	*Anthus campestris*	Tawny Pipit
☑	*Anthus similis*	Long-billed Pipit
☑	*Anthus pratensis*	Meadow Pipit
☑	*Anthus trivialis*	Tree Pipit
☐	*Anthus hodgsoni*	Olive-backed Pipit
☐	*Anthus gustavi*	Pechora Pipit
☑	*Anthus cervinus*	Red-throated Pipit
☑	*Anthus rubescens*	Buff-bellied Pipit
☑	*Anthus spinoletta*	Water Pipit
☑	*Anthus petrosus*	European Rock Pipit
☑	*Anthus berthelotii*	Berthelot's Pipit
☐	**Family Fringillidae**	**Finches**
☑	*Fringilla coelebs*	Common Chaffinch
☑	*Fringilla teydea*	Blue Chaffinch
☑	*Fringilla montifringilla*	Brambling
☐	*Mycerobas carnipes*	White-winged Grosbeak
☐	*Hesperiphona vespertina*	Evening Grosbeak
☑	*Coccothraustes coccothraustes*	Hawfinch
☑	*Pinicola enucleator*	Pine Grosbeak
☑	*Pyrrhula pyrrhula*	Eurasian Bullfinch
☑	*Pyrrhula murina*	Azores Bullfinch
☑	*Rhodopechys sanguineus*	Eurasian Crimson-winged Finch
☑	*Rhodopechys alienus*	African Crimson-winged Finch
☑	*Bucanetes githagineus*	Trumpeter Finch
☐	*Bucanetes mongolicus*	Mongolian Finch
☑	*Carpodacus erythrinus*	Common Rosefinch

☑	*Carpodacus rubicilla*	Great Rosefinch
☑	*Carpodacus synoicus*	Sinai Rosefinch
☐	*Carpodacus sibiricus*	Long-tailed Rosefinch
☐	*Carpodacus roseus*	Pallas's Rosefinch
☑	*Chloris chloris*	European Greenfinch
☐	*Rhodospiza obsoleta*	Desert Finch
☑	*Rhynchostruthus percivali*	Arabian Grosbeak
☐	*Crithagra rothschildi*	Arabian Serin
☑	*Crithagra menachensis*	Yemen Serin
☑	*Linaria flavirostris*	Twite
☑	*Linaria cannabina*	Common Linnet
☐	*Linaria yemenensis*	Yemen Linnet
☑	*Acanthis flammea*	Common Redpoll
☑	*Acanthis hornemanni*	Arctic Redpoll
☑	*Loxia pytyopsittacus*	Parrot Crossbill
☑	*Loxia scotica*	Scottish Crossbill
☑	*Loxia curvirostra*	Common Crossbill
☑	*Loxia leucoptera*	Two-barred Crossbill
☑	*Carduelis carduelis*	European Goldfinch
☐	*Carduelis citrinella*	Citril Finch
☑	*Carduelis corsicana*	Corsican Finch
☑	*Serinus pusillus*	Red-fronted Serin
☑	*Serinus serinus*	European Serin
☐	*Serinus syriacus*	Syrian Serin
☑	*Serinus canaria*	Atlantic Canary
☑	*Spinus spinus*	Eurasian Siskin
☐	**Family Parulidae**	**New World warblers**
☐	*Seiurus aurocapilla*	Ovenbird
☐	*Parkesia noveboracensis*	Northern Waterthrush
☐	*Vermivora chrysoptera*	Golden-winged Warbler
☐	*Vermivora cyanoptera*	Blue-winged Warbler
☐	*Mniotilta varia*	Black-and-white Warbler
☐	*Leiothlypis peregrina*	Tennessee Warbler
☐	*Geothlypis trichas*	Common Yellowthroat
☐	*Setophaga citrina*	Hooded Warbler
☐	*Setophaga ruticilla*	American Redstart
☐	*Setophaga tigrina*	Cape May Warbler
☐	*Setophaga cerulea*	Cerulean Warbler
☐	*Setophaga americana*	Northern Parula

☐	*Setophaga magnolia*	Magnolia Warbler
☐	*Setophaga castanea*	Bay-breasted Warbler
☐	*Setophaga fusca*	Blackburnian Warbler
☑	*Setophaga aestiva*	American Yellow Warbler
☐	*Setophaga pensylvanica*	Chestnut-sided Warbler
☐	*Setophaga striata*	Blackpoll Warbler
☐	*Setophaga caerulescens*	Black-throated Blue Warbler
☐	*Setophaga palmarum*	Palm Warbler
☐	*Setophaga coronata*	Myrtle Warbler
☐	*Setophaga dominica*	Yellow-throated Warbler
☐	*Setophaga discolor*	Prairie Warbler
☐	*Setophaga virens*	Black-throated Green Warbler
☐	*Cardellina canadensis*	Canada Warbler
☐	*Cardellina pusilla*	Wilson's Warbler
☐	**Family Icteridae**	**New World orioles, blackbirds and allies**
☐	*Xanthocephalus xanthocephalus*	Yellow-headed Blackbird
☐	*Dolichonyx oryzivorus*	Bobolink
☑	*Icterus galbula*	Baltimore Oriole
☐	*Molothrus ater*	Brown-headed Cowbird
☐	*Quiscalus quiscula*	Common Grackle
☐	**Family Emberizidae**	**Buntings, New World sparrows and allies**
☑	*Emberiza calandra*	Corn Bunting
☑	*Emberiza citrinella*	Yellowhammer
☑	*Emberiza leucocephalos*	Pine Bunting
☑	*Emberiza cia*	Rock Bunting
☐	*Emberiza stewarti*	White-capped Bunting
☑	*Emberiza buchanani*	Grey-necked Bunting
☑	*Emberiza cineracea*	Cinereous Bunting
☑	*Emberiza hortulana*	Ortolan Bunting
☑	*Emberiza caesia*	Cretzschmar's Bunting
☑	*Emberiza cirlus*	Cirl Bunting
☑	*Emberiza striolata*	Striolated Bunting
☑	*Emberiza sahari*	House Bunting
☑	*Emberiza tahapisi*	Cinnamon-breasted Bunting
☑	*Emberiza goslingi*	Gosling's Bunting
☐	*Emberiza fucata*	Chestnut-eared Bunting
☑	*Emberiza pusilla*	Little Bunting
☐	*Emberiza chrysophrys*	Yellow-browed Bunting
☑	*Emberiza rustica*	Rustic Bunting

☐	*Emberiza aureola*	Yellow-breasted Bunting
☐	*Emberiza rutila*	Chestnut Bunting
☑	*Emberiza melanocephala*	Black-headed Bunting
☐	*Emberiza bruniceps*	Red-headed Bunting
☐	*Emberiza spodocephala*	Black-faced Bunting
☐	*Emberiza pallasi*	Pallas's Reed Bunting
☑	*Emberiza schoeniclus*	Common Reed Bunting
☐	*Passerella iliaca*	Red Fox Sparrow
☐	*Melospiza melodia*	Song Sparrow
☐	*Melospiza lincolnii*	Lincoln's Sparrow
☑	*Zonotrichia leucophrys*	White-crowned Sparrow
☑	*Zonotrichia albicollis*	White-throated Sparrow
☑	*Junco hyemalis*	Dark-eyed Junco
☐	*Passerculus sandwichensis*	Savannah Sparrow
☐	*Chondestes grammacus*	Lark Sparrow
☐	*Pipilo erythrophthalmus*	Eastern Towhee
☐	**Family Calcariidae**	**Longspurs and snow buntings**
☑	*Calcarius lapponicus*	Lapland Longspur
☑	*Plectrophenax nivalis*	Snow Bunting
☐	**Family Cardinalidae**	**Cardinals, grosbeaks and allies**
☐	*Piranga rubra*	Summer Tanager
☐	*Piranga olivacea*	Scarlet Tanager
☐	*Pheucticus ludovicianus*	Rose-breasted Grosbeak
☐	*Spiza americana*	Dickcissel
☐	*Passerina cyanea*	Indigo Bunting
☐	**Indeterminate species recorded in region**	
☐	*Terpsiphone paradisi/affinis/incei*	Indian/Oriental/Amur Paradise Flycatcher
☐	*Phylloscopus tenellipes/borealoides*	Pale-legged/Sakhalin Leaf Warbler
☐		
☐		
☐		
☐		
☐		
☐		
☐		
☐		
☐		
☐		
☐		

REFERENCES

A

Adamian, M S, and Klem Jr, D. 1999. *Handbook of the Birds of Armenia*. American University of Armenia, Oakland.

Ahmed, R. 2011. Subspecific identification and status of Cattle Egret. *Dutch Birding* 33: 294-304.

Albayrak, T, Gürsoy, A, and Kirwan, G M. 2002. *International Species Action Plan for the Cinereous Bunting* (*Emberiza cineracea*). BirdLife International, Wageningen.

Alfrey, P, and Ahmad, M. 2007. Short-billed Gull on Terceira, Azores, in February-March 2003 and identification of the 'Mew Gull complex'. *Dutch Birding* 29: 201-212.

Alfrey, P, Buckell, S, Legrand, V, Monticelli, D, and Birding Azores. 2010. Birding on Corvo, Azores, and Nearctic vagrants in 2005-09. *Dutch Birding* 32: 299-315.

Alfrey, P, Crochet, P-A, and Spittle, D. 2012. Three White-eyed Vireos on Corvo, Azores, in 2005-09. *Dutch Birding* 34: 38-41.

Alfrey, P. 2006. Eastern Black Redstart at Kazbegi, Georgia, in May 2003. *Dutch Birding* 28: 98.

Aliabadian, M, Kaboli, M, Förschler, M I, Nijman, V, Chamani, A, Tillier, A, Prodon, R, Pasquet, E, Ericson, P G P, and Zuccon, D. 2012. Convergent evolution of morphological and ecological traits in the open-habitat chat complex (Aves, Muscicapidae: Saxicolinae). *Mol Phylogenet Evol* 65: 35-45.

Alieslam, A. 2014. First record for Iran of Black-naped Monarch *Hypothymis azurea*, near Jask city, Hormozgan province. *Sandgrouse* 36: 61-62.

Al-Obeidi, L A. 2016. First breeding record of Sedge Warbler *Acrocephalus schoenobaenus* and apparent breeding of White-crowned Wheatear *Oenanthe leucopyga*, for Iraq. *Sandgrouse* 38: 156-159.

Al-Saghier, O, and Porter, R F. 1997. The first African Spoonbill *Platalea alba* in Yemen and Arabia. *Sandgrouse* 19: 141.

Al-Sirhan, A. 2015. KORC Annotated Checklist of Birds. http://birdsofkuwait.com/annotated_checklist.shtml.

Alström, P, and Mild, K. 2003. *Pipits and Wagtails of Europe, Asia and North America*. Christopher Helm, London.

Alström, P, Barnes, K N, Olsson, U, Barker, F K, Bloomer, P, Khan, A A, Quresh, M A, Guillaumet, A, Crochet, P-A, and Ryan, P G. 2013. Multilocus phylogeny of the avian family Alaudidae (larks) reveals complex morphological evolution, non-monophyletic genera and hidden species diversity. *Mol Phylogenet Evol* 69: 1043-1056.

Alström, P, Colston, P, and Lewington, I. 1991. *A Field Guide to the Rare Birds of Britain and Europe*. HarperCollins, London.

Alström, P, Ericson, P G P, Olsson, U, and Sundberg, P. 2006. Phylogeny and classification of the avian superfamily Sylvioidea. *Mol Phylogenet Evol* 38: 381-397.

Alström, P, Fjeldså, J, Fregin, S, and Olsson, U. 2011. Gross morphology betrays phylogeny: the Scrub Warbler *Scotocerca inquieta* is not a cisticolid. *Ibis* 153: 87-97.

Alström, P, Fregin, S, Norman, J A, Zuccon, D, Ericson, P G P, Christidis, L, and Olsson, U. 2011. Multilocus analysis of a taxonomically densely sampled dataset reveal extensive non-monophyly in the avian family Locustellidae. *Mol Phylogenet Evol* 58: 513-526. DOI: 10.1016/j.ympev.2010.12.012.

Alström, P, Höhna, S, Gelang, M, Ericson, P G P, and Olsson, U. 2011. Non-monophyly and intricate morphological evolution in the avian family Cettiidae revealed by multi-locus analysis of a taxonomically densely sampled dataset. *BioMed Central*. http://www.biomedcentral.com/1471-2148/11/352.

Alström, P, Olsson, U, Lei, F, Wang, H-t, Gao, W, and Sundberg, P. 2008. Phylogeny and classification of the Old World Emberizini (Aves, Passeriformes). *Mol Phylogenet Evol* 47: 960-973.

Alström, P, Saitoh, T, Williams, D, Nishiumi, I, Shigeta, Y, Ueda, K, Irestedt, M, Bjorklund, M, and Olsson, U. 2011. The Arctic Warbler *Phylloscopus borealis* – three anciently separated cryptic species. *Ibis* 153: 395-410.

Amezian, M, Radi, M, Tattou, M I, Khayya, M L, Samlali, M L, and Qninba, A. 2011. Cricket Longtail breeding in southern Morocco. *Dutch Birding* 33: 229-232.

Ananian, V, Aghababyan, K, Tumanyan, S, Janoyan, G, and Bildstein, K. 2010. Shikra *Accipter badius* breeding in Armenia. *Sandgrouse* 32: 151-155.

Ananian, V, and Busuttil, S. 2003. Further observations and probable breeding of Mongolian Trumpeter Finch *Bucanetes mongolicus* in Armenia. *Sandgrouse* 25: 148-150.

Andrews, I J, Khoury, F, and Shirihai, H. 1999. Jordan Bird Report 1995-97. *Sandgrouse* 21: 10-35.

Andrews, I J. 1995. *The Birds of the Hashemite Kingdom of Jordan*. Privately published, Musselburgh.

Anon. 2004. The *British Birds* list of English names for Western Palearctic birds. *Br Birds* 97: 2-5.

Anon. 2006. Full species list of the Madeiran archipelago (the Salvage Islands excluded). http://madeira.seawatching.net/lists/Madeira_fullist.pdf.

Anon. 2008. Fauna and flora [of the Selvagens]. http://selvagens.seawatching.net/faunaflora.html.

Anon. 2012a. Annex II. Estatus de l'Avifauna Balear. *Anu Ornitol Balear* 27: 217-228.

Anon. 2012b. Bird photos from Madeira. http://madeira.seawatching.net/systematic.html.

Anon. 2014. Artenliste Österreich. Club300. www.club300.at/birds/aut.

Anon. 2015. Linnuharulduste komisjoni (HK) poolt käsitletavate liikide kõik aktsepteeritud vaatlused seisuga 12.11.2015. http://www.eoy.ee/yhing/hk/hk_aktsept.pdf.

Ararat, K. 2016. First record of Lesser Spotted Eagle *Clanga pomarina*, first breeding record of Eurasian Penduline Tit *Remiz pendulinus* and first records of Eastern Bonelli's Warbler *Phylloscopus orientalis*, Olive-tree Warbler *Hippolais olivetorum* and Goldcrest *Regulus regulus* for Iraq. *Sandgrouse* 38: 106-109.

Arbabi, T, Gonzalez, J, Witt, H-H, Klein, R, and Wink, M. 2014. Mitochondrial phylogeography of the Eurasian Reed Warbler *Acrocephalus scirpaceus* and the first genetic record of *A. s. fuscus* in Central Europe. *Ibis* 156: 799-811.

Arcos, J M (compiler). 2011. *International species action plan for the Balearic shearwater,* Puffinus mauretanicus. SEO/BirdLife and BirdLife International.

Arkhipov, V, and Blair, M. 2007. Skua (*Catharacta, Stercorarius*) occurrence in the OSME region. *Sandgrouse* 29: 183-204.

Arkhipov, V Y, and Ławicki, Ł. 2016. Nearctic passerines in Russia. *Dutch Birding* 38: 201-214.

Arkhipov, V. 2006. Status and distribution of selected bird species on the Russia-Kazakhstan border northwest of the Caspian Sea. *Sandgrouse* 28: 156-160.

Armada, R. 2010. The Red-footed Booby in Spain. *Birding World* 23: 524-529.

Arnaiz-Villena, A, Gomez-Prieto, P, and Ruiz-del-Valle, V. 2009. Phylogeography of Finches and Sparrows. In *Animal Genetics* edited by L J Rechi. Nova Science Publishers, New York.

Arthur, R W. 1963. Summer Tanager on Bardsey, Caernarvonshire (1957). *Br Birds* 56: 49-51.

Ashoori, A, Khani, A, Ghasemi, M, Rabiee, K, Mansoori, M, Musavi, S B, Hashemi, A, and Eskandari, F. 2013. Recent records and status of the Sociable Lapwing *Vanellus gregarius* in Iran. *Sandgrouse* 35: 14-19.

ASPB. 2015. Birds of Armenia Checklist. http://aspbirds.org/index.php/en/ecotourism/birds-of-armenia-checklist.

Atkinson, P W, and Caddick, J A. 2013. Checklist of the birds of Libya. Downloaded from www.africanbirdclub.org, 24 November 2013.

Aumüller, R, and Noeske, A. 2006. Der Taigazilpzalp in Europa Teil 2: Räumlich-zeitliches Vorkommen von *tristis*. www.club300.de.

Ayé, R, Roth, T and Stalling, T. 2014. Paddyfield Pipit (*Anthus rufulus*) in south-eastern Iran: a new species to the Middle East. *Zool Middle East* 60: 183-185.

Ayé, R, Schweizer, M, and Roth, T. 2012. *Birds of Central Asia*. Christopher Helm, London.

Ayé, R, Stalling, T, and Roth, T. 2014. Himalayan Vultures in Iran in April 2007 and December 2010. *Dutch Birding* 36: 159-162.

Ayé, R. 2008. Saxaul Sparrows in Iran in April 2004. *Dutch Birding* 30: 315-316.

Azazaf, H, Feltrup-Azafaf, C, Dlensi, H, and Isenmann, P. 2015. Oiseaux des Tunisie – Birds of Tunisia 2005-2014. *Alauda* 83: 7-28.

Azores Bird Club. 2014. *Azores Rare and Scarce Bird Report No 1: 2013*.

Azores Bird Club. 2015. *Azores Rare and Scarce Bird Report No 2: 2014*.

Azzopardi, J. 2006. A review of the status of Black-eared Wheatear in the Maltese Islands. *Br Birds* 99: 481-495.

B

Babbington, J, and Campbell, O. 2016. Recent status and occurrence of Crested Honey Buzzards *Pernis ptilorhynchus* in the Arabian peninsula, with emphasis on Saudi Arabia and the United Arab Emirates. *Sandgrouse* 38: 12-22.

Babbington, J. 2015. Saudi Arabia list. http://www.birdsofsaudiarabia.com/p/saudi-arabia-bird-list.html.

Baccetti, N, Fracasso, F, and Gotti, C. 2014. La lista CISO-COI degli uccelli italiani - Parte seconda: le specie naturalizzate (cat. C) e le categorie "di servizio" (cat. D, E, X). *Avocetta* 38: 1-21.

Baccetti, N, Massa, B, and Violani, C. 2007. Proposed synonymy of *Sylvia cantillans moltonii* Orlando, 1937, with *Sylvia cantillans subalpina* Temminck, 1820. *Bull Brit Ornithol Club* 127: 107-110.

Baccetti, N, Spagnesi, M, and Zenatello, M. 1997. Historical overview on introduced birds in Italy. *Supplemento alle Ricerche di Biologia della Selvaggina* 27: 299-316.

Baczó, Z, Kovács, S, and Szabó, A. 2009. Important ornithological observations in Romania. *Scientific Annals of DDI* 15: 7-10.

Baha el Din, S, and Baha el Din, M. 2000. The occurrence of the North African subspecies of Mourning Wheatear *Oenanthe lugens halophila* in Egypt. *Sandgrouse* 22: 109-112.

Baha el Din, S, and Baha el Din, M. 2001. The first Black Bush Robin *Cercotrichas podobe* in Egypt. *Sandgrouse* 23: 62-64.

Baha El Din, S, Baha El Din, M, and Bensch, S. 2010. Mangrove Olivaceous Warblers in Egypt – new to the Western Palearctic. *Birding World* 23: 215-217.

Baha El Din, S. 1984. The Black Wheatear (*Oenanthe leucura*) in Egypt. *Courser* 1: 5-7.

Baha El Din, S. 1990. Dunn's Lark breeding in Egypt. *Bull Ornithol Soc Middle East* 25: 27-28.

Baker, A J, Dennison, M D, Lynch, A, and Legrand, G. 1990. Genetic divergence in peripherally isolated populations of chaffinches in the Atlantic islands. *Evolution* 44: 981-999.

Baker, J M, and Omland, K E. 2006. Canary Island Ravens *Corvus corax tingitanus* have distinct mtDNA. *Ibis* 148: 174-178.

Baker, K. 1997. *Warblers of Europe, Asia and North Africa*. Christopher Helm, London.

Ball, R M, and Avise, J C. 1992. Mitochondrial DNA phylogeographic differentiation among avian populations and the evolutionary significance of subspecies. *Auk* 109: 626-636.

Balmer, D, Gillings, S, Caffrey, B, Swann, B, Downie, I, and Fuller, R. 2013. *Bird Atlas 2007-11: the breeding and wintering birds of Britain and Ireland*. BTO Books, Thetford.

Bangjord, G. 2016a. Registered bird species checklist for Jan Mayen. http://www.svalbardbirds.com/artsliste-jm.htm.

Bangjord, G. 2016b. Registered bird species checklist for Svalbard. http://svalbardbirds.com/artsliste-web.htm.

Banks, A N, Wright, L J, Maclean, I M D, Hann, C, and Rehfisch, M M. 2008. *Review of the Status of Introduced Non-Native Waterbird Species in the Area of the African-Eurasian Waterbird Agreement: 2007 Update*. BTO Research Report No 489. British Trust for Ornithology, Thetford.

Banks, K, Clark, H, Mackay, I R K, Mackay, S G, and Sellers, R M. 1989. Ageing, sexing and racing Snow Buntings in winter plumage. *Ringers' Bulletin* 7: 84-87.

Banks, R C, Chesser, R T, Cicero, C, Dunn, J L, Kratter, A W, Lovette, I J, Rasmussen, P C, Remsen Jr, J V, Rising, J D, and Stotz, D F. 2007. Forty-eighth supplement to the American Ornithologists' Union *Check-List of North American Birds*. *Auk* 124: 1109-1115.

Banks, R C, Chesser, R T, Cicero, C, Dunn, J L, Kratter, A W, Lovette, I J, Rasmussen, P C, Remsen Jr, J V, Rising, J D, Stotz, D F, and Winker, K. 2008. Forty-ninth supplement to the American Ornithologists' Union *Check-List of North American Birds*. *Auk* 125: 758-768.

Banks, R C, Cicero, C, Dunn, J L, Kratter, A W, Rasmussen, P C, Remsen Jr, J V, Rising, J D, and Stotz, D F. 2004. Forty-fifth supplement to the American Ornithologists' Union *Check-List of North American Birds*. *Auk* 121: 985-995.

Banks, R C, Cicero, C, Dunn, J L, Kratter, A W, Rasmussen, P C, Remsen Jr, J V, Rising, J D, and Stotz, D F. 2005. Forty-sixth supplement to the American Ornithologists' Union *Check-List of North American Birds*. *Auk* 122: 1026-1031.

Banks, R C, Cicero, C, Dunn, J L, Kratter, A W, Rasmussen, P C, Remsen Jr, J V, Rising, J D, and Stotz, D F. 2006. Forty-seventh supplement to the American Ornithologists' Union *Check-List of North American Birds*. *Auk* 123: 926-936.

Bannon, J K, and Richardson, C. 1993. White-capped Buntings *Emberiza stewarti* in the United Arab Emirates: a new bird for Arabia. *Emirates Bird Rep* 17: 3-5.

Barbet-Massin, M, and Jiguet, F. 2011. Back from a predicted climatic extinction of an island endemic: a future for the Corsican Nuthatch. *PLoS ONE* 6(3): e18228. Doi: 10.1371/journal.pone.001822.

Barezzani, R, and Ebels, E B. 2012. Mugimaki Flycatcher in Brescia, Italy, in October 2011. *Dutch Birding* 34: 41-44.

Barone, R, and Delgado, G. Iago Sparrows *Passer iagoensis* in the mirror. *Bull Afr Bird Club* 18: 226-227.

Barreiros, J P, Elias, R B, Gabriel, R, Rodrigues, P, Barcelos, L D, Branco, J O, and Borges, P A V. 2014. The Yellow-crowned Night Heron *Nyctanassa violacea* (Aves: Pelecaniformes: Ardeidae) in the Azores and Madeira Archipelagos: a new species for the Western Palearctic. *Arquipelago* 31: 37-43.

Basly, J-P. 2007. Le Léiothrix jaune *Leiothrix lutea* en Béarn: répartition, effectifs et comportement. *Ornithos* 14: 370-375.

Bathel, P H, and Helbig, A J. 2006. *Checklist of the Birds of Germany*. Limicola Verlag, Göttingen.

Batty, C. 2010. Pied Crows in Western Sahara, Morocco. *Dutch Birding* 32: 329-332.

Bauer, H-G, and Woog, F. 2008. Non-native and naturalized bird species (neozoa) in Germany, part I: occurrence, population size and status. *Vogelwarte* 46: 157-194.

Beaman, M. 1990. Identification of Black Noddy and validity of December 1984 record in Mauritania. *Dutch Birding* 12: 245-248.

Beaman, M. 1994. *Palearctic Birds. A Checklist of the Birds of Europe, North Africa and Asia north of the foothills of the Himalayas*. Harrier Publications, Stonyhurst.

Beaubrun, P C, Thévenot, M, and Lévêque, R. 1978. Le Dendrocygne fauve *Dendrocygna bicolor* au Maroc. *Alauda* 46: 177-178.

Beemster, N, and Klop, E. 2013. Risk assessment of the Black swan (*Cygnus atratus*) in the Netherlands. A&W-report 1978. Altenburg & Wymenga ecologisch onderzoek, Feanwâlden.

Bekel-Kastrup, H, Kobialka, H, Wegener, D, and Ebels, E. 2013. Hooded Wheatear at Nestos Delta, Greece, in May 2012 and status in Europe. *Dutch Birding* 35: 384-386.

Belbachir, F. 2000. Première observation d'une colonie de Capucin bec-d'argent *Euodice cantans* dans le Sahara central algérien. *Alauda* 68: 149-151.

Belgian Rare Birds Committee. 2015. Belgian official checklist. Published online: http://www.belgianrbc.be/#.

Belik, V P. 2005. The Sociable Lapwing in Eurasia: what does the future hold? *Br Birds* 98: 476-485.

Bell, J, and Håland, A. 2008. Visible migration of divers along the Norwegian Coast. In: Håland, A (ed). The Loon & Diver Workshop 2007/2008: 1: 47. Vardø, Norway.

van Bemmelen, R S A, Eggenhuizen, A H V, and Faveyts, W. 2012. Invasion of White-headed Long-tailed Tits in the Netherlands in 2010-11: progression, numbers and distribution. *Dutch Birding* 34: 137-150.

van den Berg, A B, and Haas, M. 2012. WP reports. *Dutch Birding* 34: 320-335.

van den Berg, A B, and de Roever, J W. 1984. Plumages and bare parts of Desert Sparrow. *Dutch Birding* 6: 139-140.

van den Berg, A B, and The Sound Approach. 2005. Olivaceous warblers in the western Mediterranean and their identification. *Dutch Birding* 27: 302-307.

van den Berg, A B, and The Sound Approach. 2006. Phenology and identification of Atlas and Iberian Pied Flycatchers. *Dutch Birding* 28: 1-6.

van den Berg, A B, and The Sound Approach. 2009. Calls, identification and taxonomy of Siberian Chiffchaff: an analysis. *Dutch Birding* 31: 79-85.

van den Berg, A B, Bekir, S, de Knijff, P, and The Sound Approach. 2010. Rediscovery, biology, vocalisations and taxonomy of fish owls in Turkey. *Dutch Birding* 32: 287-298.

van den Berg, A B. 2004. Population growth and vagrancy potential of Ross's Goose. *Dutch Birding* 26: 107-111.

van den Berg, A B. 2009. Yellow-billed Kite in High Atlas, Morocco, in April 2008. *Dutch Birding* 31: 172-173.

van den Berg, A B. 2011. Groene Reiger te Amsterdam en Zaandam in 2006-09. *Dutch Birding* 33: 163-176.

van den Berg, A B. 2013a. Green Heron in the Netherlands in 2006-09. *Dutch Birding* 35: 34.

van den Berg, A B. 2013b. Lijst van Nederlandse vogelsoorten. http://www.dutchbirding.nl/content/page/files/webprog20130301-85.pdf.

van den Berg, A B. 2014. Dutch Birding Bird Names. http://www.dutchbirding.nl/content/page/files/DBupdateWPchecklistArnoud20140110.pdf.

Bergier, P, and Thévenot, M. 2010. Liste des oiseaux du Maroc. Mise à jour février 2010 (rév. 3.0). *Go-South Bull* 7: 15-55.

Bergier, P, Franchimont, J, and the CHM. 2013. Les oiseaux rares au Maroc. Rapport de la Commission d'Homologation Marocaine, Numéro 18. *Go-South Bull* 10: 1-16.

Bergier, P, Franchimont, J, and the CHM. 2014. Les oiseaux rares au Maroc. Rapport de la Commission d'Homologation Marocaine, Numéro 19. *Go-South Bull* 11: 1-13.

Bergier, P, Franchimont, J, Thévenot, M, and the CHM. 2004. Les oiseaux rares au Maroc. Rapport de la Commission d'Homologation Marocaine, Numéro 9. *Go-South Bull* 1: 13-17.

Bergier, P, Franchimont, J, Thévenot, M, and the CHM. 2005. Les oiseaux rares au Maroc. Rapport de la Commission d'Homologation Marocaine, Numéro 10. *Go-South Bull* 2: 23-30.

Bergier, P, Franchimont, J, Thévenot, M, and the CHM. 2006. Les oiseaux rares au Maroc. Rapport de la Commission d'Homologation Marocaine, Numéro 11. *Go-South Bull* 3: 31-42.

Bergier, P, Franchimont, J, Thévenot, M, and the CHM. 2007. Les oiseaux rares au Maroc. Rapport de la Commission d'Homologation Marocaine, Numéro 12. *Go-South Bull* 4: 77-97.

Bergier, P, Franchimont, J, Thévenot, M, and the CHM. 2008. Les oiseaux rares au Maroc. Rapport de la Commission d'Homologation Marocaine, Numéro 13. *Go-South Bull* 5: 48-58.

Bergier, P, Franchimont, J, Thévenot, M, and the CHM. 2009. Les oiseaux rares au Maroc. Rapport de la Commission d'Homologation Marocaine, Numéro 14. *Go-South Bull* 6: 76-91.

Bergier, P, Franchimont, J, Thévenot, M, and the CHM. 2010. Les oiseaux rares au Maroc. Rapport de la Commission d'Homologation Marocaine, Numéro 15. *Go-South Bull* 7: 1-14.

Bergier, P, Franchimont, J, Thévenot, M, and the CHM. 2011. Les oiseaux rares au Maroc. Rapport de la Commission d'Homologation Marocaine, Numéro 16. *Go-South Bull* 8: 1-20.

Bergier, P, Franchimont, J, Thévenot, M, and the CHM. 2012. Les oiseaux rares au Maroc. Rapport de la Commission d'Homologation Marocaine, Numéro 17. *Go-South Bull* 9: 13-32.

Bergier, P, Thévenot, M, and Qninba, A. 2010. Liste des oiseaux du Sahara Atlantique marocain. Mise à jour février 2010 (rév. 1.0). *Go-South Bull* 7: 109-120.

Bergier, P, Thévenot, M, van den Berg, A B, and The Sound Approach. 2013. Distribution, habitat, identification and taxonomy of Streaked Scrub Warbler in Morocco. *Dutch Birding* 35: 107-121.

Bergier, P, Zadane, Y, and Qninba, A. 2009. Cape Gull: a new breeding species in the Western Palearctic. *Birding World* 22: 253-256.

Bergmann, H-H, and Schottler, B H. 2001. Tenerife Robin – a species of its own? *Dutch Birding* 23: 140-146.

Berlijn, M. 2007. Falcated Ducks in the Netherlands and the WP. *Dutch Birding* 29: 139-146.

BirdLife Cyprus. 2012. Birds of Cyprus Checklist. http://www.birdlifecyprus.org/upload/PDFs/CyChecklist_2012.pdf.

BirdLife International. 2002. *Slender-billed Curlew* Numenius tenuirostris *Bibliography*. Update No 3, December 2002. BirdLife International, Cambridge.

BirdLife International. 2014a. *IUCN Red List for Birds.* Downloaded from http://www.birdlife.org on 03/07/2014.

BirdLife International. 2014b. We have lost over 150 bird species since 1500. Presented as part of the BirdLife State of the World's Birds website. Available from: http://www.birdlife.org/datazone/sowb/casestudy/102. Checked: 03/02/2016.

Biricik, M. 2009. Unexpectedly large number of Sociable Lapwings *Vanellus gregarius* on autumn migration in Turkey and some remarks on the stopover site. *Sandgrouse* 31: 15-17.

Birkhead, T R. 1991. *The Magpies. The ecology and behaviour of black-billed and yellow-billed magpies.* T & A D Poyser, London.

Blair, M J, McKay, H, Musgrove, A J, and Rehfisch, M M. 2000. *Review of the Status of Introduced Non-Native Waterbird Species in the Agreement Area of the African-Eurasian Waterbird Agreement.* BTO Research Report No 229. British Trust for Ornithology, Thetford.

Blair, M, Preddy, S, Porter, R, and Aspinall, S. 2014. The OSME Region List of Bird Taxa – v3.1. http://osme.org/ORL.

Boere, G C, Galbraith, C A, and Stroud, D A (eds). 2006. *Waterbirds around the World.* The Stationery Office, Edinburgh.

Bolton, M. 2007. Playback experiments indicate absence of vocal recognition among temporally and geographically separated populations of Madeiran Storm-petrels *Oceanodroma castro*. *Ibis* 149: 255-263.

Bolton, M, Smith, A L, Gómez-Díaz, E, Friesen, V L, Medeiros, R, Bried, J, Roscales, J L, and Furness, R W. 2008. Monteiro's Storm-petrel *Oceanodroma monteiroi*: a new species from the Azores. *Ibis* 150: 717-727.

Bonavia, E, Borg, J J, Coleiro, C, Gauci C, Johnson, M, Raine, A, and Sultana, J. 2010. Systematic List 2000-2005. *Il-Merill* 32: 55-109.

Borg, J J, and Cachia, D. 2010. First records of the Striped Crake *Porzana marginalis* in the Maltese Islands. *Il-Merill* 32: 45.

Borras, A, Blache, S, Cabrera, J, Cabrera, T, and Senar, J C. 2005. Citril Finch (*Serinus citrinella*) populations at the north of the Pyrenees may winter in the northeast of the Iberian Peninsula. *Aves* 42: 261-265.

Borrow, N, and Demey, R. 2001. *A Guide to the Birds of Western Africa*. Princeton University Press, Princeton.

Bos, J F F P, Essetti, I, and Gilissen, N L M. 2000. Record counts of Marbled Teal in Tunisia, October 1999: consequences for population estimates and distribution. *Threatened Waterfowl Specialist Group News* 12: 49-53.

Bot, S, and van Dijk, R E. 2009. Black-headed Penduline Tits *Remiz macronyx* in Kazakhstan. *Sandgrouse* 31: 171-176.

Bot, S, Brinkhuizen, D, Pogány, Á, Székely, T, and van Dijk, R. 2011. Penduline tits in Eurasia: distribution, identification and systematics. *Dutch Birding* 33: 177-187.

Bot, S, van Bemmelen, R S A, Eggenhuizen, A H V, and Groenendijk, D. 2012. Invasion of White-headed Long-tailed Tits in the Netherlands in 2010-11: identification. *Dutch Birding* 34: 151-159.

Bourass, E, Baccetti, N, Bashimam, W, Berbash, A, Bouzainen, M, De Faveri, A, Galidan, A, Saied, A M, Yahia, J, and Zenatello, M. 2013. Results of the seventh winter waterbird census in Libya, January-February 2011. *Bull Afr Bird Club* 20: 20-26.

Bourne, Dr W R P, and Lee, Dr D S. 1994. The first record of McCormick's (or the South Polar) Skua for Europe and the Northern Hemisphere. *Sea Swallow* 43: 74-76.

Bouwman, R G. 1985. Spur-winged Goose in Morocco in June 1984. *Dutch Birding* 7: 21-22.

Boyla, K A. In prep. *Checklist of the Birds of Turkey*.

Brambilla, M, Vitulano, S, Ferri, A, Spina, F, Fabbri, E, and Randi, E. 2012. An unexpected pattern of migration revealed in the Subalpine Warbler *Sylvia cantillans* complex by mitochondrial DNA analyses. *Ibis* 154: 616-620.

Brelsford, A, and Irwin, D E. 2009. Incipient speciation despite little assortative mating: the yellow-rumped warbler hybrid zone. *Evolution* 63: 3050-3060.

Breman, F C, Jordaens, K, Sonet, G, Nagy, Z T, van Houdt, J, and Louette, M. 2013. DNA barcoding and evolutionary relationships in *Accipiter* Brisson, 1760 (Aves, Falconiformes: Accipitridae) with a focus on African and Eurasian representatives. *J Orn* 154: 265-287.

Bretagnolle, V, Carruthers, M, Cubitt, M, Bioret, F, and Cuillandre, J-P. 1991. Six captures of a dark-rumped, fork-tailed storm-petrel in the northeastern Atlantic. *Ibis* 133: 351-356.

Bretagnolle, V. 1995. Systematics of the Soft-plumaged Petrel *Pterodroma mollis* (Procellariidae): new insight from the study of vocalizations. *Ibis* 137: 207-218.

Brewer, D. 2001. *Wrens, Dippers and Thrashers*. Christopher Helm, London.

Brichetti, P A, and Fracasso, G. 2004. *Ornitologia Italiana* 2 Tetraonidae-Scolopacidae. Alberto Perdisa Editore, Bologna.

Brichetti, P, and Fracasso, G. 2015. Check-list degli uccelli italiani aggiornata al 2014. *Riv Ital Ornitol* 85: 31-50.

Bridge, E S, Jones, A W, and Baker, A J. 2005. A phylogenetic framework for the terns (Sternini) inferred from mtDNA sequences: implications for taxonomy and plumage evolution. *Mol Phylogenet Evol* 35: 459-469.

Bried, J, and Magalhães, M C. 2004. First Palearctic record of the endangered Bermuda Petrel *Pterodroma cahow*. *Bull Brit Ornithol Club* 124: 202-206.

van den Brink, B, and Leonard, P. 2011. A Palearctic Red-rumped Swallow *Cecropis daurica rufula* in Zambia. *Bull Afr Bird Club* 18: 77.

British Ornithologists' Union Records Committee. 2009. 37th Report (October 2008). *Ibis* 151: 224-230.

British Ornithologists' Union Records Committee. 2012. 40th Report (October 2011). *Ibis* 154: 212-215.

British Ornithologists' Union Records Committee. 2013. 41st Report (October 2012). *Ibis* 155: 204-207.

British Ornithologists' Union Records Committee. 2015. 43rd Report (October 2014). *Ibis* 157: 186-192.

British Ornithologists' Union Records Committee. 2015. 44th Report (January 2015). *Ibis* 157: 413.

British Ornithologists' Union Records Committee. 2016. 45th Report (October 2015). *Ibis* 158: 202-205.

Brooke, M. 2004. *Albatrosses and Petrels across the World*. Oxford University Press, Oxford.

Brown, A, and Grice, P. 2005. *Birds in England*. T & A D Poyser, London.

Brown, L H, Urban, E K, and Newman, K. 1982. *The Birds of Africa*. Volume 1. Academic Press, London.

Browne, P W P. 1981. New bird species in Mauritania. *Malimbus* 3: 62-72.

Browne, P. 2013. Proposal to remove African Palm Swift from WP list. *Dutch Birding* 35: 249-250.

Buchan, J A, Grunwell, M J, Morris, N G, and Tull, S J. The Qatar List. Qatar Bird Records Committee, 2014. Downloaded from http://www.qatarbirds.org/list.htm.

Buchanan, G, Crockford, N, and Gretton, A. 2010. The Slender-billed Curlew *Numenius tenuirostris* in Africa. *Bull Afr Bird Club* 17: 202-206.

Bundy, G, and Warr, E. 1979. *Provisional Check-list of Birds of the Arabian Gulf States*. Ornithological Society of the Middle East, Sandy.

Bundy, G, Connor, R J, and Harrison, C J O. 1989. *Birds of the Eastern Province of Saudi Arabia*. Witherby, London.

Bundy, G. 1976. *The Birds of Libya. An Annotated Checklist*. British Ornithologists' Union, London.

Burfield, I, and van Bommel, F (compilers). 2004. *Birds*

in Europe. Population Estimates, Trends and Conservation Status. BirdLife Conservation Series No. 12. BirdLife International, Cambridge.

Burg, T M, and Croxall, J P. 2004. Global population structure and taxonomy of the wandering albatross species complex. *Mol Ecol* 13: 2345-2355.

Butler, C J. 2003. Population biology of the introduced Rose-ringed Parakeet *Psittacula krameri* in the UK. PhD thesis, Edward Grey Institute of Field Ornithology. www.biology.uco.edu/PersonalPages/CButler/thesis.pdf.

Byers, C, Olsson, U, and Curson, J. 1995. *Sparrows and Buntings: a Guide to the Buntings and North American Sparrows*. Pica Press, Robertsbridge.

Byrkjedal, I, and Thompson, D B A. 1998. *Tundra Plovers: the Eurasian, Pacific and American Golden Plovers and Grey Plover*. T & A D Poyser, London.

Byrkjedal, I, Breistol, A, Mjos, A-T, and Strann, K-B. 2000. Winter habitat of White-billed and Great Northern Divers (*Gavia adamsii* and *G immer*) on the coast of Norway. *Ornis Nor* 23: 50-55.

C

Cabot, J, and Urdiales, C. 2005. The subspecific status of Sardinian Warblers *Sylvia melanocephala* in the Canary Islands with the description of a new subspecies from Western Sahara. *Bull Brit Ornithol Club* 125: 230-240.

Cade, M, and Collinson, J M. 2015. 'Stejneger's Stonechat' in Dorset: new to Britain. *Br Birds* 108: 423-428.

Cagnon, C, Lauga, B, Hémery, G, and Mouchès, C. 2004. Phylogenetic differentiation of storm petrels (*Hydrobates pelagicus*) based on cytochrome b mitochondrial DNA variation. *Marine Biology* 145: 1257-1264.

Campbell, O, Flood, R F, Al Dhaheri, K, and Talbot, G. 2013. The first confirmed records of Cory's Shearwater *Calonectris (diomedea) borealis* for the United Arab Emirates and Oman, in 2011. *Sandgrouse* 35: 126-131.

Campbell, O. 2012. The status of Isabelline Shrike taxa in the United Arab Emirates. *Br Birds* 105: 417-420.

Carlos, C J, Roselaar, C S, and Voisin, J-F. 2012. A replacement name for *Charadrius leschenaultii crassirostris* (Severtzov, 1873), a subspecies of Greater Sand Plover. *Bull Brit Ornithol Club* 132: 63-65.

Castelli, G, D'Amelio, P, and Haas, M. 2016. Lista ornitica del Palearctico Occidentale – EBN Italia. Vers 1.1. http://www.ebnitalia.it/easyNews/NewsLeggi.asp?NewsID=29.

Catley, G. 2009. A Soft-plumaged Petrel in Arctic Norway - the first record for the North Atlantic region. *Birding World* 22: 249-252.

Ceia, R S, Ramos, J A, Heleno, R H, Hilton, G, and Marques, T A. 2011. Status assessment of the Critically Endangered Azores Bullfinch *Pyrrhula murina*. *Bird Conserv Int* 21: 477-489.

Celmins, A, and Baumanis, J. 1997. Report on rare birds in Latvia in 1995. *Putni Dabā* 7: 27-32.

Celmins, A, and Baumanis, J. 1998. Report on rare birds in Latvia in 1996-97. *Putni Dabā* 8: 1-10.

Celmins, A, and Baumanis, J. 2000. Report on rare birds in Latvia in 1998-99. *Putni Dabā* 10: 12-21.

Chantler, P, and Driessens, G. 2000. Swifts. *A Guide to the Swifts and Treeswifts of the World*. Second edition. Pica Press, Robertsbridge.

Charles, D. 2008. Ring-billed Gull breeding with Common Gull on Copeland Island, Co. Down. The first confirmed breeding record for Ring-billed Gull in the Western Palearctic. *Northern Ireland Bird Report XVIII*.

Cheke, R A, Mann, C F, and Allen, R. 2001. *Sunbirds. A Guide to the Sunbirds, Flowerpeckers, Spiderhunters and Sugarbirds of the World*. Christopher Helm, London.

Cheraghi, S, and Tohdifar, M. 2010. First record of Ultramarine Flycatcher *Ficedula superciliaris* in Iran and the Middle East. *Sandgrouse* 32: 37-38.

Cherakaoui, I, and Bouajaja, A. 2014. Liste des Oiseaux du Maroc. http://grepom.org/wp-content/uploads/Liste-des-Oiseaux-du-Maroc_2014.pdf.

Chesser, R T, Banks, R C, Barker, F K, Cicero, C, Dunn, J L, Kratter, A W, Lovette, I J, Rasmussen, P C, Remsen Jnr, J V, Rising, J D, Stotz, D F, and Winker, K. 2009. Fiftieth Supplement to the American Ornithologists' Union *Check-List of North American Birds*. *Auk* 126: 705-714.

Chesser, R T, Banks, R C, Barker, F K, Cicero, C, Dunn, J L, Kratter, A W, Lovette, I J, Rasmussen, P C, Remsen Jnr, J V, Rising, J D, Stotz, D F, and Winker, K. 2010. Fifty-first Supplement to the American Ornithologists' Union *Check-List of North American Birds*. *Auk* 127: 726-744.

Chesser, R T, Banks, R C, Barker, K F, Cicero, C, Dunn, J L, Kratter, A W, Lovette, I J, Rasmussen, P C, Remsen Jnr, J V, Rising, J D, Stotz, D F, and Winker, K. 2011. Fifty-second Supplement to the American Ornithologists' Union *Check-List of North American Birds*. *Auk* 128: 600-613.

Chesser, R T, Banks, R C, Barker, K F, Cicero, C, Dunn, J L, Kratter, A W, Lovette, I J, Rasmussen, P C, Remsen Jnr, J V, Rising, J D, Stotz, D F, and Winker, K. 2012. Fifty-third Supplement to the American Ornithologists' Union *Check-List of North American Birds*. *Auk* 129: 573-588.

Chesser, R T, Banks, R C, Barker, K F, Cicero, C, Dunn, J L, Kratter, A W, Lovette, I J, Rasmussen, P C, Remsen Jnr, J V, Rising, J D, Stotz, D F, and Winker, K. 2013. Fifty-fourth Supplement to the American Ornithologists' Union *Check-List of North American Birds*. *Auk* 130: 1-14.

Chesser, R T, Banks, R C, Burns, K J, Cicero, C, Dunn, J L, Kratter, A W, Lovette, I J, Navarro-Sigüenza, A G, Rasmussen, P C, Remsen Jnr, J V, Rising, J D, Stotz, D F, and Winker, K. 2015. Fifty-sixth Supplement to the American Ornithologists' Union *Check-List of North American Birds*. *Auk* 132: 748-764.

Chesser, R T, Banks, R C, Cicero, C, Dunn, J L, Kratter, A W, Lovette, I J, Navarro-Sigüenza, A G, Rasmussen, P C, Remsen Jnr, J V, Rising, J D, Stotz, D F, and Winker, K. 2014. Fifty-fifth Supplement to the American Ornithologists' Union *Check-List of North American Birds*. *Auk* 131: CSi-CSxv. DOI: 10.1642/AUK-14-124.1.

Christensen, J S, and Rasmussen, P A F. 2015. Revideret status for sjældne fugle i Danmark før 1965. *Dansk Orn Foren Tidsskr* 109: 41-112

Christidis, L, and Boles, W E. 2008. *Systematics and*

Taxonomy of Australian Birds. CSIRO Publishing, Collingwood.

Cibois, A. 2003. *Sylvia* is a babbler: taxonomic implications for the families Sylviidae and Timaliidae. *Bull Brit Ornithol Club* 123: 257-261.

Clark, W S. 1992. The taxonomy of Steppe and Tawny Eagles with criteria for separation of museum specimens and live eagles. *Bull Brit Ornithol Club* 112: 150-156.

Clark, W S. 2005. Steppe Eagle *Aquila nipalensis* is monotypic. *Bull Brit Ornithol Club* 125: 149-153.

Clarke, T. 2006. *Birds of the Atlantic Islands*. Christopher Helm, London.

Cleere, N. 2010. *Nightjars, Potoos, Frogmouths, Oilbird and Owlet-nightjars of the World*. WildGuides, Old Basing.

Cleeves, T. 2002. Slender-billed Curlew in Northumberland: new to Britain and Ireland. *Br Birds* 95: 272-278.

Clement, P, and Hathway, R. 2000. *Thrushes*. Christopher Helm, London.

Clement, P, and Rose, C. 2015. *Robins and Chats*. Christopher Helm, London.

Clement, P, Harris, A, and Davis, J. 1993. *Finches and Sparrows: an Identification Guide*. Christopher Helm, London.

Clement, P. 1995. Southern and eastern Great Grey Shrikes in northwest Europe. *Birding World* 8: 300-309.

Clements, A. 1990. A record of Banded Martin *Riparia cincta* from Egypt. *Sandgrouse* 12: 55-56.

Clements, J F. 2000. *Birds of the World: a Checklist*. Pica Press, Robertsbridge.

Clouet, M, and Wink, M. 2000. The buzzards of Cape Verde *Buteo (buteo) bannermani* and Socotra *Buteo (buteo)* spp: first results of a genetic analysis based on nucleotide sequences of the cytochrome *b* gene. *Alauda* 68: 55-58.

Cohen, A, Granit, B, and Perlman, Y, on behalf of the Israeli Rarities and Distribution Committee (IRDC). 2015. Bulletin 9:01 on Rare Birds in Israel. http://www.israbirding.com/irdc/bulletins/bulletin_9/.

Collar, N J, and Pilgrim, D. 2007. Species-level changes proposed for Asian birds, 2005-2006. *Birding Asia* 8: 14-30.

Collar, N J. 2006. A partial revision of the Asian babblers (Timaliidae). *Forktail* 22: 85-112.

Collinson, J M, and McGowan, R Y. 2012. A genetic analysis of the first British Siberian Stonechat. *Br Birds* 105: 318-331.

Collinson, J M, Knox, A G, Parkin, D T, and Sangster, G. 2003. Specific status of taxa within the Greenish Warbler complex. *Br Birds* 96: 327-331.

Collinson, J M, McGowan, R Y, and Irestedt, M. 2014. First British records of 'Eastern' and 'Western' Subalpine Warblers. *Br Birds* 102: 282-285.

Collinson, J M, Parkin, D T, Knox, A G, Sangster, G, and Helbig, A. 2006. Species limits within the genus *Melanitta*, the scoters. *Br Birds* 99: 183-201.

Collinson, J M, Parkin, D T, Knox, A G, Sangster, G, and Svensson, L. 2008. Species boundaries in the Herring and Lesser Black-backed Gull complex. *Br Birds* 101: 340-363.

Collinson, J M, Rowlands, A, Steele, J G, McInerny, C J, and Hudson, N. 2014. The review of the record of Slender-billed Curlew at Druridge Bay, Northumberland. *Br Birds* 107: 389-404.

Collinson, J M, Waite, S, and McGowan, R Y. 2013. British records of 'Eastern Yellow Wagtail'. *Br Birds* 106: 36-41.

Collinson, M. 2001. Evolution of the Atlantic-island Chaffinches. *Br Birds* 94: 121-124.

Collinson, M. 2001. Greenish Warbler, 'Two-barred Greenish Warbler', and the speciation process. *Br Birds* 94: 278-283.

Collinson, M. 2006. Splitting headaches? Recent taxonomic changes affecting the British and Western Palearctic lists. *Br Birds* 99: 306-323.

Comité d'Homologation National. 2015. Base de données du CHN. http://www.chn-france.com/chn_donnees.php.

Commission de l'Avifaune Française. 2007. Liste officielle des Oiseaux de France (Catégories A, B et C). *Ornithos* 14: 234-246.

Constantine, M, Hopper, N, and The Sound Approach. 2012. *Catching the Bug: a Sound Approach Guide to the Birds of Poole Harbour*. The Sound Approach, Poole.

Cooper, J H, and Voous, K. 1999. Birds, bones and biogeography: Iberian azure-winged magpies come in from the cold. *Br Birds* 92: 659-665.

Cooper, J H. 2000. First fossil record of azure-winged magpie *Cyanopica cyanus* in Europe. *Ibis* 142: 150-151.

Copete, J L, Ferrer, X, Bigas, D, Sanpera, C, and Vieties, D R. 2011. Reed Cormorant collected in Catalunya, Spain, in c 1855. *Dutch Birding* 33: 40-43.

Copete, J L, Lorenzo, J A, Amengual, E, Bigas, D, Fernández, P, López-Velasco, D, Rodríguez, G, and García-Tarrasón, M – Comité de Rarezas de la Sociedad Española de Ornitología. 2015. Observaciones de aves raras en España, 2012 y 2013. *Ardeola* 62: 453-508.

Copete, J L. 2008. Distribution and identification of Iberian Chiffchaff. *Br Birds* 101: 378-379.

Corell, M, Averland, M, Bister, H, and Strömberg, N. 2015. Nya Svenska fågellistan/The complete checklist of Swedish birds. http://c3.igoterra.com/index.asp.

Cornwallis, L, and Porter, R F. 1982. Spring observations on the birds of North Yemen. *OSME Bulletin* 4: 1-36.

Correia-Fagundes, C, and Romano, H. 2011. Black-bellied Storm-petrel *Fregatta tropica* in Madeira – first record from the Western Palearctic. *Bocagiana* 233: 1-6.

Correia-Fagundes, C, Romano, H, Zino, F, and Biscoito, M. 2011. Additions and corrections to the 2010 checklist of the birds of the archipelagos of Madeira and the Selvagens. *Bol Mus Mun Funchal* 61: 23-28.

Correia-Fagundes, C, Romano, H, Zino, F, and Biscoito, M. 2011. First record of Yellow-crowned Night Heron *Nyctanassa violacea* (Aves: Pelecaniformes: Ardeidae) from the island of Madeira (NE Atlantic Ocean). *Bol Mus Mun Funchal* 61: 5-11.

Corso, A, and Dennis, P. 1998. Amur Falcon in Italy, a new Western Palearctic bird. *Birding World* 11: 259-260.

Corso, A, Jansen, J, and Kókay, S. 2014. A review of the identification criteria and variability of the Slender-billed Curlew. *Br Birds* 107: 339-370.

Corso, A, Penna, V, Gustin, M, Maiorano, I, and Ferrandes, P. 2012. Annotated checklist of the birds from Pantellaria Island (Sicilian Channel, Italy): a summary of the most relevant data, with a new species for the site and for Italy. *Biodivers J* 3: 407-428.

Corso, A. 2009. Tristan Albatross in the WP: a cautionary note. *Dutch Birding* 31: 307-308.

Corso, A. 2010. Sicilian Rock Partridge: identification and taxonomy. *Dutch Birding* 32: 79-96.

Corso, A. 2012. Additional comments on Rock Partridge morphology: *Alectoris graeca orlandoi*. *Dutch Birding* 34: 97-99.

Corso, A. 2012. African Reed Warblers in Tunisia. *Birding World* 25: 333-335.

Cox, G W. 1985. The evolution of avian migration systems between temperate and tropical regions of the New World. *The American Naturalist* 126: 451-474.

Cramp, S (ed). 1977. *Handbook of the Birds of Europe, the Middle East and North Africa – the Birds of the Western Palearctic*. Vol I: Ostrich to Ducks. Oxford University Press, Oxford.

Cramp, S (ed). 1980. *Handbook of the Birds of Europe, the Middle East and North Africa – the Birds of the Western Palearctic*. Vol II: Hawks to Bustards. Oxford University Press, Oxford.

Cramp, S (ed). 1983. *Handbook of the Birds of Europe, the Middle East and North Africa – the Birds of the Western Palearctic*. Vol III: Waders to Gulls. Oxford University Press, Oxford.

Cramp, S (ed). 1985. *Handbook of the Birds of Europe, the Middle East and North Africa – the Birds of the Western Palearctic*. Vol IV: Terns to Woodpeckers. Oxford University Press, Oxford.

Cramp, S (ed). 1988. *Handbook of the Birds of Europe, the Middle East and North Africa – the Birds of the Western Palearctic*. Vol V: Tyrant Flycatchers to Thrushes. Oxford University Press, Oxford.

Cramp, S (ed). 1992. *Handbook of the Birds of Europe, the Middle East and North Africa – the Birds of the Western Palearctic*. Vol VI: Warblers. Oxford University Press, Oxford.

Cramp, S, and Perrins, C M (eds). 1993. *Handbook of the Birds of Europe, the Middle East and North Africa – the Birds of the Western Palearctic*. Vol VII: Flycatchers to Shrikes. Oxford University Press, Oxford.

Cramp, S, and Perrins, C M (eds). 1994a. *Handbook of the Birds of Europe, the Middle East and North Africa – the Birds of the Western Palearctic*. Vol VIII: Crows to Finches. Oxford University Press, Oxford.

Cramp, S, and Perrins, C M (eds). 1994b. *Handbook of the Birds of Europe, the Middle East and North Africa – the Birds of the Western Palearctic*. Vol IX: Buntings and New World Warblers. Oxford University Press, Oxford.

Cranswick, P A, Raducescu, L, Hilton, G M, and Petkov, N. 2012. International Single Species Action Plan for the Conservation of the Red-breasted Goose (*Branta ruficollis*). AEWA Technical Series No 46, Bonn.

Cresswell, W, Donald, P, and Khrokov, V. 2005. Ongoing research on Sociable Lapwings – an update. *Br Birds* 98: 496-497.

Crochet, P-A, and Haas, M. 2008. Western Palearctic list update: deletion of White-bellied Storm Petrel. *Dutch Birding* 30: 17-18.

Crochet, P-A, and Haas, M. 2008. Western Palearctic list update: deletion of Cape Gannet. *Dutch Birding* 30: 103-104.

Crochet, P-A, and Haas, M. 2010. Western Palearctic list updates: African Palm Swift. *Dutch Birding* 32: 131-132.

Crochet, P-A, and Haas, M. 2013. Western Palearctic list updates: re-evaluation of five species from continental Mauritania. *Dutch Birding* 35: 28-30.

Crochet, P-A, and Joynt, G. 2012. *AERC List of Western Palearctic Birds*. December 2012 version.

Crochet, P-A, and Spaans, B. 2008. Spur-winged Geese at Banc d'Arguin, Mauritania, in December 2004. *Dutch Birding* 30: 101-102.

Crochet, P-A, Barthel, P H, Bauer, H-G, van den Berg, A B, Bezzel, E, Collinson, J M, Dietzen, C, Dubois, P J, Fromholtz, J, Helbig, A J†, Jiguet, F, Jirle, E, Knox, A G, Krüger, T, Le Maréchal, P, van Loon, A J, Päckert, M, Parkin, D T, Pons, J-M, Raty, L, Roselaar, C S, Sangster, G, Steinheimer, F D, Svensson, L, Tyrberg, T, Votier, S C, and Yésou, P. 2011. AERC TAC's taxonomic recommendations: 2011 report. http://www.aerc.eu/tac.html.

Crochet, P-A, Didner, E, Geniez, P, and Forsyth, E. 2012. Village Weaver at Abu Simbel, Egypt, in May 2006. *Dutch Birding* 34: 316-319.

Crochet, P-A. 2008. A Taiga Merlin on the Azores: an overlooked vagrant to Europe? *Birding World* 21: 114-116.

Crottini, A, Galimberti, A, Boto, A, Serra, L, Liu, Y, Yeung, C, Yang, X, Barbuto, M, and Casiraghi, M. 2010. Toward a resolution of a taxonomic enigma: First genetic analyses of *Paradoxornis webbianus* and *Paradoxornis alphonsianus* (Aves: Paradoxornithidae) from China and Italy. *Mol Phylogenet Evol* 57: 1312-1318.

Curson, J, Quinn, D, and Beadle, D. 1994. *New World Warblers*. Christopher Helm, London.

D

Darby, C. 2011. Eastern Common Terns in Suffolk and Belgium. *Birding World* 24: 511-512.

Daróczi, J S, Bugariu, S, Fântână, C, Kelemen, A M, Kovács, I, Ölvedi, S-Z, and Szabó, J. 2015. Report of the Romanian Rarities Committee for the period 2010-2014 (Part I), Nr. 1/2015. http://rombird.ro/uploaded/tiny/files/rrc_report_2010-2014,_part_i.pdf.

de Juana, E, and Garcia, E. 2015. *The Birds of the Iberian Peninsula*. Christopher Helm, London.

de Juana, E, y Comité de Rarezas de la Sociedad Española de Ornitología. 2005. Report on rare birds in Spain in 2003. *Ardeola* 52: 185-206.

de Juana, E. 2011. The Sociable Lapwing in Europe. *Br Birds* 104: 84-90.

de Knijff, P, van der Spek, V, and Fischer, J. 2012. Genetic identity of grey chiffchaffs trapped in the Netherlands in autumns of 2009-11. *Dutch Birding* 34: 386-392.

de la Perche, N. 1992. Moroccan Double-spurred Francolin. *Dutch Birding* 14: 10-12.

de Rouck, K, and Colin, D. 2012. Two Mourning Collared Doves at Abu Simbel, Egypt, from December 2010. *Dutch Birding* 34: 104-107.

de Smet, G, Adriaens, P, Vandegehuchte, M, and members of the Belgisch Avifaunistisch Homologatiecomité. 2006. *Lijst van de Belgische vogels*. BAHC, Oostkamp.

Dean, A R, Bradshaw, C, Martin, J, Stoddart, A, and Walbridge, G. 2010. The status in Britain of 'Siberian Chiffchaff'. *Br Birds* 103: 320-338.

del Hoyo, J, and Collar, N J, 2014. *HBW and BirdLife International Illustrated Checklist of the Birds of the World*. Volume 1: Non-passerines. Lynx Edicions, Barcelona.

del Hoyo, J, Elliott, A, Sargatal, J, Christie, D A, and de Juana, E (eds). 2016. *Handbook of the Birds of the World Alive*. Lynx Edicions, Barcelona. Retrieved from http://www.hbw.com.

Delaney, H. 2012. The Eastern Kingbird in County Galway – a new Western Palearctic bird. *Birding World* 25: 430-432.

Dennison, M D, and Baker, A J. 1991. Morphometric variability in continental and Atlantic island populations of chaffinches (*Fringilla coelebs*). *Evolution* 45: 29-39.

Deutsche Seltenheitenkommission. 2008. Seltene Vogelarten in Deutschland von 2001 bis 2005. *Limicola* 22: 249-339.

Deutsche Seltenheitenkommission. 2009. Seltene Vogelarten in Deutschland von 2006 bis 2008. *Limicola* 23: 257-334.

Deutsche Seltenheitenkommission. 2010. Seltene Vogelarten in Deutschland 2009 (mit Nächtragen 2001-2008). *Limicola* 24: 233-286.

Deutsche Seltenheitenkommission. 2012. *Seltene Vögel in Deutschland 2010*. Dachverband Deutscher Avifaunisten, Münster.

Deutsche Seltenheitenkommission. 2013. *Seltene Vögel in Deutschland 2011/12*. Dachverband Deutscher Avifaunisten, Münster.

Dhora, D. 2010. *Register of Species of the Fauna of Albania*. Universiteti i Shkodrës, Shkoder.

Dickinson, E C, and Dekker, R W R J. 2001. Systematic notes on Asian birds. 11. A preliminary review of the Alaudidae. *Zool Verh Leiden* 335: 61-84.

Dickinson, E C, Eck, S, and Martens, J. 2004. Systematic notes on Asian birds 44. A preliminary review of the Corvidae. *Zool Verh Leiden* 350: 85-109.

Dickinson, E C. 2003. *The Howard and Moore Complete Checklist of the Birds of the World*. Third edition. Christopher Helm, London.

Dickinson, E C. 2008. The name of the easternmost population of Common Nightingale *Luscinia megarhynchos*. *Bull Brit Ornithol Club* 128: 141-142.

van Diek, H, Felix, R, Hornman, M, Meininger, P L, Willems, F, and Zekhuis, M. 2004. Bird counting in Iran in January 2004. *Dutch Birding* 287-296.

Dierschke, J. 2007. The status of Black-headed and Red-headed Buntings on Helgoland. *Br Birds* 100: 552-557.

Dies, J I, Lorenzo, J A, Gutiérrez, R, García, E, Gorospe, G, Martí-Aledo, J, Gutiérrez, P, and Vidal, C. 2007. Observaciones de aves raras en España, 2005. *Ardeola* 54: 405-446.

Dietzen, C, García-del-Rey, E, Castro, G D, and Wink, M. 2008. Phylogeography of the Blue Tit (*Parus teneriffae*-group) on the Canary Islands based on mitochondrial DNA sequence data and morphometrics. *J Ornithol* 149: 1-12.

Dietzen, C, Michels, J P, and Wink, M. 2015. Formal description of a new subspecies of the European Robin from Gran Canaria Island, Spain (Aves: Muscicapidae: *Erithacus rubecula marioniae* subsp. Nov.). *Open Ornithol J* 8: 39-42.

Dijkstra, K-D. 1997. Striated Heron *Butorides striata* breeding in Egyptian Nile Valley. *Sandgrouse* 19: 140.

Dobson, A D M, and Clarke, M L. 2011. Inconsistency in the taxonomy of Hen and Northern Harriers: causes and consequences. *Br Birds* 104: 192-201.

Donald, P, Collar, N, Marsden, S, and Pain, D. 2013. *Facing Extinction. The world's rarest birds and the race to save them*. Second edition. Bloomsbury, London.

Dowsett, R J, Atkinson, P W, and Caddick, J A. 2014. Checklist of the birds of Chad. Downloaded from www.africanbirdclub.org 8 March 2014.

Dowsett, R J, Atkinson, P W, and Caddick, J A. 2014. Checklist of the birds of Niger. Downloaded from www.africanbirdclub.org 8 March 2014.

Drovetski, S, Zink, R M, Rohwer, S, Fadeev, I V, Nesterov, E V, Karagodin, I, Koblik, E A, and Red'kin, Y A. 2004. Complex biogeographic history of a Holarctic passerine. *Proc R Soc Lond [Biol.]* 271: 545-551.

Dubois, P J, Duquet, M, Le Maréchal, P, Olioso, G, and Yésou, P. 2014. Deuxième mise à jour du nouvel inventaire des oiseaux de France. *Ornithos* 21: 169-213.

Dubois, P J, Le Maréchal, P, Olioso, P, and Yésou, P. 2008. *Nouvel Inventaire des Oiseaux de France*. Delachaux and Niestlé, Paris.

Dubois, P J. 2002. Probable breeding of pure American Black Duck pair in the Azores in 2000. *Dutch Birding* 24: 12.

Dubois, P J. 2007. Les oiseaux allochtones en France: statut et interactions avec les espèces indigènes. *Ornithos* 14: 329-364.

Dudley, S P. 2010. BOU Proceedings – the Impacts of Non-native Species. http://www.bou.org.uk/bouproc-net/non-natives/dudley20100531.pdf.

Dudley, S. 2009. *A Birdwatching Guide to Lesvos*. Arlequin Press, Shrewsbury.

Dufour, P, Jones, J, and Crochet, P-A. 2016. Occurrence of multiple Elegant Terns confirmed in Western Europe. www.birdguides.com/webzine/article.asp?a=5856.

Dufourny, H. 1999. White-tailed Tropicbird in Cape Verde Islands in February 1999. *Dutch Birding* 21: 254-255.

Dunn, J, and Garrett, K. 1997. *Peterson Field Guides: Warblers*. Houghton Mifflin, New York.

Dutch Birding & Commissie Dwaalgasten Nederlandse Avifauna. 2015. Overzicht van alle vogels waargenomen in Nederland. http://www.dutchavifauna.nl/list.

E

Ebels, E B. 2003. Speciation in *Pica* magpies. *Dutch Birding* 25: 103-116.

Ebels, E, van Duivendijk, N, van der Luit, P, and Steijn, L. 2014. Ship-assisted passage by Iago Sparrows from Cape Verde Islands to Madeira and the Netherlands in May 2013. *Dutch Birding* 36: 167-171.

Eck, S, and Martens, J. 2006. Systematic notes on Asian birds. 49. A preliminary review of the Aegithalidae, Remizidae and Paridae. *Zool Med Leiden* 80-5: 1-63.

Eck, S. 1996. Die Palaearktischen Vögel – Geospezies und Biospezies. *Zool Abhandlungen (Dresden)* 49 (Suppl): 1-104.

Efe, M A, Tavares, E S, Baker, A J, and Bonatto, S L. 2009. Multigene phylogeny and DNA barcoding indicate that the Sandwich Tern complex (*Thalasseus sandvicensis*, Laridae, Sternini) comprises two species. *Mol Phylogenet Evol* 52: 263-267.

Egyptian Ornithological Rarities Committee. 2013. Checklist of the birds of Egypt. http://www.chn-france.org/upload_content/eorc_boe_2011.pdf.

van Eijk, P. 2013. Presumed second locality for Omani Owl. *Dutch Birding* 35: 386-388.

Elgvin, T O, Hermansen, J S, Fijarczyk, A, Bonnet, T, Borge, T, Saether, S A, Voje, K L, and Saetre, G-P. 2011. Hybrid speciation in sparrows II: a role for sex chromosomes? *Mol Ecol* **20**: 3823-3837.

Elorriaga, J, and Muñoz, A R. 2010. First breeding record of North African Long-legged Buzzard *Buteo rufinus cirtensis* in continental Europe. *Br Birds* 103: 399-401.

van Els, P, and Brady, M L. 2014. Specimen of Indian Pitta *Pitta brachyura* from the Islamic Republic of Iran. *Bull Brit Ornithol Club* 134: 160-162.

Elts, J, Leito, A, Leivits, A, Luigujõe, L, Mägi, E, Nellis, R, Nellis, R, Ots, M, and Pehlak, H. 2013. Eesti lindude status, pesitsusaegne ja talvine arvukus 2008-2012. *Hirundo* 26: 80-112.

Engelmoer, M, and Roselaar, C S. 1998. *Geographical Variation in Waders*. Kluwer Academic Publishers, Dordrecht.

Ericson, P G P, and Johansson, U S. 2003. Phylogeny of Passerida (Aves: Passeriformes) based on nuclear and mitochondrial sequence data. *Mol Phylogenet Evol* 29: 126-138.

Eriksen, H, and Eriksen, J. 1999. The first records of Yellow Bittern *Ixobrychus sinensis* in Oman and Arabia. *Sandgrouse* 21: 178-179.

Eriksen, H, Eriksen, J, Al-Saadi, S, and Sargeant, D E. 2001. The discovery of Oriental White-eye *Zosterops palpebrosa* in Arabia. *Sandgrouse* 23: 130-133.

Eriksen, H. 2004. The first Cape Gannet *Sula capensis* in Oman and the Middle East. *Sandgrouse* 26: 146-148.

Eriksen, J, and Victor, R. 2013. *Oman Bird List*. Edition 7. Sultan Qaboos University, Muscat.

Eriksen, J, and Victor, R. 2015. Additions and corrections to the *Oman Bird List*, Edition 7. http://www.birdsoman.com/obl7-update.html.

Eriksen, J. 2015. Oman Bird List, Edition 7.3. http://www.birdsoman.com/Oman_Bird_List_v7-3.pdf.

Erritzøe, J, Mann, C F, Brammer, F P, and Fuller, R A. 2012. *Cuckoos of the World*. Christopher Helm, London.

Ertan, K T. 2006. The evolutionary history of Eurasian redstarts, *Phoenicurus*. *Acta Zool Sinica* 52 (Supplement): 310-313.

Estonian Ornithological Society. 2015. List of Estonian birds. http://www.eoy.ee/en/node/313.

Etayeb, K S, Yahia, J, Berbash, A, Wattier, R, and Brochet, A-L. 2014. First breeding evidence of Marbled Duck (*Marmaronetta angustirostris*) in Libya. *Waterbirds* 37: 107-110.

F

Fabre, P-H, Irestedt, M, Fjeldså, J, Bristol, R, Groombridge, J J, Irham, M, and Jønsson, K A. 2012. Dynamic colonization exchanges between continents and islands drive diversification in paradise-flycatchers (*Terpsiphone*, Monarchidae). *J Biogeogr* 39: 1900-1918.

Fagundes, I, Pérez, C, and Moniz, F. 2012. Cape Verde Shearwater in Selvagens in April 2012. *Dutch Birding* 34: 310-311.

Fahy, K, on behalf of the Irish Rare Birds Committee. 2009. *Irish Rare Bird Report 2008*. BirdWatch Ireland, Kilcoole.

Fahy, K, on behalf of the Irish Rare Birds Committee. 2010. *Irish Rare Bird Report 2009*. BirdWatch Ireland, Kilcoole.

Fahy, K, on behalf of the Irish Rare Birds Committee. 2011. *Irish Rare Bird Report 2010*. BirdWatch Ireland, Kilcoole.

Fahy, K, on behalf of the Irish Rare Birds Committee. 2012. *Irish Rare Bird Report 2011*. BirdWatch Ireland, Kilcoole.

Fahy, K, on behalf of the Irish Rare Birds Committee. 2013. *Irish Rare Bird Report 2012*. BirdWatch Ireland, Kilcoole.

Fahy, K, on behalf of the Irish Rare Birds Committee. 2014. *Irish Rare Bird Report 2013*. BirdWatch Ireland, Kilcoole.

Farrelly, W, and Charles, D. 2011. The Dresser's Eider in County Donegal – a new Western Palearctic bird. *Birding World* 23: 62-64.

Fasola, M, and Gariboldi, A. 1985. Status del Colino della Virginia *Colinus virginianus* in Italia. *Proc III Conv Ital Ornithol Pavia*: 25-26.

Fasola, M, and Gariboldi, A. 1987. Il Colino della Virginia, *Colinus virginianus*, in Italia. *Riv ital Orn* 57: 3-13.

Faveyts, W, Elst, J, en de leden van het BAHC. 2012. Zeldzame vogels in België in 2010. Veertigste rapport van het Belgisch Avifaunistisch Homologatiecomité. *Natuur.oriolus* 78: 41-54.

Faveyts, W, Valkenburg, M, and Granit, B. 2011. Crested Honey Buzzard: identification, western occurrence and hybridisation with European Honey Buzzard. *Dutch Birding* 33: 149-162.

Faveyts, W, Vandegehuchte, M, en de leden van het BAHC. 2013. Zeldzame vogels in België in 2011. Veertigste rapport van het Belgisch Avifaunistisch Homologatiecomité. *Natuur.oriolus* 79: 44-57.

Faveyts, W, Vandegehuchte, M, en de leden van het BAHC. 2014. Zeldzame vogels in België in 2012. Veertigste rapport van het Belgisch Avifaunistisch Homologatiecomité. *Natuur.oriolus* 80: 91-103.

Feare, C, and Craig, A. 1998. *Starlings and Mynas*. Christopher Helm, London.

Fenech, N. 2010. *A Complete Guide to the Birds of Malta*. Midsea Books, Valletta.

Ferguson-Lees, I J, and Christie, D A. 2005. English names for Western Palearctic bids, and the role of the IOC. *Br Birds* 98: 207-211.

Ferguson-Lees, J, and Christie, D A. 2001. *Raptors of the World*. Christopher Helm, London.

Festari, I, Janni, O, and Rubolini, D. 2002. Moltoni's Warbler breeding in mainland Italy. *Dutch Birding* 24: 88-90.

Firouz, E. 2005. *The Complete Fauna of Iran*. Taurus, London.

Flade, M, and Lachmann, L. 2008. International Species Action Plan for the Aquatic Warbler *Acrocephalus paludicola*. http://ec.europa.eu/environment/nature/conservation/wildbirds/action_plans/docs/acrocephalus_paludicola.pdf.

Flade, M, Diop, I, Haase, M, Le Nevé, A, Oppel, S, Tegetmeyer, C, Vogel, A, and Salewski, V. 2011. Distribution, ecology and threat status of the Aquatic Warblers *Acrocephalus paludicola* wintering in West Africa. *J Ornithol* 152 (supplement 1): 129-140.

Flint, P R, Stewart, P F, and Kirwan, G M. 1997. A review of claimed records of Lesser Sand Plover *Charadrius mongolus* from Cyprus and Turkey. *Sandgrouse* 19: 133-139.

Flint, P, and Stewart, P. 1992. *The Birds of Cyprus*. BOU Checklist No. 6. Second Edition. British Ornithologists' Union, Tring.

Flint, P, Whaley, D, Kirwan, G M, Charalambides, M, Schweizer, M, and Wink, M. 2015. Reprising the taxonomy of Cyprus Scops Owl *Otus (scops) cyprius*, a neglected island endemic. *Zootaxa*. doi: http://dx.doi.org/10.11646/zootaxa.4040.3.3.

Flood, B, and Danzenbaker, M. 2009. Identification of Trindade Petrel and its status in the North Atlantic. *Birding World* 22: 162-166.

Flood, B, and Fisher, A. 2011. *Multimedia Identification Guide to North Atlantic Seabirds: Storm-petrels and Bulwer's Petrel*. Pelagic Birds and Birding Multimedia Identification Guides, Scilly.

Flood, B, and Fisher, A. 2013. *Multimedia Identification Guide to North Atlantic Seabirds: Pterodroma Petrels*. Pelagic Birds and Birding Multimedia Identification Guides, Scilly.

Flood, B, Romano, H, and Correia-Fagundes, C. 2013. The first confrmed South Polar Skua for Madeira. *Birding World* 26: 248-252.

Flood, B. 2007. Madeiran Storm-petrel 28 July 2007 – first live record for Britain. *Isles of Scilly Bird and Natural History Review 2007*: 112-116.

Flood, R L. 2009. 'All-dark' Oceanodroma storm-petrels in the Atlantic and neighbouring seas. *Br Birds* 102: 365-385.

Flood, R L. 2011. Black-bellied Storm-petrel first for WP. http://scillypelagics.blogspot.com/2011/08/black-bellied-storm-petrel-first-for-wp.html.

Foekens, E, and Schelvis, J. 2006. Two River Terns in Golestan, Iran, in January 2005. *Dutch Birding* 28: 21–23.

Fok, K W, Wade, C M, and Parkin, D T. 2002. Inferring the phylogeny of disjunct populations of the azure-winged magpie *Cyanopica cyanus* from mitochondrial control region sequences. *Proc R Soc Lond [Biol.]* 269: 1671-1679.

Forrester, R W, Andrews, I J, McInerny, C J, Murray, R D, McGowan, R Y, Zonfrillo, B, Betts, M W, Jardine, D C, and Grundy, D S. Eds. 2007. *The Birds of Scotland*. Vols I and II. The Scottish Ornithologists' Club, Aberlady.

Förschler, M I, Khoury, F, Bairlein, F, and Aliabadian, M. 2010a. Phylogeny of the mourning wheatear *Oenanthe lugens* complex. *Mol Phylogenet Evol* 56: 758-767.

Förschler, M I, Khoury, F, Bairlein, F, and Aliabadian, M. 2010b. Corrigendum to 'Phylogeny of the mourning wheatear *Oenanthe lugens* complex'. [*Mol Phylogenet Evol* 56: 758-767]. *Mol Phylogenet Evol* 57: 483-484.

Förschler, M I, Randler, C, Dierschke, J, and Bairlein, F. 2010. Morphometric diagnosability of Cyprus Wheatears *Oenanthe cypriaca* and an unexpected occurrence on Helgoland Island. *Bird Study* 57: 396-400.

Forsman, D. 2014. Wahlberg's Eagle – a new bird for the Western Palearctic list. http://www.dickforsman.com/2014/04/22/wahlbergs-eagle-a-new-bird-for-the-western-palaearctic-list/.

Fox, A D, Hobson, K A, Ekins, G, Grantham, M, and Green, A J. 2010. Isotope forensic analysis does not support vagrancy for a Marbled Duck shot in Essex. *Br Birds* 103: 464-467.

Fracasso, F, Baccetti, N, and Serra, L. 2009. La lista CISO-COI degli Uccelli italiani – Parte prima: liste A, B e C. *Avocetta* 35: 5-24.

Fracasso, G. 2012. Photographs of Mugimaki Flycatcher collected at Treviso, Italy, in October 1957. *Dutch Birding* 34: 109-110.

Fraser, P A, Rogers, M J, and the Rarities Committee. 2007. Report on rare birds in Great Britain in 2005. Part 1: non-passerines. *Br Birds* 100: 16-61.

Fraser, P A, Rogers, M J, and the Rarities Committee. 2007. Report on rare birds in Great Britain in 2005. Part 2: passerines. *Br Birds* 100: 72-104.

Fray, R, Pennington, M, Riddington, R, Meek, E, Higson, P, Forsyth, A, Leitch, A, Scott, M, Marr, T, ap Rheinallt, T, and Olofson, S. 2012. An unprecedented influx of Iceland Gulls in the northeastern Atlantic in January/February 2012. *Br Birds* 105: 263-272.

Frede, M, and Langbehn, H. 1997. A contribution to the identification and distribution of the Armenian Gull. *Alula* 3: 102-108.

Fregin, S, Haase, M, Olsson, U, and Alström, P. 2009. Multi-locus phylogeny of the family Acrocephalidae (Aves: Passeriformes) – the traditional taxonomy overthrown. *Mol Phylogenet Evol* 52: 866-878.

Frich, A S, Kristensen, A B, Ortvad, T E, and Schwalbe, M. 2009. Sjældne fugle I Danmark øg Grønland I 2008. Fugleåret 2008. *Dansk Ornitol Foren*, Copenhagen.

Friesen, V L, Platt, J F, and Baker, A J. 1996. Evidence from Cytochrome *B* sequences and allozymes for a 'new' species of alcid: the Long-billed Murrelet (*Brachyramphus perdix*). *Condor* 98: 681-690.

Fry, H, Fry, K, and Harris, A. 1992. *Kingfishers, Bee-eaters and Rollers*. Christopher Helm, London.

Fuchs, J, and Pons, J-M. 2015. A new classification of the Pied Woodpeckers assemblage (Dendropicini, Picidae) based on a comprehensive multi-locus phylogeny. *Mol Phylogenet Evol* 88: 28-37.

Fuisz, T, and de Kort, S R. 2007. Habitat-dependent call divergence in the common cuckoo: is it a potential signal for assortative mating? *Proc Biol Sci* 274 (1622): 2093-2097.

Furness, R W. 1987. *The Skuas*. T & A D Poyser, Calton.

G

Gallo-Orsi, U, and Boere, G C. 2001. The Slender-billed Curlew *Numenius tenuirostris*: threats and conservation. *Acta Ornithologica* 36: 73-77.

Gallo-Orsi, U. 2002. *International Action Plan for the Slender-billed Curlew (*Numenius tenuirostris*)*. BirdLife International, Cambridge.

Gamauf, A, and Haring, E. 2004. Molecular phylogeny and biogeography of Honey-buzzards (genera *Pernis* and *Henicopernis*). *J Zool Syst Evol Res* 42: 145-153.

García-del-Rey, E, and Cresswell, W. 2005. Density estimates, microhabitat selection and foraging behaviour of the endemic Blue Chaffinch *Fringilla teydea teydea* on Tenerife (Canary Islands). *Ardeola* 52: 305-317.

García-del-Rey, E, and Vargas, F J G. 2013. *Rare Birds of the Canary Islands*. Lynx Edicions, Barcelona.

García-del-Rey, E. 2011. *Field Guide to the Birds of Macaronesia: Azores, Madeira, Canary Islands, Cape Verde*. Lynx Edicions, Barcelona.

Garcia-R, J C, and Trewick, S A. 2015. Dispersal and speciation in purple swamphens (Rallidae: *Porphyrio*). *Auk* 132: 140-155.

Garcia-R, J C, Gibb, G C, and Trewick, S A. 2014. Deep global evolutionary radiation in birds: Diversification and trait radiation in the cosmopolitan bird family Rallidae. *Mol Phylogenet Evol* 81: 96-108.

Gariboldi, A, Meriggi, A, and Cesaris, C. 1992. The Bobwhite Quail (*Colinus virginianus*), an introduced species in the Ticino Natural Park, northern Italy. *Trans Intl Union Game Biol* 17: 263-266.

Garner, M, and friends. 2008. *Frontiers in Birding*. BirdGuides, Sheffield.

Garner, M, and Millington, R. 2010. The forms of Common Eider: their identification, taxonomy and vagrancy. *Birding World* 23: 65-82.

Garner, M, Lewington, I, and Crook, J. 2007. Identification of American Sandwich Tern. *Dutch Birding* 29: 273-287.

Garner, M, Lewington, I, and Slack, R. 2003. Mongolian and Lesser Sand Plovers: an identification overview. *Birding World* 16: 377-385.

Garner, M. 2002. Identification and vagrancy of American Merlins in Europe. *Birding World* 15: 468-480.

Garner, M. 2014. Bermuda *Phylloscopus* warbler: another look. http://birdingfrontiers.com/2014/04/18/bermuda-phylloscopus-warbler-another-look.

Gaskell, J. 2005. Recent changes in the status and distribution of birds in Libya. *Sandgrouse* 27: 126-138.

Gaston, A J, and Jones, I L. 1998. *Auks*. Oxford University Press, Oxford.

Gauci, M, and Mallia, M. 2010. First record of the Siberian Rubythroat *Luscinia calliope* in the Maltese Islands. *Il-Merill* 32: 40.

Gaunt, A, Greer, R, and Martin, J. 2014. A *Fregetta* storm-petrel at Severn Beach, Avon – new to Britain. *Br Birds* 107: 83-91.

Gavrilov, E I, and Gavrilov, A E. 2000. Possible ringing recoveries of Relict Gull in Bulgaria and Turkey. *Dutch Birding* 22: 219-221.

Gay, L, Defos du Rau, P, Mondain-Monval, J-Y, and Crochet, P-A. 2004. Phylogeography of a game species: the red-crested pochard (*Netta rufina*) and consequences for its management. *Mol Ecol* 13: 1035-1045.

Gedeon, K, Grüneberg, C, Mitschke, A, Sudfeldt, C, Eikhorst, W, Fischer, S, Flade, M, Frick, S, Geiersberger, I, Koop, B, Kramer, M, Krüger, T, Roth, N, Ryslavy, T, Stübing, S, Sudmann, S R, Steffens, R, Vökler, F, and Witt, K. 2014. *Atlas Deutscher Brutvogelarten. Atlas of German Breeding Birds*. Stiftung Vogelmonitoring Deutschland und Dachverband Deutscher Avifaunisten, Münster.

Gee, J P. 1984. The birds of Mauritania. *Malimbus* 6: 31-66.

Genovart, M, Juste, J, and Oro, D. 2005. Two sibling species sympatrically breeding: a new conservation concern for the critically endangered Balearic Shearwater. *Conserv Genet* 6: 601-606.

Genovart, M, Oro, D, Juste, J, and Bertorelle, G. 2007. What genetics tell us about the conservation of the critically endangered Balearic Shearwater? *Biol Conserv* 137: 283-293.

Ghasemi, M, and Khaleghizadeh, A. 2015. New record of Red-necked Stint *Calidris ruficollis* for Iran. *Sandgrouse* 37: 160.

Gibbs, D, Barnes, E, and Cox, J. 2001. *Pigeons and Doves. A Guide to the Pigeons and Doves of the World*. Pica Press, Robertsbridge.

Gibson, R, and Baker, A. 2012. Multiple gene sequences resolve phylogenetic relationships in the shorebird suborder Scolopaci (Aves: Charadriiformes). *Mol Phylogen Evol* 64: 66-72.

Gill, F, and Donsker, D (eds). 2014. IOC World Bird List (v 4.4). doi: 10.14344/IOC.ML.4.4.

Gill, R E, Canevari, P, and Iversen, E H. 1998. Eskimo Curlew (*Numenius borealis*). In: Poole, A, and Gill, F (eds). *The Birds of North America*. No 347, pp 1-28. The Academy of Natural Sciences and the American Ornithologists' Union, Philadelphia and Washington DC.

Giradoux, P, Degauquier, R, Jones, P J, Weigel, J, and Isenmann, P. 1988. Avifaune du Niger: etat des connaissances en 1986. *Malimbus* 10: 1-140.

Golovatin, M G, and Paskhalny, S P. Where are western bounds of area *S. h. minussensis*? *Open Ornithol J* 6: 52-54.

González, J, Siow, M, García-del-Rey, E, Delgado, G, and Wink, M. 2008. Phylogenetic relationships of the Cape Verde Sparrow based on mitochondrial and nuclear DNA. Poster presentation.

Gonzalez, J, Wink, M, García -del-Rey, E, and Delgado Castro, G. 2008. Evidence from DNA nucleotide sequences and ISSR profiles indicates paraphyly in subspecies of the Southern Grey Shrike (*Lanius meridionalis*). *J Ornithol* 149: 495-506.

González, L M. 1996. Action plan for the Blue Chaffinch *Fringilla teydea*. In: Heredia, B, Rose, L, and Painter, M (eds). *Globally Threatened Birds in Europe: Action Plans*. Council of Europe, and BirdLife International, Strasbourg.

González, L M. 2008. Origin and formation of the Spanish Imperial Eagle *Aquila adalberti*. *J Ornithol* 149: 151-159.

Goodman, S M, and Meininger, P L (eds). 1989. *The Birds of Egypt*. Oxford University Press, Oxford.

Gorman, G. 2014. *Woodpeckers of the World: the Complete Guide*. Christopher Helm, London.

Gosler, A. 1993. *The Great Tit*. Hamlyn, London.

Gottschling, M, Fuhrmann, K, and Beichle, U. 2005. Ein alter Beleg des Schmuckreiher *Egretta thula* als Er-

stnachweis fuer Deutschland und die Westpalaearktis. *Limicola* 19: 225-232.
Grant, P R. 1979. Evolution of the Chaffinch, *Fringilla coelebs*, on the Atlantic islands. *Biol J Linn Soc* 11: 301-332.
Grapputo, A, Pilastro, A, and Marin, G. 1998. Genetic variation and bill size dimorphism in a passerine bird, the Reed Bunting *Emberiza schoeniclus*. *Mol Ecol* 7: 1173-1182.
Grapputo, A, Pilastro, A, Baker, A J, and Marin, G. 2001. Molecular evidence for phylogenetic relationships among buntings and American sparrows (Emberizidae). *J Avian Biol* 32: 95-101.
Green, A J. 1996. International action plan for the Marbled Teal (*Marmaronetta angustirostris*). In: Heredia, B, Rose, L, and Painter, M (eds). *Globally Threatened Birds in Europe: Action Plans*. Council of Europe, and BirdLife International, Strasbourg.
Gregory G. 2007. Shikra in the Western Palearctic. *Birding World* 20: 159-161.
Gregory, G. 2002. The captive origin of bird species in Kuwait. *Sandgrouse* 24: 122-129.
Gregory, G. 2005. *The Birds of the State of Kuwait*. Privately published, Skegness.
Gregory, S M S, and Dickinson, E C. 2012. *Clanga* has priority over *Aquiloides* (or how to drop a clanger). *Bull Brit Ornithol Club* 132: 135-136.
Gretton, A, Yurlov, A. K, and Boere, G C. 2002. Where does the Slender-billed Curlew nest, and what future does it have? *Br Birds* 95: 334-344.
Grieve, A, and Kirwan, G M. 2012. Studies of Socotran birds VII. Forbes' Watson's Swift *Apus berliozi* in Arabia – the answer to the mystery of the 'Dhofar swift'. *Bull Brit Ornithol Club* 132: 194-206.
Grishchenko, V. 2004. Checklist of the birds of Ukraine. *Berkut* 13: 141-154.
Griswold, C K, and Baker, A J. 2002. Time to the most recent common ancestor and divergence times of populations of common chaffinches (*Fringilla coelebs*) in Europe and North Africa: insights into pleistocene refugia and current levels of migration. *Evolution* 56: 143-153.
Gronningsaeter, E. 2005. White-rumped Sandpipers in Arctic Norway. *Birding World* 18: 349-350.
Groth, J G. 1993. Evolutionary differentiation in morphology, vocalizations and allozymes among nomadic sibling species in the North American red crossbill (*Loxia curvirostra*) complex. *Univ Calif Publ Zool* 127: 1-143.
Gruppo Ornitologico Sardo. 2008. Checklist of the birds of Sardinia. http://gruppoornitologicosardo.blogspot.co.uk/2008/11/checklist-of-birds-of-sardinia-by.html.
Grussu, M. 2009. Frigatebird off Sardinia, Italy, in 1967-68 and records in Mediterranean. *Dutch Birding* 31: 28.
Grussu, M. 2012. Occurrence of the Eagle Owl in the Mediterranean. *Br Birds* 105: 100.
Guillaumet, A, Pons, J-M, Godelle, B, and Crochet, P-A. 2006. History of the crested lark in the Mediterranean region as revealed by mtDNA sequences and morphology. *Mol Phyl Evol* 39: 645-656.
Guillaumet, A, Crochet, P-A, and Pons, J-M. 2008. Climate-driven dispersal in two widespread *Galerida* larks. *BMC Evol Biol* 8: 32.
Guillaumet, A, Ferdy, J-B, Desmarais, E, Godelle, B, and Crochet, P-A. 2008. Testing Bergmann's rule in the presence of potentially confounding factors: a case study with three species of *Galerida* larks in Morocco. *J Biogeogr* 35: 579-591.
Gustad, J R, and Schjolberg, K. 2002. The first Red-knobbed Coot *Fulica cristata* in Oman and the Middle East. *Sandgrouse* 24: 65-67.
Gutiérrez, C, Copete, J L, Crochet, P-A, Qninba, A, and Garrido, H. 2011. History, status and distribution of Andalusian Buttonquail in the WP. *Dutch Birding* 33: 75-93.
Gutiérrez, R, and Bigas, D. 2005. Naumann's Thrush at Delta de l'Ebre, November 2005. First for Spain. www.rarebirdspain.net/arbsf048.htm.
Gutiérrez, R, and Figuerola, J. 1995. Wintering distribution of the Balearic Shearwater (*Puffinus Yelkouan mauretanicus*, Lowe 1921) off the northeastern coast of Spain. *Ardeola* 42: 161-166.
Gutiérrez, R, de Juana, E, and Lorenzo, J A. 2012. Lista de las aves de España. Version online 1.0. Sociedad Española de Ornitología (SEO/BirdLife). http://www.rarebirdspain.net/esplis12.pdf.
Gutiérrez, R, Lorenzo, J A, Elorriaga, J, Gorospe, G, López-Velasco, D, Martí-Aledo, J, Rodríguez, G, and Sales, S. 2013. Observaciones de aves raras en España, 2011. *Ardeola* 60: 437-506.
Gutiérrez, R. 2003. Occurrence of Rüppell's Griffon Vulture in Europe. *Dutch Birding* 25: 289-303.
Gutiérrez, R. 2008. The Fork-tailed Flycatcher in Spain: a new Western Palearctic bird. *Birding World* 21: 325-328.

H

Haas, M, and Crochet, P-A. 2009. Western Palearctic list updates: Cape Petrel. *Dutch Birding* 31: 24-28.
Haas, M, and Crochet, P-A. 2010. Western Palearctic list updates: Denham's Bustard deleted. *Dutch Birding* 32: 257-258.
Haas, M, and Crochet, P-A. 2012. Western Palearctic list updates: Von Schrenck's Bittern. *Dutch Birding* 34: 168-171.
Haas, M, and Slaterus, R. 2014. Rare birds in the Netherlands in 2013. *Dutch Birding* 36: 365-393.
Haas, M, Crochet, P-A, Koerkamp, G G, Arkhipov, V Y, and Loskot, V M. 2013. Occurrence of Pallas's Rosefinch in the Western Palearctic. *Dutch Birding* 35: 169-179.
Haas, M. 2009. Tristan Albatross collected in Sicily, Italy, in October 1957. *Dutch Birding* 31: 180.
Haas, M. 2012. *Extremely Rare Birds in the Western Palearctic*. Lynx Edicions, Barcelona.
Haas, M. In prep. Additions to *Extremely Rare Birds in the Western Palearctic: 2009-2013*.
Habib, M. 2014. Saunders's Terns breeding at Ras Sudr, Egypt, in 2012-13. *Dutch Birding* 36: 20-24.
Habib, M. 2015. Surveys of Chestnut-bellied Sandgrouse in Egypt in 2013-14. *Dutch Birding* 37: 95-97.
Habib, M I. 2016. Update on breeding status of Red Sea Spoonbill in Egypt. *Dutch Birding* 38: 69-74.
Hagemeijer, W J M, and Blair, M J (eds). 1997. *The EBCC Atlas of European Breeding Birds: their Distribution and Abundance*. T & A D Poyser, London.

Hale, W G, Scarton, F, and Valle, R. 2005. The taxonomic status of the Redshank *Tringa totanus* in Italy. *Bull Brit Ornithol Club* 125: 261-275.

Hall, B P, and Moreau, R E. 1970. *An Atlas of Speciation in African Passerine Birds*. British Museum (Natural History), London.

Hamedanian, A. 1997. Observations of Pleske's Ground Jay *Podoces pleskei* in Central Iran. *Sandgrouse* 19: 88-91.

Hamza, A, and Yahia, J. 2014. First documented record of Semi-collared Flycatcher *Ficedula semitorquata* for Libya. *Bull Afr Bird Club* 21: 83-85.

Hancock, J A, Kushlan, J A, and Kahl, M P. 1992. *Storks, Ibises and Spoonbills of the World*. Academic Press, London.

Hancock, J, and Kushlan, J. 1984. *The Herons Handbook*. Croom Helm, London.

Handrinos, G, and Akriotis, T. 1997. *The Birds of Greece*. Christopher Helm, London.

Hansbro, P, and Sargeant, D. 1999. The first Red-billed Oxpecker *Buphagus erythrorhynchus* in Yemen and the Middle East. *Sandgrouse* 21: 181-182.

Hanson, H C. 1997. *The Giant Canada Goose*. Revised edition. Southern Illinois University Press, Urbana.

Hanžel, J, and Šere, D. 2012. The list of birds of Slovenia with an overview of rare species. *Acrocephalus* 32: 143-203.

Hanžel, J. 2013. Rare birds in Slovenia in 2012 – Slovenian Rarities Committee's report. *Acrocephalus* 34: 83-91.

Hanžel, J. 2014. Rare birds in Slovenia in 2013 – Slovenian Rarities Committee's report. *Acrocephalus* 35: 59-72.

Haraldsson, T. 2008. The first Namaqua Dove *Oena capensis* for Lebanon? *Sandgrouse* 30: 90-91.

Haring, E, Däubl, B, Pinsker, W, Kryukov, A, and Gamauf, A. 2012. Genetic divergences and intraspecific variation in corvids of the genus *Corvus* (Aves: Passeriformes: Corvidae) – a first survey based on museum specimens. *J Zool Syst Evol Res* 50: 230-246.

Haring, E, Gamauf, A, and Kryukov, A. 2007. Phylogeographic patterns in widespread corvid birds. *Mol Phylogenet Evol* 45: 840-862.

Harrap, S, and Quinn, D. 1996. *Tits, Nuthatches and Treecreepers*. Christopher Helm, London.

Harris, T, and Franklin, K. 2000. *Shrikes and Bush-Shrikes*. Christopher Helm, London.

Harrison, C. 1982. *An Atlas of the Birds of the Western Palaearctic*. Collins, London.

Harrison, I, and Grieve, A. 2012. Around the region. *Sandgrouse* 34: 189-203.

Harrison, I, and Sargeant, D. 2005. First breeding of Bay-backed Shrike in Arabia. *Pheonix* 21: 28.

Harrison, I. 2013a. From the Rarities Committees. *Sandgrouse* 35: 82-86.

Harrison, I. 2013b. From the Rarities Committees. *Sandgrouse* 35: 161-164.

Harrison, P. 1985. *Seabirds: an Identification Guide*. Revised edition. Croom Helm, Beckenham.

Harrop, A H J, and McGowan, R Y. 2009. Britain's first Baikal Teal. *Br Birds* 102: 691-696.

Harrop, A H J, Collinson, J M, Dudley, S P, Kehoe, C, and the British Ornithologists' Union Records Committee (BOURC). 2013. The British List: A Checklist of Birds of Britain (8th edition). *Ibis* 155: 635-676.

Harrop, A H J. 2002. The Ruddy Shelduck in Britain: a review. *Br Birds* 95: 123-128.

Harrop, A H J. 2004. The 'soft-plumaged petrel' complex: a review of the literature on taxonomy, identification and distribution. *Br Birds* 97: 6-15.

Harrop, A H J. 2008. The rise and fall of Bulwer's Petrel. *Br Birds* 101: 676-681.

Harrop, A H J. 2010. Records of Hawk Owls in Britain. *Br Birds* 103: 276-283.

Harvey, P. 2010. Brown Flycatcher on Fair Isle: new to Britain. *Br Birds* 103: 651-657.

van der Have, T M, and Moerbeek, D J. 1984. Occurrence of the Harlequin Duck in Europe. *Dutch Birding* 6: 40-44.

Hayman, P, Marchant, J, and Prater, T. 1986. *Shorebirds – an Identification Guide to the Waders of the World*. Croom Helm, Beckenham.

Hazevoet, C J, Haas, M, and Crochet, P-A. 2011. Knob-billed Duck at Banc d'Arguin, Mauritania, in December 1984. *Dutch Birding* 33: 247-248.

Hazevoet, C J, Monteiro, L R, and Ratcliffe, N. 1999. Rediscovery of the Cape Verde Cane Warbler *Acrocephalus brevipennis* on São Nicolau in February 1998. *Bull Brit Ornithol Club* 119: 68-71.

Hazevoet, C J. 1999. Fourth report on birds from the Cape Verde Islands, including notes on conservation and records of 11 taxa new to the archipelago. *Bull Zool Mus Univ Amsterdam* 17: 19-32.

Hazevoet, C J. 1993. On the history and type specimens of the Cape Verde Cane Warbler *Acrocephalus brevipennis* (Keulemans, 1866) (Aves, Sylviidae). *Bijdragen tot de Dierkunde* 62: 249-253.

Hazevoet, C J. 1995. *The Birds of the Cape Verde Islands*. BOU Check-list No 13. British Ornithologists' Union, Tring.

Hazevoet, C J. 1996. Ornithological news from the Cape Verde Islands in 1995, including records of species new to the archipelago. *Bull Zoöl Mus* 15: 22-27.

Hazevoet, C J. 1997. Notes on distribution, conservation, and taxonomy of birds from the Cape Verde Islands, including records of six species new to the archipelago. *Bull Zoöl Mus* 15: 89-100.

Hazevoet, C J. 1998. Third report on birds from the Cape Verde Islands, including records of seven taxa new to the archipelago. *Bull Zoöl Mus* 16: 65-71.

Hazevoet, C J. 1999a. Fourth report on birds from the Cape Verde Islands, including notes on conservation and records of 11 taxa new to the archipelago. *Bull Zoöl Mus* 17: 19-32.

Hazevoet, C J. 1999b. Notes on birds from the Cape Verde Islands in the collection of the Centro de Zoologia, Lisbon, with comments on taxonomy and distribution. *Bull Brit Ornithol Club* 119: 25-31.

Hazevoet, C J. 2003. Fifth report on birds from the Cape Verde Islands, including records of 15 taxa new to the archipelago. *Arquivos do Museu Bocage* Vol III: 503-528.

Hazevoet, C J. 2010. Sixth report on birds from the Cape Verde Islands, including records of 25 taxa new to the archipelago. *Zool Caboverdiana* 1: 3-44.

Hazevoet, C J. 2012. Seventh report on birds from the Cape Verde Islands, including records of nine taxa new to the archipelago. *Zool Caboverdiana* 3: 1-28.

Hazevoet, C J. 2014. Eighth report on birds from the Cape Verde Islands, including records of nine taxa new to the archipelago. *Zool Caboverdiana* 5: 29-56.

Heath, M F, and Evans, M I (eds). 2004. *Important Bird Areas in Europe. Priority Sites for Conservation.* BirdLife Conservation Series No. 8. BirdLife International, Cambridge.

Heggøy, O, and Olsen, T A. 2013. Sjeldne fugler i Norge i 2011 og 2012. *Rapport fra Norsk sjeldenhetskomité for fugl (NSKF).* http://www.birdlife.no/organisasjonen/nskf/publikasjoner/NSKF2011_2012.pdf.

Heiss, M, and Gauger, K. 2009. The rediscovery of breeding Shikras *Accipiter badius* in the Western Palaearctic. *Sandgrouse* 31: 134-137.

Helbig, A J, Martens, J, Seibold, I, Henning, F, Schottler, B, and Wink, M. 1996. Phylogeny and species limits in the Palearctic chiffchaff *Phylloscopus collybita* complex: mitochondrial genetic differentiation and bioacoustic evidence. *Ibis* 138: 650-666.

Helbig, A J, Seibold, I, Martens, J, and Wink, M. 1995. Genetic differentiation and phylogenetic relationships of Bonelli's Warbler *Phylloscopus bonelli* and Green Warbler *P nitidus. J Avian Biol* 26: 139-153.

Hellenic Rarities Committee. 2006. Annual Report – 2005. http://files.ornithologiki.gr/docs/rarities/annual_report_2005_en.pdf

Hellenic Rarities Committee. 2007. Annual Report – 2006. http://files.ornithologiki.gr/docs/rarities/annual_report_2006_en.pdf

Hellenic Rarities Committee. 2008. Annual Report – 2007. http://files.ornithologiki.gr/docs/rarities/annual_report_2007_en.pdf

Hellenic Rarities Committee. 2009. Annual Report – 2008. http://files.ornithologiki.gr/docs/rarities/annual_report_2008_en.pdf

Hellenic Rarities Committee. 2010. Annual Report – 2009. http://files.ornithologiki.gr/docs/rarities/annual_report_2009_en.pdf

Hellenic Rarities Committee. 2011. Annual Report – 2010. http://files.ornithologiki.gr/docs/rarities/annual_report_2010_en.pdf

Hellenic Rarities Committee. 2012. Annual Report – 2011. http://files.ornithologiki.gr/docs/rarities/annual_report_2011_en.pdf

Hellenic Rarities Committee. 2013. Annual Report – 2012. http://files.ornithologiki.gr/docs/rarities/annual_report_2012_en.pdf

Hellenic Rarities Committee. 2014. Annual Report – 2013. http://files.ornithologiki.gr/docs/rarities/annual_report_2013_en.pdf

Hellenic Rarities Committee. 2015. List of the Birds of Greece. http://www.ornithologiki.gr/page_list.php?lID=3&sp=no&st=no&sf=yes&ss=yes&loc=en.

Hellenic Rarities Committee. 2015. The List of the Rare Birds of Greece. http://www.ornithologiki.gr/page_cn.php?aID=928&tID=1895.

Hellquist, A. 2014. Identification of Northern Eider. *Dutch Birding* 36: 221-231.

Hellström, M, and Norevik, G. 2014. The uppertail-covert pattern of 'Stejneger's Stonechat'. *Br Birds* 107: 692-700.

Hellström, M, and Wærn, M. 2011. Field identification and ageing of Siberian Stonechats in spring and summer. *Br Birds* 104: 236-254.

Hellyer, P. 2000. The first Yellow-billed Stork *Mycteria ibis* in Qatar and the Arabian Gulf. *Sandgrouse* 22: 125-126

Henderson, I. 2013. *UK Ruddy Duck Eradication Programme.* National Wildlife Management Centre, Reigate.

Hering, J, and Fuchs, E. 2009. The Cape Verde Warbler: distribution, density, habitat and breeding biology on the island of Fogo. *Br Birds* 102: 17-24.

Hering, J, and Fuchs, E. 2011. First breeding record for Little Bittern *Ixobrychus minutus* in Libya. *Bull Afr Bird Club* 18: 218-220.

Hering, J, and Hering, H. 2005. Discovery of Cape Verde Warbler *Acrocephalus brevipennis* on Fogo, Cape Verde Islands. *Bull Afr Bird Club* 12: 147-149.

Hering, J, Barthel, P H, Eilts, H-J, Frommolt, K-H, Fuchs, E, Heim, W, Müller, K, and Päckert, M. 2013. Die Chinadommel *Ixobrychus sinensis* am Roten Meer in Ägypten – erste Nachweise eines übershenen westpaläarktischen Brutvogels. *Limicola* 26: 253-278.

Hering, J, Brehme, S, Fuchs, E, and Winkler, H. 2010. African Reed Warblers and Mangrove Reed Warblers in Libya and Egypt – both new to the Western Palearctic. *Birding World* 23: 218-219.

Hering, J, Flade, M, and Wink, M. 2015. Records of Paddyfield Warbler *Acrocephalus agricola* in Turkey and evidence for a monotypic species. *Sandgrouse* 37: 130-136.

Hering, J, Fuchs, E, Heim, W, Eilts, H-J, and Ibrahim, H. 2013. New information on Three-banded Plover in Egypt. *Dutch Birding* 35: 23-27.

Hering, J, Winkler, H, and Steinheimer, F D. 2016. A new subspecies of Eurasian Reed Warbler *Acrocephalus scirpaceus* in Egypt. *Bull Brit Ornithol Club* 136: 101-128.

Hering, J. 2012. Asian Desert Warblers in south-eastern Libya in January 2011. *Dutch Birding* 34: 172-174.

Hermansen, J S, Sæther, S A, Elgvin, T O, Borge, T, Hjelle, E, and Sætre, G-P. 2011. Hybrid speciation in sparrows I: phenotypic intermediacy, genetic admixture and barriers to gene flow. *Mol Ecol* 20: 3812-3822.

Hernández, M A, Campos, F, Gutiérrez-Corchero, F, and Amezcua, A. 2004. Identification of *Lanius* species and subspecies using tandem repeats in the mitocondrial DNA control regions. *Ibis* 146: 227-230.

Hille, S M, and Collar, N J. 2009. The taxonomic and conservation status of *Milvus* kites in the Cape Verde archipelago: further (and final?) reflections. *Bull Brit Ornithol Club* 129: 217-221.

Hille, S, and Thiollay, J M. 2000. The imminent extinction of the kites *Milvis milvus fasciicauda* and *Milvus m migrans* on the Cape Verde Islands. *Bird Conserv Int* 10: 361-369.

Hilton, G M. 2003. A stable isotopic search for the Slender-billed Curlew. *Wader Study Group Bulletin*: 23.

Hirschfeld, E, and Symens, P. 1992. First records of Dead Sea Sparrow *Passer moabiticus* in Arabia. *Sandgrouse* 14: 48-51.

Hirschfeld, E, Roselaar, C S, and Shirihai, H. 2000. Identification, taxonomy and distribution of Greater and Lesser Sand Plovers. *Br Birds* 93: 162-189.

Hirschfeld, E. 1995. *Birds in Bahrain: a Study of their Migration Patterns*. Hobby Publications, Dubai.
Hoath, R. 2000. The first Three-banded Plover *Charadrius tricollaris* in Egypt and the Western Palearctic. *Sandgrouse* 22: 67-68.
Hoek, D, Ibrahim, H, and Noby, K. 2010. Ferruginous Duck *Aythya nyroca*, a new breeding species for Egypt in a temporary wetland near Aswan. *Sandgrouse* 32: 163-166.
Hofland, R, and Keijl, G. 2008. *Syrian Sociable Lapwing survey*. Foundation Working Group International Waterbird and Wetland Research, Beek-Ubbergen.
Holden, D, and Newsome, M. 2011. Eastern Crowned Warbler in Co Durham: new to Britain. *Br Birds* 104: 303-311.
Holling, M, and the Rare Breeding Birds Panel. 2011. Non-native breeding birds in the United Kingdom in 2006, 2007 and 2008. *Br Birds* 104: 114-138.
Hollyer, J N. 1970. The invasion of Nutcrackers in autumn 1968. *Br Birds* 63: 353-373.
Holyoak, D T. 2001. *Nightjars and their Allies*. Oxford University Press, Oxford.
Holzapfel, C, Levin, N, Hatzofe, O, and Kark, S. 2006. Colonisation of the Middle East by the invasive Common Myna *Acridotheres tristis* L, with special reference to Israel. *Sandgrouse* 28: 44-51.
Hölzinger, J. 1995. The Cinereous Bunting *Emberiza cineracea* breeding on Skyros (Greece). *Zool Middle East* 11: 31-36.
Hope, D. 2006. The first record of Temminck's Horned lark *Eremophila bilopha* in Cyprus. *Cyprus Bird Rep 2005*: 161-162.
Howell, S N G, Lewington, I, and Russell, W. 2014. *Rare Birds of North America*. Princeton University Press, Woodstock.
Howell, S N G, Patteson, J B, Sutherland, K, and Shoch, D T. 2010. Occurrence and identification of the Band-rumped Storm-petrel (*Oceanodroma castro*) complex off North Carolina. *North American Birds* 64: 196-207.
Howell, S N G. 2010. Identification and taxonomy of White-bellied Storm-petrels, with comments on WP report in August 1986. *Birding* 32: 36-42.
Howell, S N G. 2012. *Petrels, Albatrosses and Storm-Petrels of North America – a Photographic Guide*. Princeton University Press, Woodstock.
Howell, S. 2002. A Black-capped Petrel off the Bay of Biscay: the fourth for the Western Palearctic. *Birding World* 15: 219.
Hudson, N, and the Rarities Committee. 2008. Report on rare birds in Great Britain in 2007. *Br Birds* 101: 516-577.
Hudson, N, and the Rarities Committee. 2009. Report on rare birds in Great Britain in 2008. *Br Birds* 102: 528-601.
Hudson, N, and the Rarities Committee. 2010. Report on rare birds in Great Britain in 2009. *Br Birds* 103: 562-638.
Hudson, N, and the Rarities Committee. 2011. Report on rare birds in Great Britain in 2010. *Br Birds* 104: 557-629.
Hudson, N, and the Rarities Committee. 2012. Report on rare birds in Great Britain in 2011. *Br Birds* 105: 556-627.
Hudson, N, and the Rarities Committee. 2013. Report on rare birds in Great Britain in 2012. *Br Birds* 106: 570-641.
Hudson, N, and the Rarities Committee. 2014. Report on rare birds in Great Britain in 2013. *Br Birds* 107: 579-653.
Hughes, B, Robinson, J A, Green, A J, Li, Z W D, and Mundkur, T. 2006. *International Single Species Action Plan for the Conservation of the White-headed Duck* Oxyura leucocephala. CMS/AEWA, Bonn, Germany.
Huhtinen, H. 2013. Ruostepääskyjä Itä-Aasiasta – jo kaksi havaintoa Suomesta. *Linnut* 48: 22-25.
Hume, R A. 1993. Brown Shrike in Shetland: new to Britain and Ireland. *Br Birds* 86: 600-604.

I

Iñigo, A, Graza, V, Tella, J L, Laiolo, P, Suárez, F, and Barov, B. 2008. Action plan for the Dupont's Lark *Chersophilus duponti* in the European Union. BirdLife International for the European Commission.
Irestedt, M, Ohlson, J I, Zuccon, D, Källersjö, M, and Ericson, P G P. 2006. Nuclear DNA from old collections of avian study skins reveals the evolutionary history of the Old World suboscines (Aves, Passeriformes). *Zoologica Scripta* 35: 567-580.
Irish Rare Birds Committee. 2013. The Irish List (as of 31 December 2013). http://www.irbc.ie/topbar/IrishList/IRBC_IrishList(31122013).pdf.
Irwin, D E, Alström, P, Olsson, U, and Benowitz-Fredericks, Z M. 2001. Cryptic species in the genus *Phylloscopus* (Old World leaf warblers). *Ibis* 143: 233-247.
Irwin, D E, and Hellström, M. 2007. Green Warbler *Phylloscopus (trochiloides) nitidus* recorded at Ottenby, Öland: a first record for Scandinavia. *Ornis Svecica* 17: 75-80.
Irwin, D E, Bensch, S, and Price, T D. 2001. Speciation in a ring. *Nature* 409: 333-337.
Irwin, D E, Irwin, J H, and Smith, T B. 2011. Genetic variation and seasonal migratory connectivity in Wilson's warblers (*Wilsonia pusilla*): species-level differences in nuclear DNA between western and eastern populations. *Mol Ecol* 20: 3102-3115.
Isenmann, P, and Moali, A. 2000. *Oiseaux d'Algerie – Birds of Algeria*. SEOF Éditions, Paris.
Isenmann, P, Benmergui, M, Browne, P, Ba, A D, Diagana, C H, Diawara, Y, and Sidaty, Z E A O. *Oiseaux de Mauritanie – Birds of Mauritania*. SEOF Éditions, Paris.
Isenmann, P, Gaultier, T, Hili, A H, Azafzat, H, Dlensi, H, and Smart, M. 2005. *Oiseaux d'Tunisie – Birds of Tunisia*. SEOF, Paris.
Isenmann, P, Hering, J, Brehme, S, Essghaier, M, Etayeb, K, Bourass, E, and Azafzaf, H. 2016. *Oiseaux de Libye – Birds of Libya*. SEOF Éditions, Paris.
Isenmann, P. 2006. *The Birds of the Banc d'Arguin*. Parc National du Banc d'Arguin/FIBA.
Israeli Rarities and Distribution Committee (IRDC). 2002. Bulletin 2:02 on Rare Birds in Israel. http://www.israbirding.com/irdc/bulletins/bulletin_2-2/.
IUCN. 2012. *Guidelines for Application of IUCN Red List Criteria at Regional and National Levels*. Version 4.0. Gland, Switzerland and Cambridge, UK.

IUCN. 2014. *IUCN Red List of Threatened Species*. Version 2014.2. www.iucnredlist.org.

Ivanov, B, Iankov, P, Boev, Z, Georgiev, D, Profirov, L, and Dimitrov, M. 2015. *List of the Birds recorded in Bulgaria*. BUNARCO. http://www.bunarco.org/bg/dokladi.html.

J

James, P C, and Robertson, H A. 1985. First record of Swinhoe's storm petrel *Oceanodroma monorhis* in the Atlantic ocean. *Ardea* 73: 105-106.

Jännes, J. 1995. Paksunookka-kerttunen: kerttuseksi helppo tuntea. *Alula* 1: 18-20.

Janni, O, and Fracasso, G. 2009. Commissione Ornitologica Italiana (COI) - Report 22. *Avocetta* 33: 117-146.

Janni, O, and Fracasso, G. 2011. Commissione Ornitologica Italiana (COI) - Report 23. *Avocetta* 34: 71-106.

Janni, O, and Fracasso, G. 2012. Commissione Ornitologica Italiana (COI) - Report 24. *Avocetta* 36: 81-88.

Janni, O, and Fracasso, G. 2013. Commissione Ornitologica Italiana (COI) - Report 25. *Avocetta* 37: 67-74.

Jara, J, Alfrey, P, Costa, H, Matias, R, Moore, C C, Santos, J L, and Tipper, R. 2010. Aves de ocorrência rara ou acidental em Portugal. Relatório do Comité Português de Raridades referente aos anos de 2008 and 2009. *Anuário Ornitológico* 7: 3-71.

Jara, J, Costa, H, Matias, R, Moore, C C, Noivo, C, and Tipper, R. 2008. Aves de ocorrência rara ou acidental em Portugal. Relatório do Comité Português de Raridades referente aos anos de 2006 and 2007. *Anuário Ornitológico* 6: 1-45.

Jaramillo, A, and Burke, P. 1999. *New World Blackbirds: the Icterids*. Christopher Helm, London.

Jennings, M C. 1979. A record of Solitary Snipe *Gallinago solitaria* from Arabia. *OSME Bulletin* 3: 14-16.

Jennings, M C. 2010. Atlas of the Breeding Birds of Arabia. *Fauna of Arabia* Vol 25.

Jensen, J-K. 2006. Are dark morph Rough-legged Hawks overlooked in Europe? *Birding World* 19: 208-209.

Jensen, J-K. 2008. The diary of a Rough-legged Hawk in Europe. *Birding World* 21: 202.

Jensen, J-K. 2012. An Atlantic Yellow-nosed Albatross off the Faeroe Islands. *Birding World* 25: 328.

Jesus, J, Menzes, D, Gomes, S, Oliveira, P, Nogales, M, and Brehm, A. 2009. Phylogenetic relationships of gadfly petrels *Pterodroma* spp. from the Northeastern Atlantic Ocean: molecular evidence for specific status of Bugio and Cape Verde petrels and implications for conservation. *Bird Conserv Int* 19: 1-16.

Jiguet, F, and du Rau, P D. 2004. A Cape Gull in Paris – a new European bird. *Birding World* 17: 62-70.

Jiguet, F, and Zucca, M. 2005. The American Barn Swallows on the Azores – a new Western Palearctic bird. *Birding World* 18: 475-478.

Jiguet, F, Baha el Din, M, Baha el Din, S, Bonser, R, Crochet, P-A, Grieve, A, Hoath, R, Haraldsson, T, Riad, A, and Megalli, M. 2011. First report of the Egyptian Ornithological Rarities Committee – 2010. www.chn-france.org/upload_content/eorc_report_2.pdf.

Jiguet, F, Baha el Din, M, Baha el Din, S, Bonser, R, Crochet, P-A, Grieve, A, Hoath, R, Haraldsson, T, Riad, A, and Megalli, M. 2012. Second report of the Egyptian Ornithological Rarities Committee – 2011. www.chn-france.org/upload_content/eorc_report_8.pdf.

Jiguet, F, Baha el Din, S, Ibrahim, A L W, Bonser, R, Corso, A, Crochet, P-A, Grieve, A, Hoath, R, and Schweizer, M. 2014. Third report of the Egyptian Ornithological Rarities Committee – 2012 and 2013. http://www.chn-france.com/upload_content/eorc_report_9.pdf.

Jiguet, F, Capainolo, P, and Tennyson, A. 2012. Taxonomy of the Kelp Gull *Larus dominicanus* Lichtenstein revisited with sex-separated analyses of biometrics and wing tip patterns. *Zoological Studies* 51: 881-892.

Jiguet, F, Doxa, A, and Robert, A. 2008. The origin of out-of-range pelicans in Europe: wild bird dispersal or zoo escapes? *Ibis* 150: 606-618.

Jimenez Armesto, M J, Boehm, C, and Bowden, C (compilers). 2006. *International Single Species Action Plan for the Conservation of the Northern Bald Ibis* Geronticus eremita. AEWA Technical Series No. 10. Bonn, Germany.

Jirle, E. 2015. Västpalearktislistan version 2. http://www.sofnet.org/tk/vastpalearktislistan-ver-2-vp-2/vastpalearktislistan-i-webbformat/.

Johansson, U S, Ekman, J, Bowie, R C K, Halvarsson, P, Ohlson, J I, Price, T D, and Ericson, P G P. 2013. A complete multilocus species phylogeny of the tits and chickadees (Aves: Paridae). *Mol Phylogenet Evol* 69: 852-860.

Johnsen, A, Rindal, E, Ericson, P G P, Zuccon, D, Kerr, K C R, Stoeckle, M Y, and Lifjeld, J T. 2010. DNA barcoding of Scandinavian birds reveals divergent lineages in trans-Atlantic species. *J Ornithol* 151: 565-578.

Johnson, J A, Watson, R T, and Mindell, D P. 2005. Prioritizing species conservation: does the Cape Verde Kite exist? *Proc R Soc Lond [Biol.]* 272: 1365-1371.

Jones, T, Martin, K, Barov, B, and Nagy, S. 2008. *International Single Species Action Plan for the Conservation of the Western Palearctic Population of the Lesser White-fronted Goose* Anser erythropus. AEWA Technical Series No 36, Bonn, Germany.

Jonker, R M, Kraus, R H S, Zhang, Q, van Hooft, P, Larsson, K, van der Jeugd, H P, Kurvers, R H J M, van Wieren, S E, Loonen, M J J E, Crooijmans, R P M A, Ydenberg, R C, Groenen, M A M, and Prins, H H T. 2013. Genetic consequences of breaking migratory traditions in barnacle geese *Branta leucopsis*. *Mol Ecol* 22: 5835-5847.

Jønsson, K A, Bowie, R C K, Moyle, R G, Irestedt, M, Christidis, L, Norman, J A, and Fjeldså, J. 2010. Phylogeny and biogeography of Oriolidae (Aves: Passeriformes). *Ecography* 33: 232-241.

Jönsson, O. 2011. Great Black-backed Gulls breeding at Khniffis Lagoon, Morocco and the status of Cape Gull in the Western Palearctic. *Birding World* 24: 68-76.

Jusys, V, Pakštytė, E, Raudonikis, L, Karalius, S, and Stanevičius, V. 2015. Checklist of birds of Lithuania. http://www.birdlife.lt/upload/user_uploads/Dokumentai/LOFK_dokumentai/sistematinis_2015-08-14_UK.pdf.

K

Kamp, J, Oppel, S, Ananin, A A, Durnev, Y A, Gashev, S N, Hölzel, N, Mischenko, A L, Pessa, J, Smirenski, S M, Strelnikov, E G, Timonen, S, Wolanska, K, and Chan, S. 2015. Global population collapse in a superabundant migratory bird and illegal trapping in China. *Conserv Biol* 29: 1684-1694.

Kampe-Persson, H. 2010. Naturalised geese in Europe. *Ornis Svecica* 20: 155-173.

Kayser, Y, Paepegaey, B, and the CHN. 2014. Les oiseaux rares en France en 2012. 30th rapport du Comité d'Homologation National. *Ornithos* 21: 65-107.

Kayser, Y, Verneau, A, and the CHN. 2015. Les oiseaux rares en France en 2013. 31st rapport du Comité d'Homologation National. *Ornithos* 22: 113-145.

Kazakov, B A. 2004. Principal compiler. [*Gavia adamsii*. In: *Birds of the Northern Caucasus*. Vol 1. Gaviiformes, Podicipediformes, Procellariiformes, Pelecaniformes, Ciconiiformes, Phoenicopteriformes, Anseriformes]. Rostov State Pedagogical University, Rostov-on-Don.

Kear, J (ed). 2005. *Ducks, Geese and Swans*. Volume 1: General chapters, and Species Accounts (*Anhima* to *Salvadorina*). Oxford University Press, Oxford.

Kear, J (ed). 2005. *Ducks, Geese and Swans*. Volume 2: Species Accounts (*Cairina* to *Mergus*). Oxford University Press, Oxford.

Kehoe, C. 2006. Racial identification and assessment in Britain: a report from the RIACT subcommittee. *Br Birds* 99: 619-645.

Kelly, A, Leighton, K, and Newton, J. 2010. Using stable isotopes to investigate the provenance of an Eagle Owl found in Norfolk. *Br Birds* 103: 213-222.

Kennedy, M, and Spencer, H G. 2003. Phylogenies of the frigatebirds (Fregatidae) and tropicbirds (Phaethonidae), two divergent groups of the traditional order Pelecaniformes, inferred from mitochondrial DNA sequences. *Mol Phylogenet Evol* 31: 31-38.

Kennerley, P R, and Prŷs-Jones, R P. 2006. Occurrences of Gray's Grasshopper Warbler in Europe, including a further case of Meinertzhagen fraud. *Br Birds* 98: 506-516.

Kennerley, P, and Pearson, D. 2010. *Reed and Bush Warblers*. Christopher Helm, London.

Kerr, K C R, and Dove, C J. 2013. Delimiting shades of gray: phylogeography of the Northern Fulmar, *Fulmarus glacialis*. *Ecol Evol* 3: 1915-1930.

Kerr, K C R, Birks, S M, Kalyakin, M V, Red'kin, Y A, Koblik, E A, and Hebert, P D N. 2009. Filling the gap – COI barcode resolution in eastern Palearctic birds. *Frontiers in Zoology* 6: 29.

Kerr, K C R, Stoeckle, M Y, Dove, C J, Weigt, L A, Francis, C M, and Hebert, P D N. 2007. Comprehensive DNA barcode coverage of North American birds. *Mol Ecol Notes* 7: 535-543.

Khaleghizadeh, A, Scott, D A, Tohidifar, M, Musavi, S B, Ghasemi, M, Sehhatisabet, M E, Ashoori, A, Khani, A, Bakhtiari, P, Amini, H, Roselaar, C, Ayé, R, Ullman, M, Nezami, B, and Eskandari, F. 2011. Rare Birds in Iran in 1980-2010. *Podoces* 6: 1-48.

Khani, A, Rezayi, M, and Ali Poor, M J. 2016. A Long-tailed Shrike *Lanius schach* in Khorasan-e Razavi province, northeastern Iran. *Sandgrouse* 38: 5-6.

Khil, L, and Albegger, E. 2014. Records of rare and remarkable bird species in Austria 2010-2011. Seventh report of the Avifaunistic Commission of BirdLife Austria. *Egretta* 53: 10-28.

Khil, L, Boetzel, M, Geburzi, J, Trobitz, M, Werner, M, Weinrich, C, and Zegula, T. 2012. Rediscovery of Chestnut-bellied Sandgrouse in Egypt in March 2012. *Dutch Birding* 34: 213-218.

Khoury, F, Förschler, M I, Janaydeh, M, Aliabadian, M, and Al-Hmoud, A-R. 2010. Distribution, habitat and differentiation of the poorly-known black morph of Mourning Wheatear *Oenanthe lugens lugens* in Jordan. *Sandgrouse* 32: 113-119.

Khoury, F, Haddad, M, Al-Jbour, S, and Rahahleh, F. 2013. A mixed pair of pale and black morph Mourning Wheatears *Oenanthe lugens lugens* in the southern highlands of Jordan. *Sandgrouse* 35: 134-137.

King, B F. 1997. *Checklist of the Birds of Eurasia*. Ibis Publishing Company, Vista.

King, B. 2005. The taxonomic status of the three subspecies of *Cuculus saturatus*. *Bull Brit Ornithol Club* 125: 48-55.

King, H. 2015. Bahrain Systematic List. http://www.hawar-islands.com/checklist.html.

Kirwan, G M, and Gregory, S M S. 2005. A new genus for the Mongolian Finch *Bucanetes mongolicus* (Swinhoe, 1870). *Bull Brit Ornithol Club* 125: 68-80.

Kirwan, G M, and Grieve, A. 2007. Studies of Socotran birds II. One, two or three species: towards a rational taxonomy for the Golden-winged Grosbeak *Rhynchostruthus socotranus*. *Bull Afr Bird Club* 14: 159-169.

Kirwan, G M, and Grieve, A. 2013. Unravelling Meinertzhagen-generated confusion concerning the occurrence of Pale Sand Martin *Riparia diluta* in Egypt and the Near East, with a review of the species' status in the Middle East. *Sandgrouse* 35: 114-125.

Kirwan, G M, and Shirihai, H. 2007. Species limits in the House Bunting complex. *Dutch Birding* 29: 1-19.

Kirwan, G M, Atkinson, P W, van den Berg, A B, and Shirihai, H. 2006. Taxonomy of the Crimson-winged Finch *Rhodopechys sanguineus*: a test case for defining species limits between disjunct taxa? *Bull Afr Bird Club* 13: 136-146.

Kirwan, G M, Boyla, K, Castell, P, Demirci, B, Özen, M, Welch, H, and Marlow, T. 2008. *The Birds of Turkey*. Christopher Helm, London.

Kirwan, G M, Özen, M, and Demirci, B (compilers). 2008. Turkey Bird Report 2002-06. *Sandgrouse* 30: 166-189.

Kirwan, G M, Özen, M, Ertuhan, M, and Atahan, A (compilers). 2014. Turkey Bird Report 2007-2011. *Sandgrouse* 36: 146-175.

Kirwan, G M, Schweizer, M, and Copete, J L. 2015. Multiple lines of evidence confirm that Hume's Owl *Strix butleri* (A. O. Hume, 1878) is two species, with description of an unnamed species (Aves: Non-Passeriformes: Strigidae). *Zootaxa* 3904: 028-050.

Kirwan, G M, Schweizer, M, Ayé, R, and Grieve, A. 2009. Taxonomy, identification and status of Desert Sparrows. *Dutch Birding* 31: 139-158.

Kirwan, G M. 2004. The taxonomic position of the Afghan Scrub Sparrow *Passer (moabiticus) yatii. Sandgrouse* 26: 105-111.

Kirwan, G M. 2007. Two specimens of Red-necked Stint from Iran collected in July 1941. *Dutch Birding* 29: 92-93.

Kirwan, G, Porter, R, and Scott, D. 2015. Chronicle of an extinction? A review of Slender-billed Curlew records in the Middle East. *Br Birds* 108: 669-682.

Klaassen, R H G, Alerstam, T, Carlsson, P, Fox, J W, and Lindström, Å. 2011. Great flights by great snipes: long and fast non-stop migration over benign habitats. *Biology Letters* 7: 833-835.

Klassert, T E, Hernández, M A, Campos, F, Infante, O, Almeida, T, Suárez, N M, Pestano, J, and Hernández, M. 2008. Mitochondrial DNA points to *Lanius meridionalis* as a polyphyletic species. *Mol Phylogenet Evol* 47: 1227-1231.

Knox, A G, Collinson, J M, Parkin, D T, Sangster, G, and Svensson, L. 2008. Taxonomic recommendations for British birds: fifth report. *Ibis* 150: 833-835.

Knox, A G, Helbig, A J, Parkin, D T, and Sangster, G. 2001. The taxonomic status of Lesser Redpoll. *Br Birds* 94: 260-267.

Knox, A G. 1988. The taxonomy of redpolls. *Ardea* 76: 1-26.

Knox, A G. 1990. The sympatric breeding of Common and Scottish Crossbills *Loxia curvirostra* and *L. scotica* and the evolution of crossbills. *Ibis* 132: 454-466.

Knystautas, A. 1987. *The Natural History of the USSR*. Century, London.

Kok, D. 2001. Sneeuwuilen uit Canada en Groenland per schip naar Noordwest-Europa. *Dutch Birding* 23: 396-398.

Komisja Faunistyczna. 2011. Rzadkie ptaki obserwowane w Polsce w roku 2010. *Ornis Polonica* 52: 117-149.

Komisja Faunistyczna. 2012. Rzadkie ptaki obserwowane w Polsce w roku 2011. *Ornis Polonica* 53: 105-140.

Komisja Faunistyczna. 2013. Rzadkie ptaki obserwowane w Polsce w roku 2012. *Ornis Polonica* 54: 109-150.

Komisja Faunistyczna. 2014. Bird species recorded in Poland. http://komisjafaunistyczna.pl/?page_id=2654&lang=en.

Komisja Faunistyczna. 2014. Rewizja współczesnych i historycznych stwierdzeń rzadkich ptaków w Polsce. *Ornis Polonica* 55: 115-134.

Komisja Faunistyczna. 2014. Rzadkie ptaki obserwowane w Polsce w roku 2013. *Ornis Polonica* 55: 181-218.

König, C, and Weick, F. 2008. *Owls of the World*. Second edition. Christopher Helm, London.

Koopman, M E, McDonald, D B, Hayward, G D, Eldegard, K, Sonerud, G A, and Sermach, S G. 2005. Genetic similarity among Eurasian subspecies of boreal owls *Aegolius funereus*. *J Avian Biol* 36: 179-183.

Kopp, M, Peter, H-U, Mustafa, O, Lisovski, S, Ritz, M S, Phillips, R A, and Hahn, S. 2011. South polar skuas from a single breeding population overwinter in different oceans though show similar migration patterns. *Mar Ecol Prog Ser* 435: 263-267.

Kovalik, P, Topercer, J, Karaska, D, Danko, S, and Šrank, V. 2010. A checklist of birds of Slovakia to April 7, 2010. *Tichodroma* 22: 97-108.

Kralj, J. 2005. Rare birds in Croatia. Second report of the Croatian Rarities Committee. *Larus* 49: 37-51.

Kralj, J, and Barišić, S. 2013. Rare birds in Croatia. Third report of the Croatian Rarities Committee. *Nat Croat* 22: 375-396.

Kralj, J, and Radović, D. 2002. Rare birds in Croatia. First report of the Croatian Rarities Committee. *Larus* 48: 85-94.

Kren, J. 2000. *Birds of the Czech Republic*. Christopher Helm, London.

Kryukov, A P. 1995. Systematics of small Palearctic shrikes of the "*cristatus* group". *Proc Western Found Vert Zool* 6: 22-25.

Kushlan, J A, and Hancock, J A. 2005. *The Herons*. Oxford University Press, Oxford.

Kushlan, J A. 2015. *Bibliography of the Herons of the World*. http://www.heronconservation.org/conservation-tools-resources/heron-bibliography. Accessed 12 April 2015.

Kvetko, R, and FK SOS/BirdLife Slovensko. 2013. 13. správa Faunistickej komisie Slovenskej ornitologickej spoločnosti/BirdLife Slovensko. *Tichodroma* 25: 85-93.

Kvetko, R, and FK SOS/BirdLife Slovensko. 2014. 14. správa Faunistickej komisie Slovenskej ornitologickej spoločnosti/BirdLife Slovensko. *Tichodroma* 26: 97-106.

Kvetko, R, and Šrank V. 2012. 12. správa Faunistickej komisie Slovenskej ornitologickej spoločnosti/BirdLife Slovensko. *Tichodroma* 24: 102-108.

Kvist, L, Ponnikas, S, Belda, E J, Encabo, I, Martínez, E, Onrubia, A, Hernández, J M, Vera, P, Neto, J M, and Monrós, J S. 2011. Endangered subspecies of the Reed Bunting (*Emberiza schoeniclus witherbyi* and *E s lusitancia*) in Iberian Peninsula have different genetic structures. *J Ornithol* 152: 681-693.

L

van der Laan, J, and CDNA. 2008. Occurrence and identification of 'isabelline shrikes' in the Netherlands in 1985-2006 and records in Europe. *Dutch Birding* 30: 78-92.

Ławicki, Ł, Abramčuk, A V, Domashevsky, S V, Paal, U, Solheim, R, Chodkiewicz, T, and Woźniak, B. 2013. Range extension of Great Grey Owl in Europe. *Dutch Birding* 35: 145-154.

Ławicki, Ł, Corso, A, and Khil, L. 2013. Long-legged Buzzards in Europe – Britain next? *Birding World* 26: 332-343.

Ławicki, Ł, Khil, L, and de Vries, P. 2012. Expansion of Pygmy Cormorant in central and western Europe and increase of breeding population in southern Europe. *Dutch Birding* 34: 273-288.

Ławicki, Ł. 2012. Azure Tits and hybrids Azure x European Blue Tit in Europe. *Dutch Birding* 34: 219-231.

Ławicki, Ł. 2013. Japanese Waxwing vagrancy to Europe. *Birding World* 26: 43-44.

Leader, P J. 2011. Taxonomy of the Pacific Swift *Apus pacificus* Latham, 1802, complex. *Bull Brit Ornithol Club* 131: 81-93.

Leafloor, J O, Moore, J A, and Scribner, K T. 2013. A hybrid zone between Canada Geese (*Branta canadensis*) and Cackling Geese (*B. hutchinsii*). *Auk* 130: 487-500.

Lee, P L M, Richardson, L J, and Bradbury, R B. 2001. The phylogenetic status of the Corn Bunting *Milaria calandra* based on mitochondrial control-region DNA sequences. *Ibis* 143: 299-303.

Lees, A C, and Ball, A. 2011. Shades of grey: 'eastern' Skylarks and extralimital subspecies identification. *Br Birds* 104: 660-666.

Lees, A C, and Gilroy, J J. 2004. Pectoral Sandpipers in Europe: vagrancy patterns and the influx of 2003. *Br Birds* 97: 638-646.

Lees, A C, and Mahood, S P. 2011. Jacobin Cuckoo in Finland in September 1976: a plea for reassessment. *Dutch Birding* 33: 325-328.

Lees, A C, and Moores, R D. 2006. Identification and status of Dunn's Lark in northwest Africa. *Br Birds* 1999: 481-495.

Lefranc, N, and Worfolk, T. 1997. *Shrikes. A Guide to Shrikes of the World*. Pica Press, Robertsbridge.

Lehikoinen, A, Aalto, T, Huhtinen, H, Lampila, P, Lehikoinen, P, Normaja, J, Soilevaara, K, and Velmala, W. 2012. Rare birds in Finland in 2012. *Linnut Vuosikirja* 2012: 122-137.

Lehikoinen, A, Aalto, T, Huhtinen, H, Lampila, P, Normaja, J, Soilevaara, K, and Velmala, W. 2011. Rare birds in Finland in 2011. *Linnut Vuosikirja* 2011: 88-103.

Lehikoinen, A, Aalto, T, Nikander, P J, Normaja, J, Rauste, V, Soilevaara, K, and Velmala, W. 2009. Rariteettikomitean hyväksymät vuoden 2009 harvinaisuushavainnot. *Linnut Vuosikirja* 2009: 6-16.

Leibak, E, Lilleleht, V, and Veromann, H (eds). 1994. *Birds of Estonia. Status, Distribution and Numbers*. Estonian Academy Publishers, Tallinn.

Leisler, B, and Schulze-Hagen, K. 2011. *The Reed Warblers – diversity in a uniform bird family*. KNNV Publishing, Zeist.

Leonardo, G. 2015. *The Lanner Falcon*. Privately published.

Lepage, D. 2015. Avibase – the world bird database. http://avibase.bsc-eoc.org/avibase.jsp.

Lerner, H R L, and Mindell, D. 2005. Phylogeny of eagles, Old World vultures, and other Accipitridae based on nuclear and mitochondrial DNA. *Mol Phylogenet Evol* 37: 327-346.

Lerner, H R L, Klaver, M C, and Mindell, D P. 2008. Molecular Phylogenetics of the Buteonine Birds of Prey (Accipitridae). *Auk* 125: 304-315.

Lever, C. 2005. *Naturalised Birds of the World*. T & A D Poyer, London.

Lever, C. 2013. *The Mandarin Duck*. T & A D Poyer, London.

Liebers, D, de Knijff, P, and Helbig, A J. 2004. The herring gull complex is not a ring species. *Proc R Soc Lond B* 271: 893-901.

Limiñana, R, Romero, M, Mellone, U, and Urios, V. 2012. Mapping the migratory routes and wintering areas of Lesser Kestrels *Falco naumanni*: new insights from satellite telemetry. *Ibis* 154: 389-399.

Lindolm, A, and Lindén, A. 2003. Oriental Cuckoo in Finland. *Alula* 4: 122-133.

Lindsell, J A, Serra, G, Peske, L, Abdullah, M S, Qaim, G, al Kanani, A, and Wondafrash, M. 2009. Satellite tracking reveals the migration route and wintering area of the Middle East population of Critically Endangered Northern Bald Ibis *Geronticus eremita*. *Oryx* 43: 329-335.

Lislevand, T, Chutný, B, Byrkjedal, I, Pavel, V, Briedis, M, Adamik, P, and Hahn, S. 2015. Red-spotted Bluethroats *Luscinia s. svecica* migrate along the Indo-European flyway: a geolocator study. *Bird Study* 62: 508-515.

Liukkonen-Anttila, T, Uimaniemi, L, Orell, M, and Lumme, J. 2002. Mitochondrial DNA variation and the phylogeography of the grey partridge (*Perdix perdix*) in Europe: from Pleistocene history to present day populations. *J Evol Biol* 15: 971-982.

Livezey, B C. 1995. Phylogeny and evolutionary ecology of modern seaducks (Anatidae: Mergini). *Condor* 97: 233-255.

Livezey, B C. 2010. Phylogenetics of modern shorebirds (Charadriiformes) based on phenotypic evidence: analysis and discussion. *Zool J Linn Soc* 160: 567-618.

Lopes, R J, Hortas, F, and Wennerberg, L. 2008. Geographical segregation in Dunlin *Calidris alpina* populations wintering along the East Atlantic migratory flyway – evidence from mitochondrial DNA analysis. *Divers Distrib* 14: 732-741.

López Suárez, P, Varo Cruz, N, Hazevoet, C J, and López Jurado, L F. 2005. Restricted nesting habitat and reproductive failure of Magnificent Frigatebirds *Fregata magnificens* in the Cape Verde Islands. *Atlantic Seabirds* 7: 107-120.

Lorenzo, J A. 2002. The African Crake on Tenerife – a new Western Palearctic bird. *Birding World* 15: 60-61.

Loskot, V M. 2001. Taxonomic revision of the Hume's Whitethroat *Sylvia althaea* Hume, 1878. *Avian Ecol Behav* 6: 41-42.

Loskot, V M. 2005. Morphological variation and taxonomic revision of five south-eastern subspecies of Lesser Whitethroat *Sylvia curruca* (L.) (Aves: Sylviidae). *Zool Med Leiden* 79: 157-165.

Loskot, V M. 2006. Systematic notes on Asian birds. 61. New data on taxonomy and nomenclature of the Common Sand Martin *Riparia riparia* (Linnaeus, 1758) and the Pale Sand Martin *R. diluta* (Sharpe & Wyatt, 1893). *Zool Med Leiden* 80: 213-223.

Loskot, V. 2014. The Hyrcanian Chickadee *Poecile hyrcanus* (Zarudnyj & Loudon, 1905), an endemic species of broadleaved forests of Talysh and Elburz mountains. *J Nat Mus (Prague), Nat Hist Ser* 183: 65-88.

Louchart, A. 2011. Paleornithology, *Bubo insularis* and deletion of putative records of Brown Fish Owl in the western Mediterranean. *Dutch Birding* 33: 251-252.

Løvenskiold, H L. 1964. Avifauna Svalbardensis. *Norsk Polarinstitutt Skrifter* 129: 1-460.

Lovette, I J, Perez-Emen, J L, Sullivan, J P, Banks, R C, Fiorentino, I, Cordoba-Cordoba, S, Echeverry-Galvis, M, Barker, F K, Burns, K J, Klicka, J, Lanyon, S M, and Bermingham, E. 2010. A comprehensive multilocus phylogeny for the wood-warblers and

a revised classification of the Parulidae (Aves). *Mol Phylogenet Evol* 57: 753-770.
Lundberg, A, and Alatalo, R V. 1992. *The Pied Flycatcher*. T & A D Poyser, London.
Luoni, F. 2009. Eco-ethological characterisation of the alien species Vinous-throated Parrotbill (*Paradoxornis webbianus*) and evaluation of its effects on indigenous species. Doctoral thesis, Università degli Studi dell'Insubria. http://hdl.handle.net/10277/141.
Lysenko, V I. 1974. [About rare waders of Melitopol district]. *Ornitologiya Moscow* 11: 391-392.

M

Madge, S, and Burn, H. 1988. *Wildfowl: an Identification Guide to the Ducks, Geese and Swans of the World*. Christopher Helm, Bromley.
Madge, S, and Burn, H. 1993. *Crows and Jays*. Christopher Helm, London.
Madge, S, and Scofi, P. 2002. *Pheasants, Partridges and Grouse*. Christopher Helm, London.
Manegold, A. 2012. On the name of the Canary Blue Tit *Cyanistes teneriffae* from Gran Canaria. *Bull Brit Ornithol Club* 132: 68.
Manly, B, Arbogast, B S, Lee, D S, and van Tuinen, M. 2013. Mitochondrial DNA analysis reveals substantial population structure within the endangered Black-capped Petrel (*Pterodroma hasitata*). *Waterbirds* 36: 228-233.
Marchant, J. 2014. Factsheet: Bar-headed Goose. GB Non-Native Species Secretariat. http://www.nonnativespecies.org/factsheet/factsheet.cfm?speciesId=252.
Marks, B D, Weckstein, J D, and Moyle, R G. 2007. Molecular phylogenetics of the bee-eaters (Aves: Meropidae) based on nuclear and mitochondrial DNA sequence data. *Mol Phylogenet Evol* 45: 23-32.
Marova, I M, Fedorov, V V, Shipilina, D A, and Alekseev, V N. 2009. Genetic and vocal differentiation in hybrid zones of passerine birds: Siberian and European Chiffchaffs (*Phylloscopus [collybita] tristis* and *Ph. [c.] abietinus*) in the southern Urals. *Dokl Biol Sci* 427: 384-386.
Marques, D A, Thoma, M, Knaus, P, and Vallotton, L. 2013. Oiseaux rares et observations inhabituelles en Suisse en 2012. 22e rapport de la Commission de l'Avifaune Suisse. *Nos Oiseaux* 60: 215-236.
Marquiss, M, Hobson, K A, and Newton, I. 2008. Stable isotope evidence for different regional source areas of common crossbill *Loxia curvirostra* irruptions into Britain. *J Avian Biol* 39: 30-34. doi: 10.1111/j.0908-8857.2008.04212.x.
Marquiss, M, Newton, I, Hobson, K A, and Kolbeinsson, Y. 2012. Origins of irruptive migrations by Common Crossbills *Loxia curvirostra* into northwestern Europe revealed by stable isotope analysis. *Ibis* 154: 400-409.
Marshall, H D, and Baker, A J. 1999. Colonization history of Atlantic island common chaffinches (*Fringilla coelebs*) revealed by mitochondrial DNA. *Mol Phylogenet Evol* 11: 201-212.
Marthinsen, G, Wennerberg, L, and Lifjeld, J T. 2008. Low support for separate species within the redpoll complex (*Carduelis flammea–hornemanni–cabaret*) from analyses of mtDNA and microsatellite markers. *Mol Phylogenet Evol* 47: 1005-1017.
Martin, J R, and Mikkola, H. 2014. The changing face of Britain's Tawny Owls. *Brit Wildlife* 25: 391-399.
Martinez, M, Al-Assad, A, Matti, D, Roth, T, Stalling, D, and Stalling, T. 2016. First documented record of White-crowned Wheatear *Oenanthe leucopyga* in Syria and records there of Desert Lark *Ammomanes deserti annae* and Basalt Wheatear *Oenanthe lugens warriae*, all 2009. *Sandgrouse* 38: 161-164.
Martínez-Cruz, B, and Godoy, J A. 2007. Genetic evidence for a recent divergence and subsequent gene flow between Spanish and Eastern imperial eagles. *BMC Evolutionary Biology* 7: 170. doi: 10.1186/1471-2148-7-170.
Martins, R P, and Hirschfeld, E. 1998. Comments on the limits of the Western Palearctic in Iran and the Arabian Peninsula. *Sandgrouse* 20: 108-134.
Martins, R P, Bradshaw, C G, Brown, A, Kirwan, G M, and Porter, R F. 1996. The status of passerines in southern Yemen and the records of the OSME survey in spring 1993. *Sandgrouse* 17: 54-72.
Mason, N A, and Taylor, S A. 2015. Differentially expressed genes match bill morphology and plumage despite largely undifferentiated genomes in a Holarctic songbird. *Mol Ecol* 24: 3009-3025.
Masseti, M, and Bruner, E. 2009. The primates of the western Palearctic: a biogeographical, historical, and archeozoological review. *J Anthropol Sci* 87: 33-91.
Matessi, G. 1999. Evolutionary patterns in European populations of reed bunting (*Emberiza schoeniclus* ssp). PhD thesis, Padova.
Mather, J R. 2010. Pacific Diver in Yorkshire: new to Britain and the Western Palearctic. *Br Birds* 103: 539-545.
Matias, R, Alfrey, P, Costa, H, Jara, J, Moore, C C, Santos, J L, and Tipper, R. 2011. Adições e alterações à lista das aves de Portugal Continental: primeira actualização. *Anuário Ornitológico* 8: 105-117.
Matias, R, Catry, P, Costa, H, Elias, G, Jara, J, Moore, C C, and Tomé, R. 2007. Lista sistemática das aves de Portugal Continental. *Anuário Ornitológico* 5: 74-132.
Matias, R. 2009. Removal of Black Crake *Amaurornis flavirostris* from the Western Palearctic list, and the first record of Lesser Moorhen *Gallinula angulata* for Madeira. *Bull Brit Ornithol Club* 129: 116-119.
Matias, R. 2010. On a Green Heron *Butorides virescens* specimen at a museum in Madeira. *Bocagiana* 231: 1-8.
Matias, R. 2011. Aves exóticas em Portugal: anos de 2009 e 2010. *Anuário Ornitológico* 8: 94-104.
Matthysen, E, and Quinn, D. 1998. *The Nuthatches*. T & A D Poyser, London.
Maumary, L, and Knaus, P. 2000. Marbled Murrelet in Switzerland: a Pacific Ocean auk new to the Western Palearctic. *Br Birds* 93: 190-199.
Maumary, L, Vallotton, L, and Knaus, P. 2007. *Die Vögel der Schweiz*. Schweizerische Vogelwarte, Sempach, und Nos Oiseaux, Montmollin.
Mayr, G. 2011. The phylogeny of charadriiform birds (shorebirds and allies) – reassessing the conflict be-

tween morphology and molecules. *Zool J Linnean Soc* 161: 1096-3642.
McCracken, K G, and Sheldon, F H. 1998. Molecular and osteological heron phylogenies: sources of incongruence. *Auk* 115: 127-141.
McGowan, R Y, and Kitchener, A C. 2001. Historical and taxonomic review of the Iceland Gull *Larus glaucoides* complex. *Br Birds* 94: 191-195.
McKay, C R. 2000. Cedar Waxwing in Shetland: new to the Western Palearctic. *Br Birds* 93: 580-587.
McShane, C. 1996. Eastern Phoebe in Devon: new to the Western Palearctic. *Br Birds* 89: 103-107.
Meadows, B S. 2003. Additional distributional records from the central Hejaz, western Arabia – an addendum to Baldwin & Meadows (1988). *Bull Brit Ornithol Club* 123: 154-177.
Meadows, B S. 2010. On the status of Isabelline *Lanius isabellinus*, Turkestan *L phoenicuroides* and Red-backed Shrikes *L collurio* in the Eastern Province of Saudi Arabia. *Bull Brit Ornithol Club* 130: 215-218.
Meeth, P. 1969. Notes on sea-birds. 28. Giant Petrel (*Macronectus giganteus*) in West European waters. *Ardea* 57: 92.
Meininger, P L, Duiven, P, Marteijn, E C L, and van Spanje, T. 1990. Notable Bird Observations from Mauritania. *Malimbus* 12: 19-24.
Meirinho, A, Barros, N, Oliveira, N, Catry, P, Lecoq, M, Paiva, V, Geraldes, P, Granadeiro, J P, Ramírez, I, and Andrade, J. 2014. Atlas das Aves Marinhas de Portugal. Sociedade Portuguesa para o Estudo das Aves. http://www.atlasavesmarinhas.pt.
Melling, T. 2009. Should Red-necked Nightjar be on the British List? *Br Birds* 102: 110-115.
Melling, T. 2010. The Eskimo Curlew in Britain. *Br Birds* 103: 80-92.
Menzie, S, Gil-Velasco, M, and Collinson, J M. 2015. First genetically confirmed Eastern Subalpine Warbler *Sylvia cantillans* for Sweden. *Ornis Svecica* 25: 40-44.
Meriggi, A, Gariboldi, A, and Magnani, B. 1992. Habitat requirements of the bobwhite quail in northern Italy. *Bolletino di Zoologia* 59: 73-78.
Mikkola, H. 2012. *Owls of the World: a Photographic Guide*. Christopher Helm, London.
Milá, B, McCormack, J E, Castaneda, G, Wayne, R K, and Smith, T B. 2007. Recent postglacial range expansion drives the rapid diversification of a songbird lineage in the genus *Junco*. *Proc R Soc Lond [Biol.]* 274: 2653-2660.
Milá, B, Smith, T B, and Wayne, R K. 2007. Speciation and rapid phenotypic differentiation in the Yellow-rumped Warbler *Dendroica coronata* complex. *Mol Ecol* 16: 159-173.
Milchev, B P, and Boev, Z N. 2014. First record of Naumann's Thrush (*Turdus naumanni* Temminck, 1820, *sensu lato*) in Bulgaria. *Acta Zool Bulgar* 66: 559-561.
Miles, W, Parnaby, D, Rosser, B, Moss, J, and Collinson, J M. 2015. 'Eastern Grasshopper Warbler' on Fair Isle: new to Britain. *Br Birds* 108: 231-236.
Miller, J M, Hallager, S, Monfort, S L, Newby, J, Bishop, K, Tidmus, S A, Black, P, Houston, B, Matthee, C A, and Fleischer, R C. 2010. Phylogeographic analysis of nuclear and mtDNA supports subspecies designations in the ostrich (*Struthio camelus*). *Conserv Genet* 12: 423-431.
Milne, P, and McAdams, D G, on behalf of the Irish Rare Birds Committee. 2007. Irish Rare Bird Report 2005. http://www.irbc.ie/reports/irbr/2005_IRBR.pdf.
Milne, P, and McAdams, D G, on behalf of the Irish Rare Birds Committee. 2008. Irish Rare Bird Report 2006. http://www.irbc.ie/reports/irbr/2006_IRBR.pdf.
Milne, P, and McAdams, D G, on behalf of the Irish Rare Birds Committee. 2009. Irish Rare Bird Report 2007. *Irish Birds* 8: 583-610.
Milvus Group. 2015. Checklist of the Birds of Romania. http://www.milvus.ro/images/PDF/Checklist%20of%20the%20Birds%20of%20Romania.pdf.
Mitchell, D, and Young, S. 1999. *Photographic Handbook of the Rare Birds of Britain and Europe*. New Holland, London.
Mitchell, D. 2011. *Birds of Britain: Subspecies Checklist*. v1.1. https://www.birdwatch.co.uk/userfiles/file/Birdwatch/Birds%20of%20Britain%20-%20subspecies%20checklist%20v1_1.pdf.
Mjølsnes, K R, Bunes, V, Olsen, T A, and Solbakken, K A. 2006. Sjeldne fugler i Norge i 2004. Rapport fra Norsk sjeldenhetskomite for fugl (NSKF). *Ornis Nor* 29: 68-109.
Mlíkovský, J. 2012. Correct name for the Asian Brown Flycatcher (Aves: Muscicapidae, Muscicapa). *Zootaxa* 3393: 53-56.
MME Nomenclator Bizottság (in press): Az MME Monenclator Bizottság 2011. évi jelentése a Magyarországon ritka madárfajok előfordulásáról. http://www.birding.hu/doc/NB2011.pdf.
MME Nomenclator Bizottság (in press): Az MME Monenclator Bizottság 2012. évi jelentése a Magyarországon ritka madárfajok előfordulásáról. http://www.birding.hu/doc/NB2012.pdf.
MME Nomenclator Bizottság (in press): Az MME Monenclator Bizottság 2013. évi jelentése a Magyarországon ritka madárfajok előfordulásáról. http://www.birding.hu/doc/NB_2013.pdf.
MME Nomenclator Bizottság. 2008. *Magyarország Madarainak Névjegyzéke*. Nomenclator avium Hungariae. Magyar Madártani és Természetvédelmi Egyesület, Budapest.
MME Nomenclator Bizottság. 2010. The 2008 report of the Hungarian Checklist and Rarities Committee on rare birds in Hungary. *Aquila*. http://www.birding.hu/doc/NB2008.pdf.
MME Nomenclator Bizottság. 2010. The 2009 report of the Hungarian Checklist and Rarities Committee on rare birds in Hungary. *Aquila*. http://www.birding.hu/doc/NB2009.pdf.
MME Nomenclator Bizottság: Az MME Monenclator Bizottság 2010. évi jelentése a Magyarországon ritka madárfajok előfordulásáról. *Aquila* 120: 61-73. http://www.birding.hu/doc/NB2010.pdf.
Monroe, B L Jr, and Sibley C G. 1993. *A World Checklist of Birds*. Yale University Press, New Haven.
Montalto, J A, on behalf of the members of the Rarities Committee. 2010. National Rarities Committee – Malta. 1st Report. *Il-Merill* 32: 47-54.
Monteiro, L W, and Furness, R W. 1998. Speciation through temporal segregation of Madeiran storm petrel (*Oceanodroma castro*) populations in the Azores? *Philosophical Transactions of the Royal Society of London B* 353: 945-953.

Mori, A, Baldaccini, N E, Baratti, M, Caccamo, C, Dessì-Fulgheri, F, Grasso, R, Nouira, S, Ouni, R, Pollonara, E, Rodriguez-Godoy, F, Spena, M T, and Giunchi, D. 2014. A first assessment of genetic variability in the Eurasian Stone-curlew *Burhinus oedicnemus*. *Ibis* 156: 687-692.

Morozov, V V. 2004. Displaying Swinhoe's Snipe in eastern European Russia: a new species for Europe. *Br Birds* 97: 134-138.

Muchaxo, J, Alfrey, P, Costa, H, Jara, J, Matias, R, Moore, C C, Santos, J L, and Tipper, R. 2011. Aves de ocorrência rara ou acidental em Portugal. Relatório do Comité Português de Raridades referente ao ano de 2010. *Anuário Ornitológico* 8: 3-52.

Mullarney, K, and Murphy, J. 2005. The Rough-legged Hawk in Ireland. *Birding World* 18: 503-504.

Mullarney, K. 2006. A chaffinch resembling African Chaffinch in Ireland. *Birding World* 19: 109-112.

Muñoz-Fuentes, V, Green, A J, Sorenson, M D, Negro, J J, and Vilã, C. 2006. The ruddy duck *Oxyura jamaicensis* in Europe: natural colonisation or human introduction? *Mol Ecol* 15: 1441-1453.

Muñoz-Fuentes, V, Green, A J, Negro, J J, and Sorenson, M D. 2005. Population structure and loss of genetic diversity in the endangered white-headed duck, *Oxyura leucocephala*. *Conserv Genet* 6: 999-1015.

Muñoz-Fuentes, V, Vila, C, Green, A J, Negro, J J, and Sorenson, M D. 2007. Hybridization between white-headed ducks and introduced ruddy ducks in Spain. *Mol Ecol* 16: 629-638.

Murdoch, D A, and Betton, K F. 2008. A Checklist of the Birds of Syria. *Sandgrouse* Supplement 2.

Musílek, J. 1936. První zjištění buřňáka šedého *Puffinus kuhlii* (Boie) u nás. *Sylvia* 1: 55-57.

N

Natur & Ëmwelt absl. 2010. Liste der Vögel Luxemburgs. http://www.luxnatur.lu/lnv0051.htm.

Naturschutzzenter Lëtzebuerg. 2014. Birds of Luxembourg. http://www.luxnatur.lu/div/checklist2014.htm.

Nelson, J B. 2005. *Pelicans, Cormorants and their Relatives*. Oxford University Press, Oxford.

Netfugl.dk. 2015. Bird list of Denmark. http://www.netfugl.dk/dklist.php?language=uk.

Netfugl.dk. 2015. The birds of the Western Palearctic. http://www.netfugl.dk/wplist.php.

Neves, V C, Griffiths, K, Savory F, Furness, R W, and Mable, B. 2010. Are European starlings breeding in the Azores genetically distinct from birds breeding in mainland Europe? *Eur J Wildlife Res* 56: 95-100. DOI: 10.1007/s10344-009-0316-x.

Newell, D. 2013. Recent records of southern skuas in Britain. *Br Birds* 101: 439-441.

Newton, I. 2003. *The Speciation and Biogeography of Birds*. Academic Press, London.

Newton, S F. 2006. A further mangrove white-eye (*Zosterops*) population in the Arabian Sea. *Phoenix*: 16-17.

Nielsen, H H. 2000. Første danske fund af Ørkengærdesanger *Sylvia curruca*, ssp *minula* eller *jaxartica*. *Dansk Ornitol Foren Tidsskr* 94: 171-172.

Nightingale, B. 2005. The status of Lady Amherst's Pheasant in Britain. *Br Birds* 98: 20-25.

Nijman, V, and Aliabadian, M. 2013. DNA barcoding as a tool for elucidating species delineation in wide-ranging species as illustrated by owls (Tytonidae and Strigidae). *Zoolog Sci* 30: 1005-1009.

Norsk Ornitologisk Forening. 2014. Norgeslisten. http://www.birdlife.no/organisasjonen/nskf/norgeslisten.php.

Nyári, A, Ryall, C, and Peterson, A T. 2006. Global invasive potential of the house crow *Corvus splendens* based on ecological niche modelling. *J Avian Biol* 37: 306-311.

O

Occhiato, D. 2003. Pine Bunting in Italy: status and distribution. *Dutch Birding* 25: 32-39.

Olofson, S K K, and Sørenson, S. 2013. *Liste over Færøernes fugle – særligt med henblik på de sjældnere arter*. Privately published.

Olofson, S K K. 2012. Prothonotary Warbler. https://birdingfaroes.wordpress.com/2012/04/25/prothonotary-warbler/.

Olsen, K M, and Larsson, H. 1995. *Terns of Europe and North America*. Christopher Helm, London.

Olsen, K M, and Larsson, H. 1997. *Skuas and Jaegers, A Guide to the Skuas and Jaegers of the World*. Pica Press, Robertsbridge.

Olsen, K M, and Larsson, H. 2003. *Gulls of Europe, Asia and North America*. Christopher Helm, London.

Olsen, T A, and Mjølsnes, K R. 2007. Sjeldne fugler i Norge i 2005. Rapport fra Norsk sjeldenhetskomite for fugl (NSKF). *Ornis Nor* 30: 68-115.

Olsen, T A, and Mjølsnes, K R. 2009. Sjeldne fugler i Norge i 2007. Rapport fra Norsk sjeldenhetskomite for fugl (NSKF). *Ornis Nor* 32: 4-41.

Olsen, T A, Bunes, V, Egeland, Ø, Gullberg, A, Mjølsnes, K R, and Tveit, B O. 2010. Sjeldne fugler i Norge i 2008. Rapport fra Norsk sjeldenhetskomite for fugl (NSKF). *Ornis Nor* 33: 4-48.

Olsen, T A, Tveit, B O, Bunes, V, and Mjølsnes, K R. 2008. Sjeldne fugler i Norge i 2006. Rapport fra Norsk sjeldenhetskomite for fugl (NSKF). *Ornis Nor* 31: 48-97.

Olsen, T A. 2011. Sjeldne fugler i Norge i 2009 og 2010. *Rapport fra Norsk sjeldenhetskomité for fugl (NSKF)*. http://www.birdlife.no/organisasjonen/nskf/publikasjoner/NSKF2009_2010.pdf.

Olson, S L, and Reveal, J L. 2009. Nomenclatural history and a new name for the Blue-winged Warbler (Aves: Parulidae). *Wilson J Ornithol* 121: 618-620.

Olsson, P. 2015. The first Ruddy-breasted Crake *Porzana fusca* for Oman and the Middle East. *Sandgrouse* 37: 57-59.

Olsson, S L. 1985. The Italian specimen of *Bulweria fallax* (Procellariidae). *Bull Brit Ornithol Club* 105: 29-30.

Olsson, U, Alström, P, Svensson, L, Aliabadian, M, and Sundberg, P. 2010. The *Lanius excubitor* (Aves, Passeriformes) conundrum – taxonomic dilemma when molecular and non-molecular data tell different stories. *Mol Phylogenet Evol* 55: 347-357.

Olsson, U, Leader, P J, Carey, G J, Khan, A A, Svensson, L, and Alstrom, P. 2013. New insights into the intricate taxonomy and phylogeny of the *Sylvia curruca* complex. *Mol Phylogenet Evol* 67: 72-85.

Olsson, U, Rguibi-Idrissi, H, Copete, J L, Matos, J L A, Provost, P, Amezian, M, Alström, A, and Jiguet, F. 2016. Mitochondrial phylogeny of the Eurasian/African reed warbler complex (*Acrocephalus*, Aves). Disagreement between morphological and molecular evidence and cryptic divergence: a case for resurrecting *Calamoherpe ambigua* Brehm 1857. *Mol Phylogenet Evol* 102: 30-44.

Olsson, U, Yosef, R, and Alström, P. 2013. Assessment of species limits in African 'brown buntings' (Emberiza, Passeriformes) based on mitochondrial and nuclear sequence data. *Ibis* 155: 534-543.

Omland, K E, Tarr, C L, Boarman, W I, Mazluff, J M, and Fleischer, R C. 2000. Cryptic generation and paraphyly in ravens. *Proc R Soc Lond* B. 267: 2475-2482.

Onley, D, and Scofield, P. 2007. *Albatrosses, Petrels and Shearwaters of the World*. Christopher Helm, London.

van Oosten, H H, and Emtsev, A A. 2015. Putative segregation of two Yellow Wagtail taxa by breeding habitat in Western Siberia: possible implications for *Motacilla flava* taxonomy. *Ardea* 101: 65-70.

Oreel, G J. 2004. Origin of presumed African Chaffinch on Maasvlakte in April 2003. *Dutch Birding* 26: 46-47.

Ortvad, T E, Christensen, J H, Hansen, T H, Kristensen, A B, Olsen, K, Pedersen, K, Rasmussen, P A F, and Schwalbe, M. 2015. Revision af den danske fugleliste pr. 2013. Rapport nr. 45 fra Sjældenhedsudvalget. *Dansk Orn Foren Tidsskr* 109: 24-35.

Ortvad, T E, Christiansen, S S, Göller, O Z, Kristensen, A B, Neergaard, R S, Olsen, K, Pedersen, K, and Strack, R. 2014. Sjældne fugle i Danmark og Grønland i 2014. Rapport nr. 46 fra Sjældenhedsudvalget. *Fugleåret* 2014: 111-137.

Ortvad, T E, Christiansen, S S, Klein, S, Kristensen, A B, Pedersen, K, and Strack, R. 2014. Sjældne fugle i Danmark og Grønland i 2013. Rapport nr. 44 fra Sjældenhedsudvalget. *Fugleåret* 2013: 114-139.

Ots, M, and Paal, U. 2013. Linnuharuldused Eestis 2012. Eesti linnuharulduste komisjoni aruanne nr. 11. *Hirundo* 26: 26-50.

Ots, M, and Paal, U. 2014. Linnuharuldused Eestis 2013. Eesti linnuharulduste komisjoni aruanne nr. 12. *Hirundo* 27: 44-59.

Ottens, G, and Ryall, C. 2003. House Crows in the Netherlands and Europe. *Dutch Birding* 25: 312-319.

Ottens, G. 2003. Background and development of the Dutch population of House Crows *Corvus splendens*. *Limosa* 76: 69-74.

Ottens, G. 2006. African Darter in Israel in May 2004 and its WP occurrence. *Dutch Birding* 28: 214-218.

Ottens, G. 2013. Rosse Waaierstaart bij Camperduin in September 2013. *Dutch Birding* 35: 380-384.

Ottosson, U, Bensch, S, Svensson, L, and Waldenström, J. 2005. Differentiation and phylogeny of the olivaceous warbler *Hippolais pallida* species complex. *J Ornithol* 146: 127-136.

Ottvall, R, Bensch, S, Walinder, G, and Lifjeld, J T. 2002. No evidence of genetic differentiation between lesser redpolls *Carduelis flammea cabaret* and common redpolls *Carduelis f. flammea*. *Avian Sci* 2: 237–244.

Outlaw, R K, Voelker, G, and Bowie, R C K. 2010. Shall we chat? Evolutionary relationships in the genus *Cercomela* (Muscicapidae) and its relation to *Oenanthe* reveals extensive polyphyly among chats distributed in Africa, India and the Palearctic. *Mol Phyl Evol* 55: 284-92.

Ovaa, A, Groenendijk, D, Berlijn, M, and CDNA. 2009. Rare birds in the Netherlands in 2008. *Dutch Birding* 31: 331-352.

Ovaa, A, Groenendijk, D, Berlijn, M, and CDNA. 2010. Rare birds in the Netherlands in 2009. *Dutch Birding* 32: 363-383.

Ovaa, A, Groenendijk, D, Berlijn, M, and CDNA. 2011. Rare birds in the Netherlands in 2010. *Dutch Birding* 33: 357-376.

Ovaa, A, Groenendijk, D, Berlijn, M, and CDNA. 2012. Rare birds in the Netherlands in 2011. *Dutch Birding* 34: 349-377.

Ovaa, A, Groenendijk, D, Haas, M, Berlijn, M, and CDNA. 2013. Rare birds in the Netherlands in 2012. *Dutch Birding* 35: 357-379

Ovaa, A, van der Laan, J, Berlijn, M, and CDNA, 2008. Rare birds in the Netherlands in 2007. *Dutch Birding* 30: 369-389.

Özkan, K. 2011. An avifaunal survey of the Istranca mountains, Turkish Thrace: novel breeding bird records including the first breeding record of Wood Warbler *Phylloscopus sibilatrix* in Turkey. *Sandgrouse* 33: 163-173.

P

Päckert, M, and Martens, J. 2008. Taxonomic pitfalls in tits – comments on the Paridae chapter of the *Handbook of the Birds of the World*. *Ibis* 150: 829-831.

Päckert, M, Feigl, A, Wink, M, and Tietze, D T. 2011. Molecular phylogeny and historical biogeography of Swifts (Apodidae: Apus, Tachymarptis). 5th IBS conference, 7–11.01.2011, Irakleion, Crete, Greece.

Päckert, M, Martens, J, Eck, S, Nazarenko, A A, Valchuk, O P, Petri, B, and Veith, M. 2005. *Parus major* – a misclassified ring species. *Biol J Linn Soc* 86: 153-174.

Päckert, M, Martens, J, Hering, J, Kvist, L, and Illera, J C. 2013. Return flight to the Canary Islands – The key role of peripheral populations of Afrocanarian blue tits (Aves: *Cyanistes teneriffae*) in multi-gene reconstructions of colonization pathways. *Mol Phylogenet Evol* 67: 458-467.

Paige, J P. 1965. Field identification and winter range of the Asiatic Dowitcher *Limnodromus semipalmatus*. *Ibis* 107: 95-97.

Panov, E N. 2005. *Wheatears of Palearctic. Ecology, Behaviour and Evolution of the Genus Oenanthe*. Pensoft, Sofia.

Panov, E N. 2009. On the nomenclature of the so-called Isabelline Shrike. *Sandgrouse* 31: 163-170.

Panov, E N. 2011. *The True Shrikes (Laniidae) of the World: Ecology, Behaviour and Evolution*. Pensoft, Sofia.

Panov, E, and Bannikova, A A. 2010. On the validity of

the 'Steppe Grey Shrike' as an independent species. *Sandgrouse* 32: 141-146.

Parkin, D T, and Knox, A G. 2010. *The Status of Birds in Britain and Ireland*. Christopher Helm, London.

Parkin, D T, Collinson, M, Helbig, A J, Knox, A G, and Sangster, G. 2003. The taxonomic status of Carrion and Hooded Crows. *Br Birds* 96: 274-290.

Parkin, D T, Collinson, M, Helbig, A J, Knox, A G, Sangster, G, and Svensson, L. 2004. Species limits in *Acrocephalus* and *Hippolais* warblers from the Western Palearctic. *Br Birds* 97: 276-299.

Paterson, A M, and Riddiford, N J. 1990. Does the Cape Gannet enter European waters? *Br Birds* 83: 519-526.

Patrikeev, M. 2004. *The Birds of Azerbaijan*. Pensoft, Sofia.

Patten, M A. 2001. The roles of habitat and signaling in speciation: evidence from a contact zone of two Song Sparrow (*Melodia melospiza*) subspecies. PhD thesis. University of California, Riverside.

Pavia, M. 2009. Popolazione autosostentata di anatra muta, *Cairina moschata* (Linnaeus 1758), in Piemonte. *Riv Ital Ornitol* 79: 73-74.

Pavlova, A, Zink, R M, Drovetski, S V, Red'kin Y, Rohwer, S, and Sheldon, F. 2003. Phylogeographic patterns in *Motacilla flava* and *Motacilla citreola*: species limits and population history. *Auk* 120: 744-758.

Payne, R B. 2005. *The Cuckoos*. Oxford University Press, Oxford.

Pearson, D J. 1981. Field identification of Isabelline Shrike. *Dutch Birding* 3: 119-122.

Pearson, D J. 2000. The races of the Isabelline Shrike *Lanius isabellinus* and their nomenclature. *Bull Brit Ornithol Club* 120: 22-27.

Pearson, D, Svensson, L, and Frahnert, S. 2012. Further on the type series and nomenclature of the Isabelline Shrike *Lanius isabellinus*. *Bull Brit Ornithol Club* 132: 270-276.

Pedersen, T, and Aspinall, S J (eds). 2015. *EBRC Annotated Checklist of the Birds of the United Arab Emirates*. EAD, Abu Dhabi.

Pegoraro, K, Föger, M, and Parson, W. 2001. First evidence of DNA sequence differences between Northern Bald Ibises (*Geronticus eremita*) of Moroccan and Turkish origin. *Journal für Ornithologie* 142: 425-428.

Pellegrino, I, Negri, A, Cucco, M, Mucci, N, Pavia, M, Šálek, M, Boano, G, and Randi, E. 2014. Phylogeography and Pleistocene refugia of the Little Owl *Athene noctua* inferred from mtDNA sequence data. *Ibis* 156: 639-657.

Pennington, M G, Riddington, R, and Miles, W T S. 2012. The Lapland Bunting influx in Britain and Ireland in 2010/11. *Br Birds* 105: 654-673.

Pentzold, S, Tritsch, C, Martens, J, Tietze, D T, Giacalone, G, Valvo, M L, Nazarenko, A A, Kvist, L, and Päckert, M. 2013. Where is the line? Phylogeography and secondary contact of western Palearctic coal tits (*Periparus ater*: Aves, Passeriformes, Paridae). *Zoologischer Anzeiger* 252: 367-382.

Pérez, C. 2005. Mountain Chiffchaff *Phylloscopus sindianus* Brooks, 1880, in Gibraltar. *Iberis* 1: 3-10.

Pérez, C, and Bensusan, K. 2012. Tropical Mockingbird (*Mimus gilvus*) at Gibraltar: 1st for the area and the WP? http://www.rarebirdspain.net/arbsf067.htm.

Perez, C, and Cortes, J. 2003. Passage of the Iberian Chiffchaff *Phylloscopus ibericus* at Gibraltar. *Gibraltar Bird Rep* 2001: 29-31.

Perktaş, U, and Quintero, E. 2013. A wide geographical survey of mitochondrial DNA variation in the great spotted woodpecker complex, *Dendrocopos major* (Aves: Picidae). *Biol J Linn Soc* 108: 173-188.

Perktaş, U, Barrowclough, G F, and Groth, J G. 2011. Phylogeography and species limits in the green woodpecker complex (Aves: Picidae): multiple Pleistocene refugia and range expansion across Europe and the Near East. *Biol J Linn Soc* 104: 710-723.

Perlman, Y, and Israeli, N. 2013. Black-shouldered Kite *Elanus caeruleus* breeing in Israel. *Sandgrouse* 35: 25-27.

Perlman, Y, and Meyrav, J. 2009. *Checklist of the Birds of Israel*. Israel Ornithological Center and Society for the Protection of Nature in Israel, Tel-Aviv.

Perlman, Y. 2008. The natural history of the Nubian Nightjar *Caprimulgus nubicus* in Israel. *Sandgrouse* 30: 117-124.

Peronace, V, Cecere, J G, Gustin, M, and Rondinini, C. 2012. Lista Rossa 2011 degli Uccelli Nidificanti in Italia. *Avocetta* 36: 11-58.

Perttula, P, and Tenovuo, J. 2002. Great Grey Shrike subspecies *excubitor*, *sibiricus* and *borealis*. *Alula* 8: 54-60.

Petersen, Æ. 1992. Amerísk smyrill finnst hérlendis [an American Merlin found]. *Bliki* 12: 11-14.

Pétursson, G. 2010. Two records of Great Blue Heron in Iceland. *Br Birds* 103: 359.

Pétursson, G. 2014. Skrá yfir íslenska fugla. https://notendur.hi.is//~yannk/1111.pdf.

Pietri, C. 1993. *Le Colin de Californie,* Callipepla californica *(Shaw) 1798 – Historique, distribution et habitat en Corse*. Mémoire de Diplôme EPHE, École Pratique des Hautes Études, Montpellier.

Pineau, O, Kayser, Y, Sall, M, Gueye, A, and Hafner, H. 2001. The Kelp Gull at Banc d'Arguin – a new Western Palearctic bird. *Birding World* 14: 110-111.

Piot, B, and Vallotton, L. 2012. Oiseaux rares et observations inhabituelles en Suisse en 2011. 21e rapport de la Commission de l'Avifaune Suisse. *Nos Oiseaux* 59: 207-228.

Piskunov, V V, and Antonchikov, A N. 2007. Flock of 14 Relict Gulls in southern Russia and northwestern Kazakhstan in May 2000. *Dutch Birding* 29: 94-95.

Pitches, A, and Cleeves, T. 2005. *Birds New to Britain 1980-2004*. T & A D Poyser, London.

Poelstra, J. 2010. Speciation in shades of grey: the great grey shrike complex. *Dutch Birding* 32: 258-264.

Poelstra, J. 2015. Hybridization and speciation: the case of Italian Sparrow. *Dutch Birding* 37: 251-254.

Pons, J-M, Kirwan, G M, Porter, R F, and Fuchs, J. 2013. A reappraisal of the systematic affinities of Socotran, Arabian and East African scops owls (*Otus*, Strigidae) using a combination of molecular, biometric and acoustic data. *Ibis* 155: 518-533.

Pons, J-M, Olioso, G, Cruaud, C, and Fuch, J. 2011. Phylogeography of the Eurasian green woodpecker (*Picus viridis*). *J Biogeogr* 38: 311-325.

Pons, J-M, Thibault, J-C, Aymí, R, Grussu, M, Muntaner, J, Olioso, G, Sunyer, J R, Touihri, M, and Fuchs, J. 2015. The role of western Mediterranean islands in the evolutionary diversification of the Spotted Flycatcher (*Muscicapa striata*), a long-distance migratory passerine species. *J Avian Biol.* http://dx.doi.org/10.1111/jav.00859.

Porter, R F, Martins, R P, Shaw, K D, and Sorenson, U. 1996. The status of non-passerines in southern Yemen and the records of the OSME survey in spring 1993. *Sandgrouse* 17: 22-53.

Porter, R, and Aspinall, S. 2010. *Birds of the Middle East*. Second edition. Christopher Helm, London.

Porter, R, and Kirwan, G. 2010. Studies of Socotran birds VI. The taxonomic status of the Socotra Buzzard. *Bull Brit Ornithol Club* 130: 116-131.

Porter, R, and Stanton, D. 2007. *Working Checklist of the Birds of Yemen (including Socotra)*. Unpublished draft.

Porter, R, and Stanton, D. 2011. Observations of Clamorous Reed Warblers *Acrocephalus stentoreus brunnescens* and Mangrove Reed Warblers *Acrocephalus (scirpaceus) avicenniae* in mangroves of the Yemen Red Sea coast. *Sandgrouse* 33: 134-138.

Porter, R F. 2016. First breeding record of Lesser Grey Shrike *Lanius minor* in Iraq. *Sandgrouse* 38: 71-72.

Potapov, E, and Sale, R. 2005. *The Gyr Falcon*. T & A D Poyser, London.

Potapov, R, and Sale, R. 2013. *Grouse of the World*. New Holland, London.

Pratt, H D. 2011. Observations on species limits in the Great Egret (*Ardea alba*) complex. *J Heron Biol Conserv* 1: 5. www.HeronConservation.org/vol1/art5.

Proctor, B, and Donald, C. 2003. Yellow-headed Blackbirds in Britain and Europe. *Birding World* 16: 69-81.

Prommer, M, Bagyura, J, Chavko, J, and Uhrin, M. 2012. Migratory movements of Central and Eastern European Saker Falcons (*Falco cherrug*) from juvenile dispersal to adulthood. *Aquila* 119: 111–134.

Prommer, M, Milobog, Y, Gavrilyuk, M, Vetrov, V, and Bagyura, J. 2012. Relationship between central and east European Saker Falcon popuations based on results of ringing schemes and satellite-tracking. 4th International Conference on Birds of Prey and Owls in North Eurasia. http://sakerlife2.mme.hu/sites/default/files/Satellite-tracked_Sakers_in_Ukraine.pdf.

Pullainen, T, Tohmo, T, and Suomen Rariteettikomitea (FRC). 2015. RK:n hyväksymät havainnot Suomesta. http://www.birdlife.fi/havainnot/rk/rk-kokodata.shtml.

Puzović, S, Simić, D, Saveljić, D, Gergelj, J, Tucakov, M, Stojnić, Hulo, L, Ham, I, Vizi, O, Šćiban, M, Ružić, M, Vučanović, M, and Jovanović, T. 2003. Ptice Srbije I Crne Gore – veličine gnezdilišnih populacija I trendovi: 1990-2002. *Ciconia* 12: 35-120.

Q

Quaintenne, G, and species co-ordinators. 2014. Les oiseaux nicheurs rares et menacés en France en 2013. *Ornithos* 21: 297-331.

R

Rabbitts, B. 2009. Hooded Merganser on North Uist: a return to the British List. *Br Birds* 102: 122-129.

Radde, G. 1884. [Ornithological fauna of the Caucasus (*Ornis Caucasica*)]. Caucasian Museum Press, Tiflis (in Russian).

Ramadan-Jaradi, G, Barra, T, and Ramadan-Jaradi, M. 2008. Revised checklist of the birds of Lebanon. *Sandgrouse* 30: 22-69.

Ramadan-Jaradi, G. 2011. Climate variation impact on birds of Lebanon – assessment and identification of main measures to help the birds to adapt to climate change. *Lebanese Sci J* 12: 25-32.

Ramírez, I, Paiva, V H, Menezes, D, Silva, I, Philips, R A, Ramos, J A, and Garthe, S. 2013. Year-round distribution and habitat preferences of the Bugio petrel. *Mar Ecol-Prog Ser* 476: 269-284.

Ramos, J A. 1993. Status and ecology of the Priolo or Azores Bullfinch. PhD thesis. University of Oxford, Oxford.

Randler, C, Förschler, M I, Gonzalez, J, Aliabadian, M, Bairlein, F, and Wink, M. 2012. Phylogeography, pre-zygotic isolation and taxonomic status in the endemic Cyprus Wheatear *Oenanthe cypriaca*. *J Ornithol* 153: 303-312.

Ranner, A, and Khil, L. 2008. Records of rare and remarkable bird species in Austria 2001 to 2006 - fifth report of the Avifaunistic Commission of BirdLife Austria. *Egretta* 50: 51-75.

Ranner, A, and Khil, L. 2011. Records of rare and remarkable bird species in Austria 2007-2009. Sixth report of the Avifaunistic Commission of BirdLife Austria. *Egretta* 52: 13-32.

Ranner, A. 2003. Nachweise seltener und bemerkenswerter Vogelarten in Österreich 1999-2000. 4. Bericht der Avifaunistischen Kommission von BirdLife Österreich. *Egretta* 46: 109-135.

Ranner, A. 2014. *List of Austrian Bird Species*. Avifaunistic Commission of BirdLife Austria. http://www.birdlife-afk.at/Bird%20species%20list%20Austria_Dec2014_english.pdf.

Rannila, V-P. 2003. The Oriental Plover in Finland – a new Western Palearctic bird. *Birding World* 16: 209.

Rasmussen, P C, and Anderton, J C. 2012. *Birds of South Asia. The Ripley Guide*. Vols 1 and 2. Second edition. National Museum of Natural History – Smithsonian Institution, Michigan State University and Lynx Edicions, Washington DC, Michigan and Barcelona.

Red'kin, Y, and Konovalova, M. 2006. Systematic notes on Asian birds: 63. The eastern Asiatic races of *Sitta europaea*, Linnaeus 1758. *Zool Med Leiden* 80: 241-261.

Reeber, S, Kayser, Y, and the CHN. 2015. Les oiseaux rares en France en 2014. 32nd rapport du Comité d'Homologation National. *Ornithos* 22: 281-311.

Reeber, S, and the CHN. 2011. Les oiseaux rares en France en 2010. 28th rapport du Comité d'Homologation National. *Ornithos* 18: 325-367.

Reeber, S, and the CHN. 2012. Les oiseaux rares en France en 2011. 29th rapport du Comité d'Homologation National. *Ornithos* 19: 353-395.

Reeber, S. 2015. *Wildfowl of Europe, Asia and North America*. Christopher Helm, London.

Rees, E C, and Beekman, J H. 2010. Northwest European Bewick's Swans: a population in decline. *Br Birds* 103: 640-650.

Reeves, A B, Drovetski, S V, and Fadeev, I V. 2008. Mitochondrial DNA data imply a stepping-stone colonization of Beringia by arctic warbler *Phylloscopus borealis*. *J Avian Biol* 39: 569-575.

Reneerkens, J. 2012. African Crake at Banc d'Arguin, Mauritania, in November 2011. *Dutch Birding* 34: 95-96.

Restall, R. 1996. *Munias and Mannikins*. Pica Press, Robertsbridge.

Riad, A. 1995. First record of Seebohm's Wheatear *Oenanthe oenanthe seebohmi* in Egypt. *Orn Soc Middle East Bull* 35: 36-37.

Richardson, C (ed). 2010. *Cyprus Bird Report 2009*. BirdLife Cyprus, Nicosia.

Richardson, C. 1990. *The Birds of the United Arab Emirates*. Hobby Publications, Warrington.

Riddington, R, and Reid, J. 2000. Lesser Frigatebird in Israel: new to the Western Palearctic. *Br Birds* 93: 22-27.

Rising, J D. 1996. *A Guide to the Identification and Natural History of the Sparrows of the United States and Canada*. Academic Press, London.

Ristow, D, and Wink, M. 1992. Distribution of non-breeding Eleonora's Falcon *Falco eleonorae*. *Il Merill* 28: 1-10.

Robb, M S, van den Berg, A B, and Constantine, M. 2013. A new species of *Strix* owl from Oman. *Dutch Birding* 35: 275-310.

Robb, M S. 2000. Introduction to vocalizations of crossbills in north-western Europe. *Dutch Birding* 22: 61-107.

Robb, M, and The Sound Approach. 2015. *Undiscovered Owls*. The Sound Approach, Poole.

Robb, M, Mullarney, K, and The Sound Approach. 2008. *Petrels Night and Day*. The Sound Approach, Poole.

Robb, M S, Sangster, G, Aliabadian, M, van den Berg, A B, Constantine, M, Irestedt, M, Khani, A, Musavi, S B, Nunes, J M G, Willson, M S, and Walsh, A J. The rediscovery of *Strix butleri* (Hume, 1878) in Oman and Iran, with molecular resolution of the identity of *Strix omanensis* Robb, van den Berg and Constantine, 2013. bioRxiv. doi: http://dx.doi.org/10.1101/025122

Robel, D. 1997. Records of the Grey Hornbill (*Tockus nasutus*), Crimson-rumped Waxbill (*Estrilda rhodopyga*) and African Silverbill (*Lonchura cantans*) in Egypt. *Zool Middle East* 15: 33-35.

Robinson, T J, and Matthee, C A. 1999. Molecular genetic relationships of the extinct Ostrich, *Struthio camelus syriacus*: consequences for Ostrich introduction into Saudi Arabia. *Anim Conserv* 2: 165-171.

Rocha, R. 2013. Laughing Dove *Spilopelia senegalensis* (Columbiformes: Columbidae): first record for the Madeira Archipelago, Portugal. *Check List* 9: 432-433.

Rodebrand, S. 2015. Checklist of the birds of the Azores including 2011. Azores Bird Club. http://birdingazores.com/?page=2011checklist.

Rodrigues, P, Lopes, R J, Reis, S, Resendes, R, Ramos, J A, and Tristão da Cunha, R. 2013. Genetic diversity and morphological variation of the common chaffinch *Fringilla coelebs* in the Azores. *J Avian Biol* 45: 167-178. http://dx.doi.org/10.1111/j.1600-048X.2013.00229.x.

Rodríguez, B, Siverio, F, Siverio, M, and Rodríguez, A. 2011. Variable plumage coloration of breeding Barbary Falcons *Falco (peregrinus) peregrinoides* in the Canary Islands: do other Peregrine Falcon subspecies also occur in the archipelago? *Bull Brit Ornithol Club* 131: 140-153.

Rogers, M J, and the Rarities Committee. 2004. Report on rare birds in Great Britain in 2003. *Br Birds* 97: 558-625.

Rogers, M J, and the Rarities Committee. 2005. Report on rare birds in Great Britain in 2004. *Br Birds* 98: 628-694.

Romano, H, Correia-Fagundes, C, Zino, F, and Biscoito, M. 2010. Birds of the archipelagos of Madeira and the Selvagens II – new records and checklist update (1995-2010). *Bol Mus Mun Funchal* 60: 5-44.

Romay, C D, and Roselaar, C S. 2013. American Hawk-Owl caught off Las Palmas, Gran Canaria, Canary Islands, in October 1924. *Dutch Birding* 35: 1-6.

Roselaar, C S, and Aliabadian, M. 2009. Review of rare birds in Iran, 1860s-1960s. *Podoces* 4: 1-27.

Roselaar, C S, and Castricum, V. 2006. Steenortolaan bij Castricum in oktober 2004. *Dutch Birding* 28: 284-291.

Roselaar, C S, van den Berg, A B, van Loon, A J, and Maassen, E. 2006. Siberische Sprinkhaanzangers in Noord-Holland in September 2002-05. *Dutch Birding* 28: 273-283.

Roselaar, C S. 2006. The boundaries of the Palearctic region. *Br Birds* 99: 602-618.

Roselaar, C S. 2014. Are '*rubrifasciata*' crossbills of hybrid origin? *Dutch Birding* 36: 96-107.

Roth, T, Ayé, R, Burri, R, and Schweizer, M. 2005. Bird observations from Iran in February-March 2001, including a new species for the Middle East. *Sandgrouse* 27: 63-68.

Rouco, M, and the Comité de Rarezas de SEO/BirdLife. 2016. Reassessment of the Louisiana Waterthrush (*Parkesia motacilla*) record from La Palma, Canary Islands, in November 1991. http://www.reservoirbirds.com/Articles/RBAR_000020.pdf.

Rowlands, A, Kidner, P, and Condon, P. 2010. Yellow-nosed Albatross: new to Britain. *Br Birds* 103: 376-384.

Rowlands, A. 2010. Proposed criteria for BBRC assessment of claims of 'Amur Wagtail'. *Br Birds* 103: 268-275.

Ruggieri, L, and Festari, I. 2005. *A Birdwatcher's Guide to Italy*. Lynx Edicions, Barcelona.

Ruokonen, M, and Aarvak, T. 2011. Typology revisited: historical taxa of the bean goose-pink-footed goose complex. *Ardea* 99: 103-112.

Ruokonen, M, Litvin, K, and Aarvak, T. 2008. Taxonomy of the bean goose-pink-footed goose. *Mol Phylogenet Evol* 48: 554-562.

Ryabitsev, V K, and Wilson, M. 1999. Range extension of Long-tailed Rosefinch into the Western Palearctic. *Br Birds* 92: 498-503.

Ryall, C. 2002. Further records of range extension in the House Crow *Corvus splendens*. *Bull Brit Ornithol Club* 122: 231-240.

Ryall, C. 2003. Notes on the ecology and behavior of

House Crows at Hoek van Holland. *Dutch Birding* 25: 167-172.
Ryall, C. 2010. Further records and updates of range extension in House Crow *Corvus splendens*. *Bull Brit Ornithol Club* 130: 246-254.

S

Sætre, G-P, Borge, T, and Moum, T. 2001a. A new bird species? The taxonomic status of 'the Atlas Flycatcher' assessed from DNA sequence analysis. *Ibis* 143: 494-497.
Sætre, G-P, Borge, T, Lindell, J, Moum, T, Primmer, C R, Sheldon, B C, Haavie, J, Johnsen, A, and Ellegren, H. 2001. Speciation, introgressive hybridization and nonlinear rate of molecular evolution in flycatchers. *Mol Ecol* 10: 737–749.
Saitoh, T, Nishiumi, I, Shigeta, Y, and Ueda, K. 2012. Re-examination of the taxonomy of the Arctic Warbler *Phylloscopus borealis* (Blasius): three separate species within the *Phylloscopus* [*borealis*] superspecies. *Jap J Ornithol* 61: 46-59.
Salewski, V, and Herremans, M. 2006. Phenology of Western Olivaceous Warbler *Hippolais opaca* and Eastern Olivaceous Warbler *Hippolais pallida reiseri* on stopover sites in Mauritania. *Ring Migr* 23: 15-20.
Salewski, V, Schmaljohann, H, and Herremans, M. 2005. New bird records from Mauritania. *Malimbus* 27: 19-32.
Salewski, V, Stark, H, and Leisler, B. 2009. Olivaceous Warblers in southeast Morocco. *Br Birds* 102: 116-121.
Salim, M A, Al-Sheikhly, O F, Majeed, K A, and Porter, R F. 2012. An annotated checklist of the birds of Iraq. *Sandgrouse* 34: 4-43.
Salzburger, W, Martens, J, and Sturmbauer, C. 2002. Paraphyly of the Blue Tit (*Parus caeruleus*) suggested from cytochrome *b* sequences. *Mol Phylogenet Evol* 24: 19-25.
Sánchez, B, González, L, and Barov, B. 2008. Action plan for the Spanish Imperial Eagle *Aquila adalberti* in the European Union. BirdLife International for the European Commission.
Sangster G, Hazevoet, C J, van den Berg, A B, and Roselaar, C S. 1998. Dutch avifaunal list: species concepts, taxonomic instability, and taxonomic changes in 1998. *Dutch Birding* 20: 22-32.
Sangster, G, Alström, P, Forsmark, E, and Olsson, U. 2010. Multi-locus phylogenetic analysis of Old World chats and flycatchers reveals extensive paraphyly at family, subfamily and genus level (Aves: Muscicapidae). *Mol Phylogenet Evol* 57: 380-392.
Sangster, G, and Oreel, G J. 1996. Progress in taxonomy of Taiga and Tundra Bean Geese. *Dutch Birding* 18: 310-316.
Sangster, G, Collinson, J M, Crochet, P-A, Kirwan, G M, Knox, A G, Parkin, D T, and Votier, S C. 2015. Taxonomic recommendations for Western Palaearctic birds: 10th report. *Ibis* 157: 193-200.
Sangster, G, Collinson, J M, Crochet, P-A, Knox, A G, Parkin, D T, Svensson, L, and Votier, S C. 2005. Taxonomic recommendations for British birds: third report. *Ibis* 147: 821-826.
Sangster, G, Collinson, J M, Crochet, P-A, Knox, A G, Parkin, D T, Svensson, L, and Votier, S C. 2012. Taxonomic recommendations for British birds: eighth report. *Ibis* 154: 874-883.
Sangster, G, Collinson, J M, Crochet, P-A, Knox, A G, Parkin, D T, Svensson, L, and Votier, S C. 2013. Taxonomic recommendations for Western Palearctic birds: ninth report. *Ibis* 155: 898-907.
Sangster, G, Collinson, J M, Crochet, P-A, Knox, A G, Parkin, D T, Svensson, L, and Votier, S C. 2011. Taxonomic recommendations for British birds: seventh report. *Ibis* 153: 883-892.
Sangster, G, Collinson, J M, Helbig, A J, Knox, A G, and Parkin, D T. 2004. Taxonomic recommendations for British birds: second report. *Ibis* 146: 153-157.
Sangster, G, Collinson, J M, Helbig, A J, Knox, A G, and Parkin, D T. 2002. The generic status of Black-browed Albatross and other albatrosses. *Br Birds* 95: 583-585.
Sangster, G, Collinson, J M, Helbig, A J, Knox, A G, and Parkin, D T. 2002. The specific status of Balearic and Yelkouan Shearwaters. *Br Birds* 95: 636-639.
Sangster, G, Collinson, J M, Knox, A G, Parkin, D T, and Svensson, L. 2007. Taxonomic recommendations for British birds: fourth report. *Ibis* 149: 853-857.
Sangster, G, Collinson, J M, Knox, A G, Parkin, D T, and Svensson, L. 2010. Taxonomic recommendations for British birds: sixth report. *Ibis* 152: 180-186.
Sangster, G, Collinson, M, Helbig, A J, Knox, A G, Parkin, D T, and Prater, T. 2001. The taxonomic status of Green-winged Teal *Anas carolinensis*. *Br Birds* 94: 218-226.
Sangster, G, Knox, A G, Helbig, A J, and Parkin, D T. 2002. Taxonomic recommendations for British birds. *Ibis* 144: 153-159.
Sangster, G, Rodríguez-Godoy, F, Roselaar, C S, Robb, M S, and Luksenburg, J A. 2015. Integrative taxonomy reveals Europe's rarest songbird species, the Gran Canaria Blue Chaffinch *Fringilla polatzeki*. *J Avian Biol*. DOI 10.1111/jav.00825.
Sangster, G, van den Berg, A B, van Loon, A J, and Roselaar, C S. 2009. Dutch avifaunal list: taxonomic changes in 2004-2008. *Ardea* 97: 373-381.
Sangster, G. 1996. How many species of crossbill are there? *Dutch Birding* 18: 29-32.
Sangster, G. 1996. Species limits in the Blue Tit complex: new evidence from playback studies. *Dutch Birding* 18: 85-88.
Sangster, G. 1998. Purple Swamp-hen is a complex of species. *Dutch Birding* 20: 13-22.
Sangster, G. 2000. Genetic distance as a test of species boundaries in the Citril Finch *Serinus citrinella*: a critique and taxonomic reinterpretation. *Ibis* 142: 487-490.
Sangster, G. 2008. A new genus for the waterthrushes (Parulidae). *Bull Brit Ornithol Club* 128: 212-215.
Sapir, N, on behalf of the Israeli Rarities and Distribution Committee (IRDC). 2007. Bulletin 6:01 on Rare Birds in Israel: Updated National Bird Species List. http://www.israbirding.com/irdc/bulletins/bulletin_6/.
Sapir, N, Smith, J P, and Shochat, E, on behalf of the Israeli Rarities and Distribution Committee (IRDC). 2005. Bulletin 4:01 on Rare Birds in Israel. http://www.israbirding.com/irdc/bulletins/bulletin_4/.

Saunders, C M, and Webbon, G G. 1998. The first Banded Martin *Riparia cincta* in Saudi Arabia. *Sandgrouse* 20: 55.

Saunders, P. 2012. The Pale-legged Leaf Warbler in Dorset – a new Western Palearctic bird. *Birding World* 25: 428-429.

Saveljić, D, and Jovićević, M. 2015. *Popis ptica Crne Gore sa bibliografijom*. Centar za zaštitu i proučavanje ptica, Podgorica.

Savioz, J. 2012. Premiere mention Suisse du Bruant chanteur *Melospiza melodia*. *Nos Oiseaux* 59: 53-54.

Saygili, F, Yigit, N, Çam, P, and Yüce, D. 2013. Morphometric and allozymic differences between Bearded Tit *Panurus biarmicus* (Aves: Passeriformes) subpopulations in a large wetland and a small pond in central Anatolia, Turkey. *Turk J Zool* 37: 149-156.

Scheider, J, Wink, M, Stubbe, M, Hille, S, and Wiltschko, W. 2004. Phylographic relationships of the Black Kite *Milvus migrans*. In: Chancellor, R D, and Meyburg, B-U (eds). *Raptors Worldwide*. WWGBP/MME, Budapest.

Scheider, J, Wink, M, Wiltschko, W, and Stubbe, M. 2009. Zur Phylogenie und Taxonomie der Schwarzmilan subspezies. On the Phylogeny and Taxonomy of the subspecies of the Black Kite. *Populationsekölogie Greifvogel und Eulenarten*. Bd 6: 233-241.

Schodde, R, Kirwan, G M, and Porter, R. 2012. Morphological differentiation and speciation among darters (*Anhinga*). *Bull Brit Ornithol Club* 132: 283-294.

Schweizer, M, and Shirihai, H. 2013. Phylogeny of the *Oenanthe lugens* complex (Aves, Muscicapidae: Saxicolinae): Paraphyly of a morphologically cohesive group within a recent radiation of open-habitat chats. *Mol Phylogenet Evol* 69: 450-461.

Schweizer, M, Ayé, R, and Burri, R. 2006. Afghan Babbler – a little known Western Palearctic bird. *Birding World* 19: 69-70.

Šćiban, M, Rajković, D, Radišić, D, Vasić, V, and Pantović, U. 2015. *Birds of Serbia – Critical List of Species*. Institute for Nature Conservation of Vojvodina Province and Bird Protection and Study Society of Serbia, Novi Sad.

Sclater, P L. 1858. On the general geographical distribution of the members of the class Aves. *J Proc Linn Soc* 2: 130-145.

van der Schot, W J M. 1989. Atlantic Petrel in Israel in April 1989. *Dutch Birding* 11:170-172.

Scott, D A, and Adhami, A. 2006. An updated checklist of the birds of Iran. *Podoces* 1: 1-16.

Scott, D A. 2008. Rare birds in Iran in the late 1960s and 1970s. *Podoces* 3: 1-30.

Scott, D A. 2014. *Updates to the Iranian List since Scott and Adhami (2006)*. Unpublished paper.

Scott, M, Siddle, J, and Shirihai, H. 1999. Little known Western Palearctic birds: Mountain Chiffchaff. *Birding World* 12: 163-167.

Scuderi, A, and Corso, A. 2011. Crested Honey Buzzard in Europe. *Birding World* 24: 252-256.

Sehhatisabet, M. 2007. A preliminary survey of the distribution of Pleske's Ground Jay *Podoces pleskei* in Iran, with some information on numbers. *Podoces* 2: 37-51.

Seibold, I, Helbig, A J, Meyburg, B-U, Negro, J J, and Wink, M. 1996. Genetic differentiation and molecular phylogeny of European *Aquila* eagles according to cytochrome b nucleotide sequences. *Eagle Studies* 1-15. WWGBP, Berlin.

Seoane, J, Kouri, A, Illera, J C, Palomino, D, Alonso, C L, and Carrascal, L M. 2010. New data on the population, distribution and habitat preferences of the Canary Islands stonechat *Saxicola dacotiae*. *Ardeola* 57: 387-405.

Serra, G, Abdallah, M, Assaed, A, Abdallah, A, Al Qiam, G, Fayad, T, and Williamson, D. 2004. Discovery of a relict breeding colony of Northern Bald Ibis *Geronticus eremita* in Syria. *Oryx* 38: 106-108.

Shaw, D N. 2008. Chestnut-eared Bunting on Fair Isle: new to Britain. *Br Birds* 101: 235-240.

Shealer, D. 1999. Sandwich Tern (*Thalasseus sandvicensis*). *The Birds of North America Online* (A. Poole, Ed). Cornell Lab of Ornithology, Ithaca; retrieved from the Birds of North America Online. http://bna.birds.cornell.edu.

Sheldon, F H, Whittingham, L A, Moyle, R G, Slikas, B, and Winkler, D W. 2005. Phylogeny of swallows (Aves: Hirundinidae) estimated from nuclear and mitochondrial genes. *Mol Phylogenet Evol* 35: 253-270.

Sheldon, F H. 1987. The phylogeny of herons estimated from DNA-DNA hybridization data. *Auk* 104: 33-43.

Shields, G F. 1990. Analysis of mitochondrial DNA of Pacific Black Brant. *Auk* 107: 620-623.

Shirihai, H, and Golan, Y. 1994. First records of Long-tailed Shrike *Lanius schach* in Israel and Turkey. *Sandgrouse* 16: 36-40.

Shirihai, H, Gargallo, G, and Helbig, A J. 2001. *Sylvia Warblers. Identification, taxonomy and phylogeny of the genus Sylvia*. Christopher Helm, London.

Shirihai, H, Gargallo, G, Helbig, A J, Harris, A, and Cottridge, D. 2002. Identification, plumages, taxonomy and biology of Arabian Warbler *Sylvia leucomelaena*. *Sandgrouse* 24: 88-104.

Shirihai, H, Kirwan, G M, and Helbig, A J. 2011. A new taxon in the Mourning Wheatear complex. *Bull Brit Ornithol Club* 131: 270-291.

Shirihai, H. 1996. *The Birds of Israel*. Academic Press, London.

Shirihai, H. 1999. Fifty species new to Israel: their discovery and documentation, with tips on identification. *Sandgrouse* 21: 45-105.

Shirihai, H. 2000. The first Southern Pochard *Netta erythrophthalma* in Israel and the Western Palearctic. *Sandgrouse* 22: 131-133.

Shirihai, H. 2001. The Mascarene Shearwater. *Birding World* 14: 78-85.

Shirihai, H. 2012. Correcting the identification of two rare wheatear records in Israel. *Bull Brit Ornithol Club* 132: 226-235.

Shochat, E, and Lerman, S B, on behalf of the Israeli Rarities and Distribution Committee (IRDC). 2012. Bulletin 8:01 on Rare Birds in Israel. http://www.israbirding.com/irdc/bulletins/bulletin_8/.

Shochat, E, and Sapir, N, on behalf of the Israeli Rarities and Distribution Committee (IRDC). 2004. Bulletin 3:01 on Rare Birds in Israel (1995-2001). http://www.israbirding.com/irdc/bulletins/bulletin_3/.

Shochat, E, and Smith, J P, on behalf of the Israeli Rarities and Distribution Committee (IRDC). 2010. Bulletin 7:01 on Rare Birds in Israel. http://www.israbirding.com/irdc/bulletins/bulletin_7/.

Sibley, D. 2014. *The North American Bird Guide*. Second edition. Bloomsbury, London.

Silva, M C, Matias, R, Ferreira, V, Catry, P, and Granadeiro, J P. Searching for a breeding population of Swinhoe's Storm-petrel at Selvagem Grande, NE Atlantic, with a molecular characterization of occurring birds and relationships within the Hydrobatinae. *J Ornithol*. DOI 10.1007/s10336-015-1257-7.

Simeonov, P, Nagendran, M, Michels, E, Possardt, E, and Vangeluwe, D. 2014. Red-breasted Goose: satellite tracking, ecology and conservation. *Dutch Birding* 36: 73-86.

Simmons, R E. 2000. *Harriers of the World*. Oxford University Press, Oxford.

Simon, P. 1965. Synthèse de l'avifaune du massif montagneux du Tibesti, et distribution géographique de ces espèces en Afrique du nord et environs. *Gerfaut* 55: 26-71.

Sinclair, I, and Ryan, P. 2003. *Birds of Africa south of the Sahara*. Struik, Cape Town.

Slack, R. 2009. *Rare Birds Where and When*. Volume 1: sandgrouse to New World orioles. Rare Bird Books, York.

Slager, D L, Battey, C J, Bryson Jr, R W, Voelker, G, and Klicka, J. 2014. A multilocus phylogeny of a major New World avian radiation: The Vireonidae. *Mol Phylogenet Evol* 80: 95-104.

Slaterus, R, and van der Spek, V. 2013. Recente meldingen. *Dutch Birding* 35: 263-274.

Slaterus, R. 2013. Common Grackle migrating over Kamperhoek in the Netherlands in April 2013. www.ortolaantje.nl/dutchcommongrackle.html.

Slaterus, R. 2014. Glanstroepiaal over Kamperhoek in april 2013. *Dutch Birding* 36: 312-315.

Sluys, R. 1983. Geographical variation and distribution of the bearded tit *Panurus biarmicus* (Linnaeus, 1758). *Bijdragen tot de Dierkunde* 53: 13-36.

Smiles, M. 2016. Large-billed Leaf Warbler *Phylloscopus magnirostris* in the United Arab Emirates – first record for Arabia. *Sandgrouse* 38: 201-204.

Smith, J P, and the Israeli Rarities and Distribution Committee. 2001. Bulletin 1:02 on Rare Birds in Israel (1995-2001). http://www.israbirding.com/irdc/bulletins/bulletin_2/.

Smith, J P, and the Israeli Rarities and Distribution Committee. 2002. Bulletin 2:01 on Rare Birds in Israel (1995-2002). http://www.israbirding.com/irdc/bulletins/bulletin_2/.

Smith, J P. 2001. Bulletin on Rare Birds in Israel (1989-2000). http://www.israbirding.com/irdc/bulletins/bulletin_1/.

Smith, L. 2012. Two Shikra *Accipiter badius* records from Turkey. *Sandgrouse* 34: 69-71.

Smith, R D. 1996. Racial composition of breeding and wintering Snow Buntings *Plectrophenax nivalis* in the north-east Scottish uplands. *Ring Migr* 17: 123-136.

Snow, D W (ed). 1978. *An Atlas of Speciation in African Non-passerine Birds*. British Museum (Natural History), London.

Snow, D W, and Perrins, C M. 1998a. *The Birds of the Western Palearctic*. Concise Edition. Vol 1: Non-Passerines. Oxford University Press, Oxford.

Snow, D W, and Perrins, C M. 1998b. *The Birds of the Western Palearctic*. Concise Edition. Vol 2: Passerines. Oxford University Press, Oxford.

Soldaat, E, Leopold, M F, Meesters, E H, and Robertson, C J R. 2009. Albatross mandible at archeological site in Amsterdam, the Netherlands, and WP records of *Diomedea* albatrosses. *Dutch Birding* 31: 1-16.

Šrank, V. 2010. 10. správa Faunistickej komisie Slovenskej ornitologickej spoločnosti/BirdLife Slovensko. *Tichodroma* 22: 131-134.

Šrank, V. 2011: 11. správa Faunistickej komisie Slovenskej ornitologickej spoločnosti/BirdLife Slovensko. *Tichodroma* 23: 77-79.

Stanton, D B. 1998. The first Speckled Pigeon *Columba guinea* in Yemen and the Middle East. *Sandgrouse* 20: 145.

Stanton, D B. 2000. The first Fulvous Whistling-Duck *Dendrocygna bicolor* in Yemen and the Middle East. *Sandgrouse* 22: 130-131.

Stanton, D B. 2005. The first Black-billed Wood Doves *Turtur abyssinicus* in Yemen and the Middle East. *Sandgrouse* 27: 147.

Stattersfield, A J, Crosby, M J, Long, A J, and Wege, D. 1998. *Endemic Bird Areas of the World. Priorities for Biodiversity Conservation*. BirdLife Conservation Series No. 7. BirdLife International, Cambridge.

Steele, J. 2006. From the Rarities Committee's files: do we know what British 'soft-plumaged petrels' are? *Br Birds* 99: 404-419.

van Steenis, J H. 2008. *Galerida* galore – speciation in 'crested larks'. *Dutch Birding* 30: 238-239.

van Steenis, J H. 2014. Taxonomy of 'horned larks'. *Dutch Birding* 36: 188-194.

Steffen, B. 2010. African Openbill at Luxor, Egypt, in May 2009. *Dutch Birding* 32: 254-256.

Steijn, L B. 2005. Eastern Black Redstarts at Ijmuiden, the Netherlands, and on Guernsey, Channel Islands, in October 2003, and their identification, distribution and taxonomy. *Dutch Birding* 27: 171-194.

Steijn, L. 2012. White-tailed Tropicbird off Santiago, Cape Verde Islands, in May 2011. *Dutch Birding* 34: 312-313.

Stewart, J R. 2007. The fossil and archaeological record of the Eagle Owl in Britain. *Br Birds* 100: 481-486.

Stoddart, A. 2011. An apparent Hudsonian Dunlin in Norfolk. *Birding World* 24: 208-210.

Stoddart, A. 2013. Redpolls: a review of their taxonomy, identification and British status. *Br Birds* 106: 708-736.

Stoddart, A. 2014. Assessing and recording Subalpine Warblers. *Br Birds* 107: 420-424.

Strandberg, R. 2013. Ageing, sexing and subspecific identification of Osprey, and two WP records of American Osprey. *Dutch Birding* 35: 69-87.

Sturke, C. 1993. Trans-Atlantic hitch-hike of Snowy Owl. *Sea Swallow* 42: 70.

Suárez, N M, Betancor, E, Klassert, T E, Almeida, T, Hernandez, M, and Pestano, J J. 2009. Phylogeography and genetic structure of the Canarian common chaffinch (*Fringilla coelebs*) inferred with mtDNA and microsatellite loci. *Mol Phylogenet Evol* 53: 556-564.

Summers, R W, and Buckland, S T. 2011. A first survey of the global population size and distribution of the

Scottish Crossbill *Loxia scotica*. *Bird Conserv Int* 21: 186-198.

Summers, R W, Dawson, R J G, and Phillips, R E. 2007. Assortative mating and patterns of inheritance indicate that the three crossbill taxa in Scotland are species. *J Avian Biol* 38: 153-162.

Summers, R W, Jardine, D C, Marquiss, M, and Proctor, R. 1996. The biometrics of invading Common Crossbills *Loxia curvirostra* in Britain during 1990-1991. *Ring Migr* 17: 1-10.

Svensson, L, Collinson, M, Knox, A G, Parkin, D T, and Sangster, G. 2005. Species limits in the Red-breasted Flycatcher. *Br Birds* 98: 538-541.

Svensson, L, Mullarney, K, and Zetterström, D. 2009. *Collins Bird Guide*. Second edition. HarperCollins, London.

Svensson, L, Shirihai, H, Frahnert, S, and Dickinson, E. 2012. Taxonomy and nomenclature of the Stonechat complex *Saxicola torquata sensu lato* in the Caspian region. *Bull Brit Ornithol Club* 132: 260-269.

Svensson, L. 2001. Identification of Western and Eastern Olivaceous, Booted and Sykes's Warblers. *Birding World* 14: 192-219.

Svensson, L. 2001. The correct name of the Iberian Chiffchaff *Phylloscopus ibericus* Ticehurst 1937, its identification and new evidence of its winter grounds. *Bull Brit Ornithol Club* 121: 281-296.

Svensson, L. 2012. A new subspecies of Western Orphean Warbler *Sylvia hortensis* and criteria for separating Western from Eastern Orphean Warbler *S crassirostris*. *Bull Brit Ornithol Club* 132: 75-83.

Svensson, L. 2013. A taxonomic revision of the Subalpine Warbler *Sylvia cantillans*. *Bull Brit Ornithol Club* 133: 240-248.

Sveriges Ornitologiska Förening. 2015. Förteckning över Sveriges fågeltaxa. http://www.sofnet.org/rk/raritetskatalogen.

Sveriges Ornitologiska Förening. 2015. Raritetskatalogen/Rare bird catalog. http://www.sofnet.org/rk/raritetskatalogen.

Swiss Ornithological Institute. 2011. Checklist of the birds of Switzerland 2010. Downloaded from http://www.vogelwarte.ch.

Symens, P, and Werner, M. 1994. First record of Senegal Thick-knee *Burhinus senegalensis* in Arabia. *Bull Ornithol Soc Middle East* 33: 12-14.

Symens, P, Suhaibani, A, and Eichacker, X. 1994. The occurrence of Pallas's Fish Eagle *Haliaeetus leucoryphys* in Arabia. *Bull Ornithol Soc Middle East* 33: 3-6.

T

Tajkova, S U, and Red'Kin, A A. 2014. The Northern Shrike *Lanius borealis sibiricus* Bogdanov, 1881 (Aves: Laniidae) in Ukraine: a taxonomic assessment. *J Nat Mus (Prague), Nat Hist Ser* 89: 107.

Tajkova, S Û. 2012. Type specimens and taxonomic identity of *Lanius excubitor stepensis* Gavrilenko, 1928 (Aves: Laniidae). *J Nat Mus* 181: 73-93.

Takhtadzhyan, A L. 1978. *Floristicheskie oblasti Zemli*. Akad. Nauk. SSSR, Leningradskoe otd, Leningrad.

Tarsiger.com. http://www.tarsiger.com/news/index.php?sp=wp&lang=eng.

Taylor, B, and van Perlo, B. 1998. *Rails. A Guide to the Rails, Crakes, Gallinules and Coots of the World*. Pica Press, Robertsbridge.

Tella, J L, Vögelli, M, Serrano, D, and Carrete, M. 2005. Current status of the threatened Dupont's lark *Chersophilus duponti* in Spain: overestimation, decline, and extinction of local populations. *Oryx* 39: 90- 94.

The Gibraltar Ornithological and Natural History Society. 2016. *The Gibraltar Bird List: Bird Species Observed in or from the Territory of Gibraltar*. http://www.gonhs.org/Birds_000.htm.

Thévenot, M, Vernon, R, and Bergier, P. 2003. *The Birds of Morocco*. BOU Checklist Series: 20. British Ornithologists' Union and British Ornithologists' Club, Tring.

Thibault, J-C, and Bonaccorsi, G. 1999. *The Birds of Corsica*. BOU Checklist series: 17. British Ornithologists' Union, Tring.

Thorne, R and S. 2014. 'Asian Red-rumped Swallow' in Orkney and Highland: new to Britain. *Br Birds* 107: 405-412.

Thorup, K. 2001. First record of Pallid Swift *Apus pallidus* in Denmark and of ssp. *illyricus* in northern Europe. *Dansk Ornitol Foren Tidsskr* 95: 169-172.

Tietze, D T, Martens, J, and Sun, Y-H. 2006. Molecular phylogeny of treecreepers (*Certhia*) detects hidden diversity. *Ibis* 148: 477-488.

Tietze, D T, Päckert, M, Martens, J, Lehmann, H, and Sun, Y-H. 2013. Complete phylogeny and historical biogeography of true rosefinches (Aves: Carpodacus). *Zool J Linn Soc* 169: 215-234.

Tipper, R, and Wilson, K M. 2007. Moussier's Redstart at Cabo de São Vicente, Portugal in November 2006-January 2007. *Dutch Birding* 29: 297-301.

Tobias, J A, Seddon, N, Spottiswoode, C N, Pilgrim, J D, Fishpool, L D C, and Collar, N J. 2010. Quantitative criteria for species delimitation. *Ibis* 152: 724-746.

Toews, D P L, Mandic, M, Richards, J G, and Irwin, D E. 2013. Migration, mitochondria, and the Yellow-rumped Warbler. *Evolution* 68: 241-255.

Toms, M P. 2009. Eagle Owls in Britain: origins and conservation implications. *Brit Wildlife* 20: 405-412.

Töpfer, T, Haring, E, Birkhead, T R, Lopes, R J, Severinghaus, L L, Martens, J, and Päckert, M. 2011. A molecular phylogeny of bullfinches *Pyrrhula* Brisson, 1760 (Aves: Fringillidae). *Mol Phylogenet Evol* 58: 271-282.

Töpfer, T. 2006. The taxonomic status of the Italian Sparrow – *Passer italiae* (Vieillot 1817): Speciation by stabilised hybridisation? A critical analysis. *Zootaxa* 1325: 117-145.

Trimbos, K B, Doorenweerd, C, Kraaijeveld, K, Musters, C J M, Groen, N M, de Knijff, P, Piersma, T, and de Snoo, G R. 2014. Patterns in nuclear and mitochondrial DNA reveal historical and recent isolation in the Black-tailed Godwit (*Limosa limosa*). *PLoS ONE* 9: e83949. doi:10.1371/journal.pone.0083949.

Turner, A, and Rose, C. 1989. *A Handbook to the Swallows and Martins of the World*. Christopher Helm, London.

Tveit, B O. 2011. Identification of Asian Red-rumped Swallow and the first records for Norway and Britain. *Birding World* 24: 327-341.

U

Udvardy, M D F. 1975. *A Classification of the Biographical Provinces of the World*. IUCN Occasional Paper No 18. International Union for Conservation of Nature and Natural Resources, Morges.

UICN/PACO. 2010. *Parcs et* réserves du Niger: évaluation de l'efficacité de gestion des aires protégées. Ouagadougou, BF.

Ullman, M. 2000. The limits of the Western Palearctic. *Sandgrouse* 22: 74-75.

Urban, E K, Fry, C H, and Keith, S (eds). 1986. *The Birds of Africa*. Vol II. Academic Press, London.

Urban, E K, Fry, C H, and Keith, S (eds). 1997. *The Birds of Africa*. Vol V. Academic Press, London.

Urquhart, E. 2002. *Stonechats. A Guide to the Genus Saxicola*. Christopher Helm, London.

V

Valkama, J, Vepsäläinen, V, and Lehikoinen, A. 2011. *The Third Finnish Breeding Bird Atlas*. Finnish Museum of Natural History and Ministry of Environment. http://atlas3.lintuatlas.fi/english.

Vallotton, L, and Piot, B. 2010. Oiseaux rares et observations inhabituelles en Suisse en 2009. 19e rapport de la Commission de l'Avifaune Suisse. *Nos Oiseaux* 57: 283-301.

Vallotton, L, Piot, B, and Schneider, F. 2014. Oiseaux rares et observations inhabituelles en Suisse en 2013. 23e rapport de la Commission de l'Avifaune Suisse. *Nos Oiseaux* 61: 201-222.

Vandevyvre, X. 2005. Le Venturon montagnard (*Serinus citrinella*): son statut en Belgique. *Aves* 42: 265-266.

Vanhove, F, and Raty, L. 2015. Rapport de la Commission d'Homologation Années 2010 à 2013. *Aves* 52: 65-106.

Vaurie, C. 1959. *The Birds of the Palearctic Fauna*. Passeriformes. Witherby, London.

Vaurie, C. 1965. *The Birds of the Palearctic Fauna*. Non-Passeriformes. Witherby, London.

Vavřík, M, and FK ČSO. 2015. Checklist of birds of the Czech Republic. http://fkcso.cz/index-e.html.

Vavřík, M, and FK ČSO. 2015. Zpráva Faunistické komise ČSO za rok 2014. *Sylvia* 51: 97-116.

Velevski, M, Andevski, J, Uzunova, D, and Nikolov, S. 2007. *The Birds of Macedonia*. Macedonian Ecological Society, Skopje.

Velevski, M. 2012. Non-critical checklist of birds of Macedonia. http://www.mes.org.mk/PDFs/Other/Checklist%20of%20birds%20of%20Macedonia.pdf.

Viellard, J. 1976. La Sitelle kabyle. *Alauda* 44 (3, suppl. Spec): 351-352.

Viellard, J. 1978. Le Djebel Babour et sa Sitelle *Sitta ledanti* Viellard 1976. *Alauda* 46: 1-42.

Viellard, J. 1980. Remarques complémentaires sur la Sitelle kabyle *Sitta ledanti* Viellard 1976. *Alauda* 48: 139-150.

Vinicombe, K E V. 2007. The status of Red-headed Bunting in Britain. *Br Birds* 100: 540-551.

Vinicombe, K E, and Harrop, A H J. 1999. Ruddy Shelducks in Britain and Ireland 1986-1994. *Br Birds* 92: 225-255.

Violani, C G, and Massa, B. 1993. Extinction of the Andalusian Hemipode *Turnix s sylvatica* (Desf.) in the Mediterranean region. *Bull Brit Ornithol Club* 113: 225-229.

Voelker, G, and Klicka, J. 2008. Systematics of *Zoothera* thrushes, and a synthesis of true thrush molecular systematic relationships. *Mol Phylogenet Evol* 49: 377-381.

Voelker, G. 2002. Systematics and historical biogeography of wagtails: dispersal versus vicarience revisited. *Condor* 104: 725-739.

Voisin, J-F. 1994. Colin de Virginie *Colinus virginianus*. In: Yeatman-Berthelot, D, and Jarry, G. Eds. *Nouvel Atlas des Oiseaux Nicheurs de France 1985-1989*. Société Ornithologique de France, Paris: 240-241.

Voous, K H. 1977. *List of Recent Holarctic Bird Species*. British Ornithologists' Union, London.

Votier, S C, Bearhop, S, Attrill, M J, and Oro, D. 2008. Comment: is climate change the most likely driver of range expansion for a critically endangered top predator in northeast Atlantic waters? *Br Birds* 100: 392-406.

Votier, S C, Bearhop, S, Newell, R G, Orr, K, Furness, R W, and Kennedy, M. 2004. The first record of Brown Skua *Catharacta antarctica* in Europe. *Ibis* 146: 95-102.

Votier, S C, Bowen, G J, and Newton, J. 2009. Stable-hydrogen isotope analysis suggests natural vagrancy of Baikal Teal to Britain. *Br Birds* 102: 697-699.

Votier, S C, Harrop, A H J, and Denny, M. 2003. A review of status and identification of American Wigeon in Britain and Ireland. *Br Birds* 96: 2-22.

Votier, S C, Kennedy, M, Bearhop, S, Newell, R G, Griffiths, K, Whitaker, H, Ritz, M S, and Furness, R W. 2007. Supplementary DNA evidence fails to confirm presence of Brown Skuas *Stercorarius antarctica* in Europe: a retraction of Votier *et al* (2004). *Ibis* 149: 619-621.

W

Waheed, A. 2016. Egyptian record of Wahlberg's Eagle *Hieraaetus wahlbergi* is the first in the Western Palearctic and Middle East. *Sandgrouse* 38: 96-98.

Walbridge, G, Small, B, and McGowan, R Y. 2003. From the Rarities Committee's files: Ascension Frigatebird on Tiree – new to the Western Palearctic. *Br Birds* 96: 58-73.

Wallace, A R. 1876. *The Geographic Distribution of Animals*. Vols I and II. Macmillan, London.

Wallace, D I M, Bradshaw, C, and Rogers, M J, on behalf of BBRC. 2006. A review of the 1950-57 British rarities. *Br Birds* 99: 460-464.

Walther, B. 2006. The winter distribution and habitat use of the near-threatened Cinereous Bunting *Emberiza cineracea*. *Sandgrouse* 28: 52-57.

Waltho, C, and Coulson, J. 2015. *The Common Eider*. T & A D Poyser, London.

Walton, C. 2011. Eastern nominate Nightingales *Luscinia m. megarhynchos* in Cyprus in 2011. *Sandgrouse* 33: 174-176.

Warr, E. 1992. *A Provisional List of the Birds of South Yemen (formerly Aden Protectorate and People's Democratic Republic of Yemen, now part of the*

Republic of Yemen. Socotra excluded). OSME Report.

Wassink, A, and Oreel, G J. 2007. *The Birds of Kazakhstan.* De Cocksdorp, Texel.

Wassink, A, and Oreel, G J. 2008. Birds of Kazakhstan: new and interesting data. *Dutch Birding* 30: 93-100.

Wassink, A. 2009. Birds of Kazakhstan: new and interesting data, part 2. *Dutch Birding* 31: 101-110.

Wassink, A. 2013. Birds of Kazakhstan: new and interesting data, part 4. *Dutch Birding* 35: 30-34.

Wassmer, S, Haag, C, and Vallotton, L. 2011. Oiseaux rares et observations inhabituelles en Suisse en 2010. 20[e] rapport de la Commission de l'Avifaune Suisse. *Nos Oiseaux* 58: 211-232.

Watson, J. 1997. *The Golden Eagle.* T & A D Poyser, London.

Weenink, R, van Duivendijk, N, and Ebels, E B. 2011. Spaanse Keizerarend over Loozerheide in mei 2007. *Dutch Birding* 33: 94-102.

Wells, D R, and Inskipp, T P. 2012. A proposed new genus of booted eagles (tribe Aquilini). *Bull Brit Ornithol Club* 132: 70-72.

Wernham, C, Toms, M, Marchant, J, Clark, J, Siriwardena, G, and Baillie, S (eds). 2002. *The Migration Atlas: movements of the birds of Britain and Ireland.* T & A D Poyser, London.

West African Bird Database. 2015. http://www.wabdab.org/.

White, S, and Kehoe, C. 2014. Report on scarce migrant birds in Britain in 2008-2010. Part 1: non-passerines. *Br Birds* 107: 142-176.

White, S, and Kehoe, C. 2014. Report on scarce migrant birds in Britain in 2008-2010. Part 2: passerines. *Br Birds* 107: 251-281.

Whittier, J B, Leslie, D M Jnr, and van den Bussche, R A. 2006. Genetic variation among subspecies of Least Tern (*Sterna antillarum*): implications for conservation. *Waterbirds* 29: 176-184.

van Wijk, R E, Schaub, M, Tolkmitt, D, Becker, D, and Hahn, S. 2013. Short-distance migration of Wrynecks *Jynx torquilla* from Central European populations. *Ibis* 155: 886-890.

Wilson, M G, and Korovin, V A. 2003. Oriental Turtle Dove breeding in the Western Palearctic. *Br Birds* 96: 234-241.

Wink, M, and Sauer-Gürth, H. 2004. Phylogenetic relationships in diurnal raptors based on nucleotide sequences of mitochondrial and nuclear marker genes. In: Chancellor, R D, and Meyburg, B-U. *Raptors Worldwide.* WWGBP/MME, Budapest.

Wink, M, El-Sayed, A-A, Sauer-Gürth, H, and Gonzalez, J. 2009. Molecular phylogeny of owls (Strigiformes) inferred from DNA sequences of the mitochondrial cytochrome b and the nuclear RAG-1 gene. *Ardea* 97: 581-591.

Wink, M, Sauer-Gürth, H, and Gwinner, E. 2002. Evolutionary relationships of stonechats and related species inferred from mitochondrial-DNA sequences and genomic fingerprinting. *Br Birds* 95: 349-355.

Wink, M, Sauer-Gürth, H, and Witt, H-H. 2004. Phylogenetic differentiation of the Osprey *Pandion haliaetus* inferred from nucleotide sequences of the mitochondrial cytochrome *b* gene. In: Chancellor, R D, and Meyburg, B-U (eds). *Raptors Worldwide.* WWGBP/MME, Budapest.

Wink, M, Seibold, I, Lotfikhah, F, and Bednarek, W. Molecular Systematics of Holarctic Raptors (Order Falconiformes). In Chancellor, R D, Meyburg, B-U, and Ferrero, J J (eds). 1998. *Holarctic Birds of Prey.* ADENEX-WWGBP.

Wink, M. 2008. Phylogenetic and phylogeographic relationships. In: van Nieuwenhuyse, D, Génot, J-C, and Johnson, D H (eds). The Little Owl: conservation, ecology and behaviour of *Athene noctua*, Cambridge, pp39-44.

Wink, M. 2011. Evolution und Phylogenie der Vögel-Taxonomisch Konsequenzen. *Vogelwarte* 49: 17-24.

Winkel, E, Amini, H, van Roomen, M, and Schelvis, J. 2010. Bird Counting in Iran in January 2009. *Dutch Birding* 32: 171-188.

Winkel, E. 2009. 'Southern skua' off La Palma, Canary Islands, in October 2005. *Dutch Birding* 31: 20-23.

Winkel, E. 2015. Italian Sparrow: hybridization with House Sparrow in northern Italy and rufous-breasted morph. *Dutch Birding* 37: 238-246.

Winkler, H, Christie, D A, and Nurney, D. 1995. *Woodpeckers. A Guide to the Woodpeckers, Piculets and Wrynecks of the World.* Pica Press, Robertsbridge.

Winkler, R, en collaboration avec la Commission de l'Avifaune Suisse. 1999. *Avifaune de Suisse. Nos Oiseaux* Supplément 3.

Winters, R. 2013. Snow Bunting: sexing, ageing and subspecies. *Dutch Birding* 35: 7-14.

Worfolk, T. 2000. Identification of red-backed, isabelline and brown shrikes. *Dutch Birding* 22: 323-362.

Wymenga, E, Engelmoer, M, Smit, C J, and van Spanje, T M. 1990. Geographical breeding origin and migration of waders wintering in West Africa. *Ardea* 78: 83-112.

Wynn, R B, and Shaw, K D. 2009. Madeiran Storm-petrels in the Bay of Biscay. *Br Birds* 102: 28-29.

Wynn, R B, and Yésou, P. 2007. The changing status of Balearic Shearwater in northwest European waters. *Br Birds* 100: 392-406.

Y

Yates, B. 2010. Least Tern in East Sussex: new to Britain and the Western Palearctic. *Br Birds* 103: 339-347.

Yésou, P, and Clergeau, P. 2005. Sacred Ibis: a new invasive species in Europe. *Birding World* 18: 517-526.

Yésou, P, Flohart, G, and Murdoch, D. 2007. First record of Basra Reed Warbler *Acrocephalus griseldis* for Syria. *Sandgrouse* 29: 214-215.

Yosef, R, and Tryjanowski, P. 2000. Phenology and biometric measurements of Woodchat Shrike (*Lanius senator*) at Eilat, Israel. *Ring* 22: 213-217.

Yosef, R, Rydberg-Hedaen, M, Mitchell, L, and Smit, H. The first breeding record of Striated Heron *Butorides striata* in Israel. *Sandgrouse* 24: 134-135.

Young, G, Lawlor, M, Paintin, T, Atkinson, M, and Watson, J. 2014. *A Working List of the Birds of the Channel Islands.* http://www.birdsontheedge.org/wp-content/uploads/2012/02/Working-list-of-CI-birds-to-Dec-2013.pdf.

Z

Zarudny, N A. 1911. Verzeichnis der Vögel Persiens. *Journal für Ornithologie* 59: 185-241.

Zekhuis, M, and Ghasemi, M. 2009. Status and distribution of Oriental White-eye in Iran. *Dutch Birding* 31: 230-231.

Zink, R M, and Dittmann, D L. 1993. Gene flow, refugia, and evolution of geographic speciation in the Song Sparrow (*Melospiza melodia*). *Evolution* 47: 717-729.

Zink, R M, and Wechstein, J D. 2003. Recent evolutionary history of the Fox Sparrows (Genus: *Passerella*). *Auk* 120: 522-527.

Zink, R M, and Kessen, A E. 1999. Species Limits in the Fox Sparrow. *Birding* 31: 508-517.

Zink, R M, Drovetski, S V, and Rohwer, S. 2002. Phylogeographic patterns in the great spotted Woodpecker *Dendrocopos major* across Eurasia. *J Avian Biol* 33: 175-178.

Zink, R M, Drovetski, S V, and Rohwer, S. 2006. Selective neutrality of mitochondrial ND2 sequences, phylogeography and species limits in *Sitta europaea*. *Mol Phylogenet Evol* 40: 679-686.

Zink, R M, Pavlova, A, Drovetski, S V, Wink, M, and Rohwer, S. 2009. Taxonomic status and evolutionary history of the *Saxicola torquata* complex. *Mol Phylogenet Evol* 52: 769-773.

Zink, R M, Pavlova, A, Rohwer, S, and Drovetski, S V. 2006. Barn swallows before barns: population histories and intercontinental colonization. *Proc R Soc Lond [Biol.]* 273: 1245-1251.

Zink, R M, Rising, J D, Mockford, S, Horn, A G, Wright, J M, Leonard, M, and Westberg, M C. 2005. Mitochondrial DNA variation, species limits, and rapid evolution of plumage coloration and size in the savannah sparrow. *Condor* 107: 21-28.

Zink, R M, Rohwer, S, Andreev, A V, and Dittmann, D L. 1995. Trans-Beringia comparisons of mitochondrial DNA differentiation in birds. *Condor* 97: 639-49.

Zink, R M. 1994. The Geography of Mitochondrial DNA Variation, Population Structure, hybridization, and Species Limits in the Fox Sparrow (*Passerella iliaca*). *Evolution* 48: 96-111.

Zink, R, Pavlova, A, Drovetski, S, and Rohwer, S. 2008. Mitochondrial phylogeographies of five widespread Eurasian bird species. *J Ornithol* 149: 399-413.

Zino, F, Biscoito, M, and Zino, P A. 1995. Birds of the archipelago of Madeira and the Selvagens: new records and checklist. *Bol Mus Mun Funchal* 47: 63-100.

Zino, F, Brown, R, and Biscoito, M. 2008. The separation of *Pterodroma madeira* (Zino's Petrel) from *Pterodroma feae* (Fea's Petrel) (Aves: Procellariidae). *Ibis* 150: 326-334.

Zino, F, Phillips, P, and Biscoito, M. 2011. Zino's Petrel movements at sea – a preliminary analysis of datalogger results. *Birding World* 24: 216-219.

Zoological Museum of Moscow University. 2015. Breeding bird atlas of European Russia. http://zmmu.msu.ru/en/about-muzeum/divisions/division-of-the-scientific-public-oriented-projects/breeding-bird-atlas-of-european-russia.

Zuccon, D, and Ericson, P G P. 2010. A multi-gene phylogeny disentangles the chat-flycatcher complex (Aves: Muscicapidae). *Zool Scr* 39: 213-224.

Zuccon, D, and Ericson, P G P. 2010. The *Monticola* rock-thrushes: phylogeny, and biogeography revisited. *Mol Phylogenet Evol* 55: 901-910.

Zuccon, D, Prŷs-Jones, R, Rasmussen, P C, and Ericson, P G P. 2012. The phylogenetic relationships and generic limits of finches (Fringillidae). *Mol Phylogenet Evol* 62: 581-596.

Zwarts, L, Bijlsma, R G, van der Kamp, J, and Wymenga, E. 2009. *Living on the Edge: Wetlands and Birds in a Changing Sahel.* KNNV Publishing, Zeist.

Žydelis, R, Lorentson, S-H, Fox, A D, Kuresoo, A, Krasnov, Y, Goryaev, Y, Bustnes, J O, Hario, M, Nilsson, L, and Stipniece, A. 2006. Recent changes in the status of Steller's Eider *Polysticta stelleri* wintering in Europe: a decline or redistribution? *Bird Conserv Int* 16: 217-236.

INDEX OF SPECIES

A

Abdim's Stork 59
Abyssinian Roller 137
Abyssinian White-eye 188
Acadian Flycatcher 147
Acanthis flammea 226
Acanthis hornemanni 226
Accipiter badius 76
Accipiter brevipes 77
Accipiter gentilis 77
Accipiter nisus 77
Acridotheres burmannicus 192
Acridotheres cristatellus 191
Acridotheres ginginianus 191
Acridotheres tristis 191
Acrocephalus agricola 177
Acrocephalus arundinaceus 176
Acrocephalus brevipennis 175
Acrocephalus dumetorum 177
Acrocephalus griseldis 175
Acrocephalus melanopogon 176
Acrocephalus orientalis 176
Acrocephalus paludicola 176
Acrocephalus palustris 178
Acrocephalus schoenobaenus 177
Acrocephalus scirpaceus 177
Acrocephalus stentoreus 176
Actitis hypoleucos 101
Actitis macularius 101
Aegithalos caudatus 171
Aegolius funereus 133
Aegypius monachus 73
Aenigmatolimnas marginalis 84
Aethia cristatella 122
Aethia psittacula 122
Afghan Babbler 181
African Blue Tit 159
African Collared Dove 125
African Crake 83
African Crimson-winged Finch 223
African Darter 70
African Desert Warbler 184
African Fish Eagle 79
African Grey Hornbill 140
African Grey Woodpecker 141
African Olive Pigeon 125
African Openbill 58
African Palm Swift 135
African Paradise Flycatcher 153
African Pied Wagtail 218
African Pipit 219
African Sacred Ibis 60
African Silverbill 215
African Skimmer 107
African Spoonbill 61
African Stonechat 205
African Swamphen 85
Agropsar sturninus 192
Aix galericulata 31
Aix sponsa 31
Alaemon alaudipes 161
Alauda arvensis 163
Alauda gulgula 162
Alauda leucoptera 162
Alauda razae 162
Alaudala cheleensis 166
Alaudala raytal 166
Alaudala rufescens 166
Alca torda 121
Alcedo atthis 138
Alder Flycatcher 147
Alectoris barbara 43
Alectoris chukar 43
Alectoris graeca 43
Alectoris melanocephala 44
Alectoris philbyi 43
Alectoris rufa 44
Aleutian Tern 117
Alexandrine Parakeet 146
Algerian Nuthatch 189
Alle alle 120
Allen's Gallinule 85
Alopochen aegyptiaca 30
Alpine Accentor 215
Alpine Chough 154
Alpine Swift 135
Amandava amandava 214
Amandava subflava 215
Amaurornis flavirostris 250
Amaurornis phoenicurus 83
American Bittern 61
American Black Duck 33
American Cliff Swallow 170
American Coot 86
American Golden Plover 91
American Herring Gull 113
American Kestrel 143
American Redstart 230
American Robin 197
American Wigeon 32
American Woodcock 94
American Yellow Warbler 231
Ammomanes cinctura 161
Ammomanes deserti 161
Ammoperdix griseogularis 44
Ammoperdix heyi 44
Amur Falcon 144
Amur Paradise Flycatcher 241
Anas acuta 34
Anas americana 32
Anas capensis 32
Anas carolinensis 34
Anas clypeata 33
Anas crecca 34
Anas cyanoptera 249
Anas discors 33
Anas erythrorhyncha 34
Anas falcata 32
Anas formosa 34
Anas penelope 32
Anas platyrhynchos 33
Anas querquedula 34
Anas rubripes 33
Anas smithii 249
Anas strepera 32
Anastomus lamelligerus 58
Ancient Murrelet 121
Anhinga rufa 70
Anous stolidus 107
Anous tenuirostris 107
Anser albifrons 26
Anser anser 26
Anser brachyrhynchus 26
Anser caerulescens 27
Anser erythropus 27
Anser fabalis 25
Anser indicus 27
Anser rossii 27
Anser serrirostris 26
Anteater Chat 205
Anthus berthelotii 221
Anthus campestris 219
Anthus cervinus 220
Anthus cinnamomeus 219
Anthus godlewskii 219
Anthus gustavi 220
Anthus hodgsoni 220
Anthus petrosus 221
Anthus pratensis 219
Anthus richardi 218
Anthus rubescens 220
Anthus rufulus 219
Anthus similis 219
Anthus spinoletta 220
Anthus trivialis 220
Apus affinis 136
Apus alexandri 135
Apus apus 135
Apus berliozi 136
Apus caffer 136
Apus pacificus 136
Apus pallidus 136
Apus unicolor 136
Aquatic Warbler 176
Aquila adalberti 75
Aquila chrysaetos 76
Aquila fasciata 76
Aquila heliaca 75
Aquila nipalensis 75
Aquila rapax 75
Aquila verreauxii 76
Arabian Accentor 216
Arabian Babbler 182
Arabian Bustard 82
Arabian Golden Sparrow 212
Arabian Grosbeak 225
Arabian Partridge 44
Arabian Scops Owl 129
Arabian Serin 225
Arabian Warbler 184
Arabian Waxbill 214

Arabian Wheatear 208
Arabian Woodpecker 141
Aratinga nenday 146
Arctic Redpoll 226
Arctic Skua 120
Arctic Tern 118
Arctic Warbler 174
Ardea alba 65
Ardea cinerea 64
Ardea goliath 64
Ardea herodias 64
Ardea intermedia 65
Ardea melanocephala 64
Ardea purpurea 64
Ardenna carneipes 54
Ardenna gravis 54
Ardenna griseus 53
Ardenna pacificus 53
Ardeola bacchus 63
Ardeola grayii 63
Ardeola ralloides 63
Ardeotis arabs 82
Arenaria interpres 101
Armenian Gull 114
Ascension Frigatebird 67
Ashy Drongo 152
Asian Brown Flycatcher 198
Asian Desert Warbler 184
Asian Dowitcher 96
Asian House Martin 169
Asian Koel 128
Asian Short-toed Lark 166
Asio capensis 133
Asio flammeus 133
Asio otus 133
Athene brama 132
Athene noctua 132
Atlantic Canary 228
Atlantic Petrel 51
Atlantic Puffin 122
Atlantic Yellow-nosed Albatross 48
Atlas Pied Flycatcher 201
Audouin's Gull 110
Aythya affinis 37
Aythya americana 35
Aythya collaris 36
Aythya ferina 36
Aythya fuligula 36
Aythya marila 36
Aythya nyroca 36
Aythya valisineria 35
Azores Bullfinch 223
Azure Tit 159

B

Baikal Teal 34
Baillon's Crake 84
Baird's Sandpiper 103
Bald Eagle 80
Balearic Shearwater 54
Balearic Warbler 185
Baltimore Oriole 232
Banded Martin 168
Band-rumped Storm-petrel 49
Bank Myna 191
Barbary Falcon 146
Barbary Partridge 43
Bar-headed Goose 27
Barn Swallow 168
Barnacle Goose 29
Barolo Shearwater 55
Barred Warbler 182
Barrow's Goldeneye 40
Bar-tailed Godwit 96
Bar-tailed Lark 161
Bartramia longicauda 98
Basra Reed Warbler 175
Bateleur 74
Bay-backed Shrike 149
Bay-breasted Warbler 230
Bearded Reedling 161
Bearded Vulture 71
Beautiful Sunbird 210
Belted Kingfisher 138
Bermuda Petrel 52
Berthelot's Pipit 221
Bimaculated Lark 165
Black Crake 250
Black Drongo 152
Black Francolin 44
Black Grouse 42
Black Guillemot 121
Black Heron 65
Black Kite 79
Black Lark 165
Black Redstart 202
Black Scoter 39
Black Scrub Robin 197
Black Stork 59
Black Swan 29
Black Tern 119
Black Wheatear 207
Black Woodpecker 142
Black-and-white Warbler 229
Black-bellied Sandgrouse 123
Black-bellied Storm-petrel 48
Black-billed Cuckoo 128
Black-billed Wood Dove 126
Black-breasted Weaver 214
Black-browed Albatross 48
Blackburnian Warbler 230
Black-capped Petrel 52
Black-crowned Night Heron 62
Black-crowned Sparrow-Lark 162
Black-crowned Tchagra 148
Black-eared Wheatear 206
Black-faced Bunting 237
Black-headed Bunting 236
Black-headed Gull 109
Black-headed Heron 64
Black-headed Lapwing 89
Black-headed Penduline Tit 160
Black-headed Weaver 213
Black-legged Kittiwake 107
Black-naped Monarch 153
Black-naped Oriole 152
Black-necked Grebe 57
Blackpoll Warbler 231
Blackstart 207
Black-tailed Godwit 96
Black-throated Accentor 216
Black-throated Blue Warbler 231
Black-throated Diver 46
Black-throated Green Warbler 232
Black-throated Thrush 195
Black-winged Kite 71
Black-winged Pratincole 107
Black-winged Stilt 89
Blanford's Lark 164
Blue Chaffinch 222
Blue Grosbeak 251
Blue Rock Thrush 203
Blue Whistling Thrush 200
Blue-and-white Flycatcher 198
Blue-cheeked Bee-eater 139
Blue-naped Mousebird 137
Bluethroat 199
Blue-winged Teal 33
Blue-winged Warbler 229
Blyth's Pipit 219
Blyth's Reed Warbler 177
Bobolink 232
Bohemian Waxwing 157
Bolle's Pigeon 124
Bombycilla cedrorum 157
Bombycilla garrulus 157
Bombycilla japonica 251
Bonaparte's Gull 108
Bonelli's Eagle 76
Booted Eagle 74
Booted Warbler 178
Botaurus lentiginosus 61
Botaurus stellaris 61
Boyd's Shearwater 55
Brachyramphus perdix 121
Brahminy Kite 79
Brahminy Starling 192
Brambling 222
Branta bernicla 28
Branta canadensis 28
Branta hutchinsii 28
Branta leucopsis 29
Branta ruficollis 29
Brent Goose 28
Bridled Tern 117
Broad-billed Roller 138
Broad-billed Sandpiper 105
Brown Booby 68
Brown Fish Owl 131
Brown Noddy 107
Brown Shrike 148
Brown Skua 119
Brown Snake Eagle 74
Brown Thrasher 191
Brown Woodland Warbler 171
Brown-headed Cowbird 233
Brown-headed Gull 109
Brown-necked Raven 156
Brown-throated Martin 167
Bruce's Green Pigeon 127
Brünnich's Guillemot 121
Bubo africanus 131
Bubo ascalaphus 130
Bubo bubo 130
Bubo scandiacus 130

Bubulcus coromandus 64
Bubulcus ibis 63
Bucanetes githagineus 223
Bucanetes mongolicus 223
Bucephala albeola 39
Bucephala clangula 39
Bucephala islandica 40
Buff-bellied Pipit 220
Buff-breasted Sandpiper 105
Bufflehead 39
Bulwer's Petrel 56
Bulweria bulwerii 56
Bulweria fallax 56
Buphagus erythrorynchus 193
Burhinus capensis 88
Burhinus oedicnemus 87
Burhinus senegalensis 88
Bush Petronia 213
Butastur rufipennis 80
Butastur teesa 80
Buteo bannermani 81
Buteo buteo 81
Buteo hemilasius 80
Buteo lagopus 80
Buteo rufinus 81
Butorides striata 63
Butorides virescens 62

C

Cabot's Tern 116
Cackling Goose 28
Cacomantis passerinus 128
Cairina moschata 31
Calandra Lark 165
Calandrella acutirostris 164
Calandrella blanfordi 164
Calandrella brachydactyla 164
Calcarius lapponicus 239
Calidris acuminata 103
Calidris alba 102
Calidris alpina 104
Calidris bairdii 103
Calidris canutus 101
Calidris ferruginea 104
Calidris fuscicollis 103
Calidris himantopus 104
Calidris maritima 104
Calidris mauri 102
Calidris melanotos 103
Calidris minuta 102
Calidris minutilla 103
Calidris pusilla 102
Calidris ruficollis 102
Calidris subminuta 103
Calidris temminckii 102
Calidris tenuirostris 101
California Quail 41
Calliope calliope 200
Callipepla californica 41
Calonectris borealis 53
Calonectris diomedea 53
Calonectris edwardsii 53
Calonectris leucomelas 52
Campethera punctuligera 250
Canada Goose 28
Canada Warbler 232
Canarian Oystercatcher 247
Canary Islands Chiffchaff 172
Canary Islands Stonechat 203
Canvasback 35
Cape Gannet 68
Cape May Warbler 230
Cape Petrel 51
Cape Shoveler 249
Cape Teal 32
Cape Verde Buzzard 81
Cape Verde Shearwater 53
Cape Verde Storm-petrel 49
Cape Verde Swift 135
Cape Verde Warbler 175
Caprimulgus aegyptius 134
Caprimulgus europaeus 134
Caprimulgus eximius 134
Caprimulgus inornatus 135
Caprimulgus mahrattensis 134
Caprimulgus nubicus 134
Caprimulgus poliocephalus 134
Caprimulgus ruficollis 134
Cardellina canadensis 232
Cardellina pusilla 232
Carduelis carduelis 227
Carduelis citrinella 228
Carduelis corsicana 228
Carpodacus erythrinus 223
Carpodacus roseus 224
Carpodacus rubicilla 224
Carpodacus sibiricus 224
Carpodacus synoicus 224
Carpospiza brachydactyla 212
Carrion Crow 155
Caspian Gull 113
Caspian Plover 93
Caspian Snowcock 43
Caspian Tern 115
Caspian Tit 158
Catharus fuscescens 194
Catharus guttatus 194
Catharus minimus 194
Catharus ustulatus 194
Caucasian Grouse 42
Caucasian Snowcock 43
Cecropis abyssinica 169
Cecropis daurica 169
Cedar Waxwing 157
Centropus senegalensis 127
Centropus superciliosus 127
Cepphus grylle 121
Cercotrichas galactotes 197
Cercotrichas podobe 197
Certhia brachydactyla 190
Certhia familiaris 190
Cerulean Warbler 230
Ceryle rudis 139
Cetti's Warbler 170
Cettia cetti 170
Chaetura pelagica 135
Charadrius alexandrinus 92
Charadrius asiaticus 93
Charadrius dubius 91
Charadrius hiaticula 91
Charadrius leschenaultii 93
Charadrius mongolus 92
Charadrius morinellus 93
Charadrius pecuarius 92
Charadrius semipalmatus 91
Charadrius tricollaris 92
Charadrius veredus 93
Charadrius vociferus 92
Chelictinia riocourii 71
Chersophilus duponti 165
Chestnut Bunting 236
Chestnut-bellied Sandgrouse 122
Chestnut-bellied Starling 193
Chestnut-eared Bunting 235
Chestnut-headed Sparrow-Lark 162
Chestnut-sided Warbler 231
Chestnut-tailed Starling 192
Chimney Swift 135
Chinese Pond Heron 63
Chlamydotis macqueenii 82
Chlamydotis undulata 82
Chlidonias hybrida 118
Chlidonias leucopterus 118
Chlidonias niger 119
Chloris chloris 224
Chondestes grammacus 239
Chordeiles minor 133
Chroicocephalus brunnicephalus 109
Chroicocephalus cirrocephalus 109
Chroicocephalus genei 108
Chroicocephalus philadelphia 108
Chroicocephalus ridibundus 109
Chrysococcyx caprius 128
Chrysococcyx klaas 128
Chrysolophus amherstiae 46
Chrysolophus pictus 46
Chukar Partridge 43
Ciconia abdimii 59
Ciconia ciconia 59
Ciconia episcopus 59
Ciconia nigra 59
Cinclus cinclus 209
Cinereous Bunting 234
Cinereous Tit 160
Cinereous Vulture 73
Cinnamon Bittern 62
Cinnamon Teal 249
Cinnamon-breasted Bunting 235
Cinnyricinclus leucogaster 193
Cinnyris asiaticus 210
Cinnyris habessinicus 210
Cinnyris osea 210
Cinnyris pulchellus 210
Circaetus cinereus 74
Circaetus gallicus 74
Circus aeruginosus 77
Circus cyaneus 78
Circus hudsonius 78
Circus macrourus 78
Circus pygargus 78

Circus spilonotus 78
Cirl Bunting 235
Cisticola aridulus 251
Cisticola juncidis 181
Citril Finch 228
Citrine Wagtail 217
Clamator glandarius 127
Clamator jacobinus 127
Clamorous Reed Warbler 176
Clanga clanga 74
Clanga nipalensis 74
Clangula hyemalis 39
Coal Tit 157
Coccothraustes coccothraustes 222
Coccyzus americanus 128
Coccyzus erythropthalmus 128
Colaptes auratus 142
Colinus virginianus 42
Collared Flycatcher 201
Collared Kingfisher 138
Collared Pratincole 106
Coloeus dauuricus 155
Coloeus monedula 155
Columba arquatrix 125
Columba bollii 124
Columba eversmanni 124
Columba guinea 124
Columba junoniae 124
Columba livia 123
Columba oenas 124
Columba palumbus 124
Columba trocaz 124
Common Blackbird 195
Common Bulbul 167
Common Buttonquail 87
Common Buzzard 81
Common Chaffinch 221
Common Chiffchaff 171
Common Crane 87
Common Crossbill 227
Common Cuckoo 129
Common Eider 37
Common Firecrest 188
Common Goldeneye 39
Common Grackle 233
Common Grasshopper Warbler 180
Common Greenshank 99
Common Guillemot 121
Common Gull 111
Common Hawk-Cuckoo 128
Common House Martin 169
Common Kestrel 143
Common Kingfisher 138
Common Linnet 225
Common Merganser 40
Common Moorhen 85
Common Myna 191
Common Nighthawk 133
Common Nightingale 200
Common Ostrich 25
Common Pheasant 45
Common Pochard 36
Common Quail 45
Common Redpoll 226
Common Redshank 99
Common Redstart 202
Common Reed Bunting 237
Common Ringed Plover 91
Common Rock Thrush 203
Common Rosefinch 223
Common Sandpiper 101
Common Scoter 39
Common Shelduck 31
Common Snipe 95
Common Starling 192
Common Swift 135
Common Tern 117
Common Waxbill 214
Common Whitethroat 184
Common Wood Pigeon 124
Common Yellowthroat 229
Contopus virens 147
Coracias abyssinicus 137
Coracias benghalensis 137
Coracias caudatus 137
Coracias garrulus 137
Coracias naevius 137
Corn Bunting 233
Corn Crake 83
Corsican Finch 228
Corsican Nuthatch 189
Corvus albus 156
Corvus corax 156
Corvus cornix 156
Corvus corone 155
Corvus frugilegus 155
Corvus macrorhynchos 156
Corvus rhipidurus 157
Corvus ruficollis 156
Corvus splendens 155
Cory's Shearwater 53
Corythornis cristatus 138
Cotton Pygmy Goose 32
Coturnix coturnix 45
Coturnix delegorguei 45
Crab-plover 88
Cream-coloured Courser 106
Creatophora cinerea 193
Crested Auklet 122
Crested Honey Buzzard 72
Crested Lark 163
Crested Myna 191
Cretzschmar's Bunting 234
Crex crex 83
Crex egregia 83
Cricket Warbler 181
Crithagra menachensis 225
Crithagra rothschildi 225
Crowned Sandgrouse 123
Cuculus canorus 129
Cuculus optatus 128
Curlew Sandpiper 104
Cursorius cursor 106
Cursorius temminckii 106
Cyanistes caeruleus 159
Cyanistes cyanus 159
Cyanistes teneriffae 159
Cyanopica cooki 153
Cyanoptila cyanomelana 198
Cygnus atratus 29
Cygnus columbianus 30
Cygnus cygnus 30
Cygnus olor 29
Cyprus Warbler 187
Cyprus Wheatear 206
Cypsiurus parvus 135

D

Dalmatian Pelican 67
Daption capense 51
Dark Chanting Goshawk 76
Dark-eyed Junco 239
Dark-sided Flycatcher 198
Dartford Warbler 185
Daurian Jackdaw 155
Daurian Redstart 202
Daurian Shrike 148
Daurian Starling 192
Dead Sea Sparrow 211
Delichon dasypus 169
Delichon urbicum 169
Demoiselle Crane 87
Dendrocopos assimilis 141
Dendrocopos leucopterus 250
Dendrocopos leucotos 142
Dendrocopos major 141
Dendrocopos syriacus 141
Dendrocoptes dorae 141
Dendrocoptes medius 141
Dendrocygna bicolor 25
Dendrocygna javanica 25
Dendrocygna viduata 249
Dendronanthus indicus 216
Dendropicos goertae 141
Denham's Bustard 250
Desert Cisticola 251
Desert Finch 225
Desert Lark 161
Desert Owl 131
Desert Sparrow 212
Desert Wheatear 206
Desert Whitethroat 183
Desertas Petrel 52
Dickcissel 240
Dicrurus adsimilis 250
Dicrurus leucophaeus 152
Dicrurus macrocercus 152
Diederik Cuckoo 128
Diomedea dabbenena 48
Dolichonyx oryzivorus 232
Double-crested Cormorant 69
Double-spurred Francolin 44
Dromas ardeola 88
Dryobates minor 141
Dryocopus martius 142
Dumetella carolinensis 191
Dunlin 104
Dunn's Lark 165
Dunnock 216
Dupont's Lark 165
Dusky Thrush 196
Dusky Turtle Dove 125
Dusky Warbler 173
Dwarf Bittern 62

E

Eastern Bonelli's Warbler 173
Eastern Cattle Egret 64
Eastern Crowned Warbler 175
Eastern Imperial Eagle 75
Eastern Kingbird 148
Eastern Marsh Harrier 78
Eastern Olivaceous Warbler 178
Eastern Orphean Warbler 184
Eastern Phoebe 147
Eastern Rock Nuthatch 190
Eastern Subalpine Warbler 186
Eastern Towhee 239
Eastern Wood Pewee 147
Eastern Yellow Wagtail 217
Egretta ardesiaca 65
Egretta caerulea 65
Egretta garzetta 66
Egretta gularis 66
Egretta thula 66
Egretta tricolor 65
Egyptian Goose 30
Egyptian Nightjar 134
Egyptian Plover 93
Egyptian Vulture 71
Elanoides forficatus 72
Elanus caeruleus 71
Elegant Tern 115
Eleonora's Falcon 144
Emberiza aureola 236
Emberiza bruniceps 237
Emberiza buchanani 234
Emberiza caesia 234
Emberiza calandra 233
Emberiza chrysophrys 236
Emberiza cia 234
Emberiza cineracea 234
Emberiza cirlus 235
Emberiza citrinella 233
Emberiza fucata 235
Emberiza goslingi 235
Emberiza hortulana 234
Emberiza leucocephalos 233
Emberiza melanocephala 236
Emberiza pallasi 237
Emberiza pusilla 235
Emberiza rustica 236
Emberiza rutila 236
Emberiza sahari 235
Emberiza schoeniclus 237
Emberiza spodocephala 237
Emberiza stewarti 234
Emberiza striolata 235
Emberiza tahapisi 235
Empidonax alnorum 147
Empidonax minimus 147
Empidonax virescens 147
Erckel's Francolin 45
Eremalauda dunni 165
Eremophila alpestris 164
Eremophila bilopha 164
Eremopterix nigriceps 162
Eremopterix signatus 162
Erithacus rubecula 199
Esacus recurvirostris 88
Eskimo Curlew 97
Estrilda astrild 214
Estrilda rufibarba 214
Ethiopian Swallow 168
Eudynamys scolopaceus 128
Eumyias thalassinus 198
Euodice cantans 215
Euodice malabarica 215
Euplectes afer 214
Eurasian Bittern 61
Eurasian Blackcap 182
Eurasian Blue Tit 159
Eurasian Bullfinch 223
Eurasian Collared Dove 125
Eurasian Coot 86
Eurasian Crag Martin 169
Eurasian Crimson-winged Finch 223
Eurasian Curlew 98
Eurasian Dotterel 93
Eurasian Eagle-Owl 130
Eurasian Golden Oriole 152
Eurasian Hobby 145
Eurasian Hoopoe 140
Eurasian Jay 153
Eurasian Magpie 154
Eurasian Nuthatch 189
Eurasian Oystercatcher 88
Eurasian Penduline Tit 160
Eurasian Pygmy Owl 132
Eurasian Reed Warbler 177
Eurasian Scops Owl 130
Eurasian Siskin 228
Eurasian Skylark 163
Eurasian Sparrowhawk 77
Eurasian Spoonbill 60
Eurasian Stone-curlew 87
Eurasian Teal 34
Eurasian Three-toed Woodpecker 140
Eurasian Tree Sparrow 212
Eurasian Treecreeper 190
Eurasian Whimbrel 97
Eurasian Wigeon 32
Eurasian Woodcock 94
Eurasian Wren 188
Eurasian Wryneck 140
European Bee-eater 139
European Crested Tit 158
European Golden Plover 90
European Goldfinch 227
European Green Woodpecker 142
European Greenfinch 224
European Herring Gull 113
European Honey Buzzard 71
European Nightjar 134
European Pied Flycatcher 200
European Robin 199
European Rock Pipit 221
European Roller 137
European Serin 228
European Shag 69
European Stonechat 204
European Storm-petrel 49
European Turtle Dove 125
Eurynorhynchus pygmeus 105
Eurystomus glaucurus 138
Evening Grosbeak 222
Eversmann's Redstart 202
Eyebrowed Thrush 195

F

Falcated Duck 32
Falco amurensis 144
Falco biarmicus 145
Falco cherrug 145
Falco chicquera 144
Falco columbarius 144
Falco concolor 144
Falco eleonorae 144
Falco jugger 145
Falco naumanni 143
Falco pelegrinoides 146
Falco peregrinus 146
Falco rusticolus 145
Falco sparverius 143
Falco subbuteo 145
Falco tinnunculus 143
Falco vespertinus 144
Familiar Chat 251
Fan-tailed Raven 157
Far Eastern Curlew 98
Fea's Petrel 51
Ferruginous Duck 36
Ficedula albicilla 201
Ficedula albicollis 201
Ficedula hypoleuca 200
Ficedula mugimaki 201
Ficedula parva 201
Ficedula semitorquata 201
Ficedula speculigera 201
Ficedula superciliaris 202
Fieldfare 196
Fine-spotted Woodpecker 250
Finsch's Wheatear 208
Flesh-footed Shearwater 54
Forbes-Watson's Swift 136
Forest Wagtail 216
Fork-tailed Drongo 250
Fork-tailed Flycatcher 148
Forster's Tern 118
Foudia madagascariensis 214
Francolinus francolinus 44
Francolinus pondicerianus 44
Franklin's Gull 110
Fratercula arctica 122
Fratercula cirrhata 122
Fregata aquila 67
Fregata ariel 68
Fregata magnificens 67
Fregata minor 67
Fregetta grallaria 249
Fregetta tropica 48
Fringilla coelebs 221
Fringilla montifringilla 222
Fringilla teydea 222
Fulica americana 86
Fulica atra 86

Fulica cristata 86
Fulmarus glacialis 50
Fulvous Babbler 182
Fulvous Whistling Duck 25

G

Gabar Goshawk 76
Gadwall 32
Galerida cristata 163
Galerida macrorhyncha 163
Galerida theklae 163
Gallicrex cinerea 84
Gallinago delicata 95
Gallinago gallinago 95
Gallinago media 95
Gallinago megala 95
Gallinago solitaria 94
Gallinago stenura 95
Gallinula chloropus 85
Gambaga Flycatcher 198
Garden Warbler 182
Garganey 34
Garrulus glandarius 153
Gavia adamsii 47
Gavia arctica 46
Gavia immer 47
Gavia pacifica 46
Gavia stellata 46
Gelochelidon nilotica 114
Geokichla sibirica 193
Geopelia striata 126
Geothlypis trichas 229
Geronticus eremita 60
Glareola lactea 107
Glareola maldivarum 106
Glareola nordmanni 107
Glareola pratincola 106
Glaucidium passerinum 132
Glaucous Gull 112
Glaucous-winged Gull 112
Glossy Ibis 60
Goldcrest 188
Golden Eagle 76
Golden Nightjar 134
Golden Pheasant 46
Golden Pipit 218
Golden-backed Weaver 213
Golden-winged Warbler 229
Goliath Heron 64
Gosling's Bunting 235
Graceful Prinia 181
Gracupica contra 192
Grasshopper Buzzard 80
Gray's Grasshopper Warbler 181
Great Auk 247
Great Black-backed Gull 111
Great Blue Heron 64
Great Bustard 81
Great Cormorant 69
Great Crested Grebe 57
Great Egret 65
Great Frigatebird 67
Great Grey Owl 132
Great Grey Shrike 150
Great Knot 101
Great Northern Diver 47
Great Reed Warbler 176
Great Rosefinch 224
Great Shearwater 54
Great Skua 119
Great Snipe 95
Great Spotted Cuckoo 127
Great Spotted Woodpecker 141
Great Stone-curlew 88
Great Tit 160
Great White Pelican 66
Greater Crested Tern 115
Greater Flamingo 57
Greater Hoopoe-Lark 161
Greater Painted-snipe 94
Greater Sand Plover 93
Greater Scaup 36
Greater Short-toed Lark 164
Greater Spotted Eagle 74
Greater White-fronted Goose 26
Greater Yellowlegs 99
Green Bee-eater 139
Green Heron 62
Green Sandpiper 100
Green Warbler 174
Greenish Warbler 175
Green-winged Teal 34
Grey Catbird 191
Grey Francolin 44
Grey Heron 64
Grey Hypocolius 157
Grey Partridge 45
Grey Plover 91
Grey Wagtail 218
Grey-backed Fiscal 250
Grey-bellied Cuckoo 128
Grey-cheeked Thrush 194
Grey-headed Gull 109
Grey-headed Kingfisher 138
Grey-headed Lapwing 90
Grey-headed Swamphen 85
Grey-headed Woodpecker 143
Greylag Goose 26
Grey-necked Bunting 234
Grey-tailed Tattler 100
Grey-throated Martin 167
Griffon Vulture 73
Grus antigone 250
Grus canadensis 86
Grus grus 87
Grus leucogeranus 86
Grus virgo 87
Güldenstädt's Redstart 203
Gull-billed Tern 114
Gymnoris dentata 213
Gymnoris xanthocollis 213
Gypaetus barbatus 71
Gyps africanus 72
Gyps bengalensis 72
Gyps fulvus 73
Gyps himalayensis 73
Gyps rueppelli 72
Gyr Falcon 145

H

Haematopus meadewaldoi 247
Haematopus ostralegus 88
Halcyon leucocephala 138
Halcyon smyrnensis 138
Haliaeetus albicilla 80
Haliaeetus leucocephalus 80
Haliaeetus leucoryphus 79
Haliaeetus vocifer 79
Haliastur indus 79
Hamerkop 66
Harlequin Duck 38
Harlequin Quail 45
Hawfinch 222
Hazel Grouse 42
Hedydipna metallica 210
Hedydipna platura 209
Helmeted Guineafowl 41
Hen Harrier 78
Hermit Thrush 194
Hesperiphona vespertina 222
Hieraaetus pennatus 74
Hieraaetus wahlbergi 74
Hierococcyx varius 128
Himalayan Vulture 73
Himantopus himantopus 89
Hippolais icterina 179
Hippolais languida 179
Hippolais olivetorum 179
Hippolais polyglotta 179
Hirundapus caudacutus 135
Hirundo aethiopica 168
Hirundo rustica 168
Hirundo smithii 168
Histrionicus histrionicus 38
Hooded Crow 156
Hooded Merganser 40
Hooded Vulture 72
Hooded Warbler 230
Hooded Wheatear 206
Horned Grebe 57
Horned Lark 164
Houbara Bustard 82
House Bunting 235
House Crow 155
House Sparrow 210
Hudsonian Godwit 96
Hudsonian Whimbrel 97
Hume's Leaf Warbler 174
Hume's Short-toed Lark 164
Hume's Wheatear 208
Hume's Whitethroat 183
Hydrobates pelagicus 49
Hydrocoloeus minutus 109
Hydrophasianus chirurgus 94
Hydroprogne caspia 115
Hylocichla mustelina 195
Hypocolius ampelinus 157
Hypothymis azurea 153

I

Iago Sparrow 211
Iberian Chiffchaff 172

Iberian Green Woodpecker 142
Iberian Magpie 153
Iceland Gull 112
Ichthyaetus audouinii 110
Ichthyaetus hemprichii 111
Ichthyaetus ichthyaetus 110
Ichthyaetus leucophthalmus 111
Ichthyaetus melanocephalus 110
Ichthyaetus relictus 110
Icterine Warbler 179
Icterus galbula 232
Iduna aedon 178
Iduna caligata 178
Iduna opaca 179
Iduna pallida 178
Iduna rama 178
Indian Cormorant 249
Indian Golden Oriole 152
Indian Pitta 147
Indian Pond Heron 63
Indian Roller 137
Indian Scops Owl 250
Indian Silverbill 215
Indian Skimmer 107
Indian Paradise Flycatcher 241
Indigo Bunting 240
Intermediate Egret 65
Irania gutturalis 200
Iraq Babbler 181
Isabelline Wheatear 205
Italian Sparrow 211
Ivory Gull 108
Ixobrychus cinnamomeus 62
Ixobrychus eurhythmus 62
Ixobrychus exilis 61
Ixobrychus minutus 61
Ixobrychus sinensis 61
Ixobrychus sturmii 62
Ixoreus naevius 194

J

Jack Snipe 94
Jacobin Cuckoo 127
Japanese Waxwing 251
Jouanin's Petrel 56
Junco hyemalis 239
Jynx torquilla 140

K

Kelp Gull 112
Kentish Plover 92
Ketupa zeylonensis 131
Killdeer 92
King Eider 37
Kittlitz's Plover 92
Klaas's Cuckoo 128
Knob-billed Duck 30
Kordofan Lark 250
Krüper's Nuthatch 189
Kurdish Wheatear 209

L

Lady Amherst's Pheasant 46
Laggar Falcon 145
Lagonosticta senegala 214
Lagopus lagopus 43
Lagopus muta 42
Lamprotornis pulcher 193
Lanceolated Warbler 180
Lanius collurio 148
Lanius cristatus 148
Lanius excubitor 150
Lanius excubitoroides 250
Lanius isabellinus 148
Lanius meridionalis 150
Lanius minor 149
Lanius nubicus 151
Lanius pallidirostris 150
Lanius phoenicuroides 149
Lanius schach 149
Lanius senator 151
Lanius vittatus 149
Lanner Falcon 145
Lapland Longspur 239
Lappet-faced Vulture 73
Large-billed Crow 156
Large-billed Leaf Warbler 175
Lark Sparrow 239
Larus argentatus 113
Larus armenicus 114
Larus cachinnans 113
Larus canus 111
Larus delawarensis 111
Larus dominicanus 112
Larus fuscus 114
Larus glaucescens 112
Larus glaucoides 112
Larus hyperboreus 112
Larus marinus 111
Larus michahellis 113
Larus schistisagus 114
Larus smithsonianus 113
Larus thayeri 112
Larvivora cyane 199
Larvivora sibilans 199
Laughing Dove 126
Laughing Gull 109
Laurel Pigeon 124
Leach's Storm-petrel 50
Least Bittern 61
Least Flycatcher 147
Least Sandpiper 103
Least Tern 116
Leiothlypis peregrina 229
Leiothrix lutea 182
Leptoptilos crumenifer 59
Lesser Black-backed Gull 114
Lesser Crested Tern 115
Lesser Flamingo 58
Lesser Frigatebird 68
Lesser Grey Shrike 149
Lesser Kestrel 143
Lesser Moorhen 86
Lesser Noddy 107
Lesser Sand Plover 92
Lesser Scaup 37
Lesser Short-toed Lark 166
Lesser Spotted Eagle 74
Lesser Spotted Woodpecker 141
Lesser Striped Swallow 169
Lesser Whistling Duck 25
Lesser White-fronted Goose 27
Lesser Whitethroat 183
Lesser Yellowlegs 99
Leucophaeus atricilla 109
Leucophaeus pipixcan 110
Levaillant's Woodpecker 143
Levant Sparrowhawk 77
Lichtenstein's Sandgrouse 123
Lilac-breasted Roller 137
Limicola falcinellus 105
Limnodromus griseus 95
Limnodromus scolopaceus 96
Limnodromus semipalmatus 96
Limosa haemastica 96
Limosa lapponica 96
Limosa limosa 96
Linaria cannabina 225
Linaria flavirostris 225
Linaria yemenensis 226
Lincoln's Sparrow 238
Little Auk 120
Little Bittern 61
Little Blue Heron 65
Little Bunting 235
Little Bustard 82
Little Crake 83
Little Curlew 97
Little Egret 66
Little Grebe 56
Little Gull 109
Little Owl 132
Little Ringed Plover 91
Little Rock Thrush 203
Little Stint 102
Little Swift 136
Little Tern 116
Locustella certhiola 180
Locustella fasciolata 181
Locustella fluviatilis 180
Locustella lanceolata 180
Locustella luscinioides 180
Locustella naevia 180
Lonchura malacca 215
Lonchura punctulata 215
Long-billed Dowitcher 96
Long-billed Murrelet 121
Long-billed Pipit 219
Long-eared Owl 133
Long-legged Buzzard 81
Long-tailed Duck 39
Long-tailed Rosefinch 224
Long-tailed Shrike 149
Long-tailed Skua 120
Long-tailed Tit 171
Long-toed Stint 103
Lophoceros nasutus 140
Lophodytes cucullatus 40
Lophophanes cristatus 158
Louisiana Waterthrush 251
Loxia curvirostra 227
Loxia leucoptera 227

Loxia pytyopsittacus 226
Loxia scotica 227
Lullula arborea 162
Luscinia luscinia 199
Luscinia megarhynchos 200
Luscinia svecica 199
Lymnocryptes minimus 94
Lyrurus mlokosiewiczi 42
Lyrurus tetrix 42

M

Macqueen's Bustard 82
Macronectes giganteus 50
Madeira Firecrest 188
Maghreb Lark 163
Maghreb Wheatear 208
Magnificent Frigatebird 67
Magnolia Warbler 230
Malachite Kingfisher 138
Mallard 33
Mandarin Duck 31
Manx Shearwater 54
Marabou Stork 59
Marbled Duck 35
Marmaronetta angustirostris 35
Marmora's Warbler 185
Marsh Owl 133
Marsh Sandpiper 99
Marsh Tit 158
Marsh Warbler 178
Martial Eagle 249
Masked Booby 68
Masked Shrike 151
Matsudaira's Storm-petrel 50
Meadow Pipit 219
Mediterranean Gull 110
Megaceryle alcyon 138
Melanitta americana 39
Melanitta deglandi 38
Melanitta fusca 38
Melanitta nigra 39
Melanitta perspicillata 38
Melanocorypha bimaculata 165
Melanocorypha calandra 165
Melanocorypha yeltoniensis 165
Melierax metabates 76
Melodious Warbler 179
Melospiza lincolnii 238
Melospiza melodia 238
Ménétriés's Warbler 187
Mergellus albellus 40
Mergus merganser 40
Mergus serrator 40
Merlin 144
Merops albicollis 139
Merops apiaster 139
Merops orientalis 139
Merops persicus 139
Microcarbo africanus 69
Microcarbo pygmeus 69
Micronisus gabar 76
Middle Spotted Woodpecker 141
Milvus aegyptius 79
Milvus migrans 79
Milvus milvus 79
Mimus gilvus 251
Mimus polyglottos 191
Mirafra cantillans 162
Mirafra cordofanica 250
Mistle Thrush 197
Mniotilta varia 229
Molothrus ater 233
Moltoni's Warbler 186
Mongolian Finch 223
Monk Parakeet 146
Montagu's Harrier 78
Montane Nightjar 134
Monteiro's Storm-petrel 49
Monticola rufocinereus 203
Monticola saxatilis 203
Monticola solitarius 203
Montifringilla nivalis 213
Morus bassanus 68
Morus capensis 68
Motacilla aguimp 218
Motacilla alba 218
Motacilla cinerea 218
Motacilla citreola 217
Motacilla flava 216
Motacilla tschutschensis 217
Mountain Chiffchaff 172
Mourning Collared Dove 126
Mourning Dove 127
Mourning Wheatear 208
Moussier's Redstart 203
Moustached Warbler 176
Mugimaki Flycatcher 201
Muscicapa dauurica 198
Muscicapa gambagae 198
Muscicapa sibirica 198
Muscicapa striata 198
Muscovy Duck 31
Mute Swan 29
Mycerobas carnipes 222
Mycteria ibis 58
Myiopsitta monachus 146
Myophonus caeruleus 200
Myrmecocichla aethiops 205
Myrtle Warbler 231

N

Namaqua Dove 126
Nanday Parakeet 146
Naumann's Thrush 196
Necrosyrtes monachus 72
Neophron percnopterus 71
Neotis denhami 250
Neotis nuba 82
Netta erythrophthalma 35
Netta rufina 35
Nettapus coromandelianus 32
Nile Valley Sunbird 210
Northern Bald Ibis 60
Northern Bobwhite 42
Northern Flicker 142
Northern Fulmar 50
Northern Gannet 68
Northern Goshawk 77
Northern Harrier 78
Northern Hawk-Owl 132
Northern Lapwing 89
Northern Mockingbird 191
Northern Parula 230
Northern Pintail 34
Northern Raven 156
Northern Shoveler 33
Northern Waterthrush 229
Northern Wheatear 205
Nubian Bustard 82
Nubian Nightjar 134
Nucifraga caryocatactes 154
Numenius arquata 98
Numenius borealis 97
Numenius hudsonicus 97
Numenius madagascariensis 98
Numenius minutus 97
Numenius phaeopus 97
Numenius tenuirostris 97
Numida meleagris 41
Nyctanassa violacea 62
Nycticorax nycticorax 62

O

Oceanites oceanicus 47
Oceanodroma castro 49
Oceanodroma jabejabe 49
Oceanodroma leucorhoa 50
Oceanodroma matsudairae 50
Oceanodroma monorhis 50
Oceanodroma monteiroi 49
Oena capensis 126
Oenanthe albonigra 208
Oenanthe bottae 205
Oenanthe chrysopygia 209
Oenanthe cypriaca 206
Oenanthe deserti 206
Oenanthe familiaris 251
Oenanthe finschii 208
Oenanthe halophila 208
Oenanthe hispanica 206
Oenanthe isabellina 205
Oenanthe leucopyga 207
Oenanthe leucura 207
Oenanthe lugens 208
Oenanthe lugentoides 208
Oenanthe melanura 207
Oenanthe moesta 207
Oenanthe monacha 206
Oenanthe oenanthe 205
Oenanthe picata 207
Oenanthe pleschanka 206
Oenanthe seebohmi 205
Oenanthe xanthoprymna 209
Olive-backed Pipit 220
Olive-tree Warbler 179
Omani Owl 131
Onychognathus tristramii 193
Onychoprion aleuticus 117
Onychoprion anaethetus 117
Onychoprion fuscatus 117
Orange-breasted Waxbill 215
Oriental Cuckoo 128

Oriental Paradise Flycatcher 241
Oriental Plover 93
Oriental Pratincole 106
Oriental Reed Warbler 176
Oriental Skylark 162
Oriental Turtle Dove 125
Oriental White-eye 187
Oriolus chinensis 152
Oriolus kundoo 152
Oriolus oriolus 152
Ortolan Bunting 234
Otis tarda 81
Otus bakkamoena 250
Otus brucei 129
Otus pamelae 129
Otus scops 130
Ovenbird 229
Oxyura jamaicensis 41
Oxyura leucocephala 41

P

Pacific Diver 46
Pacific Golden Plover 90
Pacific Swift 136
Paddyfield Pipit 219
Paddyfield Warbler 177
Pagophila eburnea 108
Pale Crag Martin 169
Pale Martin 168
Pale Rockfinch 212
Pale-legged Leaf Warbler 241
Palestine Sunbird 210
Pallas's Fish Eagle 79
Pallas's Grasshopper Warbler 180
Pallas's Gull 110
Pallas's Leaf Warbler 173
Pallas's Reed Bunting 237
Pallas's Rosefinch 224
Pallas's Sandgrouse 122
Pallid Harrier 78
Pallid Scops Owl 129
Pallid Swift 136
Palm Warbler 231
Pandion haliaetus 70
Panurus biarmicus 161
Paragallinula angulata 86
Parakeet Auklet 122
Parkesia motacilla 251
Parkesia noveboracensis 229
Parrot Crossbill 226
Parus cinereus 160
Parus major 160
Passer ammodendri 210
Passer domesticus 210
Passer euchlorus 212
Passer hispaniolensis 211
Passer iagoensis 211
Passer italiae 211
Passer luteus 212
Passer moabiticus 211
Passer montanus 212
Passer pyrrhonotus 211
Passer simplex 212
Passer zarudnyi 212
Passerculus sandwichensis 239
Passerella iliaca 238
Passerina caerulea 251
Passerina cyanea 240
Pastor roseus 192
Pechora Pipit 220
Pectoral Sandpiper 103
Pelagodroma marina 47
Pelecanus crispus 67
Pelecanus onocrotalus 66
Pelecanus rufescens 67
Perdix perdix 45
Peregrine Falcon 146
Periparus ater 157
Perisoreus infaustus 153
Pernis apivorus 71
Pernis ptilorhyncus 72
Persian Shearwater 55
Persian Wheatear 209
Petrochelidon fluvicola 170
Petrochelidon pyrrhonota 170
Petronia petronia 212
Phaethon aethereus 58
Phaethon lepturus 58
Phalacrocorax aristotelis 69
Phalacrocorax auritus 69
Phalacrocorax carbo 69
Phalacrocorax fuscicollis 249
Phalacrocorax lucidus 70
Phalacrocorax nigrogularis 70
Phalaropus fulicarius 106
Phalaropus lobatus 105
Phalaropus tricolor 105
Pharaoh Eagle-Owl 130
Phasianus colchicus 45
Pheasant-tailed Jacana 94
Pheucticus ludovicianus 240
Philadelphia Vireo 151
Philby's Partridge 43
Philomachus pugnax 105
Phoeniconaias minor 58
Phoenicopterus ruber 57
Phoenicurus auroreus 202
Phoenicurus erythrogastrus 203
Phoenicurus erythronotus 202
Phoenicurus moussieri 203
Phoenicurus ochruros 202
Phoenicurus phoenicurus 202
Phylloscopus bonelli 172
Phylloscopus borealis 174
Phylloscopus borealoides 241
Phylloscopus canariensis 172
Phylloscopus collybita 171
Phylloscopus coronatus 175
Phylloscopus fuscatus 173
Phylloscopus humei 174
Phylloscopus ibericus 172
Phylloscopus inornatus 174
Phylloscopus magnirostris 175
Phylloscopus neglectus 172
Phylloscopus nitidus 174
Phylloscopus orientalis 173
Phylloscopus plumbeitarsus 175
Phylloscopus proregulus 173
Phylloscopus schwarzi 173
Phylloscopus sibilatrix 173
Phylloscopus sindianus 172
Phylloscopus tenellipes 241
Phylloscopus trochiloides 175
Phylloscopus trochilus 171
Phylloscopus umbrovirens 171
Pica pica 154
Picoides tridactylus 140
Picus canus 143
Picus sharpei 142
Picus squamatus 250
Picus vaillantii 143
Picus viridis 142
Pied Avocet 89
Pied Crow 156
Pied Kingfisher 139
Pied Myna 192
Pied Stonechat 205
Pied Wheatear 206
Pied-billed Grebe 56
Pine Bunting 233
Pine Grosbeak 222
Pinguinus impennis 247
Pinicola enucleator 222
Pink-backed Pelican 67
Pink-footed Goose 26
Pin-tailed Sandgrouse 122
Pin-tailed Snipe 95
Pipilo erythrophthalmus 239
Piranga olivacea 240
Piranga rubra 240
Pitta brachyura 147
Plain Leaf Warbler 172
Plain Nightjar 135
Plain Swift 136
Platalea alba 61
Platalea leucorodia 60
Plectrophenax nivalis 240
Plectropterus gambensis 30
Plegadis falcinellus 60
Pleske's Ground Jay 154
Ploceus benghalensis 214
Ploceus cucullatus 213
Ploceus galbula 213
Ploceus jacksoni 213
Ploceus manyar 214
Ploceus melanocephalus 213
Pluvialis apricaria 90
Pluvialis dominica 91
Pluvialis fulva 90
Pluvialis squatarola 91
Pluvianus aegyptius 93
Podiceps auritus 57
Podiceps cristatus 57
Podiceps grisegena 56
Podiceps nigricollis 57
Podilymbus podiceps 56
Podoces pleskei 154
Poecile cinctus 159
Poecile hyrcanus 158
Poecile lugubris 158
Poecile montanus 158
Poecile palustris 158
Polemaetus bellicosus 249

Polysticta stelleri 37
Pomarine Skua 120
Porphyrio alleni 85
Porphyrio madagascariensis 85
Porphyrio martinica 85
Porphyrio poliocephalus 85
Porphyrio porphyrio 84
Porzana carolina 84
Porzana fusca 84
Porzana parva 83
Porzana porzana 84
Porzana pusilla 84
Prairie Warbler 232
Prinia gracilis 181
Progne subis 168
Prothonotary Warbler 251
Protonotaria citrea 251
Prunella atrogularis 216
Prunella collaris 215
Prunella fagani 216
Prunella modularis 216
Prunella montanella 216
Prunella ocularis 216
Psittacula eupatria 146
Psittacula krameri 147
Pternistis bicalcaratus 44
Pternistis erckelii 45
Pterocles alchata 122
Pterocles coronatus 123
Pterocles exustus 122
Pterocles lichtensteinii 123
Pterocles orientalis 123
Pterocles senegallus 123
Pterodroma arminjoniana 52
Pterodroma cahow 52
Pterodroma deserta 52
Pterodroma feae 51
Pterodroma hasitata 52
Pterodroma incerta 51
Pterodroma madeira 51
Pterodroma mollis 51
Ptyonoprogne obsoleta 169
Ptyonoprogne rupestris 169
Puffinus bailloni 55
Puffinus baroli 55
Puffinus boydi 55
Puffinus mauretanicus 54
Puffinus persicus 55
Puffinus puffinus 54
Puffinus yelkouan 54
Purple Gallinule 85
Purple Heron 64
Purple Martin 168
Purple Roller 137
Purple Sandpiper 104
Purple Sunbird 210
Pycnonotus barbatus 167
Pycnonotus cafer 166
Pycnonotus leucotis 166
Pycnonotus xanthopygos 167
Pygmy Cormorant 69
Pygmy Sunbird 209
Pyrrhocorax graculus 154
Pyrrhocorax pyrrhocorax 154
Pyrrhula murina 223
Pyrrhula pyrrhula 223

Q

Quelea quelea 214
Quiscalus quiscula 233

R

Radde's Accentor 216
Radde's Warbler 173
Rallus aquaticus 83
Raso Lark 162
Razorbill 121
Recurvirostra avosetta 89
Red Avadavat 214
Red Fody 214
Red Fox Sparrow 238
Red Kite 79
Red Knot 101
Red Phalarope 106
Red Turtle Dove 126
Red-backed Shrike 148
Red-billed Chough 154
Red-billed Firefinch 214
Red-billed Leiothrix 182
Red-billed Oxpecker 193
Red-billed Quelea 214
Red-billed Teal 34
Red-billed Tropicbird 58
Red-breasted Flycatcher 201
Red-breasted Goose 29
Red-breasted Merganser 40
Red-breasted Nuthatch 189
Red-breasted Wheatear 205
Red-crested Pochard 35
Red-eyed Dove 126
Red-eyed Vireo 152
Red-flanked Bluetail 200
Red-footed Booby 68
Red-footed Falcon 144
Red-fronted Serin 228
Redhead 35
Red-headed Bunting 237
Red-knobbed Coot 86
Red-legged Partridge 44
Red-necked Falcon 144
Red-necked Grebe 56
Red-necked Nightjar 134
Red-necked Phalarope 105
Red-necked Stint 102
Red-rumped Swallow 169
Red-rumped Wheatear 207
Red-throated Diver 46
Red-throated Pipit 220
Red-throated Thrush 196
Red-vented Bulbul 166
Red-wattled Lapwing 90
Redwing 197
Reed Cormorant 69
Reeves's Pheasant 45
Regulus calendula 188
Regulus ignicapilla 188
Regulus madeirensis 188
Regulus regulus 188
Relict Gull 110
Remiz coronatus 160
Remiz macronyx 160
Remiz pendulinus 160
Rhamphocoris clotbey 161
Rhodopechys alienus 223
Rhodopechys sanguineus 223
Rhodospiza obsoleta 225
Rhodostethia rosea 109
Rhynchostruthus percivali 225
Richard's Pipit 218
Ring Ouzel 195
Ring-billed Gull 111
Ring-necked Duck 36
Riparia chinensis 167
Riparia cincta 168
Riparia diluta 168
Riparia paludicola 167
Riparia riparia 167
Rissa tridactyla 107
River Tern 117
River Warbler 180
Rock Bunting 234
Rock Dove 123
Rock Partridge 43
Rock Ptarmigan 42
Rock Sparrow 212
Rook 155
Roseate Tern 117
Rose-breasted Grosbeak 240
Rose-ringed Parakeet 147
Ross's Goose 27
Ross's Gull 109
Rostratula benghalensis 94
Rosy Starling 192
Rough-legged Buzzard 80
Royal Tern 115
Ruby-crowned Kinglet 188
Ruddy Duck 41
Ruddy Shelduck 31
Ruddy Turnstone 101
Ruddy-breasted Crake 84
Ruff 105
Rufous-tailed Robin 199
Rufous-tailed Scrub Robin 197
Rüppell's Vulture 72
Rüppell's Warbler 187
Rüppell's Weaver 213
Rustic Bunting 236
Rynchops albicollis 107
Rynchops flavirostris 107

S

Sabine's Gull 108
Saker Falcon 145
Sakhalin Leaf Warbler 241
Sand Lark 166
Sand Martin 167
Sand Partridge 44
Sanderling 102
Sandhill Crane 86
Sandwich Tern 116
Sardinian Warbler 186
Sarkidiornis melanotos 30
Sarus Crane 250
Saunders's Tern 116

Savannah Sparrow 239
Savi's Warbler 180
Saxaul Sparrow 210
Saxicola caprata 205
Saxicola dacotiae 203
Saxicola maurus 204
Saxicola rubetra 203
Saxicola rubicola 204
Saxicola stejnegeri 204
Saxicola torquata 205
Sayornis phoebe 147
Scaly-bellied Woodpecker 250
Scaly-breasted Munia 215
Scarlet Tanager 240
Scissor-tailed Kite 71
Scolopax minor 94
Scolopax rusticola 94
Scopoli's Shearwater 53
Scopus umbretta 66
Scotocerca inquieta 170
Scottish Crossbill 227
Sedge Warbler 177
Seebohm's Wheatear 205
See-see Partridge 44
Seiurus aurocapilla 229
Semicollared Flycatcher 201
Semipalmated Plover 91
Semipalmated Sandpiper 102
Senegal Coucal 127
Senegal Thick-knee 88
Serinus canaria 228
Serinus pusillus 228
Serinus serinus 228
Serinus syriacus 228
Setophaga aestiva 231
Setophaga americana 230
Setophaga caerulescens 231
Setophaga castanea 230
Setophaga cerulea 230
Setophaga citrina 230
Setophaga coronata 231
Setophaga discolor 232
Setophaga dominica 231
Setophaga fusca 230
Setophaga magnolia 230
Setophaga palmarum 231
Setophaga pensylvanica 231
Setophaga ruticilla 230
Setophaga striata 231
Setophaga tigrina 230
Setophaga virens 232
Sharp-tailed Sandpiper 103
Shikra 76
Shining Sunbird 210
Short-billed Dowitcher 95
Short-eared Owl 133
Short-toed Snake Eagle 74
Short-toed Treecreeper 190
Shy Albatross 48
Siberian Accentor 216
Siberian Blue Robin 199
Siberian Crane 86
Siberian Jay 153
Siberian Rubythroat 200
Siberian Stonechat 204
Siberian Thrush 193
Siberian Tit 159
Sinai Rosefinch 224
Sind Sparrow 211
Sind Woodpecker 141
Singing Bush Lark 162
Sinosuthora webbiana 187
Sitta canadensis 189
Sitta europaea 189
Sitta krueperi 189
Sitta ledanti 189
Sitta neumayer 190
Sitta tephronota 190
Sitta whiteheadi 189
Slaty-backed Gull 114
Slender-billed Curlew 97
Slender-billed Gull 108
Small Pratincole 107
Smew 40
Snow Bunting 240
Snow Goose 27
Snowy Egret 66
Snowy Owl 130
Sociable Lapwing 90
Socotra Cormorant 70
Soft-plumaged Petrel 51
Solitary Sandpiper 100
Solitary Snipe 94
Somateria fischeri 37
Somateria mollissima 37
Somateria spectabilis 37
Sombre Tit 158
Song Sparrow 238
Song Thrush 197
Sooty Falcon 144
Sooty Gull 111
Sooty Shearwater 53
Sooty Tern 117
Sora 84
South Polar Skua 119
Southern Giant Petrel 50
Southern Grey Shrike 150
Southern Pochard 35
Spanish Imperial Eagle 75
Spanish Sparrow 211
Speckled Pigeon 124
Spectacled Eider 37
Spectacled Warbler 185
Sphyrapicus varius 140
Spilopelia senegalensis 126
Spiloptila clamans 181
Spinus spinus 228
Spiza americana 240
Spoon-billed Sandpiper 105
Spotless Starling 193
Spotted Crake 84
Spotted Eagle-Owl 131
Spotted Flycatcher 198
Spotted Nutcracker 154
Spotted Owlet 132
Spotted Redshank 98
Spotted Sandgrouse 123
Spotted Sandpiper 101
Spotted Thick-knee 88
Spur-winged Goose 30
Spur-winged Lapwing 89
Squacco Heron 63
Stejneger's Stonechat 204
Steller's Eider 37
Steppe Eagle 75
Steppe Grey Shrike 150
Stercorarius antarcticus 119
Stercorarius longicaudus 120
Stercorarius maccormicki 119
Stercorarius parasiticus 120
Stercorarius pomarinus 120
Stercorarius skua 119
Sterna aurantia 117
Sterna dougallii 117
Sterna forsteri 118
Sterna hirundo 117
Sterna paradisaea 118
Sterna repressa 118
Sternula albifrons 116
Sternula antillarum 116
Sternula saundersi 116
Stilt Sandpiper 104
Stock Dove 124
Streaked Scrub Warbler 170
Streaked Shearwater 52
Streaked Weaver 214
Streak-throated Swallow 170
Streptopelia decaocto 125
Streptopelia decipiens 126
Streptopelia lugens 125
Streptopelia orientalis 125
Streptopelia roseogrisea 125
Streptopelia semitorquata 126
Streptopelia tranquebarica 126
Streptopelia turtur 125
Striated Heron 63
Striolated Bunting 235
Striped Crake 84
Strix aluco 131
Strix butleri 131
Strix hadorami 131
Strix nebulosa 132
Strix uralensis 131
Struthio camelus 25
Sturnia malabarica 192
Sturnia pagodarum 192
Sturnus unicolor 193
Sturnus vulgaris 192
Sudan Golden Sparrow 212
Sula dactylatra 68
Sula leucogaster 68
Sula sula 68
Summer Tanager 240
Surf Scoter 38
Surnia ulula 132
Swainson's Thrush 194
Swallow-tailed Kite 72
Swinhoe's Snipe 95
Swinhoe's Storm-petrel 50
Sykes's Nightjar 134
Sykes's Warbler 178
Sylvia althaea 183
Sylvia atricapilla 182
Sylvia balearica 185
Sylvia borin 182
Sylvia buryi 187
Sylvia cantillans 186
Sylvia communis 184

Sylvia conspicillata 185
Sylvia crassirostris 184
Sylvia curruca 183
Sylvia deserti 184
Sylvia deserticola 185
Sylvia hortensis 184
Sylvia inornata 186
Sylvia leucomelaena 184
Sylvia melanocephala 186
Sylvia melanothorax 187
Sylvia minula 183
Sylvia mystacea 187
Sylvia nana 184
Sylvia nisoria 182
Sylvia ruppeli 187
Sylvia sarda 185
Sylvia subalpina 186
Sylvia undata 185
Synthliboramphus antiquus 121
Syrian Serin 228
Syrian Woodpecker 141
Syrmaticus reevesii 45
Syrrhaptes paradoxus 122

T

Tachybaptus ruficollis 56
Tachycineta bicolor 168
Tachymarptis melba 135
Tadorna ferruginea 31
Tadorna tadorna 31
Taiga Bean Goose 25
Taiga Flycatcher 201
Tarsiger cyanurus 200
Tawny Eagle 75
Tawny Owl 131
Tawny Pipit 219
Tchagra senegalus 148
Temminck's Courser 106
Temminck's Lark 164
Temminck's Stint 102
Tengmalm's Owl 133
Tennessee Warbler 229
Terathopius ecaudatus 74
Terek Sandpiper 100
Terpsiphone affinis 241
Terpsiphone incei 241
Terpsiphone paradisi 241
Terpsiphone viridis 153
Tetrao urogallus 42
Tetraogallus caspius 43
Tetraogallus caucasicus 43
Tetrastes bonasia 42
Tetrax tetrax 82
Thalassarche cauta 48
Thalassarche chlororhynchos 48
Thalassarche melanophris 48
Thalasseus acuflavidus 116
Thalasseus bengalensis 115
Thalasseus bergii 115
Thalasseus elegans 115
Thalasseus maximus 115
Thalasseus sandvicensis 116
Thayer's Gull 112
Thekla Lark 163
Thick-billed Lark 161
Thick-billed Warbler 178
Three-banded Plover 92
Threskiornis aethiopicus 60
Thrush Nightingale 199
Tichodroma muraria 190
Tickell's Thrush 195
Tmetothylacus tenellus 218
Todiramphus chloris 138
Torgos tracheliotos 73
Toxostoma rufum 191
Tree Pipit 220
Tree Swallow 168
Treron waalia 127
Tricolored Heron 65
Tricolored Munia 215
Trindade Petrel 52
Tringa brevipes 100
Tringa erythropus 98
Tringa flavipes 99
Tringa glareola 100
Tringa melanoleuca 99
Tringa nebularia 99
Tringa ochropus 100
Tringa semipalmata 100
Tringa solitaria 100
Tringa stagnatilis 99
Tringa totanus 99
Tristan Albatross 48
Tristram's Starling 193
Tristram's Warbler 185
Trocaz Pigeon 124
Troglodytes troglodytes 188
Tropical Mockingbird 251
Tropical Shearwater 55
Trumpeter Finch 223
Tryngites subruficollis 105
Tufted Duck 36
Tufted Puffin 122
Tundra Bean Goose 26
Tundra Swan 30
Turdoides altirostris 181
Turdoides fulva 182
Turdoides huttoni 181
Turdoides squamiceps 182
Turdus atrogularis 195
Turdus eunomus 196
Turdus iliacus 197
Turdus menachensis 195
Turdus merula 195
Turdus migratorius 197
Turdus naumanni 196
Turdus obscurus 195
Turdus philomelos 197
Turdus pilaris 196
Turdus ruficollis 196
Turdus torquatus 195
Turdus unicolor 195
Turdus viscivorus 197
Turkestan Shrike 149
Turnix sylvaticus 87
Turtur abyssinicus 126
Twite 225
Two-barred Crossbill 227
Two-barred Warbler 175
Tyrannus savana 148
Tyrannus tyrannus 148
Tyto alba 129

U

Ultramarine Flycatcher 202
Upcher's Warbler 179
Upland Buzzard 80
Upland Sandpiper 98
Upupa epops 140
Ural Owl 131
Uria aalge 121
Uria lomvia 121
Urocolius macrourus 137

V

Vanellus cinereus 90
Vanellus gregarius 90
Vanellus indicus 90
Vanellus leucurus 90
Vanellus spinosus 89
Vanellus tectus 89
Vanellus vanellus 89
Variable Wheatear 207
Varied Thrush 194
Veery 194
Velvet Scoter 38
Verditer Flycatcher 198
Vermivora chrysoptera 229
Vermivora cyanoptera 229
Verreaux's Eagle 76
Village Weaver 213
Vinous-breasted Starling 192
Vinous-throated Parrotbill 187
Violet-backed Starling 193
Vireo flavifrons 151
Vireo griseus 151
Vireo olivaceus 152
Vireo philadelphicus 151
Von Schrenck's Bittern 62

W

Wahlberg's Eagle 74
Wallcreeper 190
Water Pipit 220
Water Rail 83
Watercock 84
Wattled Starling 193
Wedge-tailed Shearwater 53
Western Barn Owl 129
Western Bonelli's Warbler 172
Western Capercaillie 42
Western Cattle Egret 63
Western Jackdaw 155
Western Marsh Harrier 77
Western Olivaceous Warbler 179
Western Orphean Warbler 184
Western Osprey 70
Western Reef Egret 66
Western Rock Nuthatch 190
Western Sandpiper 102

Western Subalpine Warbler 186
Western Swamphen 84
Western Yellow Wagtail 216
Whinchat 203
Whiskered Tern 118
White Stork 59
White Wagtail 218
White's Thrush 194
White-backed Vulture 72
White-backed Woodpecker 142
White-bellied Storm-petrel 249
White-billed Diver 47
White-breasted Cormorant 70
White-breasted Waterhen 83
White-browed Coucal 127
White-capped Bunting 234
White-cheeked Tern 118
White-crowned Penduline Tit 160
White-crowned Sparrow 238
White-crowned Wheatear 207
White-eared Bulbul 166
White-eyed Buzzard 80
White-eyed Gull 111
White-eyed Vireo 151
White-faced Storm-petrel 47
White-faced Whistling-Duck 249
White-headed Duck 41
White-rumped Sandpiper 103
White-rumped Swift 136
White-rumped Vulture 72
White-spectacled Bulbul 167
White-tailed Eagle 80
White-tailed Lapwing 90
White-tailed Tropicbird 58
White-throated Bee-eater 139
White-throated Dipper 209
White-throated Kingfisher 138
White-throated Needletail 135
White-throated Robin 200
White-throated Sparrow 238
White-winged Grosbeak 222
White-winged Lark 162
White-winged Scoter 38
White-winged Snowfinch 213
White-winged Tern 118
White-winged Woodpecker 250
Whooper Swan 30
Willet 100
Willow Ptarmigan 43
Willow Tit 158
Willow Warbler 171
Wilson's Phalarope 105
Wilson's Snipe 95
Wilson's Storm-petrel 47
Wilson's Warbler 232
Wire-tailed Swallow 168
Wood Duck 31
Wood Sandpiper 100
Wood Thrush 195
Wood Warbler 173
Woodchat Shrike 151
Woodlark 162
Woolly-necked Stork 59

X

Xanthocephalus xanthocephalus 232
Xema sabini 108
Xenus cinereus 100

Y

Yelkouan Shearwater 54
Yellow Bittern 61
Yellow-bellied Sapsucker 140
Yellow-billed Cuckoo 128
Yellow-billed Kite 79
Yellow-billed Stork 58
Yellow-breasted Bunting 236
Yellow-browed Bunting 236
Yellow-browed Warbler 174
Yellow-crowned Bishop 214
Yellow-crowned Night Heron 62
Yellow-eyed Pigeon 124
Yellowhammer 233
Yellow-headed Blackbird 232
Yellow-legged Gull 113
Yellow-throated Sparrow 213
Yellow-throated Vireo 151
Yellow-throated Warbler 231
Yemen Linnet 226
Yemen Serin 225
Yemen Thrush 195
Yemen Warbler 187

Z

Zarudny's Sparrow 212
Zebra Dove 126
Zenaida macroura 127
Zino's Petrel 51
Zitting Cisticola 181
Zonotrichia albicollis 238
Zonotrichia leucophrys 238
Zoothera aurea 194
Zosterops abyssinicus 188
Zosterops palpebrosus 187